Weighted Network Analysis

Steve Horvath

Weighted Network Analysis

Applications in Genomics and Systems Biology

Springer

Steve Horvath
Professor of Human Genetics and Biostatistics
University of California, Los Angeles
Los Angeles, CA 90095-7088, USA
shorvath@mednet.ucla.edu

ISBN 978-1-4419-8818-8 e-ISBN 978-1-4419-8819-5
DOI 10.1007/978-1-4419-8819-5
Springer New York Dordrecht Heidelberg London

Library of Congress Control Number: 2011925163

Printed on acid-free paper

Springer is part of Springer Science+Business Media (www.springer.com)

To Lora, my brother Markus, my parents,
Joseph O'Brien and Joerg Zimmermann

Preface

The past decade has seen an incredible growth of network methods following publications by Laszlo Barabasi and others. Excellent text books exist on general networks and graph theory, but these books typically describe unweighted networks. This book focuses on weighted networks. In weighted networks, the pairwise connection strength between two nodes is quantified by a real number between 0 and 1. It is worth emphasizing that **most of the material also applies to unweighted networks**. Further, unweighted networks can easily be constructed from weighted networks by dichotomizing the connection strengths between nodes. While unweighted networks permit graph-theoretic visualization techniques and algorithms, weighted networks can be advantageous for many reasons including the following:

1. They preserve the continuous nature of the underlying connectivity information. For example, weighted correlation networks that are constructed on the basis of correlations between numeric variables do not require the choice of a hard threshold (Chap. 5). Dichotomizing information and (hard)-thresholding may lead to information loss.
2. They often lead to highly robust results (Zhang and Horvath 2005). In contrast, results based on unweighted networks, constructed by thresholding a pairwise association measure, often strongly depend on the threshold.
3. They can sometimes be decomposed and approximated by simpler networks. For example, networks can sometimes be approximated by "factorizable" networks (Chap. 2). Such approximations are often difficult to achieve for sparse, unweighted networks.
4. They sometimes allow for a parsimonious parametrization (in terms of modules and conformities, see Sect. 2.3).
5. They often allow one to derive simple relationships between network concepts (statistics) (Sect. 3.8 and Chap. 6). In particular, weighted *correlation* networks facilitate a geometric interpretation based on the angular interpretation of the correlation (Sect. 6.7).
6. They can be used to enhance standard data-mining methods such as cluster analysis since (dis)-similarity measures can often be transformed into weighted networks (Sect. 7.7).

Many of the applied sections in this book present analysis techniques and strategies to the wider audience of applied researchers. The book assumes little mathematical and statistical knowledge, but some sections are rather abstract. To make the book self-contained, some sections review statistical and data-mining techniques. Since several technical sections and chapters are less relevant to applied researchers, they start out with the warning that they can be skipped. I also present abstract, theoretical material since it may be useful for quantitative researchers, who carry out methodological research. In my own experience, I have found that applied researchers can be expert users of network methods and software. Of course, domain-knowledge experts have often a superior intuition about how to arrive at a meaningful analysis of their data. Many weighted network methods arose from collaborations with cancer biologists, neuroscientists, mouse geneticists, and biologists (e.g., see the acknowledgement section and references).

Although the field of weighted network analysis only began a few years ago, it is already impossible to summarize it in one book. I have tried to cite as many articles as possible, but I apologize to my colleagues for failing to cite their work. Several people are mentioned throughout the book, see the index for names and page numbers. I am acutely aware that I leave unmentioned many important ideas and techniques. My only excuse for giving too much attention to my own work is that I understand it best.

While the methods are formulated in general terms, which facilitate their application to wide variety of data, most applications involve genes, proteins, and gene expression data. It has become clear that networks have important medical and biological applications. Gene co-expression networks bridge the gap from individual genes to clinically important, emergent phenotypes. Gene networks allow one to move beyond single-gene comparisons and systematically identify biologically meaningful relationships between gene products, pathways, and phenotypes. Weighted gene co-expression network analysis (WGCNA) has been used to identify candidate disease biomarkers, to annotate genes with regard to module membership, to study the relationships between co-expression modules, and to compare the network topology of different networks. Case studies show how WGCNA can be used to screen for genes, to understand the transcriptional architecture, and to relate modules in different mouse tissues. Integrating co-expression networks with genetic marker data facilitates systems genetic applications (Sects. 11.5 and 12.3), which make use of causal testing and network edge-orienting procedures.

Freely Available R Software

This book provides an in-depth description of the `WGCNA` R package (Langfelder and Horvath 2008), which provides functions for carrying out network analysis tasks. R is a freely available, open source language and environment for statistical computing and graphics, which has become a de-facto standard in data analysis (Ihaka and Gentleman 1996; Venables and Ripley 2002; Gentleman et al. 2004,

2005; Carey et al. 2005). The R environment integrates standard data analysis and visualization techniques with packages (libraries) implementing the latest advances in data mining, statistics, and machine learning. The WGCNA package is available from the Comprehensive R Archive Network (CRAN), the standard repository for R add-on packages. To install it, type the following command into the R session:

```
install.packages("WGCNA")
```

Most of the R code and data presented in the book chapters can be downloaded from the following webpage:
www.genetics.ucla.edu/labs/horvath/CoexpressionNetwork/Book
Related scientific articles and presentations can be found at the following webpages: www.genetics.ucla.edu/labs/horvath/CoexpressionNetwork/.

Other relevant R packages mentioned throughout the book are freely available on the R CRAN package resource at
www.R-project.org
and/or on the Bioconductor webpage.

Carey VJ, Gentry J, Whalen E, Gentleman R (2005) Network structures and algorithms in bioconductor. Bioinformatics 21(1):135–136

Ihaka R, Gentleman R (1996) R: A language for data analysis and graphics. J Comput Graph Stat 5(3):299–314

Gentleman RC, Carey VJ, Bates DJ, Bolstad BM, Dettling M, Dudoit S, Ellis B, Gautier L, Ge Y, Gentry J, Hornik K, Hothorn T, Huber W, Iacus S, Irizarry R, Leisch F, Li C, Maechler M, Rossini AJ, Sawitzki G, Smith C, Smyth GK, Tierney L, Yang YH, Zhang J (2004) Bioconductor: Open software development for computational biology and bioinformatics. Genome Biol 5(10):R80

Gentleman R, Huber W, Carey V, Irizarry R, Dudoit S (2005) Bioinformatics and computational biology solutions using R and bioconductor. Springer, New York

Langfelder P, Horvath S (2008) WGCNA: An R package for weighted correlation network analysis. BMC Bioinform 9(1):559

Venables WN, Ripley BD (2002) Modern applied statistics with S, 4th edn. Springer, New York

Zhang B, Horvath S (2005) General framework for weighted gene coexpression analysis. Stat Appl Genet Mol Biol 4:17

Los Angeles, USA Steve Horvath

Acknowledgements

Many weighted network methods and R software code were developed in collaboration with colleagues, Postdoctoral researchers, and doctoral students. In particular, I mention **Peter Langfelder** who maintains the WGCNA R software package (Langfelder and Horvath 2008) and was the first author on several related publications (Langfelder and Horvath 2007; Langfelder et al. 2007, 2011). His contribution and those of others are mentioned throughout the book (see the index). **Bin Zhang** worked on the general framework for weighted gene coexpression network analysis (Zhang and Horvath 2005) and developed the first dynamic tree cutting algorithm Langfelder et al. (2007). **Jun Dong** worked on the relationships between network concepts (Dong and Horvath 2007; Horvath and Dong 2008).

Much of the work arose from close collaborations with applied researchers. In particular, weighted correlation networks were developed in joint discussions with cancer researchers **Paul Mischel** and **Stanley F. Nelson**, and neuroscientists **Daniel H. Geschwind** and **Michael C. Oldham** (Horvath et al. 2006; Carlson et al. 2006; Oldham et al. 2006, 2008; Miller et al. 2010). **Jake Lusis**, **Thomas Drake**, and **Eric Schadt** provided important mouse genetic applications and data (Ghazalpour et al. 2006; Chen et al. 2008). **Roel Ophoff, Giovanni Coppola**, and **Jeremy Miller** provided applications to neurological diseases (Saris et al. 2009; Miller et al. 2010). **Kenneth Lange** and **John Ranola** solved optimization problems. Many doctoral students have closely worked with me including the following: **Andy Yip** and **Ai Li** worked on extensions of the topological overlap matrix (Yip and Horvath 2007; Li and Horvath 2007). **Jason Aten** worked on causal anchors and the network edge orienting (NEO) (Aten et al. 2008). **Tova Fuller** worked on differential network analysis and causal analyses (Fuller et al. 2007). **Angela Presson** worked on integrating genetic markers with gene co-expression network analysis (Presson et al. 2008). **Chaochao Cai, Marc Carlson, Sudheer Doss, Charles Farber, Anatole Ghazalpour, Austin Hilliard, Wen Lin, Rui Luo, Michael Mason, Chris Plaisier, Lin Song, Yang Song, Atila Van Nas, Lin Wang, Kellen Winden, Wei Zhao, Yafeng Zhang, and Joerg Zimmermann** worked on WGCNA methods and applications (Ghazalpour et al. 2006; Carlson et al. 2006; Farber et al. 2009; van Nas et al. 2009; Mason et al. 2009).

Aten J, Fuller T, Lusis AJ, Horvath S (2008) Using genetic markers to orient the edges in quantitative trait networks: The NEO software. BMC Syst Biol 2(1):34

Carlson M, Zhang B, Fang Z, Mischel P, Horvath S, Nelson SF (2006) Gene connectivity, function, and sequence conservation: Predictions from modular yeast co-expression networks. BMC Genomics 7(7):40

Chen Y, Zhu J, Lum PY, Yang X, Pinto S, MacNeil DJ, Zhang C, Lamb J, Edwards S, Sieberts SK, Leonardson A, Castellini LW, Wang S, Champy MFF, Zhang B, Emilsson V, Doss S, Ghazalpour A, Horvath S, Drake TA, Lusis AJ, Schadt EE (2008) Variations in DNA elucidate molecular networks that cause disease. Nature 452(7186):429–435

Dong J, Horvath S (2007) Understanding network concepts in modules. BMC Syst Biol 1(1):24

Farber CR, van Nas A, Ghazalpour A, Aten JE, Doss S, Sos B, Schadt EE, Ingram-Drake L, Davis RC, Horvath S, Smith DJ, Drake TA, Lusis AJ (2009) An integrative genetics approach to identify candidate genes regulating bone density: Combining linkage, gene expression and association. J Bone Miner Res 1:105–116

Fuller TF, Ghazalpour A, Aten JE, Drake T, Lusis AJ, Horvath S (2007) Weighted gene coexpression network analysis strategies applied to mouse weight. Mamm Genome 18(6–7):463–472

Ghazalpour A, Doss S, Zhang B, Plaisier C, Wang S, Schadt EE, Thomas A, Drake TA, Lusis AJ, Horvath S (2006) Integrating genetics and network analysis to characterize genes related to mouse weight. PloS Genet 2(2):8

Horvath S, Dong J (2008) Geometric interpretation of gene co-expression network analysis. PLoS Comput Biol 4(8):e1000117

Horvath S, Zhang B, Carlson M, Lu KV, Zhu S, Felciano RM, Laurance MF, Zhao W, Shu Q, Lee Y, Scheck AC, Liau LM, Wu H, Geschwind DH, Febbo PG, Kornblum HI, Cloughesy TF, Nelson SF, Mischel PS (2006) Analysis of oncogenic signaling networks in glioblastoma identifies ASPM as a novel molecular target. Proc Natl Acad Sci USA 103(46):17402–17407

Langfelder P, Horvath S (2007) Eigengene networks for studying the relationships between co-expression modules. BMC Syst Biol 1(1):54

Langfelder P, Horvath S (2008) WGCNA: an R package for weighted correlation network analysis. BMC Bioinform 9(1):559

Langfelder P, Zhang B, Horvath S (2007) Defining clusters from a hierarchical cluster tree: The Dynamic Tree Cut library for R. Bioinformatics 24(5):719–720

Langfelder P, Luo R, Oldham MC, Horvath S (2011) Is my network module preserved and reproducible? Plos Comput Biol 7(1):e1001057

Li A, Horvath S (2007) Network neighborhood analysis with the multi-node topological overlap measure. Bioinformatics 23(2):222–231

Mason M, Fan G, Plath K, Zhou Q, Horvath S (2009) Signed weighted gene co-expression network analysis of transcriptional regulation in murine embryonic stem cells. BMC Genomics 10(1):327

Miller JA, Horvath S, Geschwind DH (2010) Divergence of human and mouse brain transcriptome highlights Alzheimer disease pathways. Proc Natl Acad Sci USA 107(28):12698–12703

van Nas A, GuhaThakurta D, Wang SS, Yehya N, Horvath S, Zhang B, Ingram-Drake L, Chaudhuri G, Schadt EE, Drake TA, Arnold AP, Lusis AJ (2009) Elucidating the role of gonadal hormones in sexually dimorphic gene coexpression networks. Endocrinology 150(3):1235–1249

Oldham MC, Horvath S, Geschwind DH (2006) Conservation and evolution of gene coexpression networks in human and chimpanzee brains. Proc Natl Acad Sci USA 103(47):17973–17978

Oldham MC, Konopka G, Iwamoto K, Langfelder P, Kato T, Horvath S, Geschwind DH (2008) Functional organization of the transcriptome in human brain. Nat Neurosci 11(11):1271–1282

Presson AP, Sobel EM, Papp JC, Suarez CJ, Whistler T, Rajeevan MS, Vernon SD, Horvath S (2008) Integrated weighted gene co-expression network analysis with an application to chronic fatigue syndrome. BMC Syst Biol 2:95

Saris C, Horvath S, van Vught P, van Es M, Blauw H, Fuller TF, Langfelder P, DeYoung J, Wokke J, Veldink J, van den Berg L, Ophoff R (2009) Weighted gene co-expression network analysis of the peripheral blood from Amyotrophic Lateral Sclerosis patients. BMC Genomics 10(1):405+

Yip A, Horvath S (2007) Gene network interconnectedness and the generalized topological overlap measure. BMC Bioinform 8(8):22

Zhang B, Horvath S (2005) General framework for weighted gene coexpression analysis. Stat Appl Genet Mol Biol 4:17

Contents

Acronyms

AF	Adjacency function
ARACNE	Algorithm for the reconstruction of accurate cellular networks
CF	ConFormity
CPA	Common pleiotropic anchor
DPI	Data processing inequality
FDR	False discovery rate
GTOM	Generalized topological overlap measure
GS	Gene (or node) Significance
kME	Connectivity based on the module eigenvector
kIM	Connectivity, intramodular
KL	Kullback–Leibler
LEO	Local edge orienting
LRT	Likelihood ratio test
ME	Module eigenvector or eigengene
MI	Mutual information
MLE	Maximum likelihood estimation
MTOM	Multinode topological overlap measure
NEO	Network edge orienting
OCA	Orthogonal causal anchor
NCF	Network concept function
PAM	Partitioning around medoids
PPI	Protein–Protein interaction
QTL	Quantitative trait locus
SEM	Structural equation model
SFT	Scale-free topology
SNP	Single nucleotide polymorphism
SVD	Singular value decomposition
TOM	Topological overlap matrix
WGCNA	Weighted gene co-expression network analysis or weighted correlation network analysis

Chapter 1
Networks and Fundamental Concepts

Abstract This chapter introduces basic terminology and network concepts. Subsequent chapters illustrate that many data analysis tasks can be addressed using network methods. Network concepts (also known as network statistics or network indices) can be used to describe the topological properties of a single network and for comparing two or more networks (e.g., differential network analysis). Dozens of potentially useful network concepts are known from graph theory, e.g., the connectivity, density, centralization, and topological overlap. Measures of node interconnectedness, e.g., based on generalizations of the topological overlap matrix, can be used in neighborhood analysis. We distinguish three types of *fundamental* network concepts: (1) whole network concepts are defined without reference to modules, (2) intramodular concepts describe network properties of a module, and (3) intermodular concepts describe relationships between two or more modules. Intermodular network concepts can be used to define networks whose nodes are modules.

1.1 Network Adjacency Matrix

Networks can be used to describe the pairwise relationships between n nodes (which are sometimes referred to as vertices). For example, we will use networks to describe the relationships between n genes. We consider networks that are fully specified by an $n \times n$ dimensional **adjacency matrix** $A = (A_{ij})$, where the entry A_{ij} quantifies the connection strength from node i to node j. For an *unweighted* network, A_{ij} equals 1 or 0 depending on whether a connection (also known as link or edge) exists from node i to node j.

For a *weighted network*, A_{ij} takes on a real number between 0 and 1. A_{ij} specifies the connection strength between node i and node j. For an undirected network, the connection strength (A_{ij}) from i to j equals the connection strength from j to i (A_{ji}), i.e., the adjacency matrix A is symmetric $(A_{ij}) = (A_{ji})$. For a directed network, the adjacency matrix is typically not symmetric (see Sect. 11.4). Unless we explicitly mention otherwise, we assume in the following that we are dealing with an undirected network. As a convention, we set the diagonal elements to 1, i.e., $A_{ii} = 1$.

S. Horvath, *Weighted Network Analysis: Applications in Genomics and Systems Biology*, DOI 10.1007/978-1-4419-8819-5_1, © Springer Science+Business Media, LLC 2011

In summary, we study networks whose adjacencies satisfy the following conditions:

$$\begin{aligned} 0 \leq A_{ij} &\leq 1, \\ A_{ij} &= A_{ji}, \\ A_{ii} &= 1. \end{aligned} \tag{1.1}$$

Many network applications use at least one node significance measure. Abstractly speaking, we define a *node significance measure* $GS = (GS_1, \dots, GS_n)$ as a vector with n components that correspond to the network nodes. For the ith node, GS_i quantifies the significance or importance with regard to a particular application. The only assumption is that $GS_i = 0$ means that node i is not significant with regard to the application under consideration. We should emphasize that node significance does not necessarily correspond to statistical significance. For example, GS_i can be an indicator variable that equals 1 if prior literature suggests that node i is known to be important and 0 otherwise. If a statistical significance level (p value) is available for each node, then a p value-based node significance measure can be defined as follows:

$$GS_i = -log(p\ value_i). \tag{1.2}$$

In this case, GS_i is proportional to the number of zeroes of the ith p value. In gene network applications, gene significance measures allow one to incorporate external gene information into the network analysis. In functional enrichment analysis, a gene significance measure could indicate pathway membership. In gene knockout experiments, gene significance could indicate knockout essentiality.

1.1.1 Connectivity and Related Concepts

The *connectivity* (also known as degree) of the ith node is defined by

$$k_i = \sum_{j \neq i} A_{ij}. \tag{1.3}$$

In unweighted networks, the connectivity k_i equals the number of nodes that are directly linked to node i. In weighted networks, the connectivity equals the sum of connection weights between node i and the other nodes.

1.1.2 Social Network Analogy: Affection Network

Since humans are organized into social networks, social network analogies should be intuitive to many readers. Therefore, we will refer to the following "affection network" throughout this book. Each individual is represented by a node in the affection network. We assume that the connection strength (adjacency) between two individuals reflects how much affection they feel for each other. To be specific, we assume that the affection (adjacency) A_{ij} equals 1 if two individuals strongly like each other, it equals 0.5 if they are neutral toward each other, and it equals 0 if they strongly dislike each other. Then the scaled connectivity K_i is a measure of relative popularity: high values of K_i indicates that the ith person is well liked by many others.

1.2 Analysis Tasks Amenable to Network Methods

Networks are useful for describing the relationships between objects (interpreted as network nodes). Networks are increasingly being used to analyze high-dimensional data sets where nodes correspond to variables (e.g., gene measurements). Networks facilitate sophisticated data analysis, which can often be described in intuitive ways. As social beings we function in social networks, which is why network language and terminology are very intuitive to us. For example, a network module can be interpreted as a social clique (e.g., a club) and highly connected hub nodes as popular people. Network methods can be used to address a variety of data analysis tasks including the following:

1. *To describe direct and indirect relationships between objects.* While the network adjacency matrix encodes direct first-order relationships, higher order relationships can be measured based on shared neighbors (see, e.g., Sect. 1.3.14)
2. *To carry out a neighborhood analysis.* Roughly speaking, a neighborhood is composed of nodes that are highly connected to a given "seed" set of nodes. Thus, neighborhood analysis facilitates a guilt-by-association screening strategy for finding nodes that are close to a given seed set of interesting nodes (see Sect. 1.4).
3. *To describe network properties using network concepts (also known as network statistics).* We describe several types of network concepts in this and subsequent chapters.
4. *To describe the module structure of a data set.* Modules (groups, clusters, cliques) of nodes can be defined in many ways. Several module detection and clustering procedures are described in Chap. 8.
5. *To define shared "consensus" modules present in multiple data sets.* By construction, consensus modules can be found in two or more networks (see Sect. 7.11.1). Consensus modules may represent fundamental preserved structural properties of the network.
6. *To identify important modules.* For example, module significance measures can be used to identify gene modules that relate to cancer survival time (Sect. 5.7). A module significance measure can be defined by averaging a node significance measure across the module genes.
7. *To measure differences in connectivity patterns between two data sets.* Differential network analysis can be used to identify changes in connectivity patterns or module structure between different conditions (Sect. 1.11). Module preservation statistics are described in Chap. 9.

8. *To find highly connected "hub" nodes*. For example, highly connected intramodular hub nodes effectively summarize or represent the module.
9. *To reduce or compress the data*. For example, focusing the analysis on modules or their representatives (e.g., intramodular hub nodes) amounts to a network-based data reduction technique. Module-based analyses greatly alleviate the multiple testing problem that plagues many statistical analyses involving large numbers of variables.
10. *To annotate objects with regard to module membership*. For example, intramodular connectivity measures can be used to annotate all network nodes with respect to how close they are to the identified modules. This can be accomplished by defining a fuzzy measure of module memberships (intramodular connectivity) that generalizes the binary module membership indicator to a quantitative measure. Fuzzy measures of module membership can be used to identify nodes that lie intermediate between (i.e., close to) two or more modules.
11. *To develop network-based or module-based node screening procedures*. For example, gene pathway-based approaches for finding biologically important genes can be defined with regard to module membership measures (intramodular connectivity). In general, node-screening criteria can be based on a variety of network concepts (e.g., based on differential network analysis).

Throughout the book, we mention additional analysis tasks that can be addressed by more specialized networks. For example, correlation networks (described in Chap. 5) are constructed on the basis of correlations between numeric variables that can be described by an $m \times n$ numeric matrix *datX*. The nodes of a correlation network correspond to the columns of the matrix *datX*. Network concepts and methods can be used to describe the correlation patterns between the variables and to reduce the data. Although other statistical techniques exist for analyzing correlation matrices, network language and concepts are particularly intuitive. Statistically speaking, networks can be used as a data exploratory techniques (similar to cluster analysis, factor analysis, or other dimensional reduction techniques), as machine learning, data mining, and variable selection techniques. While sometimes established statistical techniques can be used to address similar goals, they are often far less intuitive to applied scientists. In contrast, network methods can usually be explained using social network analogies. Often the data being analyzed correspond to network measurements, e.g., genes operate in pathways or modules. It is natural to use network methods when one tries to model pathways.

1.3 Fundamental Network Concepts

In the following, we describe existing and novel network concepts (also known as network statistics or indices) that can be used to describe local and global network properties (Dong and Horvath 2007). The prime example of a fundamental network concept is the connectivity k_i (1.3). Sometimes network concepts are defined with regard to a node significance measure GS_i. Abstractly speaking, a **fundamental network concept** is a function of the off-diagonal elements of A and/or a node significance measure GS. Below we present several network concepts including the density, maximum adjacency ratio, centralization, hub node significance, etc.

1.3.1 Matrix and Vector Notation

If M is a matrix and β is a real number, then M^β denotes the element-wise power, i.e., the ijth element of M^β is given by M_{ij}^β. Similarly, if v is a numeric vector, then the ith component of v^β is given by v_i^β. More generally, if $f()$ is a function that maps real numbers to real numbers, then $f(v)$ denotes the vector whose ith component is given by $f(v_i)$. We define $sum(M) = \sum_i \sum_j M_{ij}$ as the sum across all matrix entries, $max(M)$ as the maximum entry of matrix M, and $max(v)$ as the maximum component of the vector v. Similarly we define the minimum function $min(\cdot)$. We define the function $S_\beta(\cdot)$ for a vector v as $S_\beta(v) = \sum_i v_i^\beta = sum(v^\beta)$. Then $mean(v) = sum(v)/n$ and $variance(v) = sum(v^2)/n - (sum(v)/n)^2$. The transpose of a matrix or vector is denoted by the superscript $^\tau$. The **Frobenius matrix norm** is denoted by

$$||M||_F = \sqrt{\sum_i \sum_j m_{ij}^2} = \sqrt{sum(M^2)}. \tag{1.4}$$

Further denote by I the identity matrix and by $\mathrm{diag}(v^2)$ a diagonal matrix with its ith diagonal component given by $v_i^2, i = 1, \dots, n$.

We briefly review two types of multiplying two $n \times n$ dimensional matrices A and B. The *component-wise* product $A * B$ yields an $n \times n$ dimensional matrix whose i, jth element is given by $A_{ij} * B_{ij}$. In contrast, the *matrix multiplication* AB yields an $n \times n$ dimensional matrix whose i, jth element is given by $\sum_{l=1}^n A_{il} B_{lj}$. Note that no multiplication sign is used for the matrix multiplication. In contrast, the multiplication sign $*$ between two matrices denotes their component-wise product. The R commands for carrying out these two types of multiplication are given by `A*B` and `A\%*\%B`, respectively.

1.3.2 Scaled Connectivity

The connectivity (node degree) k_i is probably the best known fundamental network concept. Many other network concepts are functions of the connectivity. For example, the *minimum connectivity* is defined as:

$$k_{min} = min(k), \tag{1.5}$$

where $min(k)$ denotes the minimum across the n components of the vector k. The *maximum connectivity* is defined as:

$$k_{max} = max(k). \tag{1.6}$$

Consider a network concept NC_i (such as the connectivity) that depends on a node index i (where $i = 1, \dots, n$). Denote by *max*(NC) the maximum observed value across the n nodes. Then the **scaled version of the network concept** is defined as follows:

$$ScaledNC = \frac{NC}{max(NC)}. \tag{1.7}$$

For example, the **scaled connectivity** K_i of the ith node is defined by

$$ScaledConnectivity_i = \frac{k_i}{k_{max}} = K_i. \tag{1.8}$$

By definition the scaled connectivity lies between 0 and 1, i.e., $0 \leq K_i \leq 1$. Note that we distinguish the scaled from the unscaled connectivity using an uppercase "K" and a lowercase 'k', respectively. By definition $k_{max} \leq n-1$. Sometimes it is convenient to define the scaled connectivity (with a capital C) as follows:

$$C_i = \frac{k_i}{n-1}. \tag{1.9}$$

To avoid confusion, we should point out that the word "scale" has different meanings in different contexts. It has no relationships to the *scale*-free topology fitting index described in the following section.

1.3.3 Scale-Free Topology Fitting Index

Many studies have explored the frequency distribution of the connectivity, which can be defined based on the discretized connectivity vector $dk = discretize(k)$. The *discretize* function takes as input a numeric vector and outputs a vector of equal length whose components indicate the bin number into which the value falls. Denote the number of equal-width bins by *no.bins*. Then the uth component $dk_u = discretize(k, no.bins)_u$ reports the bin number $r = 1, 2, \dots, no.bins$ into which k_u falls. The *discretize* function is defined in (14.10). Denote by $p(r)$ the relative frequency of the rth bin, i.e., the proportion of components of k that fall into the rth bin. The **frequency distribution** of the connectivity can be estimated with $p(dk) = (p(1), \dots, p(no.bins))$. Using this notation, we define the **connectivity frequency** $p.Connectivity$ (sometimes denoted $p(dk)$ or $p(k)$) as follows:

$$p.Connectivity = p(dk) = p(discretize(k, no.bins)), \tag{1.10}$$

which depends on the number of bins *no.bins*. As default, we set $no.bins = 10$ when discretizing the connectivity vector *Connectivity*.

Many network theorists have studied the properties of the frequency distribution of the connectivity $p.Connectivity = p(dk)$ (Barabasi and Albert 1999; Albert and Barabasi 2000; Jeong et al. 2001; Ravasz et al. 2002; Watts 2002; Han et al. 2004; Barabasi and Oltvai 2004; Pagel et al. 2007). In many (but certainly not all) real network applications, the frequency distribution $p(dk)$ follows a power law:

$$p(r) = PositiveNumber * r^{-\gamma} \tag{1.11}$$

where $PositiveNumber = \frac{1}{\sum_{r=1}^{no.bins} r^{-\gamma}}$ and γ denote positive real numbers. In this case, the network is said to exhibit **scale-free topology** (Barabasi and Albert 1999; Barabasi and Oltvai 2004; Albert et al. 2000) with scaling parameter γ. By taking the log of both sides of (1.11), one can verify that scale-free topology implies a straight line relationship between $log(p(r))$ and $log(r)$:

$$log(p(r)) = -\gamma * log(r) + log(PositiveNumber). \tag{1.12}$$

To measure the extent of a straight line relationship between $log(p(r))$ and $log(r)$, we define the **scale-free topology fitting index**

$$ScaleFreeFit(no.bins) = cor(log(p(dk)), log(BinNo))^2 \tag{1.13}$$

as the square of the correlation coefficient (5.12) between $log(p(dk))$ and $log(BinNo)$, where $BinNo = (1, 2, \ldots, no.bins)$. We often use the following abbreviation $R^2 = ScaleFreeFit$.

Networks whose scale-free topology index R^2 is close to 1 are defined to be approximately scale free. One can visually inspect whether approximate scale-free topology is satisfied by plotting $log(p(k))$ versus $log(k)$ (see Fig. 1.5). In most real networks one observes an inverse relationship between $log(p(k))$and $log(k)$, i.e., γ is positive. Scale-free networks are extremely heterogeneous, and their topology being dominated by a few highly connected nodes (hubs) that link the rest of the less connected nodes to the system. Several models have been proposed for explaining the emergence of the power-law distribution (scale-free topology). For example, it can be explained using a network growth model in which nodes are preferentially attached to already established nodes, a property that is also thought to characterize the evolution of biological systems (Albert and Barabasi 2000). Scale-free networks display a remarkable tolerance against errors (Albert et al. 2000). Many networks satisfy the scale-free property only approximately. For example, Fig. 5.7 shows that for a yeast co-expression network, the connectivity distribution $p(r)$ is better modeled using an **exponentially truncated power law** (Csanyi and Szendroi 2004)

$$p(r) = PositiveNumber * r^{-\gamma} * exp(-\alpha r)$$

where $PositiveNumber = \frac{1}{\sum_{r=1}^{no.bins} r^{-\gamma} * exp(-\alpha r)}$, γ, and α denotes positive real numbers. On a log scale, an exponentially truncated power law is given as:

$$log(p(r)) = -\gamma * log(r) - \alpha r + log(PositiveNumber) \tag{1.14}$$

Potential Uses In Sect. 4.3, we use the scale-free topology index R^2 for formulating the scale-free topology criterion for network construction.

1.3.4 Network Heterogeneity

The *network heterogeneity* measure is based on the variance of the connectivity. Authors differ on how to scale the variance (Snijders 1981). We define it as the coefficient of variation of the connectivity distribution, i.e.,

$$Heterogeneity = \frac{\sqrt{var(k)}}{mean(k)} = \sqrt{\frac{n * sum(k^2)}{sum(k)^2} - 1}. \tag{1.15}$$

This heterogeneity measure is invariant with respect to multiplying the connectivity by a scalar.

Social Network Interpretation of the Heterogeneity: The heterogeneity can be used to measure the variation of popularity (connectivity) across the individuals.

Potential Uses of the Heterogeneity: Describing the reasons for and the meaning of the heterogeneity of complex networks has been the focus of considerable research in recent years (Albert et al. 2000; Watts 2002). As mentioned before, many complex networks have been found to exhibit approximate scale-free topology, which implies that these networks are highly heterogeneous.

1.3.5 Maximum Adjacency Ratio

For weighted networks, we define the *maximum adjacency ratio* of node *i* as follows:

$$MAR_i = \frac{\sum_{j \neq i} (A_{ij})^2}{\sum_{j \neq i} A_{ij}}, \tag{1.16}$$

which is defined if $k_i = \sum_{j \neq i} A_{ij} > 0$. One can easily verify that $0 \leq A_{ij} \leq 1$ implies $0 \leq MAR_i \leq 1$. Note that $MAR_i = 1$ if all nonzero adjacencies take on their maximum value of 1, which justifies the name "maximum adjacency ratio". By contrast, if all nonzero adjacencies take on a small (but constant) value $A_{ij} = \varepsilon$, then $MAR_i = \varepsilon$ will be small.

Social Network Interpretation of the Maximum Adjacency Ratio: $MAR_i = 1$ suggests that the ith individual does not form neutral relationships; this individual either strongly likes or dislikes others since all A_{ij} are either 0 or 1. In contrast, $MAR_i = 0.5$ suggests the ith individual forms less intense relationships with others.

Potential Uses of the Maximum Adjacency Ratio: Since $MAR_i = 1$ for all nodes in an unweighted network, the maximum adjacency ratio is only useful for weighted networks. The MAR can be used to determine whether a hub node forms moderate relationships with a lot of nodes or very strong relationships with relatively few nodes. To illustrate this point, we show in the following simple example that the *MAR* can be used to distinguish nodes that have the same connectivity. Assume a network (labeled by I) for which the adjacency between node 1 and every other node equals $A_{1,j}^{(I)} = 1/(n-1)$. Then $k_1^{(I)} = \sum_{j\neq 1} A_{1,j}^{(I)} = (n-1)/(n-1) = 1$ and $MAR_1^{(I)} = 1/(n-1)$. For a different network (labeled by II) where $A_{1,2}^{(II)} = 1$ and $A_{1,j}^{(II)} = 0$ if $j \geq 3$, the connectivity $k_1^{(II)}$ still equals 1 but $MAR_1^{(II)} = 1$.

As aside, we mention that a directed network analog of MAR_i has been used in the analysis of metabolic fluxes (Almaas et al. 2004).

1.3.6 Network Density

To simplify notation, we will make use of the function *vectorizeMatrix* which turns an $n \times n$ dimensional symmetric matrix A into a vector whose $n*(n-1)/2$ components correspond to the upper-diagonal entries of A, i.e.,

$$vectorizeMatrix(A) = (A_{12}, A_{13}, \ldots, A_{n-1,n}). \tag{1.17}$$

Using this notation, the *network density* (also known as line density (Snijders 1981)) is defined as the mean off-diagonal adjacency and is closely related to the mean connectivity.

$$\begin{aligned} Density &= mean(vectorizeMatrix(A)) \\ &= \frac{\sum_i \sum_{j>i} A_{ij}}{n(n-1)/2} \\ &= \frac{mean(k)}{n-1} \approx \frac{mean(k)}{n}, \end{aligned} \tag{1.18}$$

where $k = (k_1, \ldots, k_n)$ denotes the vector of node connectivities.

Social Network Interpretation: The density measures the overall affection among individuals. A density close to 1 indicates that all individuals strongly like each other, while a density of 0.5 suggests the presence of more ambiguous relationships.

Below, we show that many module detection (and clustering) methods aim to find subnetworks with high density.

1.3.7 Quantiles of the Adjacency Matrix

Quantiles are used to describe the distribution of a variable. The $prob = 0$ quantile of a set of numbers is the minimum, the $prob = 0.25$ quantile is the first quartile, the $prob = 0.50$ quantile is the median, and the $prob = 1.0$ quantile is the maximum. Using this terminology, we define the network concept *prob-th quantile of the adjacency* as the *prob*-th quantile of the *off-diagonal* elements of the adjacency matrix

$$quantile_{prob}(A) = quantile_{prob}(vectorizeMatrix(A)), \tag{1.19}$$

which is the quantile of the vectorized adjacency matrix (1.17). The minimum and median values across the off-diagonal elements of the adjacency matrix are denoted by $quantile_0(A) = min(A)$ and $quantile_{0.5}(A) = median(A)$, respectively. The median adjacency $quantile_{0.5}(A) = median(A)$ can be considered a robust measure of network density. In Sect. 4.5, we use general quantiles for 'calibrating' different networks.

1.3.8 Network Centralization

The *network centralization* (also known as degree centralization (Freeman 1978)) is given by

$$\begin{aligned} Centralization &= \frac{n}{n-2}\left(\frac{max(k)}{n-1} - \frac{mean(k)}{n-1}\right) \\ &= \frac{n}{n-2}\left(\frac{max(k)}{n-1} - Density\right) \\ &\approx \frac{max(k)}{n} - Density. \end{aligned} \tag{1.20}$$

The centralization is 1 for a network with star topology; by contrast, it is 0 for a network where each node has the same connectivity. Note that a regular grid network where $mean(k) = max(k)$ has centralization 0.

Social Network Interpretation of the Centralization: The centralization of the affection network is close to 1, if one individual has loving relationships with all others who in turn strongly dislike each other. In contrast, a centralization of 0 indicates that all individuals are equally popular.

Potential Uses of the Centralization: While the centralization is a widely used measure in social network studies, it has only rarely been used to describe structural differences of metabolic networks (Ma et al. 2004). We have found that the centralization can be used to describe properties of cluster trees (Dong and Horvath 2007; Horvath and Dong 2008).

1.3.9 Clustering Coefficient

The (local) *clustering coefficient* of node i is a density measure of local connections, or "cliquishness". Let us first review its definition for an unweighted network (Watts and Strogatz 1998). If the ith node is connected to a pair of other nodes (direct neighbors), these three build a triple. Thus, the number of triples equals the number of links among the neighbors of node i. If two neighbors of node i are also linked, then these three nodes build a triangle in the unweighted network. Thus, the number of triangles equals the number of links among the neighbors of node i. In an unweighted network, the ith clustering coefficient is defined as the proportion of observed triangles among all possible triangles involving node i.

$$ClusterCoef_i = \frac{\text{number of triangles involving node i}}{\text{number of triples (i.e., possible triangles)}}. \tag{1.21}$$

Algebraically, the clustering coefficient can be calculated as follows:

$$\begin{aligned} ClusterCoef_i &= \frac{\sum_{j \neq i} \sum_{k \neq i,j} A_{ij} A_{jk} A_{ki}}{\left(\sum_{j \neq i} A_{ij}\right)^2 - \sum_{j \neq i} (A_{ij})^2} \\ &= \frac{\sum_{j \neq i} \sum_{k \neq i} A_{ij} A_{jk} A_{ki} - \sum_{j \neq i} A_{ij}^2 A_{ij}}{\left(\sum_{j \neq i} A_{ij}\right)^2 - \sum_{j \neq i} (A_{ij})^2}. \end{aligned} \tag{1.22}$$

We defined the clustering coefficient for a weighted network by simply evaluating (1.22) on a weighted adjacency matrix (Zhang and Horvath 2005). One can easily prove that $0 \leq A_{ij} \leq 1$ implies that $0 \leq ClusterCoef_i \leq 1$.

Social Network Interpretation of the Clustering Coefficient: The higher the clustering coefficient of an individual, the higher is the affection among his friends. The clustering coefficient is zero if all of his friends strongly dislike each other.

Potential Uses of the Clustering Coefficient: The mean clustering coefficient has been used to measure the extent of module structure present in a network. The relationship between the clustering coefficient and the connectivity has been used to describe structural (hierarchical) properties of networks (Ravasz et al. 2002).

1.3.10 Hub Node Significance

Now we will define a network concept that makes use of a node significance measure GS_i. To measure the association between connectivity and node significance, we propose the following measure of *hub node significance*:

$$HubNodeSignif = \frac{\sum_i GS_i K_i}{\sum_i (K_i)^2}, \tag{1.23}$$

When GS_i is proportional to the scaled connectivity ($GS_i = cK_i$), the hub node significance equals the constant of proportionality: $HubNodeSignif = c$. The hub node significance equals the slope of the regression line between GS_i and the scaled connectivity K_i if the intercept term is set to 0 (see Figs. 1.5 and 5.2c).

Social Network Interpretation of the Hub Node Significance: Assume that the node significance measures the grade point average of the ith individual. Then the hub node significance can be used to assess whether there is a relationship between popularity (connectivity) and grade point average.

Potential Uses of the Hub Node Significance: Several studies have shown that the relationship between connectivity and node significance (i.e., the hub node significance) carries important biological information. For example, in the analysis of yeast networks where nodes correspond to genes, highly connected hub genes are essential for yeast survival, and hub genes tend to be preserved across species (Albert et al. 2000; Jeong et al. 2001; Albert and Barabasi 2002; Carter et al. 2004; Han et al. 2004; Oldham et al. 2006). A detailed analysis shows that the positive relationship between connectivity and knockout essentiality cannot always be observed (Carlson et al. 2006), i.e., the hub gene significance can be close to 0.

1.3.11 Network Significance Measure

We define the *network significance measure* as the average node significance of the nodes:

$$NetworkSignif = \frac{\sum_i GS_i}{n}. \tag{1.24}$$

Social Network Interpretation of the Network Significance: The network significance simply measures the average grade point average among the individuals.

1.3.12 Centroid Significance and Centroid Conformity

A **network centroid** is a suitably defined centrally located node in a network. A centroid can be defined in many different ways, e.g., based on connectivity or other centrality measures. For example, the centroid can be defined as (one of the) most highly connected node(s) in the network. If a node significance measure GS_i has been defined for the network, then the *centroid significance* is simply the node significance of the centroid:

$$CentroidSignif = GS_{i.centroid}, \tag{1.25}$$

where *i.centroid* is the node index of the centroid.

We define the *centroid conformity* of the ith node as the adjacency between the centroid and the ith node:

$$CentroidConformity_i = A_{i,i.centroid}. \tag{1.26}$$

Social Network Interpretation of the Centroid Conformity: In our affection network, we choose the most popular individual as centroid; then his or her grade point average is the centroid significance. The centroid conformity of the ith individual equals his or her affection (connection strength) with the most popular individual. The mean centroid conformity equals the average amount of affection felt by the most popular individual.

Potential Uses of the Centroid Conformity: In Chap. 6, we will characterize networks where the adjacency $A_{i,j}$ can be approximated by a product of the centroid conformities:

$$A_{i,j} \approx CentroidConformity_i CentroidConformity_j.$$

In Chap. 3, we use this insight to derive relationships among seemingly disparate network concepts.

1.3.13 Topological Overlap Measure

In an unweighted network, the number of direct neighbors of nodes i and j is given by $\sum_{l\neq i,j} A_{il}A_{jl}$. Then $numerator_{ij} = \sum_{l\neq i,j} A_{il}A_{jl} + A_{ij}$ equals the number of shared neighbors plus 1 if $A_{ij} = 1$ (i.e., if a direct link exists between nodes i and j). The *topological overlap measure* $TopOverlap_{ij}$ is a normalized version of $numerator_{ij}$. Specifically, topological overlap measure between nodes i and j is given by:

$$TopOverlap_{ij} = \begin{cases} \frac{\sum_{l\neq i,j} A_{il}A_{jl} + A_{ij}}{min\{\sum_{l\neq i} A_{il} - A_{ij}, \sum_{l\neq j} A_{jl} - A_{ij}\} + 1} & \text{if } i \neq j \\ 1 & \text{if } i = j. \end{cases} \tag{1.27}$$

Note that the denominator will never be 0 due to the addition of 1. The formula for the topological overlap matrix was first suggested in the context of unweighted networks, more specifically for protein–protein interaction networks (supplementary material of (Ravasz et al. 2002)).

We adapted the definition of the topological overlap measure (1.27) to weighted network and co-expression networks (Zhang and Horvath 2005; Carlson et al. 2006; Ghazalpour et al. 2006; Oldham et al. 2006; Horvath et al. 2006; Li and Horvath 2007). In the following, we use $0 \leq A_{ij} \leq 1$ to prove that $0 \leq TopOverlap_{ij} \leq 1$. Since $\sum_{l\neq i,j} A_{il}A_{jl} \leq \sum_{l\neq i} A_{il} - A_{ij}$ and $\sum_{l\neq i,j} A_{il}A_{jl} \leq \sum_{l\neq j} A_{jl} - A_{ij}$, we find that

$$\sum_{l\neq i,j} A_{il}A_{jl} \leq min\left\{\sum_{l\neq i} A_{il} - A_{ij}, \sum_{l\neq j} A_{jl} - A_{ij}\right\}.$$

Since $A_{ij} \leq 1$, this implies that the numerator of $TopOverlap_{ij}$ is smaller than the denominator.

Social Network Interpretation of the Topological Overlap Measure: In our affection network, two individuals have a high topological overlap if they like and dislike the same people. If two individuals share the same friends, they may be part of a clique, which we refer to as a module.

Potential Uses of the Topological Overlap Measure: Erroneous adjacencies can have a strong impact on network topological inference. To counter the effects of spurious or sparse adjacencies, it can be advantageous to use node similarity measures that are based on common interaction partners or on topological metrics (Ravasz et al. 2002; Brun et al. 2003; Chen et al. 2006; Chua et al. 2006). We will use the topological overlap measure as robust measure of interconnectedness.

The topological overlap measure has two major uses: first, it can be used in conjunction with a clustering method to define network modules as described below; second, it can be used to define the neighborhood of an initial set of nodes in a network. Intuitively speaking, a neighborhood is comprised of nodes that are highly connected to a given set of nodes. Below, we describe how neighborhood analysis facilitates a guilt-by-association screening strategy for finding nodes that interact with a given set of initial nodes.

As caveat, we point out that topological overlap-based analyses will only be useful in applications that satisfy the following basic assumption: *The more neighbors are shared between two nodes, the stronger is their relationship.* Our applications and several publications provide empirical evidence that the topological overlap matrix leads to biologically meaningful results (Ravasz et al. 2002; Zhang and Horvath 2005; Carlson et al. 2006; Ghazalpour et al. 2006; Oldham et al. 2006; Horvath et al. 2006; Li and Horvath 2007; Yip and Horvath 2007). But there will undoubtedly be situations when alternative similarity measures are preferable.

1.3.14 Generalized Topological Overlap for Unweighted Networks

Here, we describe extension of the topological overlap measure (referred to as generalized topological overlap, GTOM), which considers longer ranging relationships between nodes (Yip and Horvath 2007). Importantly, this generalization only applies unweighted networks. This work was done with **Andy Yip**.

For an unweighted network, one can define a distance measure $\mathrm{d(u,i)}$ as the length of the shortest path between nodes u and i. Define

$$N_1(i,j) = \{u \neq i,j \mid \mathrm{d(u,i)} = 1 \ \& \ \mathrm{d(u,j)} = 1\}$$

as the set of common, directly linked neighbors shared by i and j, i.e., the nodes in this set are one step away from both i and j. Similarly, define

$$N_1(i,-j) = \{u \neq i,j \mid \mathrm{d(u,i)} = 1\}$$

as the set of one-step neighbors of i excluding j. The number of elements in these sets is given by

$$|(N_1(i,j))| = \sum_{u \neq i,j} A_{iu}A_{ju}$$
$$|N_1(i,-j)| = \sum_{u \neq i,j} A_{iu}, \tag{1.28}$$

where $|\cdot|$ denotes the number of elements. Note that $|N_1(i,-j)|$ equals the connectivity k_i.

For an unweighted network, we can express the topological overlap measure as follows:

$$t_{ij} = \frac{|N_1(i,j)| + A_{ij}}{min\{|N_1(i,-j)|, |N_1(-i,j)|\} + 1}. \tag{1.29}$$

We use the notation t_{ij} instead of $TopOverlap_{ij}$ to remind the reader that these formulas only apply for an unweighted network. In the following, we extend the topological overlap measure to keep track of longer ranging interactions. Toward this end, we define multistep neighbors. For an unweighted network, one can define the $m = 2$ neighbors of node i as those nodes that can be reached within two steps. For an unweighted network, $t_{ij}^{[m]}$ keeps track of the number of shared neighbors that can be reached within m steps (Yip and Horvath 2007).

The mth order GTOM measure is constructed by (a) counting the number of m-step neighbors that a pair of nodes share and (b) normalizing it to take a value between 0 and 1. Specifically, denote by $N_m(i,j)$ (with $m > 0$) the set of nodes (excluding i,j) that are reachable from i and j within a path of length m, i.e.,

$$N_m(i,j) = \{l \neq i,j \mid d(l,i) \leq m \;\&\; d(l,j) \leq m\}. \tag{1.30}$$

Similarly, we define

$$N_m(i,-j) = \{l \neq i,j \mid d(l,i) \leq m\}. \tag{1.31}$$

Generalizing (1.28) to m-step neighbors, we define

$$t_{ij}^{[m]} = \frac{|N_m(i,j)| + A_{ij}}{min\{|N_m(i,-j)|, |N_m(-i,j)|\} + 1}. \tag{1.32}$$

We call the matrix $T^{[m]} = [t_{ij}^{[m]}]$ the *mth order generalized topological overlap matrix* (GTOMm). This quantity simply measures the agreement between the nodes that are reachable from i and from j within m steps. When $m = 1$, we obtain back the original TOM in formula (1.27). It is convenient and intuitive to define the zeroth order topological overlap matrix GTOM0 as $T^{[0]} = A$, which only considers the direct link between the pair of nodes in question. In an exercise, you are asked to derive a computational formula for $T^{[m]}$ (1.32).

Figure 1.1 shows the generalized topological overlap measure in simple examples.

Figures 1.2 and 1.3 present a simple network where GTOM1 and GTOM2 lead to different neighborhoods.

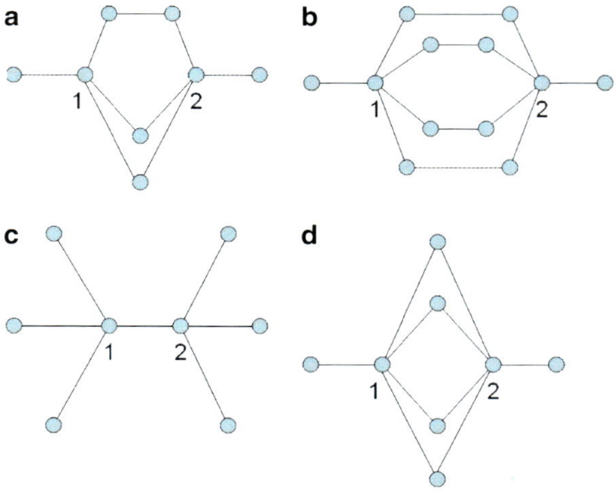

Fig. 1.1 Simple examples for illustrating the computation of the generalized topological overlap measure. (**a**) $A_{12} = 0, t_{12}^{[1]} = 0.4$, $t_{12}^{[2]} = 0.67$, (**b**) $A_{12} = 0$, $t_{12}^{[1]} = 0$, $t_{12}^{[2]} = 0.8$, (**c**) $A_{12} = 1$, $t_{12}^{[1]} = 0.25$, $t_{12}^{[2]} = 1$, (**d**) $A_{12} = 0$, $t_{ij}^{[1]} = 0.67$, $t_{12}^{[2]} = 0.8$

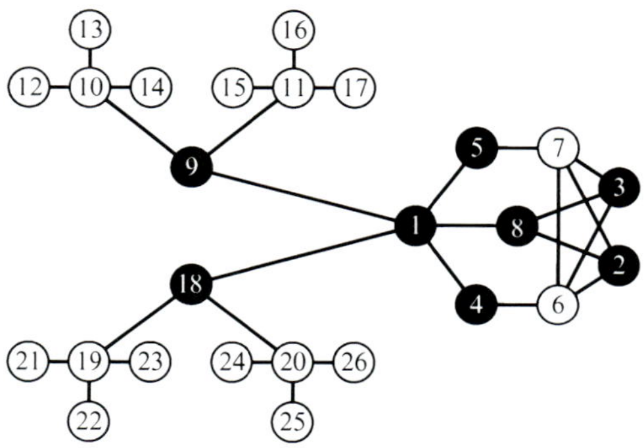

Fig. 1.2 GTOM1-based neighborhood of node 1. The eight closest neighbors of node 1 (with regard to GTOM1) are colored in black. Note that nodes 9 and 18 are part of this neighborhood

1.3.15 Multinode Topological Overlap Measure

In the following, we describe how to generalize the topological overlap measure to three or more nodes. This work was done with **Ai Li** (Li and Horvath 2007). While the standard TOM measure measures the number of neighbors shared by two nodes, the corresponding multinode measure keeps track of shared neighbors among multiple nodes. In light of formula (1.29), it is natural to define the MTOM of three different nodes i, j, k as follows:

$$t_{ijk} = \frac{|(N_1(i,j,k)| + A_{ij} + A_{ik} + A_{jk}}{min\{|N_1(i,j,-k)|, |N_1(i,-j,k)|, |N_1(-i,j,k)|\} + \binom{3}{2}}, \tag{1.33}$$

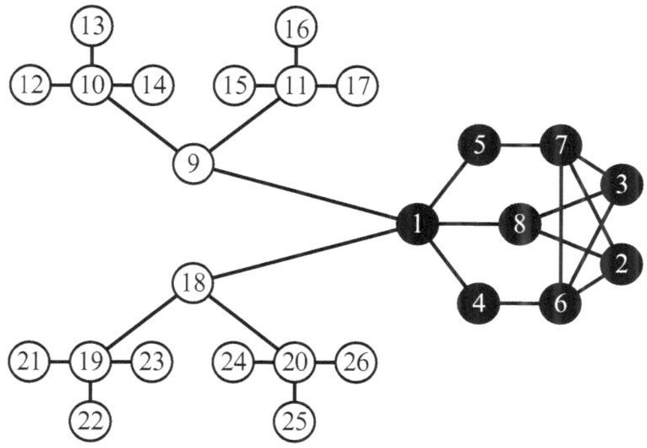

Fig. 1.3 GTOM1-based neighborhood of node 1. Note that nodes 9 and 18 are no longer among the eight closest nodes of node 1 since they do not share many of their two-step neighbors with node 1

where

$$\begin{aligned} N_1(i,j,-k) &= \{u \neq i,j,k \mid \mathrm{d(u,i)} \leq 1 \ \& \ \mathrm{d(u,j)} \leq 1\} \\ N_1(i,j,k) &= \{u \neq i,j,k \mid \mathrm{d(u,i)} \leq 1 \ \& \ \mathrm{d(u,j)} \leq 1 \ \& \ \mathrm{d(u,k)} \leq 1\}. \end{aligned}$$

Here $N(i,j,-k)$ can be regarded as the set of the neighbors shared by i and j excluding k. The following algebraic formulas can be used to calculate these quantities:

$$\begin{aligned} |(N_1(i,j,k))| &= \sum_{u \neq i,j,k} A_{iu}A_{ju}A_{ku} \\ |N_1(i,j,-k)| &= \sum_{u \neq i,j,k} A_{iu}A_{ju}. \end{aligned} \tag{1.34}$$

The binomial coefficient $\binom{3}{2} = 3$ in the denominator of (1.33) is an upper bound of $A_{ij} + A_{ik} + A_{jk}$ and equals the number of connections that can be formed between i, j, and k. Analogous to the proof for two nodes, one can prove that $0 \leq t_{ijk} \leq 1$. In the same way, we can define the MTOM of four nodes as follows:

$$t_{ijkl} = \frac{|(N_1(i,j,k,l)| + A_{ij} + A_{ik} + A_{il} + A_{jk} + A_{jl} + A_{kl}}{min\{|N_1(i,j,k,-l)|, |N_1(i,j,-k,l)|, |N_1(i,-j,k,l)| + |N_1(-i,j,k,l)|\} + \binom{4}{2}}, \tag{1.35}$$

where

$$\begin{aligned} N_1(i,j,k,-l) &= \{u \neq i,j,k,l \mid d(u,i) \leq 1 \ \& \ d(u,j) \leq 1 \ \& \ d(u,k) \leq 1\} \\ N_1(i,j,k,l) &= \{u \neq i,j,k \mid d(u,i) \leq 1 \ \& \ d(u,j) \leq 1 \ \& \ d(u,k) \leq 1 \ \& \ d(u,l) \leq 1\}. \end{aligned} \tag{1.36}$$

These quantities can be computed as follows:

$$\begin{aligned}|(N_1(i,j,k,l))| &= \sum_{u\neq i,j,k,l} A_{iu}A_{ju}A_{ku}A_{lu} \\ |N_1(i,j,k,-l)| &= \sum_{u\neq i,j,k,l} A_{iu}A_{ju}A_{ku}. \end{aligned} \tag{1.37}$$

It is straightforward to extend the definition of the topological overlap measure to more nodes. Note that the algebraic formulas for MTOM do not require that the adjacency matrix take on binary values. Since the formulas remain mathematically meaningful as long as $0 \leq A_{ij} \leq 1$, it is straightforward to use them for generalizing MTOM to **weighted networks**.

The MTOM measure is implemented in the stand-alone MTOM software, which can be downloaded from: www.genetics.ucla.edu/labs/horvath/MTOM/

1.4 Neighborhood Analysis in PPI Networks

The goal of neighborhood analysis is to find a set of nodes (the neighborhood) that is similar to an initial "seed" set of nodes. Mathematically speaking, a network interconnectedness measure is simply a node similarity measure, e.g., the adjacency matrix or the topological overlap measures could be used.

If individual network connections are susceptible to noise, it can be advantageous to use a robust interconnectedness measure, e.g., the topological overlap measure or the generalized topological overlap measure (GTOM) (Yip and Horvath 2007). A simple approach for defining a neighborhood of node i is to choose the nodes with highest adjacencies A_{ij}. In an unweighted network, this amounts to choosing the directly connected neighbors of node i. But sometimes it is advantageous to use the GTOM measures (described in Sect. 1.3.14) as we will illustrate using a protein–protein interaction network of the fly (Drosophila).

1.4.1 GTOM Analysis of Fly Protein–Protein Interaction Data

We downloaded the data from the general repository for interaction datasets (BioGRID). Protein knockout experiments in lower organisms (yeast, fly, worm) have shown that some proteins are essential for survival. In lower organisms, highly connected hub genes tend to be essential (Jeong et al. 2001, 2003; Hahn and Kern 2005; Carlson et al. 2006). Here we define neighborhoods around several seed sets of essential proteins. Specifically, we considered the 30 most highly connected essential proteins in the network and refer to them as essential hubs. The goal of the neighborhood analysis around each essential hub was to implicate other essential proteins.

The analysis assumes that a protein that is connected to an essential protein is likely to be essential as well. We used different GTOM measures to quantify whether two proteins are closely interconnected. The GTOM0 measure (i.e., the adjacency matrix) allows one to identify the directly linked neighbors of each essential hub protein. On average, each essential hub contained about 40 direct neighbors (according to GTOM0), which is why we considered a maximum neighborhood size of 40 for the GTOM1 and GTOM2 measures as well.

To assess the performance of GTOM measures, we kept track of how many essential proteins were found in the neighborhoods of each of the 30 essential hub proteins. The results were averaged over the 30 resulting neighborhood analyses. The proportions of essential proteins among the nearest neighbors of the essential hub proteins are shown in Fig. 1.4. Note that proteins that are close to essential hub proteins with respect to GTOM2 are more likely to be essential than proteins that are close with respect to GTOM1. Thus, keeping track of higher order neighbors increases the biological signal in this neighborhood analysis. But we should point out that the GTOM1 measure performs sometimes better in other neighborhood applications.

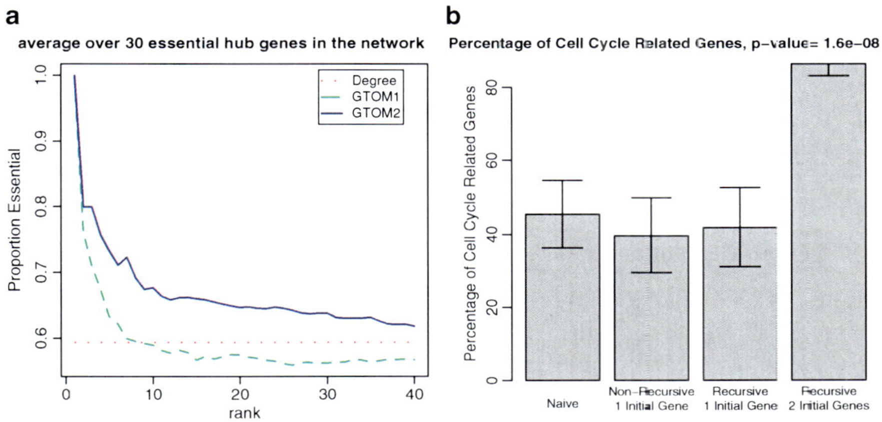

Fig. 1.4 Neighborhood analysis in protein–protein interaction networks. (**a**) Drosophila protein–protein interaction network. Proportions of essential proteins (y-axis) among the k nearest neighbors (x-axis) averaged over 30 essential hub proteins. For GTOM0 (binary adjacency matrix), the dashed horizontal line corresponds to the average proportion of essential proteins among the directly linked neighbors of the 30 essential hubs. Here GTOM2 outperforms GTOM0 or GTOM1. (**b**) Yeast Protein–Protein Physical Interaction Network (MIPS Data). Comparing the percent of cell cycle proteins R (y-axis) in different MTOM neighborhoods. Note that the recursive approach involving a seed of two cell cycle-related "hub" proteins performs better than approaches based on a single seed protein. Both recursive and the non-recursive MTOM neighborhood analysis involving a *single* seed protein tie with the naive approach of constructing a neighborhood based on the adjacency measure

1.4.2 MTOM Analysis of Yeast Protein–Protein Interaction Data

Here we illustrate how to use the multinode topological overlap measure (described in Sect. 1.3.15) in neighborhood analysis. This method is implemented in the MTOM software (Li and Horvath 2007). A limitation of many network similarity measures is that they measure pairwise similarity. While pairwise similarities are useful for clustering procedures, we show that it can be advantageous for neighborhood analysis to use a multi-node similarity measure such as MTOM.

Given an initial seed neighborhood, we consider two basic approaches for defining a neighborhood based on the concept of multi-node topological overlap. The default approach is to build the neighborhood recursively. The non-recursive alternative is computationally faster but produces less interconnected neighborhoods. The MTOM-based neighborhood analysis requires as input an initial seed neighborhood comprised of at least one node and the requested final size of the neighborhood $S \geq 1$.

1. **Recursive approach**
   a. For each node outside of the neighborhood set, compute its MTOM value with the nodes in the current neighborhood set.
   b. Add the node with the highest MTOM value to the neighborhood set.
   c. Repeat steps (a) and (b) until the neighborhood size S is reached.
2. **Non-recursive approach**
   a. For each node outside of the neighborhood set, compute its MTOM value with the nodes in the initial neighborhood set (the start seed).
   b. Choose the S nodes with the highest MTOM values as the neighborhood.

Since the recursive approach leads to neighborhoods with higher MTOM values, it is preferable over the computationally faster, non-recursive approach.

In the following, we describe an application of MTOM-based neighborhood analysis to predict cell cycle-related proteins in a yeast protein–protein interaction network (from the Munich Information Center for Protein Sequences (Guldener et al. 2006)). We restricted the analysis to the largest connected component comprised of 3,858 proteins with 7,196 direct pairwise interactions. To compare different neighborhood analysis approaches, we studied the neighborhoods of subsets of 101 cell cycle-related proteins found in the Kyoto Encyclopedia of Genes and Genomes (KEG). We considered a neighborhood size of 10. Within each neighborhood, we determined the number C of cell cycle-related proteins. We found that C is significantly correlated with the network connectivity k of the initial protein (Spearman correlation $r = 0.36$, p value ≤ 0.001). For this reason, we focused the neighborhood analysis on subsets of the 50 most highly connected "hub" cell cycle-related proteins. As can be seen from Fig. 1.4b, the neighborhoods of cell cycle genes tend to be enriched with other cell cycle genes as well. A major advantage of the MTOM-screening approach is the ability to input multiple initial nodes as seed set. Figure 1.4a shows that an initial seed neighborhood comprised of two cell cycle-related hub proteins leads to far better results than using a single protein as input. But we found that this is only true for protein pairs that have high topological overlap.

1.5 Adjacency Function Based on Topological Overlap

Let us now return to the setting of a possibly weighted adjacency matrix. In practice, an original network adjacency matrix $A^{original}$ is often transformed into new network adjacency matrix denoted by A. For example, a transformation can be used to change the topological properties of a network. We define an **adjacency function** AF as a matrix valued function that maps an $n \times n$ dimensional adjacency matrix $A^{original}$ onto a new $n \times n$ dimensional network adjacency

$$A = AF(A^{original}).$$

In the following, we will describe an important adjacency function based on the topological overlap measure (1.27). The **topological overlap matrix (TOM)-based adjacency function** AF^{TOM} maps an original adjacency matrix $A^{original}$ to the corresponding topological overlap matrix, i.e.,

$$AF^{TOM}(A^{original})_{ij} = \frac{\sum_{l \neq i,j} A_{il}^{original} A_{l,j}^{original} + A_{ij}^{original}}{min\left(\sum_{l \neq i} A_{il}^{original}, \sum_{l \neq j} A_{jl}^{original}\right) - A_{ij}^{original} + 1}. \quad (1.38)$$

The TOM-based adjacency function AF^{TOM} is particularly useful when the entries of $A^{original}$ are sparse (many zeroes) or susceptible to noise. In this case it can be advantageous to replace $A^{original}$ by $AF^{TOM}(A^{original})$. This replaces the original adjacencies by a measure of interconnected that is based on shared neighbors. In Sect. 3.10, we describe how AF^{TOM} is used to transform sparse unweighted protein–protein interaction networks into weighted networks. The topological overlap measure can serve as a filter that decreases the effect of spurious or weak connections, and it can lead to more robust networks (Li and Horvath 2007; Yip and Horvath 2007; Dong and Horvath 2007).

1.6 R Functions for the Topological Overlap Matrix

There are two different `WGCNA` functions for calculating TOM based on the adjacency matrix. The first function `TOMsimilarity` takes as input a weighted (or possibly unweighted) adjacency matrix and outputs the first order ($m = 1$) GTOM measure. This function can*not* calculate higher order ($m > 1$) generalizations of the TOM measure. Often it is more convenient to calculate the TOM-based dissimilarity measure $dissTOM = 1 - TOM$, which can be accomplished with the R function `TOMdist`.

The second function `GTOMdist` calculates $distGTOM_{ij} = 1 - t_{ij}^{[m]}$, i.e., 1 minus the generalized topological overlap matrix. It is worth repeating that the input of this function is an *unweighted* network adjacency matrix which has binary entries (1 or 0). It is straightforward to use this function to calculate the generalized topological overlap measure as follows: `TOMm=1-TOMdist(unweightedADJ,m)` The following R code shows how to calculate the minimum value of the off-diagonal elements of a generalized topological overlap matrix:

```
library(WGCNA)
# number of nodes in the network
n=500
# set the seed of the random number generator
set.seed(1)
BinaryVector=sample(c(0,1),size=n*n, prob=c(.75,.25), replace=T)
BinaryMatrix=matrix(BinaryVector,nrow=n,ncol=n)
# here we define a symmetric unweighted adjacency matrix whose
unweightedADJ=BinaryMatrix * t(BinaryMatrix)
# this is the vector of values of m
mVector=0:5
minOffDiagonal=rep(NA,length(mVector))
for (i in 1:length(mVector) ){
m=mVector[i]
# Next we calculate the GTOMm matrix
# as 1 minus the corresponding dissimilarity
GTOMm=1-GTOMdist(unweightedADJ,degree=m)
# minimum off-diagonal element of GTOMm
minOffDiagonal[i]=min(GTOMm)
}
plot(mVector,minOffDiagonal,xlab="m")
abline(h=1)
minOffDiagonal
# output
0.0000000 0.0000000 0.6710098 0.9960000 0.9960000 0.9960000
```

One can show that the entries of the topological overlap matrix increase with m. In an exercise you are asked to show that the entries of GTOMm approximate 1 if m is chosen large enough.

The multinode topological overlap measure (MTOM) is implemented in a standalone software of the same name, which can be downloaded from: www.genetics.ucla.edu/labs/horvath/MTOM/.

1.7 Network Modules

Similar to the term "cluster", there is no general agreement on the meaning of the term "module". To make our results widely applicable, we provide a very general and abstract definition: a module is a set of nodes which forms a subnetwork. The subnetwork comprised of module nodes is referred to as network module. Since a particular network module may encode a pathway or a protein complex, these special types of networks have great practical importance in biology. Assume that a module detection method (e.g., a clustering procedure) has found Q modules. Denote by $\mathcal{M}_q$ the set of node indices that correspond to the nodes in module q. We denote the adjacency matrix of the nodes inside the qth module by $A^{(q)}$. Analogously, we define $GS^{(q)}$ as the node significance measure restricted to the module nodes. Denote by $n^{(q)}$ the number of nodes inside the qth module.

Here we briefly mention an approach for defining network modules which is used in most of our applications. Typically, **we define modules as clusters of nodes that are highly interconnected** (as measured by the topological overlap measure). To simplify our notation, we introduce the **dissimilarity transformation** $D(A)$, which turns an adjacency matrix (which is a measure of node similarity) into a measure of *dissimilarity* by subtracting it from 1, i.e.,

$$D_{ij}(A) \equiv 1 - A_{ij}. \tag{1.39}$$

Note that $D(A)$ does not satisfy our definition of an adjacency matrix since its diagonal elements equal 0. Alternatively, one can use the topological overlap matrix (1.27) to define the *TOM-based dissimilarity measure*

$$\begin{aligned} dissTopOverlap_{ij} &= D_{ij}\left(AF^{TOM}(A)\right) \\ &= 1 - TopOverlap_{ij} \\ &= 1 - \frac{\sum_{u \neq i,j} A_{iu}A_{uj} + A_{ij}}{min(\sum_{u \neq i} A_{iu}, \sum_{u \neq j} A_{ju}) + 1 - A_{ij}}. \end{aligned} \tag{1.40}$$

As detailed in Sect. 8.4, we typically use the TOM-based dissimilarity as input of average linkage hierarchical clustering (Kaufman and Rousseeuw 1990), which results in a cluster tree (dendrogram) of the network. Next, network modules are defined as branches of the dendrogram. Hierarchical clustering and branch cutting are described in Sects. 8.4 and 8.6, respectively. This module detection procedure was originally proposed by Ravasz et al. 2002 for the setting of unweighted networks (Ravasz et al. 2002), but we find that it also works well for weighted networks. For example, Fig. 3.1 shows cluster trees (dendrograms) involving a gene network application. Genes or proteins of proper modules (branches) are assigned a color (e.g., turquoise, blue, etc.). Genes outside any proper module are colored gray. To date, this module detection approach has led to biologically meaningful modules in more than a hundred applications (see, e.g., Zhang and Horvath 2005; Horvath et al. 2006; Carlson et al. 2006; Ghazalpour et al. 2006; Gargalovic et al. 2006; Dong and Horvath 2007; Oldham et al. 2006, 2008).

While we often use hierarchical clustering to define modules, we emphasize that many alternative methods could be used. In the following, we assume that a module is a subnetwork inside a larger network.

1.8 Intramodular Network Concepts

*Intra*modular network concepts measure topological properties within a module.

Let us start out assuming that a module assignment Cl is available so that each node is assigned to exactly one of Q modules. For example, $Cl(i) = q$ if the ith node is in module q. The number of nodes in the qth module is denoted by $n^{(q)}$. Module assignment could be based on prior knowledge (e.g., Cl could encode groupings based on gene ontology information), or it could be the result of a clustering procedure (as described below and in Chap. 8). The following results do not depend on any particular module detection method. Instead, they are applicable for any network module, i.e., a subset of nodes which forms a subnetwork which is encoded by an $n^{(q)} \times n^{(q)}$ dimensional adjacency matrix $A^{(q)}$. Sometimes a corresponding node significance measure $GS^{(q)}$ is also available. We use the superscript (q) to denote quantities associated with the qth module. But for notational convenience, we sometimes omit superscript (q) when the context is clear.

An intramodular network concept is simply a (fundamental) network concept defined for $A^{(q)}$ and/or $GS^{(q)}$.

In the following, we present some particularly noteworthy intramodular network concepts. The **intramodular connectivity** $k_i^{(q)}$ (sometimes denoted by $\mathrm{kIM}_i^{(q)}$) is defined as the sum of connection strengths to other nodes within the same module, i.e.,

$$k_i^{(q)} = \mathrm{kIM}_i^{(q)} = \sum_{\substack{j \in \mathscr{M}_q \\ j \neq i}} a_{ij}^{(q)}, \tag{1.41}$$

where $\mathscr{M}_q$ is the set of node indices that correspond to the nodes in module q. Nodes with high intramodular connectivity are referred to as intramodular hub nodes. Intramodular connectivity has been found to be an important complementary node screening variable for finding biologically important genes (Horvath et al. 2006; Gargalovic et al. 2006; Oldham et al. 2006).

The **intramodular density** is defined as:

$$\begin{aligned} Density^{(q)} &= \sum_{i \in \mathscr{M}_q} \sum_{\substack{j \in \mathscr{M}_q \\ j \neq i}} \frac{A_{ij}^{(q)}}{n^{(q)}(n^{(q)} - 1)} \\ &= mean\left(vectorizeMatrix(A^{(q)})\right). \end{aligned}$$

The density of nodes in a subnetwork (e.g., a module) can be used to find out whether this subnetwork is tight or cohesive. The goal of many module detection methods is to find clusters of nodes with high density (see Sect. 8.1).

We refer to the network significance (1.24) of a network module $A^{(q)}$ simply as the **module significance measure**, i.e., the module significance is the average node significance of the module nodes:

$$ModuleSignif^{(q)} = NetworkSignif(A^{(q)}) = \frac{\sum_{i \in \mathcal{M}_q} GS_i^{(q)}}{n^{(q)}}. \tag{1.42}$$

The module significance measure can be used to address a major goal of gene network analysis: the identification of biologically significant modules or pathways.

The **intramodular centroid conformity** $CentroidConformity_i^{(q)}$ with regard to the centroid (or medoid) in the qth network module $A^{(q)}$ is defined as follows:

$$CentroidConformity_i^{(q)} = a_{i,i.centroid}^{(q)}. \tag{1.43}$$

In Chap. 3.10, we will provide empirical evidence that for many network modules $a_{i,j}^{(q)}$ can be approximated by a product of conformities: $a_{i,j}^{(q)} \approx CentroidConformity_i^{(q)} * CentroidConformity_j^{(q)}$.

1.9 Networks Whose Nodes Are Modules

Here we outline how to define a network among modules, i.e., each node in the network corresponds to a module. Assume that we are studying two modules denoted by q_1 and q_2, respectively. Denote by $\mathcal{M}_{q_1}$ the set of $n^{(q_1)}$ nodes inside module q_1. The adjacencies between nodes of the two modules can be represented by an $n^{(q_1)} \times n^{(q_2)}$ dimensional sub-matrix $A^{(q_1,q_2)}$ of the full adjacency matrix A.

To define a measure of adjacency between the two modules, we summarize the matrix $A^{(q_1,q_2)}$ by a number between 0 and 1:

$$\begin{aligned}
A_{q_1,q_2}^{ave} &= mean(A^{(q_1,q_2)}) = \frac{\sum_{i \in \mathcal{M}_{q_1}} \sum_{j \in \mathcal{M}_{q_2}} A_{ij}}{n^{(q_1)} n^{(q_2)}}, \\
A_{q_1,q_2}^{max} &= max(A^{(q_1,q_2)}) = max_{i \in \mathcal{M}_{q_1}, j \in \mathcal{M}_{q_2}} A_{ij}, \\
A_{q_1,q_2}^{min} &= min(A^{(q_1,q_2)}) = min_{i \in \mathcal{M}_{q_1}, j \in \mathcal{M}_{q_2}} A_{ij}.
\end{aligned} \tag{1.44}$$

Since A^{ave} uses an average, it is statistically more robust than the maximum- or the minimum-based inter-adjacency measures. But in specific applications, the minimum- and maximum-based measures can be useful as well. The inter-adjacency measures (1.44) can be used to define a network between modules, e.g.,

$$A_{q_1,q_2} = \begin{cases} A_{q_1,q_2}^{ave} & \text{if } q_1 \neq q_2 \\ 1 & \text{if } q_1 = q_2 \end{cases}. \tag{1.45}$$

Denote by $A_{modules}$ the $Q \times Q$ dimensional symmetric matrix whose q_1, q_2 element is given by $A_{q_1 q_2}$ (which measures the adjacency between the two modules). The diagonal elements of $A_{modules}$ are set to 1. Note that $A_{modules}$ can be interpreted as adjacency matrix between modules, i.e., it represents a (weighted) network whose nodes are modules. Many alternative approaches exist for defining networks whose nodes are modules (see, e.g., Sects. 2.3 and 6.5). One can also define the adjacency between modules (comprised of overlapping sets of nodes) by assessing the probability that the two index sets $\mathcal{M}_{q_1}$ and $\mathcal{M}_{q_2}$ overlap. The probability that the two index sets overlap can be calculated based on the hypergeometric distribution (i.e., Fisher's exact test) (see Sect. 14.3). In gene network applications, adjacencies between modules have also been defined by measuring the probability of overlap between gene enrichment categories (Oldham et al. 2008).

1.10 Intermodular Network Concepts

*Inter*modular network concepts measure topological properties among modules. Here we use the notation from Sect. 1.9. A *fundamental intermodular network concept* $NC(q_1, q_2) = NCF(A^{(q_1,q_2)}, A^{(q_1,q_1)}, A^{(q_2,q_2)})$ is a function of $A^{(q_1,q_2)}$, $A^{(q_1,q_1)}$, and $A^{(q_2,q_2)}$. For example, the **geometric mean density of two modules** is defined as:

$$Density(q_1, q_2) = \sqrt{Density^{(q_1)} Density^{(q_2)}}. \tag{1.46}$$

Now we will describe network concepts that can be used to measure whether two modules are separated (distinct) from one another. Our **module separability statistics** contrast *inter*modular adjacencies (1.44) with *intra*modular adjacencies. We define separability statistics as 1 minus the ratio of intermodular adjacency divided by intramodular density:

$$separability.ave(q_1, q_2) = 1 - \frac{A^{ave}_{q_1,q_2}}{Density(q_1, q_2)}, \tag{1.47}$$

$$separability.max(q_1, q_2) = 1 - \frac{A^{max}_{q_1,q_2}}{Density(q_1, q_2)}, \tag{1.48}$$

$$separability.min(q_1, q_2) = 1 - \frac{A^{min}_{q_1,q_2}}{Density(q_1, q_2)}. \tag{1.49}$$

The separability statistics take on (possibly negative) values smaller than 1. The closer a separability statistic value is to 1, the more separated (distinct) are the two modules.

1.11 Network Concepts for Comparing Two Networks

Network concepts can be used to describe differences between two networks. Assume that two $n \times n$ dimensional adjacency matrices $A^{[ref]}$ and $A^{[test]}$ are available for a set of n nodes. $A^{[ref]}$ and $A^{[test]}$ may encode the connectivity patterns among genes before and after a biological perturbation experiment. $A^{[ref]}$ may encode the gene connectivity pattern in brain tissue, while $A^{[test]}$ reports the corresponding connectivity pattern in blood tissue.

It is natural to use network concepts to describe the differences between the reference and the test network. For example, if $NC^{[ref]}$ and $NC^{[test]}$ denote the values of a network concept in the reference and test network, then one can define a **differential network concept** as follows:

$$Diff.NC = NC^{[ref]} - NC^{[test]}. \tag{1.50}$$

Based on the above-mentioned fundamental (single) network concepts, one can define the following differential network concepts:

$$\begin{aligned} Diff.K &= Diff.ScaledConnectivity_i = K_i^{[ref]} - K_i^{[test]} \\ Diff.Density &= \frac{2}{n*(n-1)} \sum_i \sum_{i \neq j} \left(A_{ij}^{[ref]} - A_{ij}^{[test]} \right) \\ Diff.ClusterCoef_i &= ClusterCoef_i^{[ref]} - ClusterCoef_i^{[test]} \\ Diff.TopOverlap_{ij} &= TopOverlap_{ij}^{[ref]} - TopOverlap_{ij}^{[ref]}. \end{aligned} \tag{1.51}$$

If the ith node is highly connected in the first network but has a low connectivity in the second network, then $Diff.ScaledConnectivity$ takes on a large positive value.

By ranking the nodes according to a suitably defined differential network concept (1.50), one can find nodes that have different connectivity patterns across two networks. Sometimes it is useful to consider two differential network concepts to screen for interesting nodes.

For example, in correlation network applications, it can be useful to plot a node significance measure GS_i (e.g., based on a Student t-test of differential expression 10.6) on the y-axis and a measure of differential connectivity $Diff.K_i$ (1.51) on the x-axis. By thresholding both GS_i and $Diff.K_i$, one can define sectors that contain nodes with high node significance and/or high differential connectivity. To find significance thresholds, one can use permutation tests as described in Fuller et al. (2007). We refer to the scatter plot along with horizontal and vertical lines (corresponding to the threshold values) as **sector plot** for differential network analysis. Differential network analysis was used to identify highly significant sectors of obesity-related genes (Fuller et al. 2007) and gender-specific genes (van Nas et al. 2009) based on mouse gene expression data. Differential network analysis has also been used to study how network connections change with age (Swindell 2007) and between primate brains (Oldham et al. 2006).

Let us now define network concepts for measuring the similarity between two networks. Consider a (single) network concept NC_i with a node index i (e.g., the scaled connectivity K_i). To measure whether $NC^{[\mathrm{ref}]}$ is correlated with $NC^{[\mathrm{test}]}$ across nodes, one can define the **network concept correlation**:

$$cor.NC = cor(NC^{[\mathrm{ref}]}, NC^{[\mathrm{test}]}). \tag{1.52}$$

The correlation coefficient (5.10) is defined in Sect. 5.1.1. For example, the **connectivity correlation** is defined as the correlation coefficient between the connectivity vectors $K[[1]]$ and $K[[2]]$:

$$cor.K = cor(K^{[\mathrm{ref}]}, K^{[\mathrm{test}]}). \tag{1.53}$$

Since the correlation coefficient is scale invariant, the connectivity correlation also equals $cor.k = cor(k^{[\mathrm{ref}]}, k^{[\mathrm{test}]})$, where k denotes the (unscaled) connectivity vector. The connectivity correlation can be used to determine whether the connectivity patterns are preserved. High values of $cor.k$ indicate that the two networks have a similar module structure.

A network concept NC_{ij} with two node indices (e.g., the adjacency matrix or the topological overlap matrix) can be vectorized and correlated between the two networks:

$$cor.NC = cor(vectorizeMatrix(NC^{[\mathrm{ref}]}), vectorizeMatrix(NC^{[\mathrm{test}]})). \tag{1.54}$$

For example, the **adjacency correlation** is defined by

$$cor.Adj = cor\left(vectorizeMatrix(A^{[\mathrm{ref}]}), vectorizeMatrix(A^{[\mathrm{test}]})\right). \tag{1.55}$$

The above-mentioned network concepts can also be applied to a subnetwork formed by the nodes of a module. For example, connectivity preservation statistics quantify how similar connectivity of a given module is between a reference and a test network. For example, module connectivity preservation can mean that, within a given module q, nodes with a high connection strength in the reference network also exhibit a high connection strength in the test network. This property can be quantified by the correlation of intramodular adjacencies in reference and test networks. Specifically, if the entries of the first adjacency matrix $A^{[\mathrm{ref}](q)}$ are correlated with those of the second adjacency matrix $A^{[\mathrm{test}](q)}$, then the adjacency pattern of the module is preserved in the second network. Therefore, we define the *adjacency correlation* of the module q network as:

$$cor.Adj^{(q)} = cor\left(vectorizeMatrix(A^{[\mathrm{ref}](q)}), vectorizeMatrix(A^{[\mathrm{test}](q)})\right). \tag{1.56}$$

High $cor.Adj^{(q)}$ indicates that adjacencies within the module q in the reference and test networks exhibit similar patterns. If module q is preserved in the second network, the highly connected hub nodes in the reference network will often be highly connected hub nodes in the test network. In other words, the intramodular connectivity $\mathrm{kIM}^{[\mathrm{ref}](q)}$ (1.41) in the reference network should be highly correlated with the corresponding intramodular connectivity $\mathrm{kIM}^{[\mathrm{test}](q)}$ in the test network. Thus, we define the correlation of intramodular connectivities as:

$$cor.kIM^{(q)} = cor\left(\mathrm{kIM}^{[\mathrm{ref}](q)}, \mathrm{kIM}^{[\mathrm{test}](q)}\right), \tag{1.57}$$

where $\mathrm{kIM}^{[\mathrm{ref}](q)}$ and $\mathrm{kIM}^{[\mathrm{test}](q)}$ are the vectors of intramodular connectivities of all nodes in module q in the reference and test networks, respectively. These and other measures will be used in Chap. 9 to define network-based measures of module preservation.

1.12 R Code for Computing Network Concepts

The function `fundamentalNetworkConcepts` in the `WGCNA` R package computes many of the above-mentioned (fundamental) network concepts based on an adjacency matrix and optionally a node significance measure. These network concepts are defined for any symmetric adjacency matrix (weighted and unweighted).

The following R code shows how to calculate network concepts for a randomly generated adjacency matrix.

```
library(WGCNA)
n=1000
# set the seed for the random number generator
set.seed(1)
# now we simulate a random n*n dimensional matrix
# whose entries lie between 0 and 1
CF=runif(n)^2
# now we define A[i,j]=CF[i]*CF[j]
A=outer(CF,CF)
# diagonal elements are set to 1
diag(A)=1
# network connectivity equals the row sum -1
Connectivity = apply(A,2,sum,na.rm=T)-1
# partition the graphics window into 2 panels
par(mfrow=c(1,2))
# now we evaluate the scale free topology fit of the network
scaleFreePlot(Connectivity,main="Scale Free Topology")
# calculate several scale free topology fitting indices
scaleFreeFitIndex(Connectivity)
# define the scaled connectivity
K=Connectivity/max(Connectivity)
# now we simulate a node significance measure
trueHubNodeSignif=0.5
GS=trueHubNodeSignif*K+rnorm(n,sd=.02)
# Fundamental network concepts
NC=fundamentalNetworkConcepts(A,GS)
```

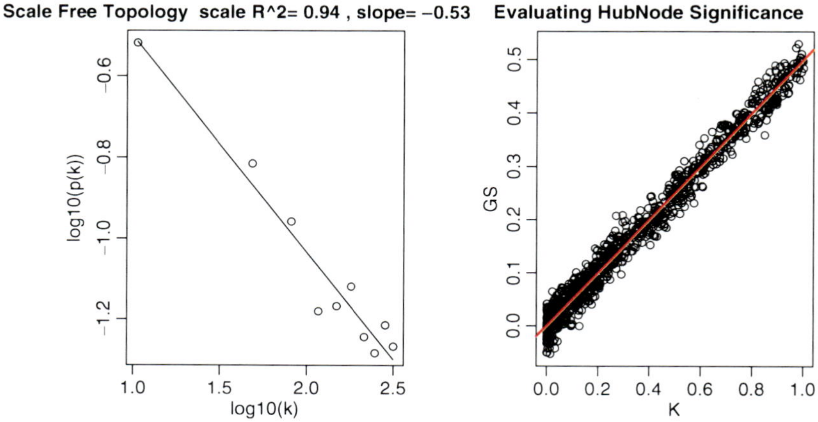

Fig. 1.5 Simulated network for illustrating scale-free fit (*left panel*) and the hub node significance (*right panel*). The R code used for generating this plot can be found in the text. Scale-free topology is approximately satisfied with $R^2 = 0.93$. The (observed) hub node significance is the slope of the line in the right figure, which results from using a linear model (without intercept term) to regress GS on K

```
NC
# scatterplot of GS versus K
plot(K,GS,main="Evaluating HubNodeSignificance")
 # fit a regression line without intercept term
lm1=lm(GS~K-1)
abline(lm1,col="red",lwd=2)
# The following output shows that the slope of the regression line
# equals the observed hub node significance (0.4990)
summary(lm1)
```

The graphical results from this R code are presented in Fig. 1.5.

1.13 Exercises

1. Exercise regarding network concepts. Consider an unweighted block diagonal adjacency matrix with two blocks. The first and second blocks contain $n^{(1)}$ and $n^{(2)}$ nodes, respectively. Each nonzero element of the first and second blocks equals b_1 and b_2, respectively. Calculate the following network concepts: connectivity k_i, scaled connectivity K_i, density, centralization, MAR, clustering coefficient, and topological overlap. Hint: Sect. 3.11. If i is in the first block, then $k_i = (n^{(1)} - 1)b_1$.

2. Exercise regarding alternative definitions of the topological overlap. Show that the following matrix satisfies the conditions of an adjacency matrix:

$$TopOverlap^{average}_{ij} = \begin{cases} \frac{\sum_{l \neq i,j} A_{il}A_{jl} + A_{ij}}{(\sum_{l \neq i} A_{il} + \sum_{l \neq j} A_{jl})/2 - A_{ij} + 1} & \text{if } i \neq j \\ 1 & \text{if } i = j. \end{cases} \quad (1.58)$$

To calculate this alternative TOM measure, use `optionTOMDenom="mean"` in function `TOMsimilarity` (implemented by P. Langfelder).

3. Exercise regarding a computational formula for the mth order generalized topological overlap matrix GTOMm (based on (Yip and Horvath 2007)). Recall the definition of the GTOMm matrix $T^{[m]} = [t^{[m]}_{ij}]$ (1.32):

$$t^{[m]}_{ij} = \frac{|N_m(i,j)| + A_{ij}}{min\{|N_m(i,-j)|, |N_m(-i,j)|\} - 1}.$$

Define the matrix $\tilde{A} = A - I$ whose diagonal elements equal 0. Let $\tilde{A}^m = \tilde{A} \ldots \tilde{A}$ denotes the matrix power based on the matrix multiplication described in Sect. 1.3.1.

(i) Argue that the i,jth element of $\tilde{A}^m$ counts the number of paths of length m connecting nodes i and j. Note that the connecting paths may contain cycles.

(ii) Argue that the matrix

$$S^{[m]} \equiv [s^{[m]}_{ij}] = \tilde{A} + \tilde{A}^2 + \ldots + \tilde{A}^m$$

(where powers denote matrix powers) gives precisely the number of distinct paths with length smaller than or equal to m connecting each pair of nodes.

(iii) Denote by $N_m(i)$ (with $m > 0$) the set of nodes (excluding i itself) that are reachable from i within a path of length m, i.e.,

$$N_m(i) = \{l \neq i \mid d(i,l) \leq m\} \quad (1.59)$$

Note that $N_m(i) \equiv \{l \neq i \mid s^{[m]}_{il} > 0\}$. Define the binary matrix $B^{[m]}$ as follows:

$$b^{[m]}_{il} = \begin{cases} 1 \text{ if } s^{[m]}_{il} > 0 \text{ and } i \neq l, \\ 0 \text{ otherwise,} \end{cases}$$

Show that $N_m(i) = \{l \neq i \mid b^{[m]}_{il} = 1\}$.

(iv) Show that the number of shared m-step neighbors, $|N_m(i,j)| = |N_m(i) \cap N_m(j)|$ can be calculated as the inner product of the ith and the jth columns of $B^{[m]}$ which equals the i,jth element of $\left(B^{[m]}\right)^2$. Hint: $B^{[m]}$ is a symmetric matrix.

(v) Show that $|N_m(i)|$ is given by the i diagonal entry of $\left(B^{[m]}\right)^2$.

(vi) Show how the above results can be used to compute $T^{[m]}$ (1.32).

(vii) Why is it computationally advantageous to compute $S^{[m]}$ recursively by the matrix product between $\tilde{A}$ and $(S^{[m-1]}+I)$?

4. Exercise regarding the asymptotic behavior of GTOMm for large m. Recall the definition of the GTOMm matrix $T^{[m]} = [t_{ij}^{[m]}]$ (1.32).

 (i) Use the R code in Sect. 1.6 to show empirically that the minimum value of the GTOMm measure increases with m.

 (ii) Consider the situation where each pair of nodes is connected by a path of length $\leq m$. In this case, show that $|N_m(i)| = n-1$ and $|N_m(i,j)| = |N_m(i) \cap N_m(j)| = n-2$ and

$$t_{ij}^{[m]} = \frac{n-2+A_{ij}+2I_{i=j}}{n-A_{ij}},$$

 where n is the number of nodes in the network.

 (iii) Show that large n implies that $T_{ij}^{[m]} \approx 1$.

 (iv) Argue that for large enough m, the generalized topological overlap measure between each pair of nodes is 1.

 (v) Argue that the GTOMm measure becomes uninformative if m is chosen too large. Comment: As default value, we choose $m = 1$. But sometimes $m = 0$ or $m = 2$ leads to more meaningful measures of interconnectedness. We expect that $m > 2$ is useful only when the unweighted input adjacency matrix contains many zeroes.

References

Albert R, Barabasi AL (2002) Statistical mechanics of complex networks. Rev Mod Phys 74:47–97

Albert R, Barabasi AL (2000) Topology of evolving networks: Local events and universality. Phys Rev Lett 85(24):5234–5237

Albert R, Jeong H, Barabasi AL (2000) Error and attack tolerance of complex networks. Nature 406(6794):378–382

Almaas E, Kovacs B, Vicsek T, Oltvai ZN, Barabasi AL (2004) Global organization of metabolic fluxes in the bacterium Escherichia coli. Nature 427:839–843

Barabasi AL, Albert R (1999) Emergence of scaling in random networks. Science 286(5439):509–512

Barabasi AL, Oltvai ZN (2004) Network biology: Understanding the cell's functional organization. Nat Rev Genet 5(2):101–113

Brun C, Chevenet F, Martin D, Wojcik J, Guenoche A, Jacq B (2003) Functional classification of proteins for the prediction of cellular function from a protein-protein interaction network. Genome Biol 5(1):R6

Carlson M, Zhang B, Fang Z, Mischel P, Horvath S, Nelson SF (2006) Gene connectivity, function, and sequence conservation: Predictions from modular yeast co-expression networks. BMC Genomics 7(7):40

Carter SL, Brechbuler CM, Griffin M, Bond AT (2004) Gene co-expression network topology provides a framework for molecular characterization of cellular state. Bioinformatics 20(14):2242–2250

Chen J, Hsu W, Lee ML, Ng S (2006) Increasing confidence of protein interactomes using network topological metrics. Bioinformatics 22:1998–2004

Chua NH, Sung W, Wong L (2006) Exploiting indirect neighbours and topological weight to predict protein function from proteinprotein interactions. Bioinformatics 22:1623–1630

Csanyi G, Szendroi B (2004) Structure of a large social network. Phys Rev 69:1–5

Dong J, Horvath S (2007) Understanding network concepts in modules. BMC Syst Biol 1(1):24

Freeman L (1978) Centrality in social networks: Conceptual clarification. Soc Networks 1:215–239

Fuller TF, Ghazalpour A, Aten JE, Drake T, Lusis AJ, Horvath S (2007) Weighted gene coexpression network analysis strategies applied to mouse weight. Mamm Genome 18(6–7):463–472

Gargalovic PS, Imura M, Zhang B, Gharavi NM, Clark MJ, Pagnon J, Yang WP, He A, Truong A, Patel S, Nelson SF, Horvath S, Berliner JA, Kirchgessner TG, Lusis AJ (2006) Identification of inflammatory gene modules based on variations of human endothelial cell responses to oxidized lipids. Proc Natl Acad Sci USA 103(34):12741–12746

Ghazalpour A, Doss S, Zhang B, Plaisier C, Wang S, Schadt EE, Thomas A, Drake TA, Lusis AJ, Horvath S (2006) Integrating genetics and network analysis to characterize genes related to mouse weight. PloS Genet 2(2):8

Guldener U, Munsterkotter M, Oesterheld M, Pagel P, Ruepp A, Mewes HW, Stumpflen V (2006) MPact: The MIPS protein interaction resource on yeast. Nucleic Acids Res 34:436–441

Hahn MW, Kern AD (2005) Comparative genomics of centrality and essentiality in three eukaryotic protein-interaction networks. Mol Biol Evol 22(4):803–806

Han JD, Bertin N, Hao T, Goldberg DS, Berriz GF, Zhang LV, Dupuy D, Walhout AJ, Cusick ME, Roth FP, Vidal M (2004) Evidence for dynamically organized modularity in the yeast protein-protein interaction network. Nature 430(6995):88–93

Horvath S, Dong J (2008) Geometric interpretation of gene co-expression network analysis. PLoS Comput Biol 4(8):e1000117

Horvath S, Zhang B, Carlson M, Lu KV, Zhu S, Felciano RM, Laurance MF, Zhao W, Shu Q, Lee Y, Scheck AC, Liau LM, Wu H, Geschwind DH, Febbo PG, Kornblum HI, Cloughesy TF, Nelson SF, Mischel PS (2006) Analysis of oncogenic signaling networks in glioblastoma identifies ASPM as a novel molecular target. Proc Natl Acad Sci USA 103(46):17402–17407

Jeong H, Mason SP, Barabasi AL, Oltvai ZN (2001) Lethality and centrality in protein networks. Nature 411:41

Jeong H, Oltvai Z, Barabasi A (2003) Prediction of protein essentiality based on genome data. ComPlexUs 1:19–28

Kaufman L, Rousseeuw PJ (1990) Finding groups in data: An introduction to cluster analysis. Wiley, New York

Li A, Horvath S (2007) Network neighborhood analysis with the multi-node topological overlap measure. Bioinformatics 23(2):222–231

Ma HW, Buer J, Zeng AP (2004) Hierarchical structure and modules in the Escherichia coli transcriptional regulatory network revealed by a new top-down approach. BMC Bioinform 5(1):199

van Nas A, GuhaThakurta D, Wang SS, Yehya N, Horvath S, Zhang B, Ingram-Drake L, Chaudhuri G, Schadt EE, Drake TA, Arnold AP, Lusis AJ (2009) Elucidating the role of gonadal hormones in sexually dimorphic gene coexpression networks. Endocrinology 150(3):1235–1249

Oldham MC, Horvath S, Geschwind DH (2006) Conservation and evolution of gene coexpression networks in human and chimpanzee brains. Proc Natl Acad Sci USA 103(47):17973–17978

Oldham MC, Konopka G, Iwamoto K, Langfelder P, Kato T, Horvath S, Geschwind DH (2008) Functional organization of the transcriptome in human brain. Nat Neurosci 11(11):1271–1282

Pagel M, Meade A, Scott D (2007) Assembly rules for protein networks derived from phylogenetic-statistical analysis of whole genomes. BMC Evol Biol 7(Suppl 1):S16

Ravasz E, Somera AL, Mongru DA, Oltvai ZN, Barabasi AL (2002) Hierarchical organization of modularity in metabolic networks. Science 297(5586):1551–1555

Snijders TA (1981) The degree variance: An index of graph heterogeneity. Soc Networks 3:163–174

Swindell W (2007) Gene expression profiling of long-lived dwarf mice: Longevity-associated genes and relationships with diet, gender and aging. BMC Genomics 8(1):353

Watts DJ (2002) A simple model of global cascades on random networks. Proc Natl Acad Sci USA 99(9):5766–5771

Watts DJ, Strogatz SH (1998) Collective dynamics of 'small-world' networks. Nature 393 (6684):440–442

Yip A, Horvath S (2007) Gene network interconnectedness and the generalized topological overlap measure. BMC Bioinform 8(8):22

Zhang B, Horvath S (2005) General framework for weighted gene coexpression analysis. Stat Appl Genet Mol Biol 4:17

Chapter 2
Approximately Factorizable Networks

Abstract In factorizable networks, the adjacency (connection strength) between two nodes can be factored into node-specific contributions, named node "conformity". Often the ith node conformity, CF_i is approximately equal to the scaled connectivity $k_i/sum(k)$ of the ith node. We describe (a) an algorithm for computing the conformity CF and for measuring the "factorizability" of a general network, and (b) a module- and CF-based decomposition of a general adjacency matrix, which can be used to arrive at a parsimonious description of a network. Approximately factorizable networks have important practical and theoretical applications, e.g., we use them to derive relationships between network concepts. Collaborative work with **Jun Dong** has shown that network modules (i.e., subnetworks comprised of module nodes) tend to be approximately factorizable (Dong and Horvath BMC Syst Biol 1(1):24, 2007).

2.1 Exactly Factorizable Networks

We define an adjacency matrix A to be exactly factorizable if, and only if, there exists a vector CF with nonnegative elements such that

$$A_{ij} = CF_i CF_j \text{ for all } i \neq j. \tag{2.1}$$

Below, we show that CF is uniquely defined if the network contains $n \geq 3$ nodes and if $A_{ij} > 0$. We call CF_i the **conformity** of node i. While the term "factorizable network" was first proposed in Dong and Horvath (2007), numerous examples of these types of networks can be found in the literature. A recent physical model for experimentally determined protein–protein interactions is exactly factorizable (Deeds et al. 2006). In that model, the 'affinity' A_{ij} between proteins i and j is the product of conformities $CF_i = exp(-K_i)$, where K_i is the number of hydrophobic residues in the ith protein. Another related example is an exactly factorizable random network model for which the edges between pairs of nodes are drawn according to a linking probability function (Servedio et al. 2004). In multi-graph models, CF_i has been used to represent the propensity for the ith node to form

S. Horvath, *Weighted Network Analysis: Applications in Genomics and Systems Biology*, DOI 10.1007/978-1-4419-8819-5_2,

an edge (Ranola et al. 2010). One can easily show that the vector CF is not unique if an exactly factorizable network contains only $n = 2$ nodes. However, for $n > 2$ the conformity is uniquely defined when dealing with a weighted network where $A_{ij} > 0$. In an exercise, you are asked to prove the uniqueness.

2.2 Conformity for a Non-Factorizable Network

Equation (2.21) provides an explicit formula for the conformity of a weighted, exactly factorizable network. For a general, non-factorizable network, we describe here how to compute the conformity by optimizing an objective function (Dong and Horvath 2007). In the following, we assume a general $n \times n$ adjacency matrix A where $n > 2$. Let $v = (v_1, v_2, \ldots, v_n)$ be a vector of length n. We define the conformity as a vector v^* that *minimizes* the following objective function:

$$f(v) = \sum_i \sum_{j \neq i} (A_{ij} - v_i v_j)^2 = ||A - \left(I - diag\left(v^2\right) + vv^\tau\right)||_F^2,$$

where $||.||_F$ denotes the Frobenius matrix norm (1.4). An equivalent but more convenient approach is to define the conformity CF as the maximizer of the following **factorizability function**:

$$\begin{aligned} F_A(v) &= 1 - \frac{f(v)}{f(0)} = 1 - \frac{\sum_i \sum_{j \neq i} (A_{ij} - v_i v_j)^2}{\sum_i \sum_{j \neq i} (A_{ij})^2} \\ &= 1 - \frac{||A - \left(vv^\tau - \mathrm{diag}\left(v^2\right) + I\right)||_F^2}{||A - I||_F^2}, \\ &= 1 - \frac{||A - A_v||_F^2}{||A - I||_F^2}, \end{aligned} \tag{2.2}$$

where $A_v = vv^\tau - \mathrm{diag}(v^2) + I$ is a matrix whose diagonal elements equal 1. One can easily show that if v^* maximizes $F_A(v)$, then $-v^*$ also maximizes $F_A(v)$. Further, all components of v^* must have the same sign, since otherwise flipping the sign of the negative components leads to a higher value of $F_A(v)$. This leads us to the following:

Definition 2.1 (Conformity, Factorizability). We define the conformity CF as the vector with nonnegative entries that maximizes $F_A(v)$. If there is more than one such maximizer, then a maximizer closest to $k/\sqrt{sum(k)}$ is chosen. Further, we define the factorizability $F(A)$ as the corresponding maximum value $F_A(CF)$.

Thus, the **network factorizability measure** is defined as follows:

$$F(A) = F_A(CF) = max_v(F_A(v)) = 1 - \frac{||A - A_{CF}||_F^2}{||A - I||_F^2}, \tag{2.3}$$

where

$$A_{CF} = CFCF^{\tau} - \text{diag}\left(CF^2\right) + I.$$

A_{CF} can be interpreted as the exactly factorizable network which best approximates A (according to the Frobenius norm). In general, the closer $F(A)$ is to 1, the better A_{CF} approximates A. Note that $F_A(0) = 0$ implies that $F(A) \geq 0$.

An algorithmic definition of the factorizability $F(A)$ is provided in (2.11). We find that for most real networks, the conformity is highly related to the first eigenvector of the adjacency matrix, i.e., the conformity vector CF is roughly equal to $\sqrt{d_1}u_1$ where d_1 is the largest eigenvalue of A and u_1 is the corresponding unit length eigenvector with positive components. The idea of using a variant of the singular value decomposition for decomposing an adjacency matrix has been proposed by several authors. However, we prefer to define the conformity as a maximizer of the factorizability function $F_A(v)$ (2.2) for the following reasons: First, the factorizability satisfies that $F_A(CF) = 1$ if, and only if, A is exactly factorizable network with $A_{ij} = CF_iCF_j$. Second, we prefer to define the conformity without reference to the diagonal elements A_{ii} of the adjacency matrix. Third, the definition naturally fits within the framework of least squares factor analysis where conformity can be interpreted as the first factor (de Leeuw and Michailidis 2000). While network analysis focuses on the adjacency matrix, factor analysis takes as input a correlation or covariance matrix.

2.2.1 Algorithm for Computing the Node Conformity

This technical section may be skipped by readers who are not interested in learning the algorithmic definition of the node conformity and factorizability. We describe an algorithm for finding a vector v (with nonnegative components) that maximizes the objective function $F_A(v)$ (2.2). In general, $F_A(v)$ may have multiple maximizers as can be demonstrated with the block diagonal simulated example (3.37) by choosing $n^{(1)} = n^{(2)}$ and $b_1 = b_2$. By forming the first derivative of the factorizability function $F_A(v)$ in terms of v_i, one can show that a local maximum satisfies

$$\sum_{j \neq i} A_{ij}CF_j = CF_i \sum_{j \neq i} CF_j^2, \tag{2.4}$$

i.e.,

$$\left(A - I + \text{diag}\left(CF^2\right)\right) CF = CF||CF||_2^2. \tag{2.5}$$

Equation (2.5) suggests that the conformity is an eigenvector of the "hat" adjacency matrix which also depends on the conformity

$$\hat{A} = A - I + \text{diag}\left(CF^2\right).$$

Denote by $CF^{(s-1)}$ an estimate of the conformity CF. Next define

$$\hat{A}^{(s-1)} = A - I + \text{diag}\left(\left(CF^{(s-1)}\right)^2\right). \tag{2.6}$$

Define a new estimate of the conformity by

$$CF^{(s)} = \sqrt{\hat{d}_1^{(s-1)}}\hat{u}_1^{(s-1)}, \tag{2.7}$$

where $\hat{d}_1^{(s-1)}$ and $\hat{u}_1^{(s-1)}$ denote the largest eigenvalue and corresponding unit length eigenvector of $\hat{A}^{(s-1)}$. One can easily show that all the components of $\hat{u}_1^{(s-1)}$ must have the same sign and we assume without loss of generality nonnegative components. Our algorithm for computing the conformity is monotonic as can be proven with the following:

Lemma 2.1. *If A denotes a symmetric real matrix with eigenvalues $d_1, \ldots, d_n$ sorted according to their absolute values, i.e., $|d_1| \geq |d_2| \geq \ldots \geq |d_n|$, and the corresponding orthonormal eigenvectors are denoted by $u_1, \ldots, u_n$, then $\|A - vv^\tau\|_F^2$ is minimized at $v^* = \sqrt{|d_1|}u_1$.*

The proof can be found in Horn and Johnson (1991).

Lemma 2.1 with $A = \hat{A}^{(s-1)}$ implies that

$$\begin{aligned}&\left\|A - I + \text{diag}\left(\left(CF^{(s-1)}\right)^2\right) - CF^{(s-1)}\left(CF^{(s-1)}\right)^\tau\right\|_F^2 \\ &\geq \left\|A - I + \text{diag}\left(\left(CF^{(s-1)}\right)^2\right) - CF^{(s)}\left(CF^{(s)}\right)^\tau\right\|_F^2.\end{aligned} \tag{2.8}$$

Considering the diagonal elements, one can easily show that

$$\begin{aligned}&\left\|A - I + \text{diag}\left(\left(CF^{(s-1)}\right)^2\right) - CF^{(s)}\left(CF^{(s)}\right)^\tau\right\|_F^2 \\ &\geq \left\|A - I + \text{diag}\left(\left(CF^{(s)}\right)^2\right) - CF^{(s)}\left(CF^{(s)}\right)^\tau\right\|_F^2.\end{aligned} \tag{2.9}$$

Thus, we arrive at the following:

$$F_A\left(CF^{(s)}\right) \geq F_A\left(CF^{(s-1)}\right), \tag{2.10}$$

which suggests a monotonic algorithm for computing CF. Equation (3.24) suggests to choose $k/\sqrt{sum(k)}$ as a starting value of the algorithm. These comments give rise to the following:

Definition 2.2 (Algorithmic Definition of Conformity, Factorizability). For a general network A, set $CF^{(1)} = k/\sqrt{sum(k)}$ and apply the monotonic iterative algorithm described by (2.6) and (2.7). If the limit $CF^{(\infty)}$ exists, we define it as the conformity $CF = CF^{(\infty)}$. Further, we define the network factorizability (2.3) as:

$$F(A) = F_A(CF) = 1 - \frac{\|A - A_{CF}\|_F^2}{\|A - I\|_F^2}. \tag{2.11}$$

Note that the conformity satisfies (2.5) by definition of convergence. The monotonic algorithm described by (2.6) and (2.7) can be derived as a majorization–minimization (MM) algorithm (Lange 2004; Ranola et al. 2010), and it can be considered as a special case of an algorithm described for fitting a least squares factor analysis model with one factor (de Leeuw and Michailidis 2000; Gifi 1990).

2.3 Module-Based and Conformity-Based Approximation of a Network

A factorizable network allows for a simple and parsimonious parametrization in terms of the components of the conformity vector. But in general, $n*(n-1)/2$ upper diagonal elements A_{ij} are needed to parameterize an adjacency matrix A. While most adjacency matrices A are *not* approximately factorizable, they may contain modules that are factorizable. For example, we argue in Sect. 3.9 (Observation 3.6) that clusters of highly interconnected nodes are often factorizable. Assume that a module assignment Cl is available so that each node is assigned to exactly one of Q modules. The module assignment $Cl(i) = q$ if the ith node is in the qth module. In the following, we outline when a general network adjacency A can be approximated with a parsimonious matrix $A^{Cl}_{CF,app}$ (2.15) that involves only $Q(Q-1)/2+n$ parameters.

Denote the module size, i.e., the number of nodes in the qth module, by $n^{(q)}$ and the $n^{(q)} \times n^{(q)}$ dimensional intramodular adjacency matrix by $A^{(q,q)}$. Assume that $A^{(q,q)}$ is approximately factorizable, i.e.,

$$A^{(q,q)}_{ij} \approx CF^{(q)}_i CF^{(q)}_j. \tag{2.12}$$

Let us now study the relationships between two modules labeled by q_1 and q_2, respectively. In this case, the $n^{(q_1)} \times n^{(q_2)}$ matrix $A^{(q_1,q_2)}$ specifies the *inter*modular adjacencies. We find that the intermodular adjacency matrix $A^{(q_1,q_2)}_{ij}$ can sometimes be factored as follows:

$$A^{(q_1,q_2)}_{ij} \approx CF^{(q_1)}_i CF^{(q_2)}_j A_{q_1 q_2}, \tag{2.13}$$

where $A_{q_1 q_2}$ is a number, i and j denote nodes in modules q_1 and q_2, respectively. In Sect. 6.4, we characterize *correlation* networks where (2.13) is approximately satisfied. Note that (2.12) and (2.13) can be combined as follows:

$$A^{(q_1,q_2)}_{ij} \approx CF^{(q_1)}_i CF^{(q_2)}_j A_{q_1 q_2}, \tag{2.14}$$

where $A_{q_1 q_2} = 1$ if $q_1 = q_2$. If $0 \le A_{q_1 q_2} \le 1$, then $A_{q_1 q_2}$ can be interpreted as an element of a $Q \times Q$ dimensional adjacency matrix $A_{modules}$ whose nodes are modules (see also Sect. 1.9).

The right-hand side of (2.14) motivates us to define an $n \times n$ dimensional matrix $A^{Cl}_{CF,app}$ whose i,jth element is given by

$$A^{Cl}_{CF,app,ij} = CF_i^{(q_1)} CF_j^{(q_2)} A_{q_1 q_2} \text{ if } i \in \text{ module } q_1, j \in \text{ module } q_2. \quad (2.15)$$

This leads us to the following:

Definition 2.3 (Module- and CF-based approximation of A). We refer to $A^{Cl}_{CF,app}$ (2.15) as the module- and CF-based approximation of A.

Note that $A^{Cl}_{CF,app}$ only involves $Q(Q-1)/2+n$ parameters (comprising the upper off-diagonal elements of $A_{modules}$ and the $n = \sum_{u=1}^{Q} n^{q_u}$ conformity parameters). If the approximation is accurate, then $A^{Cl}_{CF,app}$ provides a parsimonious parametrization of A, which is very intuitive since it is based on the underlying module structure. To quantify how well $A^{Cl}_{CF,app}$ approximates A, one can use the **module- and CF-based factorizability measure** $F(A,Cl)$ (2.19) defined below.

Let us now describe how to estimate the conformity vectors $CF^{(q_1)}, \ldots, CF^{(q_Q)}$ and the intermodular adjacencies A_{q_1,q_2}. The idea is to estimate the parameters such that the resulting matrix $A^{Cl}_{CF,app}$ is as close to A as possible. Specifically, the parameters are estimated by minimizing the following objective function:

$$f\left(v^{(q_1)}, \ldots, v^{(q_Q)}, M_{modules}\right) = \sum_{q_1=1}^{Q} \sum_{q_2=1}^{Q} \sum_{i \in \mathcal{M}_{q_1}} \sum_{j \neq i\, j \in \mathcal{M}_{q_2}} \left(A_{ij}^{(q_1,q_2)} - M_{q_1,q_2} v_i^{(q_1)} v_j^{(q_2)}\right)^2, \quad (2.16)$$

where the vector $v^{(q_1)}$ is used to estimate $CF^{(q_1)}$ and the symmetric matrix $M_{modules}$ is used to estimate $A_{modules}$. We assume that the elements of $M_{modules}$ are nonnegative and that the diagonal elements equal 1. It can be convenient to reformulate the optimization problem as a maximization problem whose objective function is given by the **module-based factorizability function**:

$$F_{A,Cl}\left(v^{(q_1)}, \ldots, v^{(q_Q)}, M_{modules}\right) = 1 - \frac{f\left(v^{(q_1)}, \ldots, v^{(q_Q)}, M_{modules}\right)}{f(0, \ldots, 0)}. \quad (2.17)$$

One can easily show that if the set of parameters $v^{(q_1)}, \ldots, v^{(q_Q)}, M_{modules}$ maximizes $F_{A,Cl}$, then $-v^{(q_1)}, \ldots, -v^{(q_Q)}, M_{modules}$ also maximizes this function. Since the elements of $M_{modules}$ are nonnegative, the components of a maximizing conformity vector $v^{(q_1)}$ must have the same sign. Otherwise, flipping the sign of the negative components leads to a higher value of $F_{A,Cl}$. Assume that the vectors $v^{*(q_1)}, \ldots, v^{*(q_Q)}$ (with nonnegative components) and the matrix $M^*_{modules}$ (with nonnegative entries) maximize $F_{A,Cl}$. If more than one set of conformity vectors maximizes $F_{A,Cl}$ (2.17), i.e., if several solutions tie, then the maximizing set closest to the set of scaled intramodular connectivities

$$\left\{ \frac{k^{(q_1)}}{\sqrt{sum\left(k^{(q_1)}\right)}}, \ldots, \frac{k^{(q_Q)}}{\sqrt{sum\left(k^{(q_Q)}\right)}} \right\} \tag{2.18}$$

is chosen. We define the conformity vectors $CF^{(q_1)}, \ldots, CF^{(q_Q)}$ by $v^{*(q_1)}, \ldots, v^{*(q_Q)}$ and the intermodular adjacency matrix $A_{modules}$ by $M^*_{modules}$. Further, we define the **module-based factorizability measure** as follows:

$$F(A, Cl) = F_{A,Cl}\left(CF^{(q_1)}, \ldots, CF^{(q_Q)}, A_{modules}\right). \tag{2.19}$$

The closer $F(A, Cl)$ is to 1, the better $A^{Cl}_{CF,app}$ (2.15) approximates A. If $F(A, Cl) = 1$, then we refer to $CF^{(q_1)}, \ldots, CF^{(q_Q)}, A_{modules}$ as the **module- and CF-based decomposition** of A.

An **algorithm for optimizing the two objective functions** (2.16 and 2.17) can be informed using the following comments.

First, it is natural to choose the scaled intramodular connectivity $\frac{k^{(q_1)}}{\sqrt{sum(k^{(q_1)})}}$ (2.18) as starting value for $v^{(q_1)}$.

Second, given estimates $\hat{v}^{(q_1)}$ of the conformity vectors one can derive the following explicit solution of the optimizing problem for the intermodular adjacencies

$$\hat{M}_{q_1,q_2} = \frac{mean\left(A^{(q_1,q_2)}\right)}{mean\left(\hat{v}^{(q_1)}\right) mean\left(\hat{v}^{(q_2)}\right)} \tag{2.20}$$

with the usual arguments from calculus (set the first derivative with respect to the conformities to zero and solve for M_{q_1,q_2}). This suggests to optimize the objective function by iterating between two steps: one step estimates $v^{(q_1)}$ and the other estimates M_{q_1,q_2} (using (2.20)).

Third, an *approximate* solution to the optimization problem can be found by ignoring intermodular adjacencies when it comes to estimating $v^{(q_1)}$. In this case, the algorithm for finding conformity vectors (in Sect. 2.2.1) can be used to find an approximate estimate $\hat{v}^{(q_1)} = \hat{CF}^{(q_1)}$ based on $A^{(q_1,q_1)}$ alone. An algorithm for finding these approximate solutions is implemented in the R function conformity Decomposition. It is worth repeating that these approximate estimates are typically different from the true solutions of the optimization problem. **John Ranola** and **Kenneth Lange** have developed a minorization–majorization (MM) algorithm for finding a solution of the optimization problem, which is implemented in the R function `propensityDecomposition` (Ranola et al. 2011). We briefly mention that one can extend the optimization problem using an objective function that includes the module assignment *Cl* as an additional parameter. Optimizing this extended objective function with regard to *Cl*, the conformities, and the intermodular adjacencies leads to a CF based clustering algorithm, which is implemented in the function `propensityClustering`. Example R code illustrating the use of `propensityDecomposition` and `propensityClustering` can be found in the help files of these functions.

2.4 Exercises

1. Exercise regarding the uniqueness of conformity in case of an exactly factorizable network. Specifically, prove the following statement. If A is an $n \times n$ ($n \geq 3$) dimensional adjacency matrix with positive entries ($A_{ij} > 0$), then the system of (2.1)
$$A_{ij} = CF_i CF_j \text{ for all } i \neq j$$
has at most one solution CF with positive entries. If the solution exists, it is given by
$$CF_i = \left(\frac{p_i}{\left(\prod_{h=1}^n p_h\right)^{1/(2(n-1))}} \right)^{\frac{1}{n-2}}, \tag{2.21}$$
where $p_i = \prod_{j=1}^n A_{ij}$ denotes the **product connectivity** of the ith node. Hint: By assumption, we have $A_{ij} = CF_i CF_j$ for a positive vector CF and $n \geq 3$. Multiplying both sides of (2.1) yields $\prod_h \prod_{l \neq h} A_{lh} = \prod_h \prod_{l \neq h} CF_l CF_h = \left(\prod_{l=1}^n CF_l\right)^{2(n-1)}$. Since $\prod_{l=1}^n CF_l$ is positive, we find
$$\prod_{l=1}^n CF_l = \left(\prod_h \prod_{l \neq h} A_{lh} \right)^{\frac{1}{2(n-1)}}.$$
Similarly, eliminating the ith row and column from A yields
$$\prod_{l \neq i} CF_l = \left(\prod_{h \neq i} \prod_{l \neq h,i} A_{lh} \right)^{\frac{1}{2(n-2)}} = \left(\prod_h \prod_{l \neq h} A_{lh} \Big/ \left(\prod_{l \neq i} A_{li} \right)^2 \right)^{\frac{1}{2(n-1)}}.$$
Since $CF_i = \prod_{l=1}^n CF_l / \prod_{l \neq i} CF_l$, we conclude that CF_i is uniquely defined by
$$CF_i = \frac{\left(\prod_h \prod_{l \neq h} A_{lh}\right)^{\frac{1}{2(n-1)}}}{\left(\prod_{h \neq i} \prod_{l \neq h,i} A_{lh}\right)^{\frac{1}{2(n-2)}}} = \left(\frac{p_i}{\left(\prod_{h=1}^n p_h\right)^{1/(2(n-1))}} \right)^{\frac{1}{n-2}}.$$

2. Exercise regarding the module- and CF-based approximation (decomposition) of a block diagonal network A. Consider an unweighted block diagonal adjacency matrix with two blocks. The first and second blocks contain $n^{(1)}$ and $n^{(2)}$ nodes, respectively. Each nonzero element of the first and second blocks equals b_1 and b_2, respectively.

(i) Calculate the module- and CF-based approximation of A if Cl indicates block (module) membership. Hint: See Sects. 3.11 and 2.3. Determine $CF^{(1)}$, $CF^{(2)}$, and the 2×2 dimensional matrix $A_{modules}$.

(ii) Is the result of (i) a decomposition of A? Hint: Determine whether the module-based factorizability measure $F(A, Cl)$ equals 1.

(iii) Determine the module- and CF-based decomposition of a general block diagonal matrix with Q blocks of sizes $n^{(1)}, \ldots, n^{(Q)}$.

References

Deeds EJ, Ashenberg O, Shakhnovich EI (2006) A simple physical model for scaling in protein–protein interaction networks. Proc Natl Acad Sci USA 103(2):311–316

Dong J, Horvath S (2007) Understanding network concepts in modules. BMC Syst Biol 1(1):24

Gifi A (1990) Nonlinear multivariate analysis. Wiley, Chichester, England

Horn RA, Johnson CR (1991) Topics in matrix analysis. Cambridge University Press, Cambridge

Lange K (2004) Optimization. Springer, New York

de Leeuw J, Michailidis G (2000) Block relaxation algorithms in statistics. J Comput Graph Stat 9:26–31

Ranola JMO, Ahn S, Sehl ME, Smith DJ, Lange K (2010) A Poisson model for random multigraphs. Bioinformatics 26(16):2004–2011

Ranola JMO, Langfelder P, Song L, Horvath S, Lange K (2011) An MM algorithm for module and propensity based decomposition of a network. UCLA Technical Report

Servedio VDP, Caldarelli G, Butta P (2004) Vertex intrinsic fitness: How to produce arbitrary scale-free networks. Phys Rev E – Stat Nonlin Soft Matter Phys 70(5):056126

Chapter 3
Different Types of Network Concepts

Abstract While some network concepts (e.g., connectivity) have found important practical applications, other network concepts (e.g., network centralization) are rarely being used. Before attempting to understand why some concepts are more interesting than others, it is important to understand how network concepts relate to each other. As a step toward this goal, we explore the meaning of network concepts in approximately factorizable networks. In these types of networks, simple relationships exist between seemingly disparate network concepts. To provide a formalism for relating network concepts to each other, we define three types of network concepts: fundamental-, conformity-, and approximate conformity-based concepts. For each type, one can define intramodular and intermodular network concepts. Fundamental concepts include the standard definitions of connectivity, density, centralization, heterogeneity, clustering coefficient, and topological overlap. Approximate conformity-based analogs of fundamental network concepts have several theoretical advantages. First, they allow one to derive simple relationships between seemingly disparate network concepts. For example, we derive simple relationships between the clustering coefficient, the heterogeneity, the density, the centralization, and the topological overlap. Second, they allow one to show that fundamental network concepts can be approximated by simple functions of the connectivity in network modules. Third, they allow one to study the effects of transforming a network. We illustrate these results using protein–protein interaction networks and using block-diagonal networks. This work reviews and extends work with **Jun Dong**.

In Chap. 1, we defined fundamental network concepts as functions of the off-diagonal elements of an adjacency matrix A and/or a node significance measure GS. Based on the sets of nodes, we distinguished whole network concepts (based on all nodes), intramodular network concepts (based on nodes inside a given module), and intermodular network concepts (defined using nodes from two different modules). Apart from these fundamental network concepts, we will also define network concepts based on different types of approximations to the adjacency matrix. Table 3.1 provides an overview of different types of network concepts, which will be explained below. CF-based network concepts are defined in Definition 3.2, approximate CF-based network concepts are defined in Definition 3.3, and eigenvector-based network concepts are defined in Definition 6.1. While

S. Horvath, *Weighted Network Analysis: Applications in Genomics and Systems Biology*, DOI 10.1007/978-1-4419-8819-5_3,

Table 3.1 Overview of different types of network concepts

Network	Type of concept	Adjacency	Input		Example	Equation
Whole	Fundamental	General	$A-I$	GS	$K_i = \frac{\sum_{j\neq i} A_{ij}}{max(\sum_{j\neq i} A_{ij})}$	(1.8)
Whole	CF-based	General	$A_{CF}-I$	GS_{CF}	$K_{CF,i} = \frac{CF_i(\sum_j CF_j - CF_i)}{max(CF_i(\sum_j CF_j - CF_i))}$	(3.48)
Whole	Approx. CF-based	General	$A_{CF,app}$	GS_{CF}	$K_{CF,app,i} = \frac{CF_i}{CF_{max}}$	(3.49)
Whole	Eigenvector -based	Correlation	A_E	GS_E	$K_{E,i} = \frac{A_{e,i}}{A_{e,max}} \approx \lvert cor(x_i, E^{(q)})\rvert^{\beta}$	(6.30)
Intramod	Fundamental	General	$A^{(q)}-I$	$GS^{(q)}$	$K_i^{(q)} = \frac{\sum_{j\neq i} A_{ij}^{(q)}}{max(\sum_{j\neq i} A_{ij}^{(q)})}$	
Intramod	CF-based	General	$A_{CF}^{(q)}-I$	$GS_{CF}^{(q)}$	$K_{CF,i}^{(q)} = \frac{CF_i^{(q)}(\sum_j CF_j^{(q)} - CF_i^{(q)})}{max(CF_i^{(q)}(\sum_j CF_j^{(q)} - CF_i^{(q)}))}$	
Intramod	Approx. CF-based	General	$A_{CF,app}^{(q)}$	$GS_{CF}^{(q)}$	$K_{CF,app,i}^{(q)} = \frac{CF_i^{(q)}}{CF_{max}^{(q)}}$	
Intramod	Eigenvector -based	Correlation	$A_E^{(q)}$	$GS_E^{(q)}$	$K_{E,i}^{(q)} \approx \lvert cor(x_i, E^{(q)})\rvert^{\beta}$	
Intermod	Fundamental	General	$A^{(q)}$	$A^{(q_1,q_2)}$	$separability.ave(q_1,q_2)$ $= 1 - \frac{mean(A^{(q_1,q_2)})}{\sqrt{Density^{(q_1)} Density^{(q_2)}}}$	(1.47)
Intermod	Approx. CF-based	General	$A_{CF,app}^{(q)}$	$A_{CF}^{(q_1,q_2)}$	$separability.ave_{CF,app}(q_1,q_2)$ $\approx 1 - A_{q_1 q_2}$	(3.44)
Intermod	Eigenvector -based	Correlation	$A_E^{(q)}$	$A_E^{(q_1,q_2)}$	$separability.ave_E(q_1,q_2)$ $\approx 1 - \lvert cor(E^{(q_1)}, E^{(q_2)})\rvert^{\beta}$	(6.45)

A network concept arises by evaluating a *network concept function* on a special type of input matrix and node significance measure. The diagonal elements of $A-I$ and $A_{CF}-I = CFCF^{\tau} - \text{diag}(CF^2)$ are zero. In contrast, the ith diagonal element of $A_{CF,app} = CFCF^{\tau}$ is given by CF_i^2. GS denotes a node significance measure, which is a vector of real numbers. The conformity-based analog of the node significance measure is defined as $GS_{CF} = CF_i * CF_t$, where CF_t is a constant. Eigenvector-based network concepts are defined in Sect. 6. Examples include concepts based on the scaled connectivity $ScaledConnectivity(M)$ (3.1) and intermodular separability measures.

fundamental network concepts are the standard concepts used in network applications, approximate CF-based and eigenvector-based concepts are very useful for theoretical derivations.

3.1 Network Concept Functions

To define different types of network concepts, we make use of network concept functions (Dong and Horvath 2007; Horvath and Dong 2008). A **network concept function** is a function of an $n \times n$ dimensional symmetric matrix $M = (M_{ij})$

(interpreted as a type of adjacency matrix) and/or a corresponding vector $G = (G_1, \ldots, G_n)$ (interpreted as node significance measure). In particular, we will use the following network concept functions:

$$
\begin{aligned}
Connectivity(M)_i &= \sum_j M_{ij}, \\
ScaledConnectivity(M)_i &= K(M)_i = \frac{Connectivity(M)_i}{max(Connectivity(M))}, \\
p.Connectivity(M, no.bins) &= p(discretize(Connectivity(M), no.bins)), \\
ScaleFreeFit(M, no.bins) &= cor(log(p.Connectivity(M, no.bins)), log(BinNo))^2, \\
Density(M) &= \frac{\sum_i \sum_j M_{ij}}{n(n-1)}, \\
MAR(M)_i &= \frac{\sum_j (M_{ij})^2}{\sum_j M_{ij}}, \\
Centralization(M) &= \frac{n}{n-2}\left(\frac{max(Connectivity(M))}{n-1} - Density(M)\right), \\
Heterogeneity(M) &= \sqrt{\frac{n(\sum_i \sum_j \sum_k M_{ij} M_{ik})}{(\sum_i \sum_j M_{ij})^2} - 1}, \\
ClusterCoef(M)_i &= \frac{\sum_j \sum_k M_{ij} M_{jk} M_{ki} - \sum_j M_{ij}^2 M_{jj}}{(\sum_j M_{ij})^2 - \sum_j M_{ij}^2}, \\
TopOverlap(M)_{ij} &= \frac{\sum_u M_{iu} M_{uj} + M_{ij}}{min\{Connectivity(M)_i, Connectivity(M)_j\} + 1 - M_{ij}}, \\
NetworkSignif(G) &= \frac{\sum_i G_i}{n}, \\
HubGeneSignif(M, G) &= \frac{\sum_i G_i ScaledConnectivity(M)_i}{\sum_i (ScaledConnectivity(M)_i)^2},
\end{aligned} \tag{3.1}
$$

where the parameter $no.bins$ in the function $p.Connectivity$ refers to the number of bins used in the discretization function and $BinNo = (1, 2, \ldots, no.bins)$.

Network concept functions allow us to define different types of network concepts. An abstract definition of a fundamental network concept is given by:

Definition 3.1 (Fundamental network concept). A **fundamental network concept** is defined by evaluating a network concept function NCF (3.1) on $A - I$ and/or a node significance measure GS, i.e.,

$$NC_{fundamental} = NCF(A - I, GS). \tag{3.2}$$

Since the diagonal elements of an adjacency matrix are set to 1, the diagonal elements of $A - I$ (where I is the identity matrix) equal 0. Thus, fundamental network concepts are functions of the off-diagonal elements of the adjacency matrix A.

For example, the fundamental network concept of (whole network) connectivity arises by evaluating by $Connectivity()$ on $A-I$, i.e.,

$$k_i = Connectivity(A-I)_i = \sum_{j \neq i} A_{ij}.$$

3.2 CF-Based Network Concepts

Let us now assume that a (unique) conformity vector CF exists for the adjacency matrix A. Then we are able to define the **CF-based adjacency matrix** as follows:

$$A_{CF} = CFCF^{\tau} - \mathrm{diag}(CF^2) + I, \tag{3.3}$$

where $\mathrm{diag}(CF^2)$ denotes the diagonal matrix with diagonal elements CF_i^2, $i = 1 \dots n$. Note that the diagonal elements of A_{CF} equal 1. Let us first assume that the network concept function $NCF(M)$ (3.1) does not depend on a node significance measure. In this case, we define the **CF-based network concept** by evaluating the network concept function on $A_{CF} - I$ (3.3), i.e.,

$$NC_{CF} = NCF(A_{CF} - I).$$

If the network concept function $NCF(M,G)$ does depend on a node significance measure, assume that the adjacency matrix and the node significance measure can be approximated as follows:

$$\begin{aligned} A_{ij} &\approx CF_i CF_j \text{ for } i \neq j \\ GS_i &\approx CF_i CF_t. \end{aligned} \tag{3.4}$$

Toward this end, a minimization algorithm can be used to define CF and CF_t as minimizing solutions v^*, v_t^* of

$$f(v,v_t) = \sum_i \sum_{j>i} (A_{ij} - v_i v_j)^2 + \sum_i (GS_i - v_i v_t)^2, \tag{3.5}$$

where we require nonnegative components for v and a nonnegative number v_t. More details are presented below. Based on the right-hand side of (3.4), we define the **CF-based node significance measure as:**

$$GS_{CF,i} = CF_i CF_t. \tag{3.6}$$

We are now ready to provide a general definition of a CF-based network concept:

Definition 3.2 (CF-based network concept). A **CF-based network concept** is defined by evaluating a network concept function (3.1) on $A_{CF} - I$ (3.3) and/or GS_{CF} (3.6), i.e.,

$$NC_{CF} = NCF(A_{CF} - I, GS_{CF}).$$

For example, the CF-based connectivity is given by

$$Connectivity_{CF,i} = CF_i sum(CF) - CF_i^2.$$

In an exercise, you are asked to derive expressions for the other network concepts.

3.3 Approximate CF-Based Network Concepts

In the following, we define *approximate* CF-based network concepts (Dong and Horvath 2007; Horvath and Dong 2008) which turn out to satisfy simple relationships. We define the **approximate CF-based adjacency matrix** as follows:

$$A_{CF,app} = CF CF^{\tau}. \tag{3.7}$$

Note that only the ith diagonal element of $A_{CF,app}$ is $A_{CF,app,ii} = (CF_i)^2$, while it is 1 for A_{CF}.

Definition 3.3 (Approximate CF-based analog of a network concept). An **approximate CF-based analog** of a network concept is defined by evaluating a network concept function (3.1) on $A_{CF,app}$ (3.7) and/or GS_{CF} (3.6), i.e.,

$$NC_{CF,app} = NCF(A_{CF,app}, GS_{CF}).$$

Denote the maximum conformity by $CF_{max} = max(CF)$. In an exercise, you are asked to derive approximate CF-based analogs of fundamental network concepts. It turns out that the approximate CF-based clustering coefficient does not depend on the i-index. This is why we sometimes omit this index and simply write $ClusterCoef_{CF,app}$.

In the following, we will demonstrate a major advantage of approximate CF-based network concepts: they reveal surprising and often simple relationships between seemingly disparate network concepts. In an exercise, you are asked to prove the following relationships:

$$\begin{aligned}
CF_i &= \frac{k_{CF,app,i}}{\sqrt{sum(k_{CF,app})}} \\
K_{CF,app,i} &= \frac{CF_i}{CF_{max}} \\
mean\left(K_{CF,app,j}\right) &\approx \frac{\sqrt{Density_{CF,app}}}{CF_{max}} \\
Centralization_{CF,app} &\approx \sqrt{Density_{CF,app}}\left(CF_{max} - \sqrt{Density_{CF,app}}\right) \\
HubNodeSignif_{CF,app} &\approx CF_{max} CF_t \\
NetworkSignif_{CF,app} &\approx \frac{\sqrt{Density_{CF,app}}}{CF_{max}} HubNodeSignif_{CF,app}.
\end{aligned} \tag{3.8}$$

In the next section, we study conditions when these simple relationships are approximately satisfied by the analogous fundamental network concepts.

We will now introduce a restrictive assumption on the conformity vector. The **maximum conformity assumption** states that

$$CF_{max} = max(CF) \approx 1. \tag{3.9}$$

The maximum conformity assumption is rarely satisfied by whole networks but is often satisfied by network modules. In particular, most correlation network modules satisfy the maximum conformity assumption as described in Sect. 6. Under the maximum conformity assumption, (3.8) implies

$$\begin{aligned} K_{CF,app,i} &= CF_i \\ Centralization_{CF,app} &\approx \sqrt{Density_{CF,app}}\left(1 - \sqrt{Density_{CF,app}}\right) \\ HubNodeSignif_{CF,app} &\approx CF_t. \end{aligned} \tag{3.10}$$

3.4 Fundamental Network Concepts Versus CF-Based Analogs

Let us first assume a network concept that only depends on the adjacency matrix. In this case, the fundamental network concept $NCF(A-I)$ is simply a function of $A-I$, and its CF-based analog is given by $NCF(A_{CF}-I)$. By definition of the factorizability measure (2.3), $A \approx A_{CF}$ is equivalent to $F(A) \approx 1$. For a continuous network concept function, $A \approx A_{CF}$ implies

$$NCF(A-I) \approx NCF(A_{CF}-I), \tag{3.11}$$

i.e., the fundamental network concept is approximately equal to its CF-based analog. Thus, a fundamental network concept can be approximated by its CF-based analog if the network is approximately factorizable.

Let us now consider the case when the fundamental network concept $NCF(A-I,GS)$ depends on a node significance measure. Then the CF-based analog is given by $NCF(A_{CF}-I,GS_{CF})$, where $GS_{CF,i} = CF_iCF_t$ (3.6). For a continuous network concept function, the fundamental network concept is equal to its CF-based analog if $A \approx A_{CF}$ and $GS \approx GS_{CF}$. A more quantitative measure can be defined using the extended factorizability measure which will be defined in the following. Assume that the entries of the node significance measure GS_i are nonnegative numbers. Let $v = (v_1, v_2, \ldots, v_n)$ be a vector of length n and v_t be a number. The conformity vector CF and the node significance conformity CF_t are defined as minimizing solutions of the following objective function:

$$f(v, v_t) = \sum_i \sum_{j>i} (A_{ij} - v_i v_j)^2 + \sum_i (GS_i - v_i v_t)^2.$$

An equivalent but more convenient approach is to define these quantities as the maximizer of the following **extended factorizability function**:

$$F_{A,GS}(v,v_t) = 1 - \frac{f(v,v_t)}{f(0,0)}. \tag{3.12}$$

Similar to Definition 2.1, the conformity vector CF is defined as the vector with non-negative entries that maximizes $F_{A,GS}(v,v_t)$, and the node significance conformity CF_t is analogously defined. Further, we define the extended factorizability $F(A,GS)$ as the corresponding maximum value $F_A(CF,CF_t)$, i.e.,

$$F(A,GS) = max_{v,v_t}(F_{A,GS}(v,v_t)) = F_{A,GS}(CF,CF_t). \tag{3.13}$$

Further, we set $A_{CF} = CFCF^\tau - \text{diag}(CF^2) + I$ and $GS_{CF} = CF_iCF_t$. In general, the closer $F(A,GS)$ is to 1, the better A_{CF} approximates A and the better GS_{CF} approximates GS. Using this notation, we arrive at the following:

Observation 3.1. *A fundamental network concept $NC_{fundamental}$ is approximately equal to its CF-based analog NC_{CF} if the network is approximately factorizable $F(A) \approx 1$. For a network concept that depends on a node significance measure, $F(A,GS) \approx 1$ implies*

$$\begin{aligned} NCF(A-I,GS) &\approx NCF(A_{CF}-I,GS_{CF}) \\ \textit{fundamental concept} &\approx \textit{CF-based concept}. \end{aligned} \tag{3.14}$$

3.5 CF-Based Concepts Versus Approximate CF-Based Analog

To determine when a CF-based network concept is close to its approximate CF-based analog, we find it useful to introduce two types of assumptions on the conformity vector CF. First, the β-th power ($\beta \le 4$) of the conformity is assumed to satisfy.

$$\frac{CF_i^\beta}{sum\left(CF^\beta\right)} \approx 0, \text{ where } \beta \le 4. \tag{3.15}$$

Since this assumption implies that each individual component CF_i^β is negligible compared to the sum over all components, we refer to it as the **negligible CF component assumption**. This weak assumption is typically satisfied by conformity vectors. The second assumption on the conformity vector is concerned with the βth sample moment $mean(CF^\beta)$ (where $\beta \le 4$). If these moments are bounded (i.e., take on finite values) then

$$MomentRatio(\beta) = \frac{mean(CF^{2*\beta})}{mean(CF^\beta)^2}$$

is typically much smaller than n. For large networks (i.e., large n), we often find that the following **bounded CF moment assumption** is satisfied:

$$MomentRatio(\beta) = \frac{1}{n} \frac{mean\left(CF^{2*\beta}\right)}{mean\left(CF^{\beta}\right)^2} \approx 0. \tag{3.16}$$

In an exercise, you are asked to prove that

$$\frac{1}{n} \leq MomentRatio(\beta) \leq 1. \tag{3.17}$$

Incidentally, the moment ratio for $\beta = 2$ has a nice interpretation in terms of the Frobenius matrix norm:

$$MomentRatio(\beta = 2) = \frac{||(A_{CF} - I) - A_{CF,app}||_F^2}{||A_{CF,app}||_F^2}.$$

In an exercise, you are asked to prove the following:

Observation 3.2 (CF-based versus approximate CF-based concepts). *CF-based network concepts are close to their approximate CF-based analogs under weak conditions on the conformity vector. The negligible CF component assumption (3.15) implies that the following CF-based network concepts are close to their approximate CF-based analogs:* $Connectivity_{CF}$, $p.Connectivity_{CF}$, $ScaleFreeFit_{CF}$, MAR_{CF}, $ClusterCoef_{CF}$, $TopOverlap_{CF}$, $NetworkSignif_{CF}$, $HubGeneSignif_{CF}$. *The bounded CF moment assumption (3.16) implies that the following CF-based network concepts are close to their approximate CF-based analogs:* $Density_{CF}$, $Centralization_{CF}$, $Heterogeneity_{CF}$. *In brief, the two conditions imply that*

$$\begin{aligned} NCF(A_{CF} - I, GS) &\approx NCF(A_{CF,app}, GS_{CF}) \\ CF\text{-}based\ concept &\approx approximate\ CF\text{-}based\ concept. \end{aligned} \tag{3.18}$$

3.6 Higher Order Approximations of Fundamental Concepts

This technical section can be skipped. It is only recommended to readers who are familiar with the multivariable Taylor expansion of a function, which is taught in calculus. In the following, we describe higher order Taylor expansions for improving the approximation between fundamental network concepts and conformity-based network concepts. Assume that the fundamental network concept $NCF(A - I)$ is defined by a differentiable network concept function $NCF(M)$ that depends on the matrix M. We now describe Taylor series expansions of $NCF()$ around $A_{CF} - I$.

The *zeroth order Taylor expansion* $Taylor^{(0)} = NCF(A_{CF} - I)$ is just the CF-based analog of the network concept. The *first-order Taylor expansion* is given by:

$$\begin{aligned} Taylor^{(1)} &= NCF(A_{CF} - I) + \sum_{i \neq j} (A_{ij} - CF_i CF_j) DNCF(A_{CF} - I)_{ij} \\ &= NCF(A_{CF} - I) + sum\left((A - A_{CF}) * DNCF(A_{CF} - I)\right), \end{aligned} \quad (3.19)$$

where $DNCF(A_{CF} - I)$ denotes a matrix of first derivatives of $NCF()$ evaluated at $A_{CF} - I$, i.e., the i,jth component of DNCF is given by

$$DNCF(A_{CF} - I)_{ij} = \left(\frac{\partial}{\partial M_{ij}} NCF \right)(A_{CF} - I). \quad (3.20)$$

An even better approximation is given by the second-order Taylor expansion:

$$\begin{aligned} Taylor^{(2)} = {} & NCF(A_{CF} - I) + \sum_{i \neq j} (A_{ij} - CF_i CF_j) DNCF(A_{CF} - I)_{ij} \\ & + \sum_{i \neq j} \sum_{i' \neq j'} (A_{ij} - CF_i CF_j)(A_{i'j'} - CF_{i'} CF_{j'}) D^2 NCF(A_{CF} - I)_{ij,i'j'}, \end{aligned} \quad (3.21)$$

where the second-order derivative of NCF with respect to M_{ij} and $M_{i'j'}$ is defined by

$$D^2 NCF_{ij,i'j'} = \left(\frac{\partial^2}{\partial M_{ij} \partial M_{i'j'}} NCF \right)(A_{CF} - I). \quad (3.22)$$

3.7 Fundamental Concepts Versus Approx. CF-Based Analogs

We relate a fundamental network concept (e.g., *Density*) to its approximate CF-based analog (e.g., $Density_{CF,app}$) in two steps. In step 1 (fundamental→CF-based), the results from Sect. 3.4 imply that fundamental network concepts are close to their CF-based analogs in approximately factorizable networks. In step 2 (CF-based→approximate CF-based), the results from Sect. 3.5 imply that CF-based network concepts are close to their approximate CF-based analogs if CF satisfies the negligible CF component and/or the bounded CF moment ratio assumption. In summary, we arrive at the following:

Observation 3.3. *A fundamental network concept is close to its approximate CF-based analog, i.e.,*

$$\textit{fundamental concept} \approx \textit{approximate CF-based analog}, \quad (3.23)$$

if the network is approximately factorizable (see Observation 3.1) and if it satisfies the negligible CF component and/or the bounded CF moment ratio assumption (see Observation 3.2).

In our real data applications, we show empirically that fundamental concepts can be approximated by their approximate CF-based analogs even in networks that satisfy the assumptions underlying Observation 3.3 only approximately. Observation 3.3 is illustrated for network density, centralization, heterogeneity, and clustering coefficients in Fig. 3.2 (Drosophila PPI network) and Fig. 3.3 (yeast PPI network).

3.8 Relationships Among Fundamental Network Concepts

An important consequence of Observation 3.3 is that the simple relationships satisfied by approximate CF-based network concepts also apply to their corresponding fundamental network concepts in approximately factorizable networks. For example, (3.8) implies that

$$CF \approx \frac{k}{\sqrt{sum(k)}}. \tag{3.24}$$

This relationship has several important consequences. First, it can be used to argue that an approximately factorizable network satisfies scale-free topology (1.11) if the same holds for the distribution of CF. Second, it can be used to derive the following:

Observation 3.4. *In approximately factorizable networks, fundamental network concepts are simple functions of the connectivity.*

Since approximate CF-based network concepts are simple functions of the conformity, substituting $\frac{k}{\sqrt{sum(k)}}$ for CF implies that fundamental concepts can be approximated by simple functions of the connectivity in approximately factorizable networks.

Let us now look at the mean clustering coefficient. In approximately factorizable networks, we find

$$mean(ClusterCoef) \approx ClusterCoef_{CF,app}. \tag{3.25}$$

Observation 3.4, (3.50), and (3.25) imply

$$mean(ClusterCoef) \approx \frac{(sum(k^2))^2}{(sum(k))^3}. \tag{3.26}$$

Further, one can derive the following

Observation 3.5 (Special relationships among network concepts). *Under the maximum conformity assumption (3.9) and the assumptions underlying Observation 3.3, the following special relationships are satisfied by fundamental network concepts:*

$$K_i \approx CF_i \tag{3.27}$$

$$HubNodeSignif \approx CF_t \quad (3.28)$$

$$NetworkSignif \approx \sqrt{Density} \times HubGeneSignif \quad (3.29)$$

$$mean(ClusterCoef) \approx \left(1 + Heterogeneity^2\right)^2 \times Density \quad (3.30)$$

$$Centralization \approx \sqrt{Density}\left(1 - \sqrt{Density}\right). \quad (3.31)$$

Equation (3.30) is illustrated in Fig. 3.4 (Drosophila PPI network). The relationship between centralization and density (3.31) is surprisingly simple, but it certainly does not hold in general networks. For a general network, one can only derive an upper bound for the centralization in terms of the density (Snijders 1981). Our empirical studies (described below) show that (3.31) is not very robust with regard to deviations from our theoretical assumptions.

3.8.1 Relationships for the Topological Overlap Matrix

In an exercise, you are asked to show that under weak assumptions, the following relationship holds in approximately factorizable networks:

$$TopOverlap_{ij} \approx \frac{max(k_i, k_j)}{n}(1 + Heterogeneity^2). \quad (3.32)$$

Equation (3.32) is illustrated in Fig. 3.4 (Drosophila PPI network). Equation (3.32) has several important consequences. To begin with, it illustrates that the topological overlap between the most highly connected node and all other nodes is approximately constant. Specifically, if we denote the index of the most highly connected node by $[1]$ and its connectivity by $k_{[1]} = max(k)$, then

$$TopOverlap_{[1]j} \approx \frac{k_{[1]}}{n}(1 + Heterogeneity^2). \quad (3.33)$$

As an aside, we briefly mention that $TopOverlap_{[1]j}$ has a simple interpretation in terms of the hierarchical clustering dendrogram that results from using $dissTopOverlap_{ij} = 1 - TopOverlap_{ij}$ as input. In this case, $TopOverlap_{[1]j}$ is related to the longest branch length in the dendrogram.

In the following, we relate $TopOverlap_{[1]j}$ to the fundamental network concept $Centralization$. According to (1.20), $\frac{max(k)}{n} \approx Centralization + Density$. Substituting this expression in (3.33) implies

$$TopOverlap_{[1]j} \approx (Centralization + Density)(1 + Heterogeneity^2). \quad (3.34)$$

Equation (3.34) is illustrated in Fig. 3.4 (Drosophila PPI network).

3.9 Alternative Expression of the Factorizability *F(A)*

In the following, we will present alternative expressions for the factorizability when the conformity vector CF of the adjacency matrix A exists.

In an exercise, you are asked to prove that the factorizability $F(A)$ (2.3) can be expressed as follows:

$$F(A) = \frac{||A_{CF} - I||_F^2}{||A - I||_F^2} = \frac{Density \text{ of } A_{CF}^2}{Density \text{ of } A^2}, \tag{3.35}$$

i.e., it can be expressed as a ratio of network densities (1.18). Further, we find the following striking relationship between the factorizability and the approximate CF-based clustering coefficient:

$$F(A) = \frac{sum(CF^2)^2 - sum(CF^4)}{Density(A^2)} = ClusterCoef_{CF,app} \frac{Density(A_{CF})}{Density(A^2)}. \tag{3.36}$$

Let us now assume that a clustering procedure (see Chap. 8) has led to clusters of densely interconnected nodes. In the context of networks, we refer to these clusters as modules. Denote by $A^{(q)}$ the adjacency matrix of the qth module and by $CF^{(q)}$ the corresponding conformity vector. Equation (3.36) shows that the module is approximately factorizable ($F(A^{(q)}) \approx 1$) if $ClusterCoef_{CF,app}^{(q)} \approx 1$ (cliquish nodes) and if $\frac{Density(A_{CF}^{(q)})}{Density((A^{(q)})^2)} \approx 1$ (densely connected nodes). Since many clustering procedures lead to clusters whose nodes are "cliquish" and are densely connected, we arrive at the following:

Observation 3.6. *For many modules defined with a clustering procedure, the sub-network comprised of the module nodes is approximately factorizable.*

Observation 3.6 motivated us to study network concepts in approximately factorizable networks. In Sect. 3.10, we provide empirical evidence for Observation 3.6. Further, in Sect. 6.2, we characterize approximately factorizable weighted correlation networks, e.g., network modules.

3.10 Approximately Factorizable PPI Modules

Approximate factorizability is a very strong structural assumption on an adjacency matrix. It certainly does not hold for general networks. Here we provide empirical evidence in support of Observation 3.6, i.e., we illustrate that many clusters (modules) of genes or proteins in real networks are approximately factorizable (Dong and Horvath 2007). Using protein–protein interaction, gene co-expression, we find that (a) many networks comprised of module nodes (i.e., network modules) are approximately factorizable and (b) in these types of networks, simple relationships exist between seemingly disparate network concepts (Dong and Horvath 2007). Here we review some of these results for protein–protein interaction networks. Approximately factorizable *correlation* networks are discussed in Sect. 6.2.

To illustrate our results, we computed network concepts in network modules based on Drosophila and yeast protein–protein interaction (PPI) networks downloaded from BioGrid (Breitkreutz et al. 2003). The original PPI networks $A^{original}$ were unweighted networks, i.e., the adjacency between two proteins is either 1 or 0 depending on whether they are known to interact.

Since these original adjacency matrices of the PPI network adjacency matrices are very sparse (i.e., they contain a lot of zeroes), we replaced them by the corresponding topological overlap matrices. Thus, we used the TOM-based adjacency function (1.38) to replace the original PPI network by $A = AF^{TOM}(A^{original})$. Thus, the i, jth element of A is given by $A_{ij} = dissTopOverlap_{ij}$. The resulting adjacency matrix A is a weighted network, which can be considered as a "smoothed out" version of the original adjacency matrix.

As described in Sect. 1.7, we use $dissTopOverlap_{ij}$ as input of average linkage hierarchical clustering to arrive at a dendrogram (clustering tree) (Kaufman and Rousseeuw 1990). As described in Sect. 8.6, we defined the modules as branches of the hierarchical clustering dendrogram (see Fig. 3.1).

For example, in Fig. 3.1, we show the dendrograms of our network applications. Genes or proteins of proper modules are assigned a color (e.g., turquoise, blue, etc.). Genes outside any proper module are colored gray. Of the 1371 proteins in

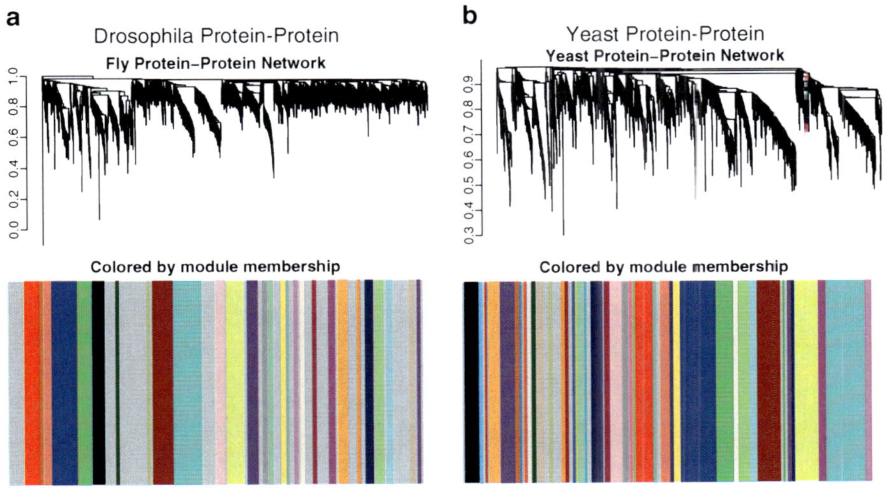

Fig. 3.1 Hierarchical clustering tree (dendrogram) and module definition. (**a**) Module of the drosophila PPI network. As described in more detail in Fig. 8.3, the dendrogram results from average linkage hierarchical clustering. The color-band below the dendrogram denotes the modules, which are defined as branches in the dendrogram. Of the 1371 proteins, 862 were clustered into 28 proper modules, and the remaining proteins are colored in *gray*; (**b**) yeast PPI network

the Drosophila PPI network, 862 were clustered into 28 modules, and the remaining proteins grouped into an improper (gray) module. The module sizes of the proper modules range from 10 to 96, mean 30.79, median 23, and interquartile range 24.

Of the 2,292 proteins in the yeast PPI network, 2,050 were clustered into 44 proper modules, and the remaining proteins grouped into an improper module. The module sizes of the proper modules range from 10 to 219, mean 46.59, median 24, and interquartile range 38.8.

Since the scope of this chapter is mathematical and topological analysis of network modules, we defined modules without regard to external gene ontology information. Also we do not provide an in-depth analysis of the biological meaning of the network modules. But we briefly mention that there is indirect evidence that most of the resulting modules are biologically meaningful. Using the functional gene annotation tools from the Database for Annotation, Visualization and Integrated Discovery (DAVID) (Dennis et al. 2003) to test for both enriched biochemical pathways and subcellular compartmentalization, we find that most modules are significantly enriched with known gene ontologies. A functional enrichment analysis for each network application can be found in the Additional Files of Dong and Horvath (2007).

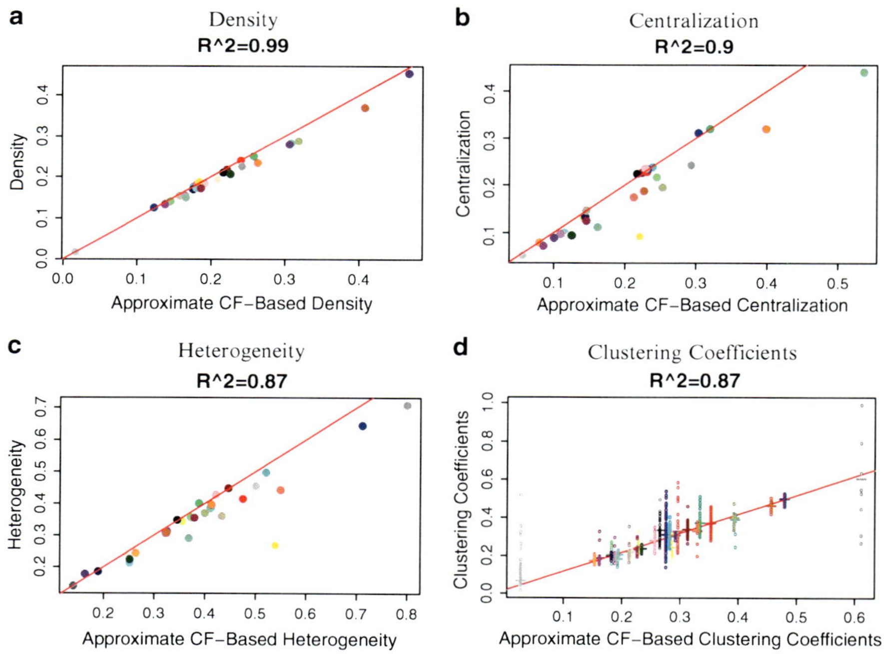

Fig. 3.2 Drosophila protein–protein interaction network modules: the relationship between fundamental network concepts $NetworkConcept(A-I)$ (y-axis) and their approximate CF-based analogs $NetworkConcept_{CF,app}$ (x-axis). In (**a–c**), each dot corresponds to a module since these network concepts summarize an entire network module. In (**d**), each dot corresponds to a node since these network concepts are node specific. A reference line with intercept 0 and slope 1 has been added to each plot

Our empirical results support Observation 3.6: for many modules defined with a clustering procedure, the subnetwork comprised of the module nodes is approximately factorizable. In the Drosophila PPI network, the mean factorizability $F(A)$ is 0.82 for 'proper' modules defined as branches in the network. In contrast, the factorizability of the subnetwork comprised of non-module (gray) nodes is only 0.17. In the yeast PPI network, the mean factorizability of proper modules is 0.85, while it equals only 0.20 for the gray module. We briefly mention that similar results are obtained for modules inside a correlation network (Dong and Horvath 2007).

In accordance with Observation 3.3, we find a close relationship ($R^2 \geq 0.6$) between the fundamental network concepts and their approximate CF-based analogs. Specifically, we relate the network density, centralization, heterogeneity, and clustering coefficients to their approximate CF-based analogs in Figs. 3.2 (Drosophila PPI network) and 3.3 (yeast PPI network).

In accordance with (3.30), we find a close relationship ($R^2 \geq 0.6$) between the mean clustering coefficient $mean(ClusterCoef)$ and $\left(1+Heterogeneity^2\right)^2 \times Density$.

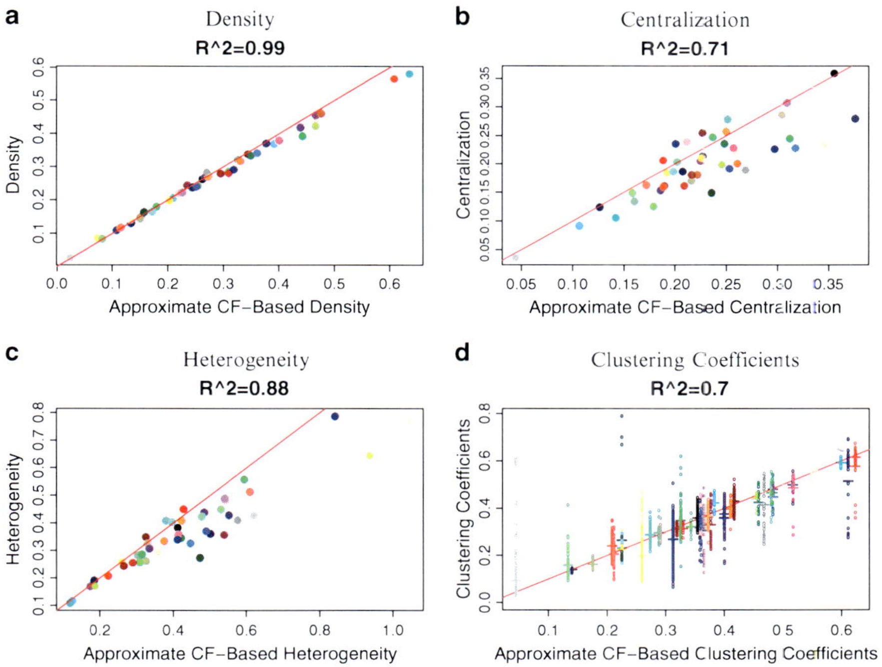

Fig. 3.3 Yeast protein–protein interaction network modules: the relationship between fundamental network concepts $NetworkConcept(A-I)$ (y-axis) and their approximate CF-based analogs $NetworkConcept_{CF,app}$ (x-axis). In (**a**–**c**), each dot corresponds to a module since these network concepts summarize an entire network module. In (**d**), each dot corresponds to a node since these network concepts are node specific. A reference line with intercept 0 and slope 1 has been added to each plot

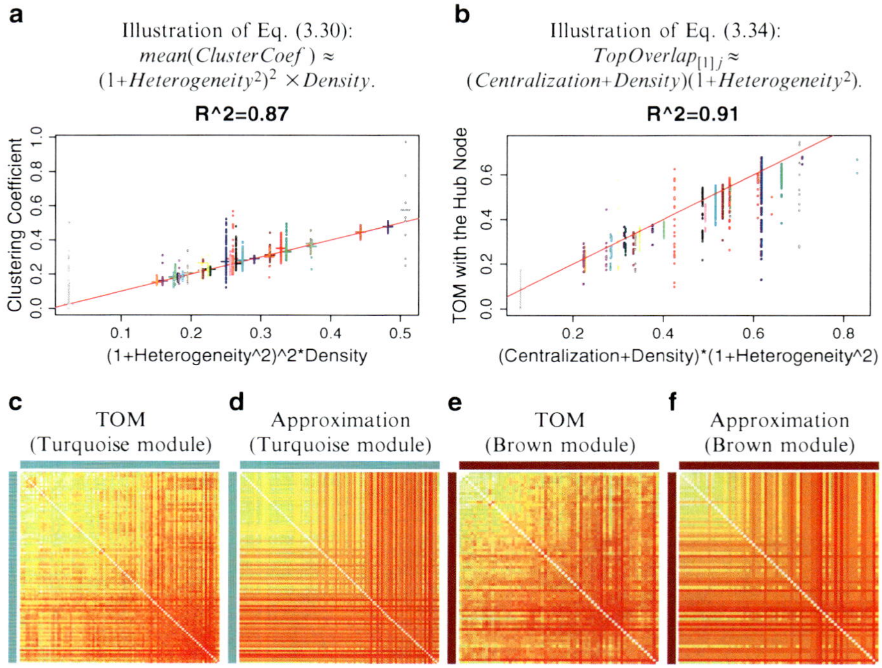

Fig. 3.4 Drosophila protein–protein interaction network modules: the relationship between fundamental network concepts. In (**a**,**b**), each dot corresponds to a node since these network concepts are node specific, and the *red line* has intercept 0 and slope 1. (**c**) is a color-coded depiction of the topological overlap matrix $TopOverlap_{ij}$ in the turquoise network module. (**d**) represents the corresponding approximation $max(k_i, k_j)(1+Heterogeneity^2)/n$ (3.32). (**e**,**f**) are their analogs for the brown module. The turquoise and the brown module represent the largest and third largest module. Analogous plots for the other modules can be found in our online supplement

In accordance with (3.34), we find a close relationship between $TopOverlap_{[1]j}$ and $(Centralization + Density)(1 + Heterogeneity^2)$ [see Fig. 3.4 (Drosophila PPI network)].

Since the number of genes inside a module (module size) varies greatly among the modules, it is natural to wonder whether the reported relationships between network concepts are due to the underlying module sizes. As detailed in an Additional File of Dong and Horvath (2007), we find that the relationship between fundamental network concepts and their approximate CF-based analogs remains highly significant even after correcting for module sizes. The same holds for the relationships between network concepts. Thus, none of the reported relationships is trivially due to module sizes. But we find that many network concepts depend on the underlying module size. We find that large modules are less factorizable than small modules: there is a strong negative correlation between module factorizability $F(A)$ and module size. We also find that fundamental network concepts (e.g., density) depend on module size in our applications. For the factorizability, density, centralization, heterogeneity, and mean clustering coefficient, the correlation coefficients with module size are −0.84, −0.46, −0.17, 0.26, and −0.36 in Drosophila PPI network modules; they are −0.55, −0.52, 0.05, 0.5, and −0.44 in yeast PPI network modules. More details can be found in Dong and Horvath (2007).

3.11 Studying Block Diagonal Adjacency Matrices

A simple, exactly factorizable network is given by an adjacency matrix A with constant adjacencies ($A_{ij} = b$, $b \in (0,1]$). The adjacency matrix is exactly factorizable since $A_{ij} = CF_i CF_j$, where $CF_i = \sqrt{b}$. This network can be interpreted as the expected adjacency matrix of an Erdös–Rényi network (Erdos and Renyi 1960). One can easily derive the following expressions for the fundamental network concepts: $Connectivity_i = (n-1)b$, $Density = b$, $Centralization = 0$, $Heterogeneity = 0$, $ClusterCoef_i = b$ and $TopOverlap_{ij} = b$.

Since A is exactly factorizable, the fundamental network concepts equal their CF-based analogs. However, the *approximate* CF-based concepts are different from their exact counterparts (see Table 3.2). For reasonably large values of n, the fundamental network concepts are very close to their approximate CF-based analogs, which illustrates Observation 3.3. The results in Table 3.2 can be used to verify our simple relationships between network concepts [e.g., (3.30) and (3.24)].

In the following, we will consider a block diagonal adjacency matrix where each block has constant adjacencies, i.e.,

$$A = \begin{pmatrix} 1 & b_1 & \cdots & b_1 & 0 & 0 & \cdots & 0 \\ b_1 & 1 & \cdots & b_1 & 0 & 0 & \cdots & 0 \\ \vdots & \vdots & \ddots & \vdots & \vdots & \vdots & \ddots & \vdots \\ b_1 & b_1 & \cdots & 1 & 0 & 0 & \cdots & 0 \\ 0 & 0 & \cdots & 0 & 1 & b_2 & \cdots & b_2 \\ 0 & 0 & \cdots & 0 & b_2 & 1 & \cdots & b_2 \\ \vdots & \vdots & \ddots & \vdots & \vdots & \vdots & \ddots & \vdots \\ 0 & 0 & \cdots & 0 & b_2 & b_2 & \cdots & 1 \end{pmatrix}. \tag{3.37}$$

Table 3.2 Network concepts in the constant Erdös–Rényi network

Network concepts	Fundamental	Approximate CF-based
$Connectivity_i$	$(n-1)b$	nb
$Density$	b	$b\frac{n}{n-1}$
$Centralization$	0	0
$Heterogeneity$	0	0
$TopOverlap_{ij}$	b	$b\frac{nb+1}{(n-1)b+1}$
$ClusterCoef_i$	b	b

We assume that the first and second blocks have dimensions $n^{(1)} \times n^{(1)}$ and $n^{(2)} \times n^{(2)}$, respectively. Such a block diagonal matrix can be interpreted as a network with two distinct modules. Setting $n^{(2)} = 0$ results in the simple constant adjacency matrix, which we considered before.

We denote by $f_1 = (1,1,\ldots,1,0,0,\ldots,0)$ the vector whose first $n^{(1)}$ components equal 1 and the remaining components equal 0. Similarly, we define $f_2 = (0,0,\ldots,0,1,1,\ldots,1) = 1 - f_1$. To simplify the calculation of the conformity, we further assume that

$$\frac{n^{(2)}\left(n^{(2)}-1\right)b_2^2}{n^{(1)}\left(n^{(1)}-1\right)b_1^2} < 1. \tag{3.38}$$

Then the conformity is uniquely defined by

$$CF = \sqrt{b_1} f_1,$$

as one can show using equations (2.2) and (2.4). Further, one can show that the factorizability is given by

$$F(A) = \frac{n^{(1)}\left(n^{(1)}-1\right)b_1^2}{n^{(1)}\left(n^{(1)}-1\right)b_1^2 + n^{(2)}\left(n^{(2)}-1\right)b_2^2}. \tag{3.39}$$

In particular, if $n^{(1)} \approx n^{(2)}$ and $b_1 = b_2$, i.e., if the adjacency matrix is comprised of two nearly identical blocks, the factorizability is $F(A) \approx 1/2$. Similarly, one can show that if the matrix A is comprised of Q identical blocks, then $F(A) \approx 1/Q$. This block diagonal network allows one to arrive at explicit formulas for fundamental-, CF-based-, and approximate CF-based network concepts (see Table 3.3).

Table 3.3 Network concepts of the block diagonal constant adjacencies. The indicator function $Ind(\cdot)$ takes on the value 1 if the condition is satisfied and 0 otherwise

Concept	Fundamental	CF based	Approx CF based
$Connectivity_i$	$\sum_{q=1}^{Q}(n^{(q)}-1)b_q Ind_{i\le n^{(q)}}$	$(n^{(1)}-1)b_1 Ind_{i\le n^{(1)}}$	$n^{(1)} b_1 Ind_{i\le n^{(1)}}$
$Density$	$\frac{n^{(1)}(n^{(1)}-1)b_1+n^{(2)}(n^{(2)}-1)b_2}{(n^{(1)}+n^{(2)})(n^{(1)}+n^{(2)}-1)}$	$\frac{n^{(1)}(n^{(1)}-1)b_1}{(n^{(1)}+n^{(2)})(n^{(1)}+n^{(2)}-1)}$	$\frac{(n^{(1)})^2 b_1}{(n^{(1)}+n^{(2)})(n^{(1)}+n^{(2)}-1)}$
$Centralization$	$\frac{n^{(2)}((n^{(1)}-1)b_1+(n^{(2)}-1)b_2)}{(n^{(1)}+n^{(2)}-1)(n^{(1)}+n^{(2)}-2)}$	$\frac{(n^{(1)}-1)n^{(2)}b_1}{(n^{(1)}+n^{(2)}-1)(n^{(1)}+n^{(2)}-2)}$	$\frac{n^{(1)}n^{(2)}b_1}{(n^{(1)}+n^{(2)}-1)(n^{(1)}+n^{(2)}-2)}$
$Heterogeneity$	$\sqrt{\frac{(n^{(1)}+n^{(2)})[\sum_{q=1}^{Q} n^{(q)}(n^{(q)}-1)^2 b_q^2]}{[\sum_{q=1}^{Q} n^{(q)}(n^{(q)}-1)b_q]^2}-1}$	$\sqrt{\frac{n^{(2)}}{n^{(1)}}}$	$\sqrt{\frac{n^{(2)}}{n^{(1)}}}$
$TopOverlap_{ij}$	$b_1 Ind_{i,j\le n^{(1)}} + b_2 Ind_{i,j>n^{(1)}}$	$b_1 Ind_{i,j\le n^{(1)}}$	$b_1 \frac{n^{(1)}b_1+1}{(n^{(1)}-1)b_1+1} Ind_{i,j\le n^{(1)}}$
$ClusterCoef_i$	$b_1 Ind_{i\le n^{(1)}} + b_2 Ind_{i>n^{(1)}}$	$b_1 Ind_{i\le n^{(1)}}$	$b_1 Ind_{i\le n^{(1)}}$

In the following, we study the relationship between fundamental network concepts and their approximate CF-based analogs in the limit when the block diagonal network becomes approximately factorizable. Specifically, we calculate network concepts in the limit $b_2 \rightarrow 0$ when $n^{(1)}$, $n^{(2)}$, and b_1 are kept fixed. Under this assumption, $b_2 \rightarrow 0$ is equivalent to $F(A) \rightarrow 1$. Then, one can easily show that

$$\begin{aligned}
Connectivity_i &\approx \frac{n^{(1)}-1}{n^{(1)}} Connectivity_{CF,app,i},\\
Density &\approx \frac{n^{(1)}-1}{n^{(1)}} Density_{CF,app},\\
Centralization &\approx \frac{n^{(1)}-1}{n^{(1)}} Centralization_{CF,app},\\
Heterogeneity &\approx Heterogeneity_{CF,app},\\
TopOverlap_{ij} &\approx \frac{(n^{(1)}-1)b_1+1}{n^{(1)}b_1+1} TopOverlap_{CF,app,ij},\\
ClusterCoef_i &\approx ClusterCoef_{CF,app,i}.
\end{aligned}$$

For reasonably large values of $n^{(1)}$ (say $n^{(1)} > 20$), these limits illustrate Observation 3.3. Similarly, one can easily verify our relationships between network concepts in the case when the factorizability $F(A)$ is close to 1 and $n^{(1)}$ is reasonably large. But note that the maximum conformity assumption (3.9) is only satisfied when b_1 is close to 1.

3.12 Approximate CF-Based Intermodular Network Concepts

Here we briefly outline how to define approximate CF-based analogs of the fundamental intermodular concepts described in Sect. 1.10. For simplicity, we assume that the intermodular concepts do not depend on a node significance measure. In this case, the fundamental intermodular concept can be found by evaluating a suitable defined network concept function

$$NCF\left(A^{(q_1,q_1)}-I, A^{(q_2,q_2)}-I, A^{(q_1,q_2)}\right)$$

on intramodular adjacency matrices $A^{(q_1,q_1)}-I$, $A^{(q_2,q_2)}-I$ and the intermodular adjacency matrix $A^{(q_1,q_2)}$.

Recall our module-based factorizability decomposition (2.14) of the network

$$A^{(q_1,q_2)} \approx A^{(q_1,q_2)}_{CF,app}, \tag{3.40}$$

where

$$A^{(q_1,q_2)}_{CF,app} = CF^{(q_1)}\left(CF^{(q_2)}\right)^{\tau} A_{q_1q_2} \tag{3.41}$$

and $A_{q_1q_2}$ denotes the adjacency between modules q_1 and q_2. We are now ready to define.

Definition 3.4 (Approximate CF-based analog of an intermodular network concept). An approximate CF-based analog of an intermodular network concept

$$NCF\left(A^{(q_1,q_2)}_{CF,app}, A^{(q_1,q_1)}_{CF,app}, A^{(q_2,q_2)}_{CF,app}\right)$$

is defined by evaluating a network concept function (3.1) on an approximate CF-based intermodular adjacency matrix and/or intramodular adjacency matrices.

For example, the approximate CF-based analog of the geometric mean of two module densities (1.46) is defined as

$$\begin{aligned} Density_{CF,app}(q_1,q_2) &= \sqrt{\frac{sum\left(CF^{(q_1)}\right)^2}{n^{(q_1)}*\left(n^{(q_1)}-1\right)} * \frac{sum\left(CF^{(q_2)}\right)^2}{n^{(q_2)}*\left(n^{(q_2)}-1\right)}} \\ &\approx mean\left(CF^{(q_1)}\right) mean\left(CF^{(q_2)}\right), \end{aligned} \tag{3.42}$$

where the latter approximation assumes that $n^{(q)}/(n^{(q)}-1) \approx 1$.

The approximate CF-based analog of the average inter-adjacency matrix is given by:

$$A^{ave}_{CF,app}(q_1,q_2) = A_{q_1q_2} mean\left(CF^{(q_1)}\right) mean\left(CF^{(q_2)}\right). \tag{3.43}$$

Using (3.42) and (3.43), we find that the approximate CF-based analog of the separability statistic (1.47) is given by separability statistics as 1 minus the ratio of intermodular adjacency divided by intramodular density:

$$separability.ave_{CF,app}(q_1,q_2) \approx 1 - A_{q_1q_2}. \tag{3.44}$$

3.13 CF-Based Network Concepts for Comparing Two Networks

In Sect. 1.11, we introduced several fundamental network concepts for comparing two networks $A^{[\text{ref}]}$ and $A^{[\text{test}]}$. Denote by $CF^{[\text{ref}]}$ and $CF^{[\text{test}]}$ the conformity vectors of the reference and test network, respectively. Then one can define the following conformity-based measure of connectivity preservation:

$$cor.CF = cor\left(CF^{[\text{ref}]}, CF^{[\text{test}]}\right). \tag{3.45}$$

In an exercise you are asked to show that for approximately factorizable reference and test networks, the following approximate relationships hold true under mild assumptions:

$$\begin{aligned} cor.K &= cor\left(K^{[\text{ref}]}, K^{[\text{test}]}\right) \approx cor.CF \\ cor.Adj &= cor\left(vectorizeMatrix(A^{[\text{ref}]}), vectorizeMatrix\left(A^{[\text{test}]}\right)\right) \approx cor.CF \\ cor.K &= \approx cor.Adj. \end{aligned} \tag{3.46}$$

These relationships are very important when studying whether a subnetwork formed by the nodes of a module is preserved across networks. Since network modules tend to be factorizable, (3.46) implies that seemingly disparate measures of intramodular connectivity preservation are highly related. We will revisit this point in Chap. 9 where we define network-based measures of module preservation.

3.14 Discussion

In this chapter, we present theoretical results that clarify the mathematical relationship between network concepts in approximately factorizable networks. A deeper understanding of network concepts may guide the data analyst on how to construct and use networks in practice. Our results apply to any network that is approximately factorizable irrespective of its definition. Importantly, we provide empirical evidence that many network modules are approximately factorizable (Dong and Horvath 2007). In Sect. 6.2, we provide a theoretical description of approximately factorizable correlation networks.

We briefly highlight the main results of this chapter.

Observation 3.1 states that a fundamental network concept $NC_{fundamental}$ is approximately equal to its CF-based analog NC_{CF} if the network is approximately factorizable. Higher order approximations of fundamental network concepts are described in Sect. 3.6.

Observation 3.2 states that CF-based network concepts are close to their approximate CF-based analogs under weak conditions on the conformity vector.

Observation 3.3 states that a fundamental network concept is close to its approximate CF-based analog if the network is approximately factorizable (see Observation 3.1) and if the conformity vector satisfies weak assumptions (see Observation 3.2).

Observation 3.4 states that fundamental network concepts are simple functions of the connectivity in approximately factorizable networks.

Observation 3.5 presents special relationships among network concepts in approximately factorizable networks. These results shed some light on the relationship between network concepts traditionally used by social scientists (e.g., centralization, heterogeneity) and concepts used in practice (e.g., clustering coefficient, topological overlap).

Observation 3.6 states that network modules (that are defined using a clustering procedure) are approximately factorizable. While modules in gene co-expression networks tend to be approximately factorizable if the corresponding expression profiles are highly correlated (see Sect. 6.2), the situation is more complicated for modules in protein–protein interaction networks: only after replacing the original adjacency matrix by a 'smoothed out' version (the topological overlap matrix), do we find that the resulting modules are approximately factorizable. In Chap. 6, we will derive further results for special types of approximately factorizable networks, namely approximately factorizable correlation networks. Our empirical data also highlight how network concepts differ between subnetworks of "proper" modules and the subnetwork comprised of improper (gray) module nodes (Dong and Horvath 2007). For all applications, we find that proper modules have high factorizability, high density, and high mean conformity. Based on our theoretical derivations, it comes as no surprise that proper modules also have a high average clustering coefficient and a high centralization when compared to the improper module. But we find no difference in heterogeneity between proper and improper network modules.

A general network is typically not approximately factorizable. But we show in Sect. 2.3 that A can often be approximated by the $n \times n$ dimensional matrix $A^{Cl}_{CF,app}$ (2.15) whose i,jth element is given by

$$A^{Cl}_{CF,app,ij} = CF_i^{(q_1)} CF_j^{(q_2)} A_{q_1 q_2} \text{ if } i \in \text{ module } q_1, j \in \text{ module } q_2.$$

$A^{Cl}_{CF,app}$ is an adequate approximation of A if a module assignment Cl can be defined that leads to approximately factorizable modules.

The idea of deriving relationships between fundamental network concepts by first deriving analogous relationships between approximate CF-based concepts often works well even if the underlying network is only approximately factorizable. However, we caution the reader that for some relationships exact factorizability is a vital assumption. For example, assume that we want to study the Spearman correlation $corkCCSpearman = cor(k, ClusterCoef)$ between the connectivity and the cluster coefficient. For exactly factorizable networks, we find the following highly accurate relationship

$$corkCCSpearman \approx 0.96 - 2.19 \frac{sum(k)^2}{nsum(k^2)} = 0.96 - 2.19 \frac{1}{1 + Heterogeneity^2}. \tag{3.47}$$

In an exercise, you are asked to show that this relationship holds for exactly factorizable networks but breaks down completely for non-factorizable networks.

3.15 R Code

Here we present R code for solving the following:

Exercise regarding the relationship between fundamental and conformity-based network concepts.

(i) Simulate a $n \times n$ dimensional adjacency matrix where $n = 500$. Answer:

```
library(WGCNA)
n=500
# set the seed for the random number generator
set.seed(1)
# now we simulate a random n*n dimensional matrix
power1=1
A=matrix(runif(n*n)^power1,nrow=n,ncol=n)
#symmetrize the matrix
A=(t(A)+A)/2
# set diagonal elements to 1
diag(A)=1
# round to 12 digits to get rid of numeric errors
A=round(A,12)
```

(ii) Compute the factorizability $F(A)$, the conformity CF, and check whether the bounded CF moment condition (3.16). Answer:

```
# compute conformity based and fundamental network concepts
   for A
NC.A=conformityBasedNetworkConcepts(A)
# this is the factorizability of A
Factorizability=NC.A$Factorizability
Factorizability # it equals 0.8580625
# this is the conformity vector of A
CF=NC.A$conformityBasedNCs$Conformity
sum(CF^4)/sum(CF^2)^2 # equals 0.002002598
sum(CF^2)/sum(CF)^2 # equals 0.002000649
```

(iii) Check whether fundamental network concepts are approximately equal to their approximate CF-based analogs $NC_{fundamental} \approx NC_{CF.app}$ (3.23 and Observation 3.3). Answer:

```
#Connectivity of node 1
NC.A$fundamentalNCs$Connectivity[1]#252.99
NC.A$approximateConformityBasedNCs$Connectivity.CF.App[1]
   #253.54
#Clustering Coefficient of node 1
NC.A$fundamentalNCs$ClusterCoef[1]# 0.5002779
NC.A$approximateConformityBasedNCs$ClusterCoef.CF.App[1]
   #0.5004
#MAR of node 1
NC.A$fundamentalNCs$MAR[1]#0.5840411
NC.A$approximateConformityBasedNCs$MAR.CF.App[1]#0.5072
# Density
NC.A$fundamentalNCs$Density#0.5000992
NC.A$approximateConformityBasedNCs$Density.CF.App# 0.50110
# Centralization
```

```
NC.A$fundamentalNCs$Centralization#0.03182366
NC.A$approximateConformityBasedNCs$Centralization.CF.App
   #0.03158
#Heterogeneity
NC.A$fundamentalNCs$Heterogeneity#0.01797534
NC.A$approximateConformityBasedNCs$Heterogeneity#0.01801197
```

(iv) Calculate the density, heterogeneity, and clustering coefficients based on A, A_{CF}, and $A_{CF,app}$. Answer:

```
Density=NC.A$fundamentalNCs$Density#0.5000992
Density.CF= NC.A$conformityBasedNCs$Density#0.5000991
Density.CF.App=
NC.A$approximateConformityBasedNCs$Density.CF.App#0.5011
Heterogeneity=NC.A$fundamentalNCs$Heterogeneity#0.01797534
Heterogeneity.CF=NC.A$conformityBasedNCs$Heterogeneity
   #0.017975
Heterogeneity.CF.App=
NC.A$approximateConformityBasedNCs$Heterogeneity.CF.App
   #0.01801
ClusterCoef=NC.A$fundamentalNCs$ClusterCoef
ClusterCoef.CF= NC.A$conformityBasedNCs$ClusterCoef
ClusterCoef.CF.App=
NC.A$approximateConformityBasedNCs$ClusterCoef.CF.App
# show that ClusterCoef.CF.App is constant=0.50042
ClusterCoef.CF.App[1:10]
```

(v) Calculate density measures based on A^2, $(A_{CF})^2$. Answer:

```
# compute network concepts for A^2
NC.A.squared=conformityBasedNetworkConcepts(A^2)
A.CF.app=outer(CF,CF)
# now we define the A.CF
A.CF=A.CF.app
diag(A.CF)=1
# compute network concepts for (A.CF)^2
NC.A.CF.squared=conformityBasedNetworkConcepts((A.CF)^2)
Density.A.squared= NC.A.squared$fundamentalNCs$Density ]
Density.A.CF.squared= NC.A.CF.squared$fundamentalNCs$Density
```

(vi) Study different expressions involving the clustering coefficient. In particular, check whether $ClusterCoef_{CF,app} \approx mean(ClusterCoef)$ (3.25), $ClusterCoef_{CF,app} = \frac{Density(A_{CF}^2)}{Density(A_{CF})}$ (3.8), and $mean(ClusterCoef) \approx (1+Heterogeneity^2)^2 \times Density$ (3.30). Answer:

```
# the mean (fundamental) ClusterCoef is given by
mean(ClusterCoef)# 0.500099
ClusterCoef.CF.App[1]
#it equals the ratio of the following densities
Density.A.CF.squared/Density.CF#0.500423
(1+Heterogeneity^2)^2*Density#0.5004224
```

(vii) Check the following relationships for the factorizability: $F(A) = \frac{Density(A_{CF}^2)}{Density(A^2)}$ (3.35), $F(A) = ClusterCoef_{CF,app} * \frac{Density(A_{CF})}{Density(A^2)}$ (3.36) and

$$F(A) \approx mean(ClusterCoef) * \frac{Density(A)}{Density(A^2)}.$$

Answer:

```
Factorizability #= 0.8580625
Density.A.CF.squared/Density.A.squared#0.8580625
ClusterCoef.CF.App[1]*Density.CF/Density.A.squared#0.85806
mean(ClusterCoef)*Density/Density.A.squared #0.85750
```

(vii) Check $TopOverlap_{ij} \approx \frac{max(k_i,k_j)}{n}(1+Heterogeneity^2)$ (3.32). Answer:

```
TopOverlap=TOMsimilarity(A)
k=NC.A$fundamentalNCs$Connectivity
TopOverlapApproximated=
1/n*outer(k,k,FUN="pmax")*(1+Heterogeneity^2)
LHS=vectorizeMatrix(TopOverlap)
RHS=vectorizeMatrix(TopOverlapApproximated)
summary(abs(LHS-RHS))#maximum abs. diff 1.866e-02
# the following plot shows a high correlation of 0.9
verboseScatterplot(LHS,RHS,xlab="TOM[i,j]",
ylab="max(k[i],k[j])/n*(1+Hetero^2)")
```

(viii) Check that the maximum conformity assumption is *not* satisfied in this example. Answer:

```
max(CF) #equals 0.751568
```

Additional R software tutorials regarding network concepts can be found at the following webpage: www.genetics.ucla.edu/labs/horvath/ModuleConformity/ModuleNetworks.

3.16 Exercises

1. Exercise regarding the big picture.
   (i) Which networks are likely (or unlikely) to be approximately factorizable?
   (ii) Which networks are likely to satisfy the maximum conformity assumption?
   (iii) Why is it useful to study approximate CF-based network concepts?
2. Exercise regarding CF-based analogs of fundamental network concepts. Derive the following expressions for the CF-based network concepts:

$$\begin{aligned}
Connectivity_{CF,i} &= CF_i sum(CF) - CF_i^2, \\
ScaledConnectivity_{CF,i} &= K_{CF,i} = \frac{CF_i(sum(CF) - CF_i)}{max(Connectivity_{CF})}, \\
p.Connectivity_{CF} &= p(discretize(Connectivity_{CF}, no.bins)), \\
ScaleFreeFit_{CF} &= cor(log(p.Connectivity_{CF}), log(BinNo))^2,
\end{aligned}$$

$$
\begin{aligned}
Density_{CF} &= \frac{sum(CF)^2 - sum\left(CF^2\right)}{n(n-1)}, \\
MAR_{CF,i} &= CF_i \frac{sum\left(CF^2\right) - CF_i^2}{sum(CF) - CF_i}, \\
Centralization_{CF} &= \frac{n}{n-2}\left(\frac{max(Connectivity_{CF})}{n-1} - Density_{CF}\right), \\
ClusterCoef_{CF,i} &= \frac{\left(sum\left(CF^2\right) - CF_i^2\right)^2 - \left(sum\left(CF^4\right) - CF_i^4\right)}{\left(sum(CF) - CF_i\right)^2 - \left(sum\left(CF^2\right) - CF_i^2\right)}, \\
NetworkSignif_{CF} &= CF_t mean(CF), \\
HubGeneSignif_{CF} &= CF_t \frac{\sum_i CF_i K_{CF,i}}{\sum_i \left(K_{CF,i}\right)^2},
\end{aligned} \tag{3.48}
$$

$$
\begin{aligned}
Heterogeneity_{CF} &= \sqrt{\frac{n\left(sum\left(CF^2\right) sum(CF)^2 - 2sum\left(CF^3\right) sum(CF) + sum\left(CF^4\right)\right)}{\left(sum(CF)^2 - sum\left(CF^2\right)\right)^2} - 1}, \\
TopOverlap_{CF,ij} &= \frac{CF_i CF_j \left(sum\left(CF^2\right) - CF_i^2 - CF_j^2\right) + CF_i CF_j}{min\left(CF_i\left(sum(CF) - CF_i\right), CF_j\left(sum(CF) - CF_j\right)\right) + 1 - CF_i CF_j}.
\end{aligned}
$$

Hint: Evaluate network concept functions (3.1) on $A_{CF} - I$ (3.3) and $GS_{CF,i} = CF_i CF_t$.

3. Exercise regarding approximate CF-based analogs of fundamental network concepts. Derive the following expressions for the approximate CF-based network concepts:

$$
\begin{aligned}
Connectivity_{CF,app,i} &= k_{CF,app,i} = CF_i sum(CF), \\
ScaledConnectivity_{CF,app,i} &= K_{CF,app,i} = \frac{CF_i}{CF_{max}}, \\
p.Connectivity_{CF,app} &= p(discretize(CF, no.bins)), \\
ScaleFreeFit_{CF,app} &= cor\left(log\left(p.Connectivity_{CF,app}\right), log(BinNo)\right)^2, \\
Density_{CF,app} &= \frac{sum(CF)^2}{n(n-1)} = \frac{n}{n-1} mean(CF)^2, \\
MAR_{CF,app,i} &= CF_i \frac{sum\left(CF^2\right)}{sum(CF)}, \\
Centralization_{CF,app} &= \frac{n * sum(CF)}{(n-1)(n-2)}\left(CF_{max} - mean(CF)\right) \\
&\approx mean(CF)\left(CF_{max} - mean(CF)\right), \\
ClusterCoef_{CF,app,i} &= \frac{sum\left(CF^2\right)^2 - sum\left(CF^4\right)}{sum(CF)^2 - sum\left(CF^2\right)}, \\
NetworkSignif_{CF,app} &= CF_t mean(CF), \\
HubNodeSignif_{CF,app} &= CF_t * CF_{max},
\end{aligned}
$$

$$Heterogeneity_{CF,app} = \sqrt{\frac{mean\left(CF^2\right)}{mean(CF)^2} - 1},$$
$$TopOverlap_{CF,app,ij} = \frac{CF_i CF_j \left(sum\left(CF^2\right)+1\right)}{min\left(CF_i, CF_j\right) sum(CF) + 1 - CF_i CF_j}. \quad (3.49)$$

4. Exercise. Derive the relationships among approximate CF-based network concepts (3.8).
5. Exercise. Show that the bounded CF moment assumption (3.16) implies the following relationships (3.50), (3.51), (3.52) involving the approximate CF-based clustering coefficient:

$$ClusterCoef_{CF,app} \approx \left(\frac{sum(CF^2)}{sum(CF)}\right)^2$$
$$ClusterCoef_{CF,app} \approx \left(1 + Heterogeneity^2_{CF,app}\right)^2 \times Density_{CF,app}$$
$$MAR_{CF,app,i} \approx CF_i \sqrt{ClusterCoef_{CF,app}}$$

6. Exercise: Prove that the moment ratio

$$MomentRatio(\beta) = \frac{mean\left(CF^{2*\beta}\right)}{mean\left(CF^{\beta}\right)^2}$$

satisfies the following inequalities (3.17):

$$\frac{1}{n} \le MomentRatio(\beta) \le 1.$$

Hint. Recall that CF_i is a nonnegative number. The first inequality follows from the Cauchy–Schwartz inequality

$$\sum_i \left((CF_i)^{\beta} 1\right)^2 \le \sum_i (CF_i)^{2\beta} \sum_i 1^2.$$

The second inequality follows from Proposition 4.1 and (4.14).
7. Exercise. Prove that under the maximum conformity assumption $CF_{max} \approx 1$ (3.9), the following relationships (3.10) are satisfied:

$$K_{CF,app,i} = CF_i$$
$$Centralization_{CF,app} \approx \sqrt{Density_{CF,app}}\left(1 - \sqrt{Density_{CF,app}}\right)$$

8. Exercise. Prove Observation 3.2 regarding CF-based versus approximate CF-based concepts. Toward this end, prove

 (i) that the negligible CF component assumption (3.15) implies that the following CF-based network concepts are close to their approximate

CF-based analogs: $Connectivity_{CF}$, $ScaledConnectivity_{CF}$, $p.Connectivity_{CF}$, $ScaleFreeFit_{CF}$, MAR_{CF}, $ClusterCoef_{CF}$, $TopOverlap_{CF}$, $NetworkSignif_{CF}$, $HubGeneSignif_{CF}$

(ii) that the bounded CF moment assumption (3.16) implies that the following CF-based network concepts are close to their approximate CF-based analogs: $Density_{CF}$, $Centralization_{CF}$, $Heterogeneity_{CF}$

9. Exercise. Under what assumptions do the following relationships hold?

$$ClusterCoef_{CF,app} \approx \left(\frac{mean\left(CF^2\right)}{mean(CF)}\right)^2 \tag{3.50}$$

$$ClusterCoef_{CF,app} \approx \left(1+Heterogeneity_{CF,app}^2\right)^2 \times Density_{CF,app} \tag{3.51}$$

$$MAR_{CF,app,i} \approx K_{CF,app,i} CF_{max} \sqrt{ClusterCoef_{CF,app}}. \tag{3.52}$$

10. Exercise. Prove that the factorizability $F(A)$ (2.3) can be expressed as follows (3.35):

$$F(A) = \frac{Density \text{ of } A_{CF}^2}{Density \text{ of } A^2} = \frac{\|A_{CF}-I\|_F^2}{\|A-I\|_F^2} = \frac{sum\left(CF^2\right)^2 - sum\left(CF^4\right)}{\|A-I\|_F^2}.$$

Hint: Since $F(A) = 1 - \frac{\|(A-I)+diag(CF^2)-CFCF^\tau\|_F^2}{\|A-I\|_F^2}$, it will be sufficient to show that $\|(A-I)-(A_{CF}-I)\|_F^2 = \|A-I\|_F^2 - \|A_{CF}-I\|_F^2$. Note that the Frobenius matrix norm of B (1.4) can also be expressed as follows: $\|B\|_F^2 = trace(B^\tau B)$, where the trace of a matrix is the sum of its diagonal elements. Thus, $\|(A-I)-(A_{CF}-I)\|_F^2 = \|A-I\|_F^2 + \|A_{CF}-I\|_F^2 - 2 \times trace((A-I)(A_{CF}-I))$. Using (2.5), we find that $trace((A-I)(A_{CF}-I)) = trace((A-I)CFCF^\tau) - trace((A-I)diag(CF^2)) = CF^\tau(A-I)CF = \|A_{CF}-I\|_F^2$. Thus, $\|(A-I)-(A_{CF}-I)\|_F^2 = \|A-I\|_F^2 - \|A_{CF}-I\|_F^2$. The remainder of the proof is straightforward.

11. Exercise. Show that under weak assumptions, the following relationship (3.32) holds in approximately networks:

$$TopOverlap_{ij} \approx \frac{max\left(k_i,k_j\right)}{n}\left(1+Heterogeneity^2\right). \tag{3.53}$$

Hint: Consider the following weak assumptions: $\frac{1}{sum(CF^2)} \approx 0$ and $\frac{1-CF_iCF_j}{min(CF_i,CF_j)sum(CF)} \approx 0$. Note that

$$\begin{aligned} TopOverlap_{CF,app,ij} &\approx max\left(CF_i,CF_j\right)\frac{sum\left(CF^2\right)}{sum(CF)} \\ &= \frac{max\left(CF_i sum(CF), CF_j sum(CF)\right)}{n}\frac{n sum\left(CF^2\right)}{sum(CF)^2} \\ &\approx \frac{max\left(k_{CF,app,i},k_{CF,app,j}\right)}{n}\left(1+Heterogeneity_{CF,app}^2\right). \end{aligned} \tag{3.54}$$

12. Exercise regarding the approximate CF-based analog of the scale-free topology fitting index. What conditions does CF have to satisfy in order that the approximate scale-free topology index equals 1?
13. R exercise regarding the R simulation in Sect. 3.15. Repeat the analysis using the following matrix:

```
power1=10; A=matrix(runif(n*n)^power1,nrow=n,ncol=n)
A=(t(A)+A)/2 ; diag(A)=1
```

Determine which relationships remain valid even though the network is no longer approximately factorizable.
14. R exercise regarding the R simulation in Sect. 3.15. Repeat the analysis with the following exactly factorizable adjacency matrix:

```
CF=runif(n); A=outer(CF,CF); diag(A)=1
```

15. Exercise regarding relationships between approximate CF-based network concepts.

    (i) Derive your own relationship between approximate CF-based network concepts.
    (ii) Generalize this relationship to fundamental network concepts in approximately factorizable networks.
    (iii) Numerically validate the relationship using R code. Hint: Sect. 3.15.
    (iv) Does the relationship still hold for networks with relatively low factorizability?

16. Exercise regarding the relationship between clustering coefficient and connectivity. Let $corkCCSpearman = cor(k, ClusterCoef)$ be the *Spearman* correlation between the connectivity and the cluster coefficient. The overarching goal of this exercise is to relate $corkCCSpearman$ to other network concepts.

    (i) For an exactly factorizable networks, the following relationship (3.47) is highly accurate ($R^2 \approx 0.97$):

$$corkCCSpearman \approx 0.96 - 2.19\frac{sum(k)^2}{nsum(k^2)}$$
$$= 0.96 - 2.19\frac{1}{1+Heterogeneity^2}.$$

    Hint: Confirm this relationship numerically.
    (ii) Check whether relationship (3.47) holds approximately for non-factorizable networks? Hint: No.

17. Exercise regarding the relationship between connectivity-based preservation statistics. Assume that two adjacency matrices $A^{[ref]}$ and $A^{[test]}$ are exactly factorizable. Use numerical studies to argue that the following relationships

(3.46) are approximately satisfied under mild assumptions:

$$cor.K = cor\left(K^{[\text{ref}]}, K^{[\text{test}]}\right) \approx cor\left(CF^{[\text{ref}]}, CF^{[\text{test}]}\right) = cor.CF$$
$$cor.Adj = cor\left(vectorizeMatrix(A^{[\text{ref}]}), vectorizeMatrix(A^{[\text{test}]})\right) \approx cor.CF$$
$$cor.K = \approx cor.Adj.$$

Hint: The following R code can be used to show that *cor.Adj* is similar to *cor.CF*.

```
library(WGCNA)
# network size (which should be varied)
n=10
no.replicates=100
# prop. of nodes whose CF differs between the networks
propNoise=0.5
# the minimum conformity value allowed in the two networks
minCF=.5
corADJ=rep(NA,no.replicates)
corCF=rep(NA,no.replicates)
for(i in 1:no.replicates){
CF=runif(n,min=minCF,max=1)
ADJ=outer(CF,CF,FUN="*"); diag(ADJ)=1
vADJ=vectorizeMatrix(ADJ)
# number of nodes with noised up conformity
n.noise=as.integer(n*propNoise)
# no. of nodes whose conformity is perfectly preserved
n.signal=n-n.noise
if (n.signal==0) {CF2=runif(n.noise,min=minCF,max=1)}
if (n.noise==0) {CF2=CF}
if (n.signal>0 & n.noise>0) {
CF2=c(CF[1:n.signal],runif(n.noise,min=minCF,max=1))}
ADJ2=outer(CF2,CF2,FUN="*"); diag(ADJ2)=1
vADJ2=vectorizeMatrix(ADJ2)
corADJ[i]=cor(vADJ,vADJ2)
corCF[i]=cor(CF,CF2) } # end of for loop
par(mfrow=c(1,1))
verboseScatterplot(corADJ, corCF,main=paste("n=",n, ",
   minCF=",
minCF, ", prop.noise=", propNoise));abline(0,1)
```

References

Breitkreutz BJ, Stark C, Tyers M (2003) The GRID: The general repository for interaction datasets. Genome Biol 4:R23

Dennis G, Sherman B, Hosack D, Yang J, Gao W, Lane H, Lempicki R (2003) DAVID: Database for annotation, visualization, and integrated discovery. Genome Biol 4(9):R60

Dong J, Horvath S (2007) Understanding network concepts in modules. BMC Syst Biol 1(1):24

Erdos P, Renyi A (1960) On the evolution of random graphs. Publ Math Inst Hung Acad Sci5: 17–60

Horvath S, Dong J (2008) Geometric interpretation of gene co-expression network analysis. PLoS Comput Biol 4(8):e1000117

Kaufman L, Rousseeuw PJ (1990) Finding groups in data: An introduction to cluster analysis. Wiley, New York

Snijders TA (1981) The degree variance: An index of graph heterogeneity. Soc Networks 3: 163–174

Chapter 4
Adjacency Functions and Their Topological Effects

Abstract An adjacency function AF can be used to transform a given (original) network into a new network, i.e., $A^{original}$ is transformed into $A = AF\left(A^{original}\right)$. We introduce several useful adjacency functions and show how they can be used to construct networks, e.g., the power adjacency function $AF^{power}()$, which raises each adjacency to a fixed power β, can be used to construct weighted correlation networks. In general, adjacency functions depend on parameter values, e.g., AF^{power} depends on the power β. The choice of these parameter values can be informed by topological criteria based on network concepts. For example, we describe the scale-free topology criterion that considers the effect of the AF parameters on the scale-free topology fitting index. The power adjacency function can also be used to calibrate one network with respect to another network, which is often a prerequisite for differential network analysis or for consensus module detection. If an adjacency function (e.g., AF^{power}) is defined via a monotonically increasing real-valued function, then it is called rank-preserving since the ranking of the original adjacencies equals that of the transformed adjacencies. The terminology of "rank-equivalence" and "threshold-equivalence" between adjacency functions, networks, and network construction methods allows us (a) to provide a general definition of a correlation network and (b) to study the relationships between different network construction methods.

4.1 Definition of Important Adjacency Functions

Recall from Sect. 1.5 that an adjacency function AF is a matrix valued function that maps an $n \times n$ dimensional adjacency matrix $A^{original}$ onto a new $n \times n$ dimensional network adjacency

$$A = AF\left(A^{original}\right).$$

Networks are often transformed to change their topological properties (measured by network concepts). For example, the density (1.18) of $A = \left(A^{original}\right)^{\beta}$ $(\beta \neq 0)$ is a monotonic function of β. A transformation can also be used to map a weighted adjacency matrix into an unweighted adjacency matrix by thresholding its adjacencies.

S. Horvath, *Weighted Network Analysis: Applications in Genomics and Systems Biology*, DOI 10.1007/978-1-4419-8819-5_4,

Consider the step function $s \longrightarrow af^{step}(s)$ which equals 1 if $s \geq \tau$ and 0 otherwise. By applying the step function to the off-diagonal elements of $A^{original}$, we define the **threshold adjacency function** (and sometimes refer to it as **hard-thresholding** or **step** adjacency function):

$$AF^{threshold}\left(A^{original}, \tau\right)_{ij} = \begin{cases} 1 \text{ if } A_{ij}^{original} \geq \tau \\ 0 \text{ if } A_{ij}^{original} < \tau \end{cases}. \tag{4.1}$$

The threshold adjacency function can be used for transforming a weighted network to an unweighted network (5.22). The parameter τ is sometimes referred to as **hard threshold**.

The **power adjacency function** $AF^{power}(*, \beta)$ is also known as **soft-thresholding function**:

$$AF^{power}\left(A^{original}, \beta\right)_{ij} = |A_{ij}^{original}|^{\beta}. \tag{4.2}$$

AF^{power} can be used to define a weighted correlation network from correlations (5.22). The parameter β is sometimes referred to as **soft threshold** or simply as power. We will use the power adjacency function repeatedly since it has the following attractive properties:

1. It satisfies $AF^{power}(0) = 0$ and $AF^{power}(1) = 1$.
2. It is a continuous function.
3. It depends on a single parameter β.
4. It preserves exact factorizability: if $A^{original}$ is exactly factorizable (2.1), then $AF^{power}(A^{original}, \beta)$ is factorizable as well.

The **Moebius adjacency function** $AF^{Moebius}$

$$AF^{Moebius}\left(A^{original}, \tau\right)_{ij} = \begin{cases} \frac{a*A_{ij}^{original}+b}{c*A_{ij}^{original}+d} & \text{if } i \neq j \\ 1 & \text{if } i = j \end{cases} \tag{4.3}$$

is based on the Moebius function $s \longrightarrow af^{Moebius}(s) = \frac{a*s+b}{c*s+d}$ whose four parameters a, b, c, d are required to satisfy the following conditions:

$$a * d - c * b \geq 0 \text{ which implies that } af^{Moebius}(s) \text{ is nondecreasing,} \tag{4.4}$$

$$\frac{b}{d} \geq 0 \text{ which implies that } af^{Moebius}(0) \geq 0, \tag{4.5}$$

$$\frac{a+b}{c+d} \leq 1 \text{ which implies that } af^{Moebius}(1) \leq 1. \tag{4.6}$$

For example, the functions $s \longrightarrow 0.5 * s + 0.5$ and $s \longrightarrow 1/(2-s)$ are increasing Moebius functions that map the unit interval onto itself.

4.2 Topological Effects of the Power Transformation AF^{power}

The power adjacency function $AF^{power}(,\beta)$ (power transformation) does not change an unweighted network since $0^{\beta} = 0$ and $1^{\beta} = 1$ for $\beta \neq 0$, but it changes the topological properties (network concepts) of weighted networks.

Consider a fundamental network concept, which is defined by evaluating a network concept function NCF (3.1) on an adjacency matrix. Denote by $NC^{original}$ and NC the value of the network concept for networks $A^{original}$ and $A = AF^{power}(A^{original}, \beta) = (A_{ij}^{original})^{\beta}$, respectively. For example, $Density^{original}$ and Density denote the density (1.18) of networks $A^{original}$ and A, respectively. In Sect. 4.2.2, we prove that $\beta \geq 1$ implies

$$MAR_i^{original} \geq MAR_i \tag{4.7}$$

$$Connectivity_i^{original} \geq Connectivity_i \tag{4.8}$$

$$Density^{original} \geq Density. \tag{4.9}$$

In an exercise you are asked to show that for the vast majority of networks

$$mean(ClusterCoef^{original}) \text{ typically } \geq mean(ClusterCoef) \tag{4.10}$$

$$Heterogeneity^{original} \text{ typically } \leq Heterogeneity \tag{4.11}$$

$$separability^{original} \text{ typically } \leq separability, \tag{4.12}$$

where the word "typically" signifies that counter examples can be constructed. For example, one can construct a (small) network $A^{original}$ for which the heterogeneity $A^{original}$ is larger than that of A (Horvath and Dong 2008). When it comes to studying the effect of adjacency function, it can be advantageous to replace fundamental network concepts by their approximate CF-based analogs. For example, (6.45) can be used to argue that the separability (1.47) between approximately factorizable modules tends to increase with $\beta > 1$. As another example of the use of CF-based analogs, we will show in Sect. 4.2.1 that the approximate CF-based heterogeneity increases with β.

For other network concepts, the effect of the power transformation is more ambiguous: the centralization (1.20) and the scale-free topology fitting index (1.13) can both increase and decrease with β. But for many weighted correlation networks $A^{original}$, the scale-free fit $ScaleFreeFit(AF^{power}(A^{original}, \beta))$ $(\beta > 1)$ of the transformed network is higher than that of the original network as will be discussed in Sect. 4.3.

4.2.1 Studying the Power AF Using Approx. CF-Based Concepts

When it comes to studying the effect of an adjacency function, it can be advantageous to replace fundamental network concepts by their approximate conformity-based analogs. To illustrate this point, we briefly describe the effect of soft thresholding (i.e., the power transformation) $AF^{power}\left(A^{original}\right)$ (4.2) on the network heterogeneity. Using extensive simulation studies, we found that for the vast majority of networks, the heterogeneity increases with the soft threshold β (Horvath and Dong 2008). Thus, for most networks, increasing β makes it easier to discern highly connected nodes from less connected nodes. However, one can construct networks for which increasing β leads to a lower heterogeneity. The situation is much simpler for the approximate conformity-based heterogeneity $Heterogeneity_{CF,app}$ (3.49).

Now we prove that the approximate conformity-based heterogeneity is a monotonically increasing function of the soft threshold β. Thus, the heterogeneity is an increasing function of β if it can be approximated by its eigenvector-based analog (Observation 6.2). Recall that

$$Heterogeneity_{CF,app} = \sqrt{\frac{nsum(CF^2)}{(sum(CF))^2} - 1}$$

which implies that it is a decreasing function of

$$S(\beta) = \frac{\left(\sum_{i=1}^{n} CF^{\beta}\right)^2}{\sum_{i=1}^{n} CF^{2\beta}}, \tag{4.13}$$

To prove that the heterogeneity increases with β, it suffices to show that $S(\beta)$ is a decreasing function of β, which can be done with Proposition 4.1 (4.14).

4.2.2 *MAR* Is a Nonincreasing Function of β

Here we prove that the jth maximum adjacency ratio $MAR_j(A) = \frac{\sum_{i\neq j}(A_{ij})^2}{\sum_{i\neq j} A_{ij}}$ (1.16) of the network $A = \left(A^{original}\right)^{\beta}$ $\beta > 1$ is smaller than or equal to that of the original network $A^{original}$. This mathematical section can be skipped at first reading. To prove the claim, it suffices to show that

$$S(\beta) = \frac{\left(\sum_{i\neq j}^{n} (A_i)^{\beta}\right)^2}{\sum_{i\neq j}^{n} (A_i)^{2\beta}}$$

satisfies $S(\beta) \leq S(1)$ if $\beta > 1$. The fact that the summation skips the jth node is not important in our proof. Without loss of generality it suffices to prove the following:

Proposition 4.1. *Let $\{A_i, i = 1, \ldots, n\}$ be a group of nonnegative numbers and $\beta > 1$ then the following inequality holds:*

$$S(\beta) = \frac{\left(\sum_{i=1}^n A_i^\beta\right)^2}{\sum_{i=1}^n A_i^{2\beta}} \leq \frac{\left(\sum_{i=1}^n A_i\right)^2}{\sum_{i=1}^n A_i^2} = S(1). \tag{4.14}$$

The prove the Proposition, we will make use of the following.

Lemma. *Let $\{u_i, i = 1, \ldots, n\}$ and $\{v_i, i = 1, \ldots, n\}$ be groups of nonnegative numbers, and θ be a number $0 \leq \theta < 1$. Then the following inequality holds:*

$$\sum_{i=1}^n u_i^\theta v_i^{1-\theta} \leq \left(\sum_{i=1}^n u_i\right)^\theta \left(\sum_{i=1}^n v_i\right)^{1-\theta}. \tag{4.15}$$

The Lemma can be proved with Hölder's inequality (5.4), which can be re-expressed as follows:

$$\sum_{i=1}^n |x_i y_i| \leq \left(\sum_{i=1}^n |x_i|^{1/\theta}\right)^\theta \left(\sum_{i=1}^n |y_i|^{1/(1-\theta)}\right)^{1-\theta}. \tag{4.16}$$

We use the Lemma with $\theta_1 = \beta/(2\beta - 1)$, $u_i = A_i$, and $v_i = A_i^{2\beta}$ to derive

$$\sum_{i=1}^n A_i^\beta \leq \left(\sum_{i=1}^n A_i\right)^{\theta_1} \left(\sum_{i=1}^n A_i^{2\beta}\right)^{1-\theta_1}.$$

Further, we use the Lemma with $\theta_2 = (2\beta - 2)/(2\beta - 1)$, $u_i = A_i$, and $v_i = A_i^{2\beta}$ to derive

$$\sum_{i=1}^n A_i^2 \leq \left(\sum_{i=1}^n A_i\right)^{\theta_2} \left(\sum_{i=1}^n A_i^{2\beta}\right)^{1-\theta_2}.$$

By squaring the first inequality and multiplying it with the second inequality, we arrive at

$$\begin{aligned}\left(\sum_{i=1}^n A_i^\beta\right)^2 \left(\sum_{i=1}^n A_i^2\right) &\leq \left(\sum_{i=1}^n A_i\right)^{2\theta_1+\theta_2} \left(\sum_{i=1}^n A_i^{2\beta}\right)^{3-(2\theta_1+\theta_2)} \\ &= \left(\sum_{i=1}^n A_i\right)^2 \left(\sum_{i=1}^n A_i^{2\beta}\right)\end{aligned}$$

since $2\theta_1 + \theta_2 = 2$ and $3 - (2\theta_1 + \theta_2) = 1$. The last inequality completes the proof since it is equivalent to inequality (4.14).

4.3 Topological Criteria for Choosing AF Parameters

Most network construction procedures depend on parameters. For example, an unweighted correlation network constructed with the threshold adjacency function $AF^{threshold}$ depends on the threshold parameter τ. A weighted correlation network constructed with the power adjacency function AF^{power} depends on the parameter β. We will now discuss several methods for choosing these parameters. The choice of the parameters determines the connectivity patterns and topological properties of the network. For example, increasing the value of τ in the threshold adjacency function decreases the number of node connections, which may eliminate spurious connections. But if τ is chosen too high, the resulting network contains too many zeroes to be informative. If A is a correlation network or another network that threshold-implies a matrix of p values, then the hard threshold τ can be chosen based on statistical significance as described in Sect. 10.17. But statistical significance is not always the same as biologic significance, and topologically meaningless networks may result when only considering statistical criteria.

In the following, we present several *topological* criteria for choosing the parameters of an adjacency function. These criteria aim to produce networks with topologically realistic or desirable properties. Since each parameter value of an adjacency function AF leads to a network with different topological properties, the choice of the parameter value can be guided by desirable topological properties. For example, the **mean connectivity criterion** selects adjacency function parameters such that the mean connectivity $mean(k) = \sum_{i=1}^{n} k_i/n$ takes on a given value. One can prove that the mean connectivity of a weighted or an unweighted network is a monotonically decreasing function of the adjacency function parameters β or τ, respectively. Thus, the higher the parameter value, the lower is the mean connectivity. Roughly speaking, a network is uninformative if the mean connectivity is very small or very large. But in many applications, it is not clear what constitutes a reasonable choice for the mean connectivity.

For some adjacency functions, the relationship between parameter choices and network concepts is well understood. For example, we showed in Sect. 4.2 that high values of the power adjacency function parameter β lead to low values of the connectivity, density, maximum adjacency ratio, and typically the clustering coefficient. For most networks, the higher the β, the larger the heterogeneity and the module separability. These insights can be used by a data analyst to rule out values of β that lead to undesirable (or unrealistic) networks.

In the following, we will describe how to use the scale-free fit index (1.13)

$$R^2 = ScaleFreeFit = cor(log(p(dk)), log(BinNo))^2$$

to choose adjacency function parameters. As mentioned in Sect. 1.3.3, many real networks have been found to exhibit approximate scale-free topology (Barabasi and Albert 1999; Albert et al. 2000; Barabasi and Oltvai 2004). The scale-free fit index R^2 (1.13) of a transformed network $A = AF(A^{original}, parameters)$ is a function of the adjacency function parameters.

In some network applications, prior knowledge suggests that the network should satisfy scale-free topology at least approximately. In this case, one can ignore adjacency function parameter values that do not lead to approximately scale-free networks. We find it expedient to use a signed version of the scale-free topology fitting index. Since it is biologically implausible that a networks contains more hub genes than non-hub genes, we multiply R^2 with -1 if the slope of the regression line between $log(p(dk))$ and $log(BinNo)$ is positive. These considerations have motivated us (Zhang and Horvath 2005) to propose the following **scale-free topology criterion**: *Only consider those parameter values that lead to a network satisfying scale-free topology at least approximately, e.g., signed* $R^2 > 0.80$. A more refined scale free topology criterion is based on the following empirical observation.

In practice, we often find that the relationship between R^2 and the threshold parameter (τ or β) follows approximately a saturation curve as can be seen in the brain cancer network (Figs. 5.5 and 5.6) and the mouse gene network (Sect. 12.2). *The refined scale-free topology criterion recommends to use the first threshold parameter value where saturation is reached as long as it is above* 0.8.

But is worth repeating that there does not have to be a monotonic relationship between the adjacency parameter and the scale-free fit. Sometimes the scale free fitting index R^2 decreases for large values of β. For example, when dealing with a correlation network constructed on the basis of two groups of very distinct arrays, the scale-free topology criterion may not work well. In this case, we recommend to choose a default values for the parameters, e.g., $\beta = 6$ for an unsigned correlation network and $\beta = 12$ for a signed correlation network.

When it comes to studying scale-free topology, several authors have studied the slope $-\hat{\gamma}$ of the regression line between $log(p(dk))$ and $log(dk)$. We often find that it takes a value between -1 and -2.

We are hesitant of formulating the scale-free topology criterion as an optimization problem because noise affects the relationship between R^2 and the AF parameters (see Figs. 5.5 and 5.6). These figures and Table 5.2 show that unlike the mean number of connections, R^2 is not a *strictly* monotonic function of τ.

4.4 Differential Network Concepts for Choosing AF Parameters

Network concepts for comparing two networks (described in Sect. 1.11) can be used to choose the parameter values of an adjacency function. This is particularly useful if a reference or "gold standard" network $A^{[\mathrm{ref}]}$ is available based on prior knowledge about the connectivity patterns between the nodes. In this case the adjacency function parameters can be chosen such that the resulting network $A^{[\mathrm{test}]} = AF(A^{original}, parameters)$ is as similar possible to $A^{[\mathrm{ref}]}$. In other words, the parameters are chosen such that

$$A^{[\mathrm{test}]} \approx A^{[\mathrm{ref}]}.$$

To measure the similarity between $A^{[\mathrm{ref}]}$ and $A^{[\mathrm{test}]}$, one can calculate the connectivity correlation $cor.K = cor(K^{[\mathrm{ref}]}, K^{[\mathrm{test}]})$ (1.53), the adjacency correlation coefficient *cor.Adj* (1.55), or other network concept correlations (1.52). Adjacency function parameters can be chosen (or at least narrowed down) by requiring high correlations. Alternatively, one can try to minimize a differential network concept $Diff.NC = NC^{[\text{ref}]} - NC^{[\text{test}]}$ (1.50), e.g., the absolute value of the differential density.

4.5 Power AF for Calibrating Weighted Networks

In practical applications, two network adjacency matrices may be highly correlated, i.e., have high adjacency correlation (1.55), but their adjacencies may be very different. To make two networks comparable, it can be advantageous to calibrate one with respect to the other. As an example, consider two networks $A^{[\text{ref}]}$ and $A^{[\text{test}]}$ which are related using the power adjacency transformation (4.2) $A^{[\text{test}]} = AF^{power}(A^{[\text{ref}]}, \beta = 10) = (A^{[\text{ref}]})^{10}$. Despite the high correlation between the adjacencies of $A^{[\text{ref}]}$ and those of $A^{[\text{test}]}$ the two networks will be differently "calibrated". By this we mean that many of the quantiles $quantile_{prob}(A^{[\text{ref}]})$ (1.19) of the first network will be different from those $quantile_{prob}(A^{[\text{test}]})$ of the second network. For the above example, one can show that $quantile_{prob}(A^{[\text{test}]}) = quantile_{prob}((A^{[\text{ref}]})^{\beta}) = quantile_{prob}((A^{[\text{ref}]}))^{\beta}$. In practice, one can use a quantile–quantile plot (R command qqplot) to compare the calibration of two networks. The power adjacency function $AF^{power}(*, \beta)$ (4.2) can be used to calibrate one network with respect to another network. For example, to enforce that test network has the same prob = 0.95 quantile as the reference networks, one can use the power

$$\beta.95 = \frac{log(quantile_{0.95}(A^{[\text{ref}]}))}{log(quantile_{0.95}(A^{[\text{test}]}))}$$

to define the calibrated version of the second adjacency matrix

$$calibrated.A^{[\text{test}]} = AF^{power}(A^{[\text{test}]}, \beta = \beta.95).$$

By construction, the prob= 0.95 quantile of the entries of $calibrated.A^{[\text{test}]}$ equals that of the first network $A^{[\text{ref}]}$.

R code for calibrating different networks using the adjacency function can be found in Sect. 12.6.2.

4.6 Definition of Threshold-Preserving Adjacency Functions

Threshold-preserving adjacency functions are very useful for relating different methods of constructing (**unweighted**) networks. **Threshold-preserving adjacency functions** can be defined in many equivalent ways. First, an adjacency function AF is threshold-preserving if, and only if,

$$A_{ij}^{original} < A_{kl}^{original} \text{ implies } AF\left(A^{original}\right)_{ij} \leq AF\left(A^{original}\right)_{kl}. \tag{4.17}$$

Second, AF is threshold-preserving if a nondecreasing function *nonDecreasingF* can be defined such that

$$vectorizeMatrix\left(AF\left(A^{original}\right)\right) = nonDecreasingF\left(vectorizeMatrix\left(A^{original}\right)\right),$$

where $vectorizeMatrix(A^{original})$ (1.17) is a vector whose $n*(n-1)/2$ components correspond to the upper-diagonal entries of $A^{original}$. Third, AF is threshold-preserving if, and only if, for each threshold value τ (between 0 and 1) another value τ_{AF} can be defined such that the following equation holds:

$$AF^{threshold}\left(A^{original}, \tau_{AF}\right) = AF^{threshold}\left(AF\left(A^{original}\right), \tau\right) \tag{4.18}$$

Fourth, AF is threshold preserving if the C-index (5.18) of the vectorized matrices equals 1, i.e.,

$$C.index\left(vectorizeMatrix\left(A^{original}\right), vectorizeMatrix\left(AF\left(A^{original}\right)\right)\right) = 1.$$

A more detailed description of threshold-preserving functions can be found in Sect. 10.11.

In the following, we describe several threshold-preserving adjacency functions. The first threshold-preserving function is defined by linearly transforming the original adjacencies as follows $AF(A^{original}) = 0.5 + 0.5 * A^{original}$. The second threshold-preserving function is defined by raising the original adjacencies to the power $\beta > 0$. $AF(A^{original})_{ij} = (A_{ij}^{original})^{\beta}$. The third threshold-preserving adjacency function is the constant adjacency function AF which maps each original adjacency to the number 1. The fourth threshold-preserving adjacency function is the threshold adjacency function $AF^{threshold}(,\tau)$.

We also find it useful to make use of the following terminology: An adjacency function is called **rank-preserving** if $A^{original}$ and $AF(A^{original})$ are rank equivalent, i.e.,

$$A_{ij}^{original} < A_{kl}^{original} \text{ is equivalent to } AF\left(A^{original}\right)_{ij} < AF\left(A^{original}\right)_{kl}. \tag{4.19}$$

A rank-preserving adjacency function (4.19) is also threshold preserving but the converse is not necessarily true. For example, the constant adjacency function that maps each original adjacency to the number 1 is threshold preserving but not rank preserving.

An adjacency function is rank preserving if an increasing function can be defined such that

$$vectorizeMatrix(AF(A^{original})) = increasingFunction(vectorizeMatrix(A^{original})).$$

The name "rank preserving" stems from the following property of rank-preserving adjacency functions:

$$rank\left(vectorizeMatrix\left(AF\left(A^{original}\right)\right)\right) = rank\left(vectorizeMatrix\left(A^{original}\right)\right), \tag{4.20}$$

where the rank function $rank()$ is defined in Sect. 5.1.2. To quantify how well an adjacency function preserves the ranks, one can use the Spearman correlation

$$spearmanCor\left(vectorizeMatrix\left(AF\left(A^{original}\right)\right), vectorizeMatrix\left(A^{original}\right)\right)$$

or a symmetrized version (7.9) of the C-index (5.18).

In the following, we describe important examples of threshold-preserving and rank-preserving adjacency functions. Any nondecreasing (increasing) real-valued function $af(s)$ that maps the unit interval $[0,1]$ into $[0,1]$ can be used to define a threshold-preserving (rank-preserving) adjacency function AF by applying it to the off-diagonal entries of $A^{original}$.

For example, the power function $af^{power}(s) = |s|^{\beta}$ $(\beta > 0)$ leads to the rank-preserving power adjacency function AF^{power}.

The Moebius adjacency function $AF^{Moebius}$ is always threshold preserving and sometimes rank preserving. Specifically, condition (4.4) implies that the first derivative

$$\frac{d}{ds} af^{Moebius}(s) = \frac{ad - cb}{(cs+d)^2}$$

is nonnegative, i.e., that $af^{Moebius}(s)$ is a nondecreasing function. Note that $ad - cb = 0$ implies that $AF^{Moebius}$ is a constant (and threshold preserving) function while $ad - cb > 0$ implies that $AF^{Moebius}$ is a rank-preserving function.

4.7 Equivalence of Network Construction Methods

Given many alternative ways of defining networks, we find it useful to study their relationships. For example, if a weighted network adjacency matrix is thresholded (using the threshold adjacency function), then the resulting unweighted network is in a certain sense "implied" by the original network. Or if two network adjacency matrices are related by the power adjacency function, they are in a certain sense "equivalent". In the following, we introduce terminology for making these statements more precise. Assume that we have two network construction methods that result in two network adjacency matrices $A^{method1}$ and $A^{method2}$. We say that **method 1 threshold-implies method 2** if a threshold-preserving adjacency function $AF^{1\to 2}$ can be defined such that

$$A^{method2} = AF^{1\to 2}(A^{method1}).$$

For example, if $A^{method2} = AF^{threshold}(A^{method1}, \tau)$, then method 1 threshold-implies method 2.

We call two network construction methods rank equivalent if a rank-preserving adjacency function $AF^{1\to 2}$ can be defined for relating the two adjacency matrices. For example, if $A^{method2} = \left(A^{method1}\right)^2$, then the two resulting construction methods are rank equivalent.

4.8 Exercises

1. Exercise regarding the effect of the power transformation on the topology. The goal of this exercise is to determine the effect of the power transformation on the following network concepts: density, mean maximum adjacency ratio, heterogeneity, mean clustering coefficient, centralization, scale-free topology fitting index. Simulate a 500×500 dimensional random network adjacency matrix $A^{original}$ and transform it with the power adjacency function $A = \left(A^{original}\right)^{\beta}$ with $\beta = 2$.

   (i) Repeat the analysis $no.replicates = 20$ times and determine how often the density is decreased, the mean maximum adjacency ratio is decreased, the heterogeneity is increased, the mean clustering coefficient is decreased, the centralization is decreased, and the scale-free topology index is increased. Hint:

   ```
   library(WGCNA)
   # power parameter of the transformation
   beta=2
   #number of replicates
   no.replicates=20
   # number of nodes
   n=500
   # the following vectors store the results
   DensityDecreased=rep(NA,no.replicates)
   MeanMARDecreased=rep(NA,no.replicates)
   HeterogeneityIncreased=rep(NA,no.replicates)
   MeanClusterCoefDecreased=rep(NA,no.replicates)
   CentralizationDecreased=rep(NA,no.replicates)
   SFTincreased=rep(NA,no.replicates)
   for (i in 1:no.replicates) {
   set.seed(i); printFlush(i)
   # matrix whose entries lie between 0 and 1
   power2=10 #change it for task ii
   M=matrix(runif(n*n)^power2,nrow=n,ncol=n);
   # symmetrize M
   A.original=(M+t(M))/2
   # get rid of numerical errors
   A.original=round(A.original,12)
   # set the diagonal elements to 1
   ```

```
diag(A.original)=1
A=(A.original)^beta
NC.original=fundamentalNetworkConcepts(A.original)
NC=fundamentalNetworkConcepts(A)
Connectivity.original=NC.original$Connectivity
Connectivity=NC$Connectivity
Rsquared.SFT.original=
scaleFreeFitIndex(Connectivity.original)$Rsquared.SFT
Rsquared.SFT=scaleFreeFitIndex(Connectivity)$Rsquared.SFT
# Comparisons
Nonstandardized[i]=NC$Density<NC.original$Density
MeanMARDecreased[i]=mean(NC$MAR)<mean(NC.original$MAR)
HeterogeneityIncreased[i]=
NC$Heterogeneity>NC.original$Heterogeneity
MeanClusterCoefDecreased[i]=
mean(NC$ClusterCoef)<mean(NC.original$ClusterCoef)
CentralizationDecreased[i]=
NC$Centralization<NC.original$Centralization
SFTincreased[i]=Rsquared.SFT>Rsquared.SFT.original
} # end of for loop
table(DensityDecreased);table(MeanMARDecreased)
table(HeterogeneityIncreased);table(MeanClusterCoefDecreased)
table(CentralizationDecreased);table(SFTincreased)
```

(ii) How do the results change if you choose `power2=1` in the above R code?

2. Exercise regarding properties of the power adjacency function (4.2) and the threshold adjacency function (4.1). Let $A^{original}$ be a possibly weighted adjacency matrix. Define

$$A^{weighted} = AF^{power}\left(A^{original}, \beta\right) = \left(A^{original}\right)^{\beta},$$
$$A^{unweighted} = AF^{threshold}\left(A^{original}, \tau\right),$$

where the soft threshold $\beta >= 1$.

(i) Prove that *Density* of $A^{weighted} \leq$ *Density* of $A^{original}$, i.e., the network density (1.18) of $A^{weighted}$ is lower than or equal to that of $A^{original}$.
(ii) Prove that *Density* of $A^{unweighted} \leq$ *Density* of $A^{original}$.
(iii) Prove that $A^{weighted}$ threshold-implies $A^{unweighted}$. Hint: C-index (Sect. 5.1.4).

3. Exercise regarding the properties of the adjacency functions defined in Chap. 4.

(i) Prove that the adjacency functions AF^{power}, $AF^{threshold}$, and $AF^{Moebius}$ map a weighted adjacency matrix onto another weighted adjacency matrix. Hint: Show $0 \leq A_{ij} \leq 1$.
(ii) Which of the above adjacency functions are rank-preserving (4.20)?

4. Exercise regarding the TOM adjacency function AF^{TOM} (1.38). Show that the TOM adjacency function is in general *not* rank preserving. Hint: Use

the simulated expression data from Sect. 6.13 to define the adjacency matrix $A^{original}$. Calculate the Spearman correlation and C-index (Sect. 5.1.4) between $vectorizeMatrix(A^{original})$ and $vectorizeMatrix\left(TOM\left(A^{original}\right)\right)$.

References

Albert R, Jeong H, Barabasi AL (2000) Error and attack tolerance of complex networks. Nature 406(6794):378–382

Barabasi AL, Albert R (1999) Emergence of scaling in random networks. Science 286(5439):509–512

Barabasi AL, Oltvai ZN (2004) Network biology: Understanding the cell's functional organization. Nat Rev Genet 5(2):101–113

Horvath S, Dong J (2008) Geometric interpretation of gene co-expression network analysis. PLoS Comput Biol 4(8):e1000117

Zhang B, Horvath S (2005) General framework for weighted gene coexpression analysis. Stat Appl Genet Mol Biol 4:17

Chapter 5
Correlation and Gene Co-Expression Networks

Abstract A correlation network is a network whose adjacency matrix is constructed on the basis of pairwise correlations between numeric vectors. The numeric vectors may represent observed quantitative measurements of variables. For example, the gene expression levels (transcript abundances) across different conditions can be represented by a numeric vector. In general, the relationship between a pair of numeric vectors can be measured in many ways, in particular, using a correlation coefficient (e.g., the Pearson-, Spearman-, or biweight mid-correlation) or using the concordance index. Mouse gene expression data are used to illustrate how network concepts can be used to describe the pairwise relationships among gene expression profiles. While cluster trees and heat maps can be used to *visualize* relationships between variables, concepts of correlation networks can be used to *quantify* them. Brain cancer gene expression data are used to illustrate the topological effects of hard- and soft-thresholding. We provide an overview of weighted gene coexpression network analysis and different gene network (re-)construction methods.

5.1 Relating Two Numeric Vectors

Before defining correlation networks, we describe different approaches for relating two numeric vectors x and y. Assume each vector has m components. We adopt the R software notation regarding the element-wise operation between numeric (quantitative) vectors and matrices. For example, $x+2$ denotes a vector whose uth element is given by x_u+2. Any function that maps a real number onto another real number is defined for matrices or vectors using element-wise operations. For example, $abs(x)$, $(x)^{\beta}$, $exp(x)$ denote vectors whose uth components are given by $|x_u|$, $(x_u)^{\beta}$ and $exp(x_u)$, respectively.

Sample statistics are defined for vectors and matrices by applying them to the set of components. For example, $max(x)$ denotes the maximum values across the set of component values, i.e., $max(x) = max(\{x_u\})$. The sum over the components of x is given by

$$sum(x) = \sum_{u=1}^{m} x_u = x_1 + x_2 + \cdots + x_m.$$

S. Horvath, *Weighted Network Analysis: Applications in Genomics and Systems Biology*,
DOI 10.1007/978-1-4419-8819-5_5, © Springer Science+Business Media, LLC 2011

The (sample) mean of x is given by

$$mean(x) = \frac{\sum_{u=1}^{m} x_u}{m}.$$

The (sample) variance is given by

$$var(x) = \frac{(\sum_{u=1}^{m} x_u - mean(x))^2}{m-1}.$$

A vector is **scaled** (also known as **standardized** or **studentized**) by subtracting its sample mean from each component and dividing the difference by the square root of the variance. Specifically, the uth component of $scale(x)$ is given by

$$scale(x)_u = \frac{x_u - mean(x)}{\sqrt{var(x)}}. \tag{5.1}$$

As a result, $mean(scale(x)) = 0$ and $var(scale(x)) = 1$. The scale transformation of a vector (or the columns of matrix) can be computed with the R function `scale`.

The p-**norm** of x is defined as:

$$||x||_p = \left(\sum_{u=1}^{m} |x_u|^p \right)^{1/p}. \tag{5.2}$$

The Euclidean norm $||x||$ of a vector equals the 2-norm, i.e., $||x|| = ||x||_2$. In the following, we will introduce notation for equal length vectors $x = (x_1, \ldots, x_m)$ and $y = (y_1, \ldots, y_m)$. $x * y$ denotes the vector whose uth component is given by $x_u * y_u$.

The **inner product** (also known as **dot product**) $\langle x, y \rangle$ between two vectors x and y (of equal length) is defined as

$$\langle x, y \rangle = \sum_{u=1}^{m} x_u y_u = sum(x * y).$$

The inequality $|\sum_{u=1}^{m} x_u * y_u| \le \sum_{u=1}^{m} |x_u * y_u|$ can be rewritten as follows:

$$|\langle x, y \rangle| \le ||x * y||_1. \tag{5.3}$$

An upper bound for $||x * y||_1$ is given by **Hoelder's inequality**

$$||x * y||_1 \le ||x||_p ||y||_q, \text{ where } 1/p + 1/q = 1 \tag{5.4}$$

and $1 \le p, q \le \infty$. The **Cauchy–Schwarz inequality** is a special case of Hoelder's inequality (with $p = q = 2$):

$$\sum_{u=1}^{m} |x_u y_u| \le \sqrt{\left(\sum_{u=1}^{m} |x_u|^2 \right) \left(\sum_{u=1}^{m} |y_u|^2 \right)}. \tag{5.5}$$

The **cosine correlation** between vectors x and y:

$$cosineCor(x,y) = \frac{\langle x,y\rangle}{||x||||y||} = \frac{\sum_{u=1}^{m} x_u y_u}{\sqrt{(\sum_{u=1}^{m} |x_u|^2)(\sum_{u=1}^{m} |y_u|^2)}}. \tag{5.6}$$

Note that inequalities (5.3) and (5.4) imply that $|\frac{\langle x,y\rangle}{||x||||y||}| \leq 1$, i.e.,

$$-1 \leq cosineCor(x,y) \leq 1. \tag{5.7}$$

The **angle** $\theta(x,y)$ between the two vectors x and y is defined as inverse cosine (denoted acos) of their cosine correlation, i.e.,

$$\begin{aligned}\theta(x,y) &= acos\left(\frac{<x,y>}{||x|| * ||y||}\right), \\ &= acos(cosineCor(x,y)). \end{aligned} \tag{5.8}$$

Note that $\langle x,y\rangle = 0$ implies $\theta = \pi/2$, i.e., the two vectors are orthogonal. Since $cos(acos(r)) = r$, (5.8) implies

$$cosineCor(x,y) = cos(\theta(x,y)). \tag{5.9}$$

5.1.1 Pearson Correlation

The **Pearson correlation** (also known as sample correlation) between two vectors x and y is defined as the cosine correlation (5.6) between their scaled version, i.e.,

$$\begin{aligned}cor(x,y) &= cosineCor(scale(x), scale(y)) \\ &= \frac{\langle scale(x), scale(y)\rangle}{||scale(x)||||scale(y)||}, \end{aligned} \tag{5.10}$$

where the scale transformation $scale(x)$ is defined in (5.1). Equations (5.10) and (5.9) can be combined to provide the following angular interpretation of the Pearson correlation:

$$cor(x,y) = cosineCor(scale(x), scale(y)) = cos(\theta(scale(x), scale(y))). \tag{5.11}$$

Thus, $cor(x,y)$ equals the cosine of the angle θ between the two scaled vectors $scale(x)$ and $scale(y)$.

A **linear transformation** of a vector x onto a new vector $xnew$ is defined as follows:

$$xnew = slope1 * x + intercept1,$$

where *slope1* and *intercept1* are real numbers. It is straightforward to show that $scale(xnew) = sign(slope1)x$, where the sign $sign()$function takes on values $+1$, $0, -1$ depending on the sign of its argument. Our definition of the correlation (as cosine correlation of scaled vectors (5.10)) implies that it is **scale invariant** with respect to linear transformations, i.e.,

$$cor(intercept1 + slope1 * x, intercept2 + slope2 * y) \\ = sign(slope1)sign(slope2)cor(x,y).$$

If there is a straight line relationship between y and x (i.e., when $y = slope * x + intercept$) then $cor(x,y) = sign(slope)$. Thus, $cor(x,y)$ equals -1 if, and only if, there is a decreasing straight line relationship between x and y. Inequality (5.7) applied to scaled vectors implies that

$$-1 \leq cor(x,y) \leq 1.$$

The correlation can be expressed as follows:

$$cor(x,y) = \frac{cov(x,y)}{\sqrt{var(x)var(y)}}, \tag{5.12}$$

where the **covariance** between the vectors is defined as follows:

$$cov(x,y) = \frac{\sum_u (x_u - mean(x))(y_u - mean(y))}{m-1}.$$

Note that the variance of x equals the covariance of x with itself, i.e., $var(x) = cov(x,x)$.

5.1.2 Robust Alternatives to the Pearson Correlation

The Pearson correlation is sensitive to outlying observations. Here we will describe alternative measures of correlation which are robust to outliers. The **Spearman correlation** between two quantitative vectors x and y is the (Pearson) correlation of their ranks, i.e.,

$$spearmanCor(x,y) = cor(rank(x), rank(y)), \tag{5.13}$$

where the **rank function** replaces the components of a quantitative vector by their ranks. For example, if $x = (-2, 1.1, 1.5, 1.7)$, then $rank(x)$ results in the vector $(1,2,3,4)$. The rank function is invariant with regard to monotonically increasing transformations of the vectors, i.e.,

$$spearmanCor(IncreasingF(x), IncreasingG(y)) = spearmanCor(x,y).$$

The rank invariance is a consequence of the rank invariance of the rank function:

$$rank(IncreasingF(x)) = rank(x).$$

This rank invariance shows that the Spearman correlation assesses how well an arbitrary monotonic function could describe the relationship between the two variables.

5.1.3 Biweight Midcorrelation

While the Spearman correlation protects against outliers, it is overly conservative in many applications. As an alternative, we recommend to use a modified version of the **biweight midcorrelation** since it combines advantages of the Pearson correlation (relatively high power) and the Spearman correlation (relatively high robustness) (Wilcox 1997; Hardin et al. 2007). In preparation of the definition of the biweight midcorrelation, we define the following three functions. The first function $robustScale(x)$ inputs a numeric vector x and outputs a vector whose uth component is given by

$$robustScale(x)_u = \frac{x_u - median(x)}{mad(x) * 9}, \tag{5.14}$$

where $mad(x) = median(|x - median(x)|)$ is the median absolute deviation of x and 9 is a number that could be changed. The second function $weight(x)$ inputs a numeric vector x and outputs a vector whose uth component is given by

$$weight(x)_u = (1 - (x_u)^2)^2 H(1 - |x_u|), \tag{5.15}$$

where $H()$ denotes the Heaviside unit step function that equals 1 if its argument is positive and zero otherwise. The third function is the median-based, weighted covariance function:

$$covMedianWeighted(x,y) = \frac{sum((x - median(x))w.x(y - median(y))w.y)}{\sqrt{sum(w.x^2)sum(w.y^2)}}, \tag{5.16}$$

where the weights are given by $w.x = weight(robustScale(x))$.

Using these function, the **biweight midcorrelation** between x and y is defined as follows:

$$bicor(x,y) = \frac{covMedianWeighted(x,y)}{\sqrt{covMedianWeighted(x,x)covMedianWeighted(y,y)}}. \tag{5.17}$$

Peter Langfelder implemented a computationally fast version of the biweight midcorrelation in the `WGCNA` function `bicor` (Langfelder and Horvath 2011). Here is some R code that illustrates the definition of bicor:

```
robustScale=function(x){
      (x-median(x,na.rm=T))/(9*mad(x,constant=1))}
weight=function(x){(1-x^2)^2*ifelse(abs(x)<=1,1,0)}

covMedianWeighted= function(x,y){
w.x=weight(robustScale(x));w.y=weight(robustScale(y))
numerator=
sum((x-median(x,na.rm=T))*w.x*(y-median(y,na.rm=T))*w.y,na.rm=T)
denominator=sqrt(sum(w.x^2,na.rm=T)*sum(w.y^2,na.rm=T))
numerator/denominator}

# this is the bicor function
bicorF=function(x,y){
denom=sqrt(covMedianWeighted(x,x)*covMedianWeighted(y,y))
covMedianWeighted(x,y)/denom}

# Here we simulate two correlated vectors
set.seed(1)
x=rnorm(50); y=x+rnorm(50)

bicorF(x,y) #=0.5902869
# same result results from the bicor function
bicor(x,y) #=0.5902869
# compare it to the Pearson correlation
cor(x,y) #=0.6339331
# and the Spearman correlation
cor(x,y,method="s") #=0.5671549
```

5.1.4 C-Index

Apart from correlation coefficients, many alternative methods exist for measuring the relationship between two numeric vectors x and y. Here we will briefly review the concordance index $C.index(x,y)$, which is also known as *C-index*. The C-index measures how well the components of x are able to "discriminate" the components of y. In general, the C-index is not symmetrical in x and y, i.e., $C.index(x,y) \neq C.index(y,x)$. For the experts, we briefly mention that the C-index can be considered a generalization of the area under the receiver operator curve (ROC) and as a type of Goodman–Kruskal gamma rank correlation. The C-index estimates the "probability of concordance" between the values of x and y. The C-index can be computed using the following steps:

1. Define the set of all distinct y-component pairs. A pair of y-components (y_u, y_w) is a distinct pair if $y_u \neq y_w$. As an example, consider $y = c(1,1,3)$. Then the set of all distinct pairs of components is given by $\{(y_1,y_3),(y_2,y_3)\}$. Denote the number of distinct pairs by *no.distinct.pairs*.
2. For each distinct pair (y_u, y_w) determine whether the corresponding values of x are concordant, i.e., determine whether $y_u > y_w$ implies $x_u > x_w$ or whether $y_u < y_w$ implies $x_u < x_w$. If so, then the distinct pair (y_u, y_w) is concordant with

the corresponding pair (x_u, x_w). Several different options exist for dealing with ties ($x_u = x_w$) in the x-values. Here we remove distinct pairs with tied x-values from the calculation of the C-index. Let *no.concordant.pairs* be the number of distinct pairs whose corresponding x-values are concordant.

3. Define the C-index by dividing the number of concordant pairs by the number of distinct pairs:

$$C.index(x,y) = \frac{no.concordant.pairs}{no.distinct.pairs}. \tag{5.18}$$

Since $no.distinct.pairs \geq no.concordant.pairs$, the C-index takes on values between 0 and 1. $C.index > 0.5$ implies positive concordance, $C.index = 0.5$ implies random concordance (no better than flipping a coin), and $C.index < 0.5$ implies opposite concordance (worse than random, but if you flip the direction it becomes a good concordance). The C-index strongly depends on how ties in x are being treated. Our choice of removing distinct pairs whose x-values tie is implemented by the option `outx=TRUE` in the command `rcorr.cens` of the R package `Hmisc` (Harrell 2001). Specifically, we compute $C.index(x,y)$ with the following R-commands:

```
library(Hmisc)
C.index=rcorr.cens(x,y,outx=TRUE)[[1]]
```

While our analysis focused on two numeric vectors, we should point out that the C-index is also defined for binary and even censored vectors (e.g., survival time).

5.2 Weighted and Unweighted Correlation Networks

A correlation network is a network whose adjacency matrix is constructed on the basis of correlations between quantitative measurements. The input data can be described by an $m \times n$ matrix $datX = [x_{ui}]$ where the row indices ($u = 1, \ldots, m$) correspond to sample measurements and the column indices correspond to network nodes ($i = 1, \ldots, n$):

$$datX = [x_{ui}] = (x_1 x_2 \cdots x_n). \tag{5.19}$$

We refer to the ith column x_i as the ith vector or **node profile across m samples**. Sometimes a quantitative measure (referred to as **sample trait**) is provided for the rows of $datX$. For example, the vector $t = (t_1, \ldots, t_m)$ could be a binary grouping variable (e.g., case-control status) or a quantitative variable (e.g., body weight). Abstractly speaking, we define a sample trait t as a vector with m components that correspond to the rows of the data matrix $datX$. A sample trait can be used to define a node significance measure. For example, a **trait-based node significance measure** can be defined as the absolute value of the correlation between the ith node profile x_i and the sample trait t:

$$GS_i = |cor(x_i, t)| \tag{5.20}$$

where the Pearson correlation coefficient is defined in (5.12). Alternatively, a correlation test p value (described in Sect. 10.3) or a regression-based p value for assessing the statistical significance between x_i and the sample trait t can be used to define a p value-based node significance measure (1.2). In the following, we will focus on the construction of the network adjacency matrix and will revisit node significance measures at a later point.

A correlation network adjacency matrix is constructed on the basis of the pairwise correlations $cor(x_i,x_j)$ (5.12) between the columns of $datX$. A weighted adjacency matrix can be defined as follows:

$$A_{ij} = (|cor(x_i,x_j)|)^{\beta}, \tag{5.21}$$

where the power parameter is required to satisfy $\beta \geq 1$. An *unweighted* correlation network can be defined by thresholding the absolute values of the correlation matrix, i.e.,

$$A_{ij} = \begin{cases} 1 \text{ if } |cor(x_i,x_j)| \geq \tau \\ 0 \text{ if } |cor(x_i,x_j)| < \tau \end{cases}. \tag{5.22}$$

In Sect. 5.3, we study general correlation networks including signed correlation networks which keep track of the sign of the correlation.

Weighted correlation networks have been used in a wide variety of settings. Although they were originally developed for gene expression data, these techniques have also been used for analyzing microRNA data (Wang et al. 2009), functional magnetic resonance imaging data (Mumford et al. 2010), DNA methylation data, and genetic polymorphisms.

A special case of weighted correlation networks is weighted gene co-expression networks, which arose in collaboration with people from different disciplines including bioinformaticist **Bin Zhang**, pathologist **Paul Mischel**, neuroscientist **Dan Geschwind**, and human geneticist **Stan Nelson**.

5.2.1 Social Network Analogy: Affection Network

Recall that each node in our affection network example represents an individual. Assume that the n individuals filled out an interest questionnaire that contained m value-based ranking questions. A typical question is the following. On a scale from 1 (not at all) to 10 (very much), how much do you like playing golf? The answers to the m questions by the ith individual can be represented by a vector $x_i = (x_{i1}, \ldots, x_{im})$ whose components record the answers to the m questions. To arrive at a similarity matrix between the individuals, we use the absolute value of the pairwise correlations $|cor(datX)| = |cor(x_i,x_j)|$. Our definition of the affection network is based on the following assumption: the stronger the interest relationship between two individuals, the more they will talk with each other and the stronger the affection they will feel for each other. More specifically, we assume that the affection (adjacency) A_{ij} between two individuals is proportional to their similarity on a logarithmic scale, i.e.,

$$log(A_{ij}) = \beta * log(|cor(x_i,x_j)|). \tag{5.23}$$

This is equivalent to using the power adjacency function for soft thresholding: $A_{ij} = (|cor(x_i,x_j)|)^\beta$. Incidentally, it would probably make more sense to define the affection network in a way that keeps track of the sign of the correlation coefficient, see (5.25).

5.3 General Correlation Networks

In Sect. 5.2, we introduced weighted and unweighted correlation networks (5.21 and 5.22) based on thresholding the absolute value of a correlation matrix. Here we generalize these correlation networks to keep track of the sign of the correlation. A general definition of correlation networks will allow us to compare different procedures for constructing networks.

As described in Sect. 5.2, a correlation network is constructed on the basis of an $m \times n$ matrix $datX$ (5.19) whose columns x_i correspond to network nodes ($i = 1,\dots,n$). If a sample trait t is available, then one can define either an unsigned trait-based node significance measure $GS_i = |cor(x_i,t)|$ (5.20) or a signed significance measure $GS_i = cor(x_i,t)$.

A correlation network adjacency matrix is constructed on the basis of the pair-wise correlations $cor(x_i,x_j)$ between the columns of $datX$. We typically use the Pearson correlation but an outlier-resistant correlation coefficient could also be used. We distinguish two types of correlation network adjacencies, signed and unsigned. The elements of an unsigned adjacency matrix can be written as a nondecreasing function of the absolute correlations, i.e., $A_{ij} = nonDecreasingF(|cor(x_i,x_j)|)$. In contrast, the elements of a signed adjacency matrix can be written as a nondecreasing function of the correlation, i.e., $A_{ij} = nonDecreasingF(cor(x_i,x_j))$. The nondecreasing function is required to yield adjacencies that continue to satisfy our conditions imposed on a general adjacency matrix (in particular, the elements are required to lie in $[0,1]$). Several forms of the nondecreasing function have been proposed in the literature. Unsigned and signed weighted correlation network adjaceny matrices are defined by choosing $nonDecreasingF(s) = af^{power}(s) = s^\beta$, i.e.,

$$A_{ij}^{unsigned,weighted} = (|cor(x_i,x_j)|)^\beta, \tag{5.24}$$

$$A_{ij}^{signed,weighted} = (0.5 + 0.5cor(x_i,x_j))^\beta, \tag{5.25}$$

respectively. Unsigned and signed *unweighted* correlation network adjaceny matrices are defined by choosing $nonDecreasingF(s) = af^{threshold}(s)$, e.g.,

$$A_{ij}^{unsigned,unweighted} = \begin{cases} 1 & \text{for } cor(x_i,x_j) \geq \tau \\ 0 & \text{for } cor(x_i,x_j) < \tau \end{cases}, \tag{5.26}$$

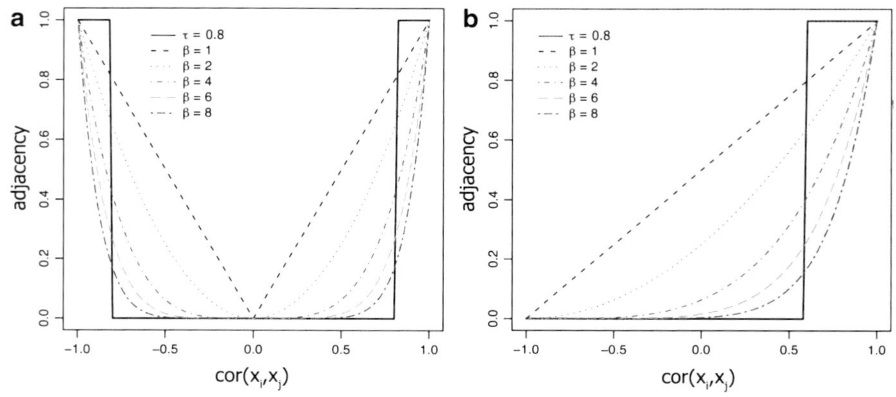

Fig. 5.1 Network adjacency (y-axis) versus the correlation coefficient (x-axis) for an unweighted network (black step function with $\tau = 0.8$) and weighted networks (*dashed lines* corresponding to different powers, β) in an unsigned network (**a**) and a signed network (**b**). Note that $cor(x_i, x_j) = -1$ leads to adjacency $= 0$ in the signed network. The weighted network preserves the continuous nature of the correlation measure, while an unweighted network dichotomizes the correlation

which depends on the threshold parameter τ. The choice of signed versus unsigned networks depends on the application; both signed (Mason et al. 2009) and unsigned (Horvath et al. 2006) correlation networks have been successfully used in genomic applications. The relationship between correlations and the adjacencies is visualized in Fig. 5.1.

With the help of adjacency functions, we provide an abstract definition of a **general correlation network** using two steps. First, an original (preliminary) adjacency matrix $A^{original}$ is defined as a function of the pairwise correlation coefficients $cor(x_i, x_j)$. Second, a threshold-preserving adjacency transformation AF (see, e.g., Chap. 4) is used to transform the original adjacency matrix

$$A = AF(A^{original}).$$

We distinguish two original adjacency matrices based on a correlation coefficient between vectors. The **unsigned correlation-based adjacency** is the absolute value of the Pearson correlation

$$A_{ij}^{unsigned,original} = |cor(x_i, x_j)|. \tag{5.27}$$

The **signed correlation-based adjacency measure** is given by

$$A_{ij}^{signed,original} = \frac{1 + cor(x_i, x_j)}{2}. \tag{5.28}$$

To change the topological properties of the original network, the original adjacency matrix can be transformed with a threshold-preserving adjacency function.

A weighted (power) adjacency matrix results when the power adjacency function (4.2) is used

$$A_{ij}^{weighted} = AF^{power}(A^{original}, \beta)_{ij} = |A_{ij}^{original}|^{\beta}. \quad (5.29)$$

An unweighted network results when $AF^{threshold}$ (4.1) is used

$$A_{ij}^{unweighted} = AF^{threshold}(A^{original}, \tau)_{ij}. \quad (5.30)$$

Of course, alternative threshold-preserving adjacency functions could also be used for defining a correlation network. For example, a Moebius adjacency transformation $AF^{Moebius}(A^{original}, a, b, c, d)_{ij}$ (4.3) such as

$$A_{ij} = 1/(2 - |cor(x_i, x_j)|)$$

defines an unsigned weighted network.

Given the profusion of methods for defining correlation networks, it is worth mentioning that according to our definition all correlation networks are threshold-implied (Sect. 4.7) by one of two very simple networks: either $A_{ij}^{unsigned,original}$ (5.27) or $A_{ij}^{signed,original}$ (5.28). This implies that each unweighted network, whose elements are given by a nondecreasing function of an underlying correlation matrix, can be defined by simply thresholding the correlation matrix.

If two correlation networks are defined using rank-preserving (strictly monotonic) adjacency functions of an underlying correlation matrix, then the two resulting correlation networks are rank-equivalent, i.e., one adjacency matrix can be written as a monotonically increasing function of the other. For example, $A_{ij}^{method1} = 1/(2 - |cor(x_i, x_j)|)$ is rank-equivalent to $A_{ij}^{method2} = |cor(x_i, x_j)|^6$.

5.4 Gene Co-Expression Networks

Correlation network methods have been widely used to analyze gene expression data. Typically, gene expression refers to the amount of messenger RNA that corresponds to the gene. Many technological platforms exist to measure gene expression data. For example, a gene expression microarray measures the expression level (messenger RNA abundance) of thousands of genes in a given condition. A gene is often represented by multiple probes on a given microarray. In this case, one can either construct a network among probes (i.e., the nodes represent probes) or "collapse" (combine) the probes into gene level measurements and then construct a network whose nodes correspond to genes. To combine multiple probes into genes, one can use the WGCNA function `collapseRows` (see the R help file of this function for more details). In most gene co-expression network applications, networks are constructed between probes. In the following, we will use the terms "gene" and "probes" interchangeably. We should also mention that many approaches exist for preprocessing ("normalizing") gene expression data. A review of different preprocessing methods with regard to gene network construction can be found in Lim et al. (2007). Microarray preprocessing techniques are beyond our scope, and we will assume in the following that the data are already normalized. Then the normalized expression data can be represented by an $m \times n$ dimensional matrix $datX$ whose ith column x_i corresponds to the ith gene (or probe). We refer to x_i as the ith gene expression profile. The m components of x_i correspond to m sample measurements (e.g., microarray arrays or experimental conditions).

Gene expression profiles across samples can be highly correlated. Genes with correlated expression profiles may correspond to protein complexes, pathways, or participate in regulatory and signaling circuits (Eisen et al. 1998). It is natural to describe the pairwise correlations between genes using correlation network language. Gene co-expression networks can be defined as weighted or unweighted correlation networks among the x_i. For example, an unweighted gene co-expression network is defined by thresholding the absolute values of the correlation matrix (5.22). To preserve the continuous nature of the co-expression information, one could simply define a weighted adjacency matrix as the absolute value of the gene expression correlation matrix i.e., $A_{ij} = |cor(x_i, x_j)|$. However, since microarray data can be noisy and the number of samples is often small, we and others have found it useful to emphasize strong correlations and to punish weak correlations. It is natural to define the adjacency between two genes as a power $\beta \geq 1$ of the absolute value of the correlation coefficient, i.e., to define a weighted correlation network (5.21) (Zhang and Horvath 2005; Horvath et al. 2006). Weighted gene co-expression networks have found important medical applications, e.g., to find brain cancer genes (Horvath et al. 2006), for characterizing obesity genes (Ghazalpour et al. 2006), for understanding atherosclerosis (Gargalovic et al. 2006), and for understanding the differences between human and chimp brain (Oldham et al. 2006). If a quantitative microarray sample trait $t = (t_1, \ldots, t_m)$ is available, it can be used to define a **trait-based gene significance measure** (5.20). If t is a binary variable encoding two groups (e.g., case versus control status), then a gene significance measure can be defined using the Student t-test (10.6) for testing differential expression between the two groups. Gene co-expression network methods have also been used for typical microarray data analysis tasks such as gene screening (filtering) (Zhang and Horvath 2005; Horvath et al. 2006; Oldham et al. 2006; Fuller et al. 2007).

In this book, we describe many alternative approaches for measuring linear and nonlinear co-expression relationships between gene expression profiles. Informally, we refer to all of these resulting networks as gene co-expression networks. Gene co-expression networks have been successfully used to describe the transcriptome in different organisms, e.g., yeast, flies, worms, plants, mice, and humans (Butte et al. 2000; Zhou et al. 2002; Steffen et al. 2002; Stuart et al. 2003; Carter et al. 2004; Zhang and Horvath 2005; Cabusora et al. 2005; Wei et al. 2006; Voy et al. 2006; Ghazalpour et al. 2006; Oldham et al. 2006; Horvath et al. 2006; Cokus et al. 2006; Swindell 2007; Weston et al. 2008; Keller et al. 2008; Shieh et al. 2008). Since co-expression networks capitalize on the statistical dependence or association between gene expression profiles, they are sometimes referred to as (statistical) dependence networks or association networks.

The **interpretation of gene co-expression relationships** depends heavily on biological context. For example, in a dataset consisting of samples from multiple tissue types, co-expression modules (i.e., modules defined by co-expression similarity) will often distinguish genes that are expressed in tissue-specific patterns (e.g., Jordan et al. 2004; Oldham et al. 2006). In a dataset consisting of samples from a single tissue type, co-expression modules may distinguish sets of genes that are preferentially expressed in distinct cell types that comprise that tissue. For example, **Mike Oldham**'s has shown that modules in human cortical networks correspond to cell types (Oldham et al. 2008). In a dataset consisting of samples from a homogeneous cellular population, co-expression modules may correspond more directly to sets of genes that work in tandem to perform various intracellular functions. In many cases, co-expression modules may not present immediate functional interpretations. However, previous work has shown that many co-expression modules are conserved across phylogeny (Stuart et al. 2003; Snel et al. 2004; Oldham et al. 2006; Miller et al. 2010), enriched with protein–protein interactions (Oldham et al. 2008; Huang et al. 2007), and enriched with specific functional categories of genes, including ribosomal, mitochondrial, synaptic, immune, hypoxic, mitotic, and many others.

Although elucidating the functional significance of identified co-expression modules requires substantial effort from biologists and bioinformaticians, the importance of co-expression modules lies not only in their functional interpretation but also in their reproducibility. Because transcriptome organization in a given biological system is highly reproducible (Oldham et al. 2008), co-expression modules provide a natural framework for comparisons between species, tissues, and pathophysiological states. This framework can reduce dimensionality by approximately three orders of magnitude (e.g., moving from 40,000 transcripts to 40 modules) (Horvath and Dong 2008), while simultaneously placing identified gene expression differences within specific cellular and functional contexts (inasmuch as the cellular and functional contexts of the modules are understood). The co-expression modules themselves are simply summaries of interdependencies that are already present in the data.

Module preservation statistics (described in Chap. 9) can be used to address an important question in co-expression module-based analyses: how to show whether the modules are robust and reproducible across data sets.

5.5 Mouse Tissue Gene Expression Data from of an F2 Intercross

We illustrate our methods using data from an F2 mouse intercross (referred to as B×H cross) (Ghazalpour et al. 2006; Cervino et al. 2005; Wang et al. 2006; Fuller et al. 2007) involving two inbred mouse strains (C57BL/6J.*Apoe* null and C3H/HeJ.*Apoe* null). The mouse gene expression data were generated by the labs of

Jake Lusis and **Eric Schadt**. The mouse strain C57BL/6J is susceptible to a variety of ailments related to atherosclerosis, diabetes, obesity, and heart disease to which the mouse strain C3H/HeJ is resistant. The offspring mice of this F2 mouse intercross are expected to show a significant spectrum of atherosclerosis and metabolic syndrome responses to a high-fat diet. A variety of physiological traits were measured, including mouse body weight, fat mass, insulin, glucose, free fatty-acid levels in the blood, and cholesterol fractions (HDL and LDL+VLDL). Since significant differences in the gene expression profiles between male and female mice have been observed (Wang et al. 2006), we analyzed each gender separately. For a detailed description of the data and the biological implications, we refer the reader to Ghazalpour et al. (2006).

The microarray data measure the expression levels in multiple tissue samples (liver, adipose, brain, muscle) from male and female mice. About 100 mice were available for each gender. The original data sets contained over 20,000 gene (probe) expression profiles (numeric variables). But in this motivational example for network concepts, we only study the pairwise correlations among 498 genes that had previously been found to form a subnetwork (i.e., a module) related to mouse body weight (Ghazalpour et al. 2006; Fuller et al. 2007).

The 498 genes form a subnetwork (i.e., network module) in female liver tissues (the Blue module described in (Ghazalpour et al. 2006)). Thus, our analysis studies intramodular network concepts in this particular network module.

Here we focus on the mathematical and topological properties of the pairwise absolute correlations $A_{ij} = |cor(x_i, x_j)|$ among the module genes. For each gender and tissue type, Fig. 5.2a depicts a hierarchical cluster tree of the genes. Figure 5.2b shows the corresponding heat maps, which color-code the absolute pairwise correlations A_{ij}. As can be seen from the color bar underneath the heat maps, red and green in the heat map indicate high and low absolute correlation, respectively. The genes in the rows and columns of each heat map are sorted by the corresponding cluster tree.

It is visually obvious that the heat maps and the cluster trees of different gender/tissue combinations can look quite different. Network theory offers a wealth of intuitive concepts for describing the pairwise relationships among genes that are depicted in cluster trees and heat maps. To illustrate this point, we describe several such concepts in the following. By visual inspection of Fig. 5.2b, genes appear to be more highly correlated in liver than in adipose (a lot of red versus green color in the corresponding heat maps). This property can be captured by the concept of network density. The density of the female liver network is 0.39, while it is only 0.23 for the female adipose network. Another example for the use of network concepts is to quantify the extent of cluster (module) structure. In this example, branches of a cluster tree (Fig. 5.2a) correspond to modules in the corresponding network. The cluster structure is also reflected in the corresponding heat maps: modules correspond to large red squares along the diagonal. Network theory provides a concept for quantifying the extent of module structure in a network: the mean clustering coefficient. The female liver, male liver, and female brain networks have high mean clustering coefficients (mean $ClusterCoef = 0.42, 0.43, 0.41$, respectively). In contrast,

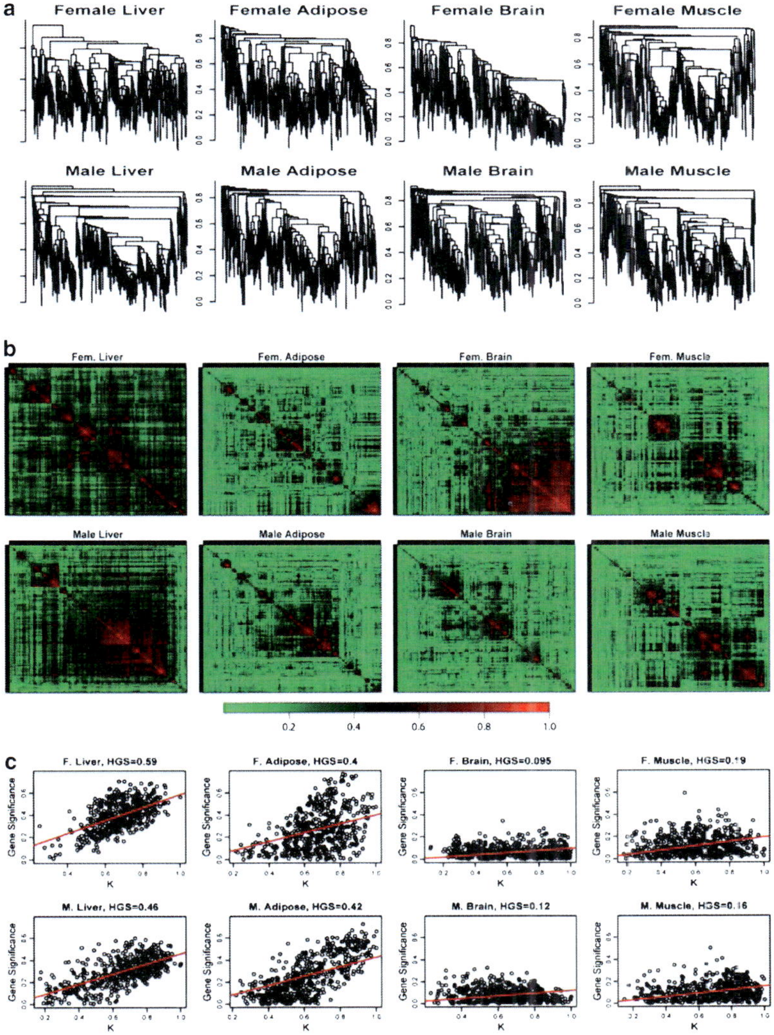

Fig. 5.2 This motivational example explores the pairwise absolute correlations $A_{ij} = |cor(x_i, x_j)|$ among 498 genes in different mouse tissues. The biological significance of this network is described in Ghazalpour et al. (2006) and Fuller et al. (2007). For each gender and tissue type (liver, adipose, brain, muscle), Figure (**a**) depicts an average linkage hierarchical cluster tree of the genes. Figure (**b**) shows the corresponding heat maps, which color-code the absolute pairwise correlations A_{ij}: *red* and *green* in the heat map indicate high and low absolute correlation, respectively. The genes in the rows and columns of each heat map are sorted by the corresponding cluster tree. Figure (**c**) depicts the relationship between gene significance GS (y-axis) and connectivity (x-axis). The gene significance of the ith gene was defined as the absolute correlation between the ith gene expression profile and mouse body weight. The hub gene significance HGS (1.23) is defined as the slope of the *red line*, which results from a regression model without an intercept term

the female adipose, male adipose, and male brain networks have lower mean clustering coefficients (mean $ClusterCoef = 0.27, 0.27, 0.25$, respectively). Difference in module structure may reflect true biological differences or they may reflect noise (e.g., technical artifacts or tissue contaminations).

As another example for the use of network concepts, compare the cluster tree of the female brain network with that of the male brain network. The cluster tree of the female network appears to be comprised of a single large branch, i.e., a highly connected hub gene at the tip of the branch forms the center in this network. In contrast, the cluster tree corresponding to the male brain network appears to split into multiple smaller branches, i.e., no single gene forms the center. To measure whether a highly connected hub gene forms the center in a network, one can use the concept of centralization. The female brain and male brain networks have centralization 0.34 and 0.21, respectively.

These examples illustrate that graph theory contains a wealth of network concepts that can be used to describe gene expression data. But we will argue that standard statistical techniques (used for gene expression data analysis) can also be used to derive network theoretic results. For example, network theorists have long studied the relationship between node (gene) significance and connectivity. Several network articles have pointed out that highly connected hub nodes are central to the network architecture (Albert et al. 2000; Jeong et al. 2001; Albert and Barabasi 2002; Han et al. 2004), but hub genes may not always be biologically significant (Carlson et al. 2006). To define a sample trait-based gene significance measure (5.20), we define the gene significance of gene i as the absolute correlation between the gene expression profile x_i and body weight t, i.e., $GS_i = |cor(x_i, t)|$. Figure 5.2c shows the relationship between this gene significance measure and connectivity in the different gender/tissue type networks. We find a strong positive relationship between gene significance and connectivity in the female and the male mouse liver networks. The positive relationship between gene significance and connectivity suggests that both variables could be used to implicate genes related to body weight. For example, we used connectivity as a variable in a systems biologic gene screening method (Fuller et al. 2007). While most network theorists would agree that connectivity is an important variable for finding important genes in a network (Zhang and Horvath 2005), the statistical advantages of combining gene significance and connectivity are not clear. Below, we use the geometric interpretation of co-expression network analysis to argue that intramodular connectivity can be interpreted as a fuzzy measure of module membership. Thus, a systems biologic gene screening method that combines a gene significance measure with intramodular connectivity amounts to a pathway-based gene screening method. Empirical evidence shows that the resulting systems biologic gene screening methods can lead to important biological insights (Horvath et al. 2006; Ghazalpour et al. 2006; Oldham et al. 2006; Fuller et al. 2007). Before combining gene significance and connectivity in a systems biologic gene screening approach, it is important to study their relationship. Toward this end, we define a measure of hub gene significance HGS as slope of a regression line (through the origin) between gene significance and scaled connectivity. As can be seen from Fig. 5.2c, the hub gene significance is high in liver and adipose tissues but it is low in brain and muscle tissues. Below, we use the geometric interpretation of co-expression networks to characterize co-expression networks that have high hub gene significance if the gene significance measure is based on a sample trait t.

In weighted correlation networks, we find empirically that the maximum adjacency ratio MAR_i is often highly correlated with the connectivity K_i (see also (3.52)). As we demonstrate in Fig. 5.3, the MAR_i is sometimes (but not always) superior to the scaled connectivity K_i when it comes to identifying biologically important intramodular hub genes.

Table 5.1 presents fundamental network concepts for the networks corresponding to different gender/tissue combinations. Note that a differential network analysis would find that many network concepts differ between networks. To interpret the changes in network concepts, it is useful to understand the relationships between network concepts. To investigate the relationships among seemingly disparate fundamental network concepts, we study conformity-based network concepts in Chap. 3. To arrive at a geometric interpretation of network concepts in correlation networks, we study eigenvectors-based network concepts in Chap. 6.

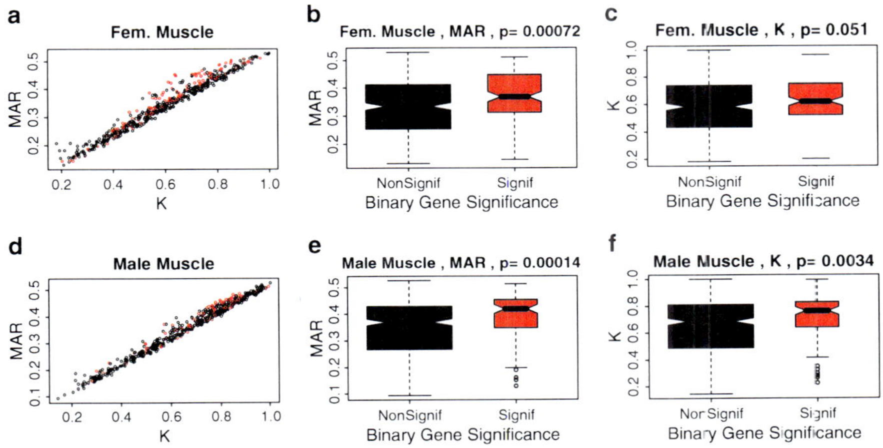

Fig. 5.3 Relationships among maximum adjacency ratio, scaled connectivity, and gene significance. Figure (**a**) depicts the relationship between MAR_i (y-axis) and scaled connectivity K_i using the female mouse muscle tissue network described in the motivational example. The genes are colored *red* or *black* depending on whether they are significantly ($p < 0.05$) related to mouse body weight. Figure (**b**) shows boxplots and a Kruskal–Wallis test p value ($p = 0.00072$) for studying whether MAR_i differs between significant (*red*) and nonsignificant (*black*) genes. Figure (**c**) shows the analogous boxplots and p value for the scaled connectivity K_i. In this female muscle tissue application, MAR_i is more significantly ($p = 0.00072$) related to GS_i than is K_i ($p = 0.051$). Figures (**d**,**e**,**f**) show the analogous relationships for male muscle. Again, the MAR_i is more significantly ($p = 0.00014$) related to GS_i than to K_i ($p = 0.0034$). As a caveat, we mention that in other applications (e.g., the yeast co-expression network (Carlson et al. 2006)), we have found that K_i is more significantly related to GS_i than MAR_i

Table 5.1 Network concept in the different mouse gender/tissue networks reported in Fig. 5.2

	Female liver		Female adipose		Female brain		Female muscle	
Network Concept	Fundam.	Eigenvector.	Fundam.	Eigen.	Fundam.	Eigen.	Fundam.	Eigen.
Factorizability	0.92	0.91	0.72	0.46	0.89	0.82	0.79	0.68
Density	0.39	0.39	0.23	0.14	0.32	0.27	0.24	0.19
Centralization	0.19	0.19	0.11	0.19	0.34	0.23	0.17	0.22
Heterogeneity	0.18	0.19	0.22	0.59	0.36	0.54	0.32	0.57
Mean Cluster Coef	0.42	0.42	0.27	0.26	0.41	0.46	0.30	0.33
	Male liver		Male adipose		Male brain		Male muscle	
Network Concept	Fundam.	Eigenvector.	Fundam.	Eigen.	Fundam.	Eigen.	Fundam.	Eigen.
Factorizability	0.93	0.92	0.76	0.59	0.73	0.46	0.76	0.48
Density	0.37	0.36	0.23	0.16	0.21	0.13	0.25	0.16
Centralization	0.19	0.21	0.15	0.21	0.21	0.20	0.12	0.21
Heterogeneity	0.28	0.32	0.28	0.65	0.27	0.67	0.30	0.64
Mean Cluster Coef	0.43	0.44	0.27	0.31	0.25	0.26	0.31	0.31

The table reports fundamental network concepts (defined in Table 3.1) and their eigenvector-based analogs (defined in Chap. 6). For each network, the table reports the network factorizability $F(A)$, the eigenvector factorizability $EF(E)$, network concepts, and their eigenvector-based analogs. Here we use a soft threshold $\beta = 1$ for constructing a weighted network. The results for higher powers β and for unweighted networks can be found (Dong and Horvath 2007).

5.6 Overview of Weighted Gene Co-Expression Network Analysis

One of the many biological applications of gene co-expression networks is the identification of pathways (modules) and centrally located genes (referred to as module centroids). In our applications, we define highly connected intramodular hub genes as module centroids. Weighted gene co-expression network analysis (WGCNA, Zhang and Horvath 2005; Horvath et al. 2006) can be considered a step-wise data reduction technique, which (a) starts from the level of thousands of variables (e.g., gene expression profiles), (b) identifies biologically interesting modules based on a node significance measure, (c) represents the modules by their centroids (e.g., eigenvectors or intramodular hubs), (d) uses intramodular connectivity (kIM or kME 6.53) as quantitative measures of module membership, and (e) combines node significance and module membership measures for identifying significant hub nodes. The module centric analysis alleviates the multiple testing problem inherent in high dimensional data, e.g., gene expression data. Instead of relating thousands of variables to a sample trait, it focuses on the relationship between a few (usually less than 10) modules and the sample trait.

An outline of WGCNA is presented in Fig. 5.4. Typically, the module definition does not make use of a priori defined gene sets. Instead, modules are constructed from the expression data using a (hierarchical) clustering procedure. Although it is advisable to relate the resulting modules to gene ontology information to assess

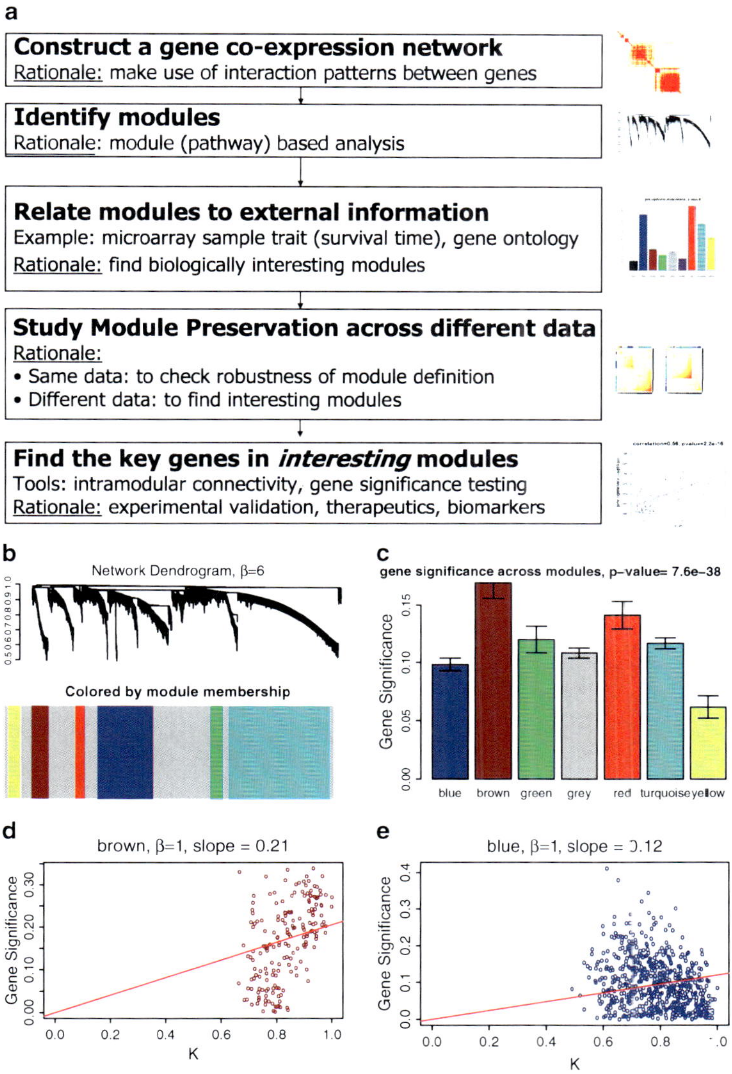

Fig. 5.4 Overview and an example application of gene co-expression network analysis. (**a**) Typical analysis flow chart. Gene co-expression network analysis aims to identify pathways (modules) and their key drivers (e.g., intramodular hub genes). (**b**) depicts the hierarchical cluster tree of genes in the brain cancer network (described in Sect. 5.7). Here modules correspond to branches of the tree. The branches and module genes are assigned a color as can be seen from the color-bands underneath the tree. *Gray* denotes genes outside of proper modules. A functional enrichment analysis of these modules can be found in Horvath et al. (2006). (**c**) Module significance (average gene significance) of the modules with regard to prognostic significance for survival time. The underlying gene significance is defined with respect to the patient survival time (5.20). (**d**) and (**e**) show scatter plots of gene significance GS (y-axis) versus scaled connectivity K (x-axis) in the *brown* and *blue* module, respectively. The hub gene significance (1.23) is defined as the slope of the *red line*, which results from a regression model without an intercept term

their biological plausibility, it is not required. Because the modules may correspond to biological pathways, focusing the analysis on modules (and corresponding centroids) amounts to a biologically motivated data reduction method.

In a high level view, gene co-expression networks can be viewed as a data reduction techniques (e.g., eigengene networks described in Sect. 6.5). In a low level view, many network analyses focus on the pairwise relationships of relatively few (hundreds) of nodes, e.g., nodes that form a single module in a larger network (see the mouse module application in Sect. 5.5). The low-level analysis of a single network module may help identify key genes that may be used as therapeutic targets or candidate biomarkers. Since intramodular hub nodes are centrally located in the module, it is natural to consider module membership measures (of kIM or kME (6.53)) when it comes to pathway-based gene screening. Intramodular hubs in significant modules lend themselves as candidates for biomarkers. Examples of biological studies that show the importance of intramodular hub genes can be found reported in (Horvath et al. 2006; Ghazalpour et al. 2006; Gargalovic et al. 2006; Oldham et al. 2006; Carlson et al. 2006). Because the expression profiles x_i of intramodular hub genes are often highly correlated (in our data, $r > 0.90$), typically dozens of candidates result. Although these candidates are statistically equivalent, they may differ in terms of biological plausibility or clinical utility. To screen for biologically important genes, one can use network screening methods based on intramodular connectivity or other types of network concepts. Of course, external knowledge (such as gene ontology information or prior publications) may also help one to select genes. Further, causal testing (described in Chap. 11 and Sect. 12.3) can also be used to screen for causal drivers.

5.7 Brain Cancer Network Application

Here we describe a weighted gene co-expression network that was constructed on the basis of 55 microarray samples of glioblastoma multiforme (brain cancer) patients. A detailed description of the data, modules, and biological implications can be found in Horvath et al. (2006), which involved a collaboration with **Paul Mischel**, **Stan Nelson**, **Tim Cloughesy**, and other UCLA researchers. An (unsigned) weighted gene co-expression network was constructed as described in Sect. 5.2. A power $\beta = 6$ was used for soft-thresholding. This power happens to be the default value but it could also be found with the scale-free topology criterion (described in Sect. 4.3 and Fig. 5.5) The adjacency matrix was transformed into the topological overlap measure (1.27), which was subsequently used as input of hierarchical clustering. Hierarchical clustering and branch cutting are described in Sect. 8.4 and Sect. 8.6, respectively. We defined six co-expression modules as branches of the cluster tree (Fig. 5.4b). These six modules (branches) are labeled by colors reflecting their module sizes (in decreasing order): turquoise, blue, brown,

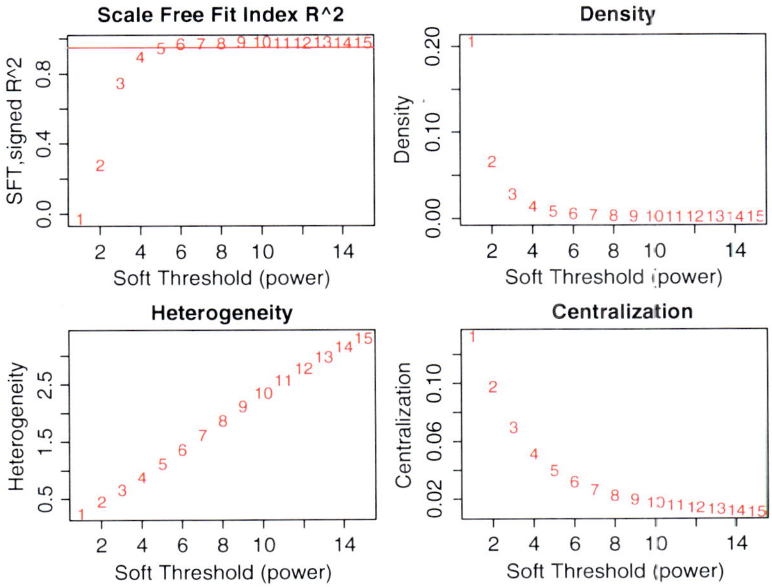

Fig. 5.5 Network concepts (y-axis) versus different values of β used in the power adjacency function (x-axis). The data are an unsigned correlation networks based on 8,000 most varying genes from a gene co-expression network based on 55 brain cancer samples (Horvath et al. 2006). Note that the scale-free topology fitting index R^2 and the heterogeneity tend to increase with β, while the density and the network centralization tend to decrease with β

yellow, green and red. Gray denotes the color of genes that were not grouped into any of the six proper modules. To allow for a comparison, we often report results for the "improper" module comprised of gray genes.

We used the patient survival time as sample trait t and defined a gene significance measure as the absolute value of the correlation between t and the gene expression profiles (5.20). The module significance was defined as average gene significance (1.24). Figure 5.4c shows that the brown module had the highest module significance.

By relating the gene significance measure GS_i to the scaled (intramodular) connectivity K_i, we arrive at a hub gene significance measure (1.23). As illustrated in Figs. 5.4d and e, the hub gene significance is defined as the slope of a regression model without intercept term. The brown module had the highest hub gene significance (Table 6.2). It turns out that the brown module is highly enriched with cell cycle and mitosis genes (Horvath et al. 2006). It is biologically sensible that high expression levels of cell cycle genes (reflecting a proliferating cancer) are negatively related to cancer survival time.

The fact that genes with high intramodular connectivity in the brown module are more likely to be prognostically significant facilitates a novel systems biologic

gene screening strategy for finding prognostic genes: Screen for genes that are (a) intramodular hub genes in the brown module and (b) that are significantly correlated with survival time. We should mention that it is essential to take a module centric view: one should only consider survival time-related modules since hub genes in other modules may have no prognostic value.

Incorporating intramodular connectivity (which reflects module membership) in gene-screening criteria facilitates systems biologic gene-screening procedures that may increase validation success. For example, we found that 68% of survival time-related intramodular hubs validated (p value smaller 0.05) in a second independent data set. A standard gene-screening approach (described in Sect. 10.10), which selects genes according to their (marginal) relationship with survival time, fails to detect many of the mitosis module genes. Only 26% of the top 300 most significant survival time-related genes validate in the second data set. In this application, considering intramodular connectivity facilitated a biologically superior gene-screening approach that allowed us to identify novel mitosis genes related to brain cancer survival (Horvath et al. 2006). The superior performance of module-based screening is no surprise in this application since the brown module was "preserved" (i.e., reproducible) in the second brain cancer data set. Obviously, intramodular connectivity should not be used as gene screening variable if the module is not preserved in the second data set. Techniques for measuring module preservation are described in Chap. 9. Methods for defining consensus modules, i.e., modules present in multiple data sets, are described in Sect. 7.11.1.

5.8 R Code for Studying the Effect of Thresholding

Here we provide R code for studying the dependence of network concepts on the soft threshold β and the hard threshold τ, i.e., the parameters of the adjacency functions AF^{power} and $AF^{threshold}$, respectively. Toward this end, we use the brain cancer gene expression data described in Sect. 5.7.

The `WGCNA` function `pickHardThreshold` was used to calculate Table 5.2, which reports the results for varying the hard threshold parameter τ for the brain cancer co-expression network (Sect. 5.7). Hard thresholding allows us to attach a significance level (p values) to each threshold (e.g., using the Fisher transformation of the correlation coefficient as described in Sect. 10.3). In our opinion, it would be difficult to use the p values to argue that a parameter value of $\tau = 0.70$ ($p = 1.9 * 10^{-9}$) is superior to $\tau = 0.50$ ($p = 8.7 * 10^{-5}$) since both are highly significant. However, the scale-free topology fitting index R^2 clearly favors $\tau \geq 0.65$. For example, $\tau = 0.65$ leads to a scale-free fitting index of $R^2 = 0.95$ and $\tau = 0.70$ leads to $R^2 = 0.97$, while $\tau = 0.50$ has a comparably low fitting index of $R^2 = 0.79$. In this application, we would choose $\tau \geq 0.65$ since this is where the R^2 curve starts to reach its "saturation" point (Fig. 5.6). It leads to a good scale-free topology fit and a reasonably high density.

Table 5.2 Brain cancer network characteristics for different hard thresholds τ

τ	p value	Signed scalefree R^2	Slope	Signed truncated exp. R^2	Mean(k)	Median(k)	*max*(k)
0.20	1.4e-01	−0.68	1.79	−0.95	3530	3580	5520
0.30	2.5e-02	0.12	0.07	−0.95	1960	1890	4200
0.40	2.3e-03	0.60	−0.84	0.94	947	787	2940
0.50	8.7e-05	0.79	−1.21	0.90	395	232	1860
0.55	1.1e-05	0.85	−1.23	0.90	243	110	1410
0.60	1.0e-06	0.84	−1.24	0.87	145	43	1080
0.65	5.9e-08	0.95	−1.11	0.95	85.9	14	795
0.70	1.9e-09	0.97	−1.05	0.97	50.2	4	616
0.75	2.9e-11	0.98	−1.01	0.98	28.9	1	480
0.80	1.4e-13	0.95	−1.03	0.94	15.7	0	383
0.85	2.2e-16	0.97	−1.03	0.97	7.2	0	262
0.90	<1e-22	0.98	−1.09	0.98	2.2	0	154
0.95	<1e-22	0.93	−1.26	0.97	0.2	0	47

The asymptotic p value for τ were calculated using the Fisher transform of the correlation coefficient. The sign of the scale-free model fitting index R^2 is determined by minus the sign of the slope. We also report the slope of the regression line used for assessing scale-free fit and the fitting index for the exponentially truncated power law (1.14). This table was generated with the `WGCNA` function command `pickHardThreshold`.

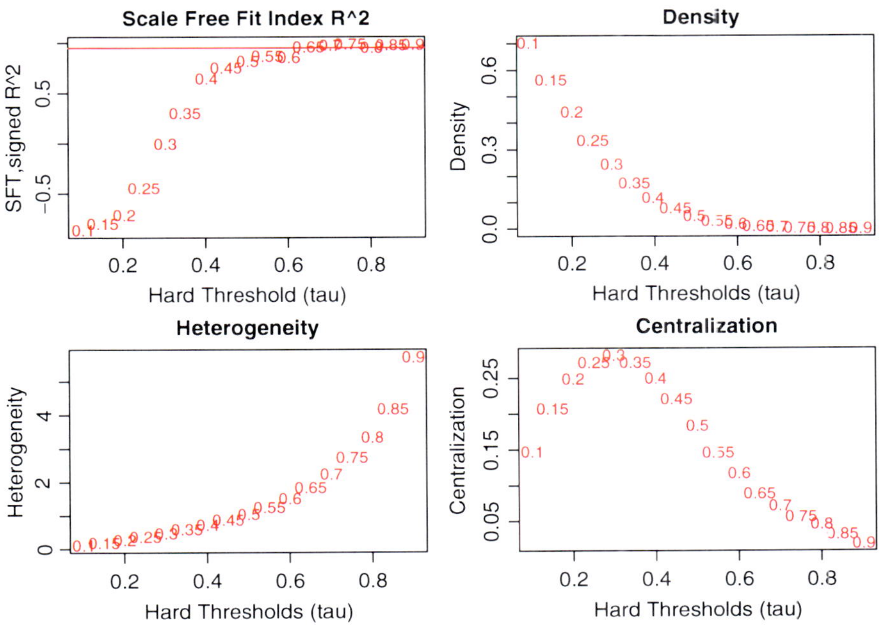

Fig. 5.6 Network concepts (y-axis) versus different values of τ used in the threshold adjacency function (x-axis). The data are an unsigned correlation networks based on 8,000 most varying genes from a gene co-expression network based on 55 brain cancer samples (Horvath et al. 2006). Note that the scale-free topology fitting index R^2 and the heterogeneity tend to increase with β, while the density and the network centralization tend to decrease with β

Figure 5.5 shows the dependence of some network concepts on the soft threshold β. The data are available from our webpage www.genetics.ucla.edu/labs/horvath/CoexpressionNetwork/Book. The following R code was used to create Fig. 5.5.

```
powers1=seq(from=1,to=15,by=1)
pst=pickSoftThreshold(datExprdataOne,powerVector=powers1,
moreNetworkConcepts=T)[[2]]
cex1=1.2;cex.axis=1.5;cex.lab=1.5;cex.main=1.5
par(mfrow=c(2,2),mar=c(4,4,3,2)+0.1)
plot(powers1,-sign(pst[,3])*pst[,2],type="n",
   xlab="Soft Threshold",
ylab="SFT,signed R^2",cex.axis=cex.axis,cex.main=cex.main,
cex.lab=cex.lab,main="Scale Free Fit Index R^2")
text(powers1,-sign(pst[,3])*pst[,2],
labels=powers1,cex=cex1,col="red")
# this line corresponds to using an R^2 cut-off of h
abline(h=0.95,col="red")
plot(powers1,pst$Density,
type="n",xlab="Soft Threshold",ylab="Density",
cex.axis=cex.axis,cex.main=cex.main,cex.lab=cex.lab,
    main="Density")
text(powers1,pst$Density,labels=powers1,cex=cex1,col="red")
plot(powers1,pst$Heterogeneity,type="n",xlab="Soft Threshold",
ylab="Heterogeneity",cex.main=cex.main,cex.lab=cex.lab,
cex.axis=cex.axis,main="Heterogeneity")
text(powers1,pst$Heterogeneity,labels=powers1,cex=cex1,col="red")
plot(powers1,pst$Centralization,type="n",xlab="Soft Threshold",
ylab="Centralization",cex.axis=cex.axis,cex.main=cex.main,
cex.lab=cex.lab,main="Centralization")
text(powers1,pst$Centralization,labels=powers1,cex=cex1,
   col="red")
```

The relationship between network concepts and different hard thresholds τ is shown in Fig. 5.6

5.9 Gene Network (Re-)Construction Methods

Correlation networks have become the benchmark approach for defining networks between numeric variables, but many alternative approaches for constructing networks based on gene expression data or other "omics" data have been proposed in the literature (reviewed in (van Someren et al. 2002; Margolin et al. 2006)). Since many network construction methods aim to recover a "true" underlying network, these methods are also known as **network reconstruction, reverse engineering (Basso et al. 2005), inference,** and **deconvolution** methods. Most gene network reconstruction methods lead to an unweighted (possibly nonsymmetrical) network adjacency matrix which can be represented by a graph whose nodes correspond to genes and edges encode direct gene regulatory interactions. For example, an edge could represent the interaction between a transcription factor and its target gene.

There is no clear agreement in the literature on how to construct *realistic* gene interaction networks. It may even be impossible to search for a perfect construction method since a method that works well for one set of data may not work well for another. To provide maximum flexibility to the data analysis, this book describes a host of methods and tools for constructing networks, e.g.,

- Chapter 7 shows how to construct networks based on input matrices
- Chapter 13 shows how to construct networks using prediction and regression models
- Chapter 14 describes association networks and mutual information networks among categorical variables
- Chapter 15 describes general association networks

Often seemingly different network construction turn out to be highly related. Methods for relating different types of networks and their construction methods are described in Sects. 10.15, 10.16, 4.6, and elsewhere.

An edge (or adjacency measure) between two genes can have very different meanings in different contexts. Edges in association networks (Butte et al. 2000; Butte and Kohane 2000; D'haeseleer et al. 2000; Perkins et al. 2005; Barrett and Palsson 2006; Smith 2006; Thakar et al. 2007; Price et al. 2007; Needham et al. 2007) simply reflect significant statistical associations between gene measurements. Thus, edges between genes often do *not* represent physical interactions between them. Many factors hamper the construction of physical interaction networks based on gene expression data. The lack of time series gene expression data precludes the inference of temporal associations and thus plausible causal relationships (reviewed in Wiggins and Nemenman 2003). Instead, gene expression measurements are typically based on cell populations collected from different individuals. The resulting gene expression data measure random steady states of the underlying biochemical dynamics. We caution the reader that steady-state statistical dependencies are not obviously linked to the underlying physical interaction model.

5.10 R Code

An R function for computing the p-norm (5.2) is defined as follows:

```
p.Norm=function(x,p){sum(abs(x)^p)^{1/p}}
```

As an aside, we mention that the R command `pnorm` has *no* relationship with the p-norm.

If x or y contain missing values, one can compute the correlation across pairwise-complete observations with the R command `cor(x,y,use="p')`. To calculate a matrix of correlations between the columns of a matrix (or data frame) $datE$, one can use the following R command:

```
MatrixCorrelations=cor(datE,use="p")
```

To speed up the calculation of correlations in R, Peter Langfelder implemented the `WGCNA` functions `corFast` and `cor1`.

To compute the ranks of a vector x, one can use the R command `rank(x)`. If some of the components of x are equal to each other (i.e., if there are ties), the user can specify different tie breaking methods using the option `ties.method` in the R function `rank`. Here, we will use the default setting `ties.method="average"`, which replaces the corresponding indices by their average. For example, the R command `rank(c(1,1,1,50,60))` yields the following vector: $(2,2,2,4,5)$.

The WGCNA function `bicor` implements a version of the biweight midcorrelation (see Sect. 5.1.3), which we have found to work very well for gene expression data.

In the following, we illustrate our theoretical derivations using the yeast gene co-expression network. This network is different from the yeast protein–protein interaction network mentioned in other sections. This microrray dataset recorded gene expression levels during different stages of cell cycles in yeasts (Spellman et al. 1998). We focused on the 4,000 most varying genes across 44 microarrays (Spellman et al. 1998). A detailed biological description of the weighted gene co-expression network (including a functional annotation of its modules) can be found in Carlson et al. (2006). The yeast gene co-expression network analysis is described in Zhang and Horvath (2005) and Carlson et al. (2006), and the data and a related R software tutorial can be found at the following webpage: www.genetics.ucla.edu/labs/horvath/GeneralFramework/.

In the following, we present R code for evaluating the scale-free topology fit. Figure 5.7 can be created with the following R code:

```
# power for the power adjacency function
beta1=7
Connectivity.soft=softConnectivity(datExpr,power=beta1)
# hard threshold for the threshold adjacency function
tau1=0.65
ADJ.hard=abs(cor(datExpr,use="p"))>0.65+0.0
diag(ADJ.hard)=0
Connectivity.hard= apply(ADJ.hard,2,sum)
# Let's create a scale free topology plot where
# the black curve corresponds to scale free topology and
# the red curve corresponds to the truncated exponential
    topology.
par(mfrow=c(2,1),mar=c(4,4,2,2))
scaleFreePlot(Connectivity.soft,
main=paste("AF.power, beta=",beta1),truncated=T)
scaleFreePlot(Connectivity.hard,
main=paste("AF.threshold, tau=",tau1),truncated=T)
#calculate several scale free topology fitting indices
scaleFreeFitIndex(Connectivity.soft)
```

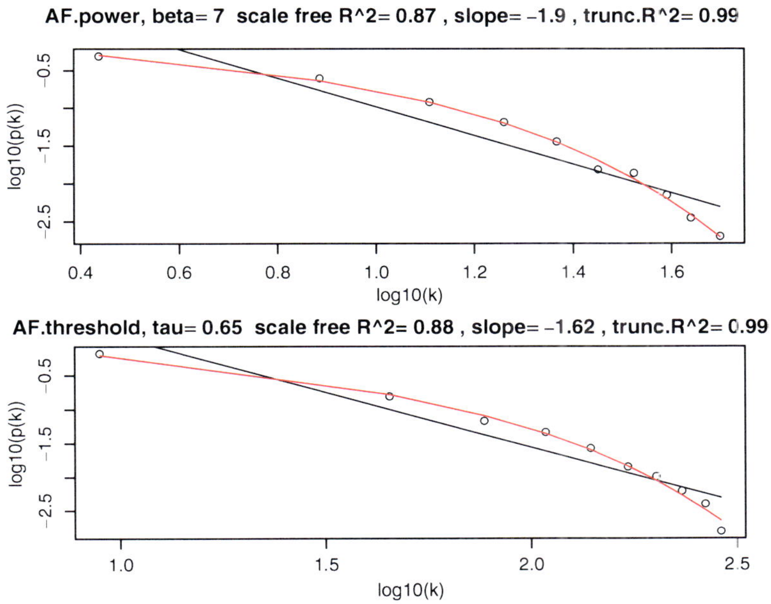

Fig. 5.7 Studying the scale-free topology of the yeast co-expression network. The *upper* and *lower panels* describe the scale-free fit of a weighted (soft threshold $\beta = 7$) and an unweighted (hard threshold $\tau = 0.65$) correlation network, respectively. This scatter plot shows the relationship between $log(p(r))$ (y-axis) and $log(r)$ (x-axis). The *black line* shows the model fit for scale-free topology and the *red line* shows the model fit an exponentially truncated power law

5.11 Exercises

1. Exercise regarding basic R commands.

   (i) Simulate a vector x of 100 numbers from a normal distribution with mean 3 and variance 5. Hint:

   ```
   # set the seed of random number generator
   # to ensure reproducibility of simulations
   set.seed(1)
   # simulate vector
   x=rnorm(100,mean=3,sd=sqrt(5))
   ```

   (ii) Compare the results of the following R commands

   ```
   var(x)
   1/(m-1)*sum((x-mean(x))^2)
   ```

   (iii) Simulate another vector y of 100 numbers from a normal distribution with mean 0 and variance 1. Verify that the covariance between two vectors x and y (of the same length) can be computed using the following R commands:

   ```
   cov(x,y)
   sum((x-mean(x))*(y-mean(y)))/(length(x)-1)
   ```

(iv) Verify that the correlation $cor(x,y)$ between two vectors can be computed using the following R commands

```
cor(x,y)
cor.test(x,y)
1/(length(x)-1)*sum(scale(x)*scale(y))
```

2. Exercise. Prove that

$$||scale(x) - scale(y)||^2 = 2(m-1)(1 - cor(x,y)).$$

Hint: The following equations may be helpful $||vector||^2 = \langle vector, vector \rangle$, $\langle v - w, v - w \rangle = \langle v, v \rangle - 2\langle v, w \rangle + \langle w, w \rangle$, and $\langle scale(x), scale(y) \rangle = (m-1)\, cor(x,y)$.

3. Exercise regarding the relationship between variance and Euclidean norm.

(i) Show that two vectors x and y with mean 0 (i.e., $mean(x) = mean(y) = 0$) have the same Euclidean length if and only if they have the same sample variance. Hint: Prove $var(x) = \frac{||x - mean(x)||^2}{m-1}$.

(ii) Argue that a set of scaled vectors $\{scale(x_1), scale(x_2), \ldots, scale(x_n)\}$ can be visualized as n points on a hyper-sphere in $m = length(x_i)$-dimensional space. What is the radius of the sphere? Hint: What is the Euclidean norm (2-norm) of $scale(x_1)$?

4. Invariance of the discretize function.

(i) Show that the discretize function is invariant with respect to linear transformations. Hint: Show $discretize(x, no.bins) = discretize(slope * x + intercept, no.bins)$ where $slope \neq 0$ and $intercept$ denote (constant) numbers. Hint: Either derive it mathematically or compare the result of the following R commands:

```
library(infotheo)
x=rnorm(100)
dxA=discretize(5*x+3,nbins=10,disc="equalwidth")[,1]
dxB=discretize(x,nbins=10,disc="equalwidth")[,1]
table(dxA,dxB)
```

(ii) Show that $discretize(x)$ is in general different from $discretize(x^2)$.

5. Exercise. Prove that the network $A_{ij}^{method1} = 1/(2 - |cor(x_i, x_j)|)$ is rank-equivalent to $A_{ij}^{method2} = |cor(x_i, x_j)|^6$.

References

Albert R, Barabasi AL (2002) Statistical mechanics of complex networks. Rev Mod Phys 74:47–97

Albert R, Jeong H, Barabasi AL (2000) Error and attack tolerance of complex networks. Nature 406(6794):378–382

Barrett CL, Palsson BO (2006) Iterative reconstruction of transcriptional regulatory networks: An algorithmic approach. PLoS Comput Biol 2(5):e52

Basso K, Margolin AA, Stolovitzky G, Klein U, Dalla-Favera R, Califano A (2005) Reverse engineering of regulatory networks in human B cells. Nat Genet 37(4):382–390

Butte AJ, Kohane IS (2000) Mutual information relevance networks: Functional genomic clustering using pairwise entropy measurments. Pac Symp Biocomput 5:418–429

Butte A, Tamayo P, Slonim D, Golub T, Kohane I (2000) Discovering functional relationships between RNA expression and chemotherapeutic susceptibility using relevance networks. Proc Natl Acad Sci USA 97:12182–12186

Cabusora L, Sutton E, Fulmer A, Forst CV (2005) Differential network expression during drug and stress response. Bioinformatics 21(12):2898–2905

Carlson M, Zhang B, Fang Z, Mischel P, Horvath S, Nelson SF (2006) Gene connectivity, function, and sequence conservation: Predictions from modular yeast co-expression networks. BMC Genomics 7(7):40

Carter SL, Brechbuler CM, Griffin M, Bond AT (2004) Gene co-expression network topology provides a framework for molecular characterization of cellular state. Bioinformatics 20(14):2242–2250

Cokus S, Rose S, Haynor D, GronbechJensen N, Pellegrini M (2006) Modelling the network of cell cycle transcription factors in the yeast Saccharomyces cerevisiae. BMC Bioinform 7:381

D'haeseleer P, Liang S, Somogyi R (2000) Genetic network inference: From co-expression clustering to reverse engineering. Bioinformatics 16(8):707–726

Dong J, Horvath S (2007) Understanding network concepts in modules. BMC Syst Biol 1(1):24

Eisen MB, Spellman PT, Brown PO, Botstein D (1998) Cluster analysis and display of genome-wide expression patterns. Proc Natl Acad Sci USA 95(25):14863–14868

Fuller TF, Ghazalpour A, Aten JE, Drake T, Lusis AJ, Horvath S (2007) Weighted gene co-expression network analysis strategies applied to mouse weight. Mamm Genome 18(6–7): 463–472

Gargalovic PS, Imura M, Zhang B, Gharavi NM, Clark MJ, Pagnon J, Yang WP, He A, Truong A, Patel S, Nelson SF, Horvath S, Berliner JA, Kirchgessner TG, Lusis AJ (2006) Identification of inflammatory gene modules based on variations of human endothelial cell responses to oxidized lipids. Proc Natl Acad Sci USA 103(34):12741–12746

Ghazalpour A, Doss S, Zhang B, Plaisier C, Wang S, Schadt EE, Thomas A, Drake TA, Lusis AJ, Horvath S (2006) Integrating genetics and network analysis to characterize genes related to mouse weight. PloS Genet 2(2):8

Han JD, Bertin N, Hao T, Goldberg DS, Berriz GF, Zhang LV, Dupuy D, Walhout AJ, Cusick ME, Roth FP, Vidal M (2004) Evidence for dynamically organized modularity in the yeast protein-protein interaction network. Nature 430(6995):88–93

Hardin J, Mitani A, Hicks L, VanKoten B (2007) A robust measure of correlation between two genes on a microarray. BMC Bioinformatics 8(1):220

Harrell F (2001) Regression modeling strategies, corrected edition. Springer, New York

Horvath S, Dong J (2008) Geometric interpretation of gene co-expression network analysis. PLoS Comput Biol 4(8):e1000117

Horvath S, Zhang B, Carlson M, Lu KV, Zhu S, Felciano RM, Laurance MF, Zhao W, Shu Q, Lee Y, Scheck AC, Liau LM, Wu H, Geschwind DH, Febbo PG, Kornblum HI, Cloughesy TF, Nelson SF, Mischel PS (2006) Analysis of oncogenic signaling networks in glioblastoma identifies ASPM as a novel molecular target. Proc Natl Acad Sci USA 103(46):17402–17407

Huang Y, Li H, Hu H, Yan X, Waterman MS, Huang H, Zhou XJ (2007) Systematic discovery of functional modules and context-specific functional annotation of human genome. Bioinformatics 23(13):i222–i229

Jeong H, Mason SP, Barabasi AL, Oltvai ZN (2001) Lethality and centrality in protein networks. Nature 411:41

Jordan IK, MarinoRamirez L, Wolf YI, Koonin EV (2004) Conservation and coevolution in the scale-free human gene coexpression network. Mol Biol Evol 21(11):2058–2070

Keller MP, Choi YJ, Wang P, Belt Davis D, Rabaglia ME, Oler AT, Stapleton DS, Argmann C, Schueler KL, Edwards S, Steinberg HA, Chaibub Neto E, Kleinhanz R, Turner S, Hellerstein MK, Schadt EE, Yandell BS, Kendziorski C, Attie AD (2008) A gene expression network model of type 2 diabetes links cell cycle regulation in islets with diabetes susceptibility. Genome Res 18(5):706–716

Langfelder P, Horvath S (2011) Fast R functions for robust correlations and hierarchical clustering. J Stat Software. In press

Lim WK, Wang K, Lefebvre C, Califano A (2007) Comparative analysis of microarray normalization procedures: Effects on reverse engineering gene networks. Bioinformatics 23(13): i282–i288

Margolin AA, Nemenman I, Basso K, Wiggins C, Stolovitzky G, Favera RD, Califano A (2006) ARACNE: An algorithm for the reconstruction of gene regulatory networks in a mammalian cellular context. BMC Bioinform 7(Suppl. 1):S7

Mason M, Fan G, Plath K, Zhou Q, Horvath S (2009) Signed weighted gene co-expression network analysis of transcriptional regulation in murine embryonic stem cells. BMC Genomics 10(1):327

Miller JA, Horvath S, Geschwind DH (2010) Divergence of human and mouse brain transcriptome highlights Alzheimer disease pathways. Proc Natl Acad Sci USA 107(28):12698–12703

Mumford JA, Horvath S, Oldham MC, Langfelder P, Geschwind DH, Poldrack RA (2010) Detecting network modules in fMRI time series: A weighted network analysis approach. NeuroImage 52(4):1465–1476

Needham CJ, Bradford JR, Bulpitt AJ, Westhead DR (2007) A primer on learning in bayesian networks for computational biology. PLoS Comput Biol 3(8):e129

Oldham MC, Horvath S, Geschwind DH (2006) Conservation and evolution of gene coexpression networks in human and chimpanzee brains. Proc Natl Acad Sci USA 103(47):17973–17978

Oldham MC, Konopka G, Iwamoto K, Langfelder P, Kato T, Horvath S, Geschwind DH (2008) Functional organization of the transcriptome in human brain. Nat Neurosci 11(11):1271–1282

Perkins TJ, Jaeger J, Reinitz J, Glass L (2005) Reverse engineering the gap gene network of Drosophila melanogaster. PLoS Comput Biol 2(5):e51

Price MN, Dehal PS, Arkin AP (2007) Orthologous transcription factors in bacteria have different functions and regulate different genes. PLoS Comput Biol 3(9):e175

Shieh G, Chen CM, Yu CY, Huang J, Wang WF, Lo YC (2008) Inferring transcriptional compensation interactions in yeast via stepwise structure equation modeling. BMC Bioinform 9(1):134

Smith GD (2006) Randomized by (your) god: Robust inference from an observational study design. J Epidemiol Community Health 60:382–388

Snel B, van Noort V, Huynen MA (2004) Gene co-regulation is highly conserved in the evolution of eukaryotes and prokaryotes. Nucleic Acids Res 32(16):4725–4731

van Someren EP, Wessels LF, Backer E, Reinders MJ (2002) Genetic network modeling. Pharmacogenomics 3(4):507–525

Spellman PT, Sherlock G, Zhang MQ, Iyer VR, Anders K, Eisen MB, Brown PO, Botstein D, Futcher B (1998) Comprehensive identification of cell cycle-regulated genes of the yeast Saccharomyces cerevisiae by microarray hybridization. Mol Biol Cell 9(12):3273–3297

Steffen M, Petti A, Aach J, D'haeseleer P, Church G (2002) Automated modelling of signal transduction networks. BMC Bioinform 3(1):34

Stuart JM, Segal E, Koller D, Kim SK (2003) A gene-coexpression network for global discovery of conserved genetic modules. Science 302(5643):249–255

Swindell W (2007) Gene expression profiling of long-lived dwarf mice: Longevity-associated genes and relationships with diet, gender and aging. BMC Genomics 8(1):353

Thakar J, Pilione M, Kirimanjeswara G, Harvill ET, Albert R (2007) Modeling systems-level regulation of host immune responses. PLoS Comput Biol 3(6):e109

Voy BH, Scharff JA, Perkins AD, Saxton AM, Borate B, Chesler EJ, Branstetter LK, Langston MA (2006) Extracting gene networks for low-dose radiation using graph theoretical algorithms. PLoS Comput Biol 2(7):e89

Wang J, Zhang S, Wang Y, Chen L, Zhang XS (2009) Disease-aging network reveals significant roles of aging genes in connecting genetic diseases. PLoS Comput Biol 5(9):e1000521

Wang S, Yehya N, Schadt EE, Drake TA, Lusis AJ (2006) Genetic and genomic analysis of fat mass trait with complex inheritance reveals marked sex specificity. PLoS Genet 2(2):e15

Wei H, Persson S, Mehta T, Srinivasasainagendra V, Chen L, Page GP, Somerville C, Loraine A (2006) Transcriptional coordination of the metabolic network in arabidopsis. Plant Physiol 142(2):762–774

Weston D, Gunter L, Rogers A, Wullschleger S (2008) Connecting genes, coexpression modules, and molecular signatures to environmental stress phenotypes in plants. BMC Syst Biol 2(1):16

Wiggins C, Nemenman I (2003) Process pathway inference via time series analysis. Exp Mech 43(3):361–370

Wilcox RR (1997) Introduction to robust estimation and hypothesis testing. Academic, San Diego, CA

Zhang B, Horvath S (2005) General framework for weighted gene coexpression analysis. Stat Appl Genet Mol Biol 4:17

Zhou X, Kao MJ, Wong WH (2002) Transitive functional annotation by shortest path analysis of gene expression data. Proc Natl Acad Sci USA 99(20):12783–12788

Chapter 6
Geometric Interpretation of Correlation Networks Using the Singular Value Decomposition

Abstract The nodes of a correlation network correspond to the columns of a numeric data matrix *datX*. Based on the singular value decomposition (SVD) of *datX*, we are able to characterize approximately factorizable correlation networks, i.e., adjacency matrices that factor into node-specific contributions. The SVD yields singular vectors that have important practical applications. For example, the first left singular vector (referred to as the eigenvector) explains the maximum amount of variation of the columns of *datX*. The eigenvector is also known as module eigengene in the context of a gene co-expression network module. Right singular vectors can be used for signal balancing, e.g., to remove batch effects and other technical artifacts. Based on the eigenvector (the first left singular vector), we define a new type of network concept, referred to as eigenvector-based network concept. Eigenvector-based concepts are analogous to approximate conformity-based network concepts but have a major advantage: they often allow for a geometric interpretation based on the angular interpretation of correlations. The underlying structure of correlation networks affects network analysis results. For example, there are geometric reasons why intramodular hub nodes in important modules tend to be important, and why hub nodes in one module cannot be hubs in another distinct module. The hub node significance of a module can be interpreted as angle between a sample trait and the module eigengene. Since the intramodular connectivity kIM_i is highly related to the module membership measure kME_i, it can be interpreted as angle between x_i and the module eigenvector ME. A short dictionary for translating between data mining- and network theory language may facilitate the communication between the two fields. Mouse and brain gene co-expression network applications are used to illustrate the results. This work reviews and extends work with **Jun Dong** (Horvath and Dong PLoS Comput Biol 4(8):e1000117, 2008).

6.1 Singular Value Decomposition of a Matrix *datX*

Assume an $m \times n$ dimensional matrix *datX* whose ith column x_i is a numeric vector with m components. Before carrying out the singular value decomposition, we typically scale the columns of *datX* so that they have mean zero and variance 1,

S. Horvath, *Weighted Network Analysis: Applications in Genomics and Systems Biology*, DOI 10.1007/978-1-4419-8819-5_6, © Springer Science+Business Media, LLC 2011

i.e., $datX = scale(datX)$ (5.1). The **singular value decomposition (SVD)** of *datX* is given by the matrix multiplication of three matrices:

$$datX = UD(V)^{\tau}, \tag{6.1}$$

where $^{\tau}$ denotes the transpose. The $m \times min(m,n)$ dimensional matrix U and the $n \times min(m,n)$ dimensional matrix V contain orthonormal columns, i.e., they satisfy:

$$(U)^{\tau}U = I(min(m,n)), \tag{6.2}$$

$$(V)^{\tau}V = I(min(m,n)), \tag{6.3}$$

where $I(min(m,n))$ denotes the $min(m,n) \times min(m,n)$ dimensional identity matrix, i.e., a matrix whose diagonal elements equal 1 and whose off-diagonal elements equal 0. The columns of the matrices U and V are referred to as **left** and **right singular vectors**, respectively. The matrices U and D are given by

$$\begin{aligned} U &= \left(u_1 u_2 \ldots u_{min(m,n)}\right) \\ D &= diag\left\{|d_1|, |d_2|, \ldots, |d_{min(m,n)}|\right\}. \end{aligned} \tag{6.4}$$

The **singular values** are defined as diagonal elements $|d_1|, |d_2|, \ldots$ of the $min(m,n) \times min(m,n)$ dimensional diagonal matrix D. We use the absolute value sign around the singular values to remind us that the singular values are *nonnegative* real numbers. In the following, we assume that the singular values are arranged in a decreasing fashion, i.e., $|d_1| \geq |d_2| \geq |d_3|, \ldots$.

In general, the singular value decomposition of a matrix is not unique. A singular value is called degenerate if it corresponds to two independent left (or right) singular vectors. If *datX* has degenerate singular values, then its singular value decomposition is not unique. Nondegenerate singular values always have unique left and right singular vectors, up to multiplication by a sign. Consequently, if all singular values of *datX* are nondegenerate and nonzero, then its singular value decomposition is unique, up to multiplication of a column of U by a sign and simultaneous multiplication of the corresponding column of V by the same sign. Often the first singular value $|d_1|$ is nondegenerate since it is strictly larger than the other singular values. In this case, u_1 and v_1 are uniquely defined up to a sign. In practice, we fix the sign of u_1 by requiring that its average correlation with the columns of *datX* is positive.

6.1.1 Signal Balancing Based on Right Singular Vectors

The matrix V (6.1) of right singular vectors has been used to "balance" the signal encoded in a numeric $m \times n$ dimensional matrix *datX*. This idea has been used for preprocessing gene expression data (Hibbs et al. 2007). The "signal balanced" matrix is defined as the transpose of the matrix of right singular vectors, i.e.,

$$datX.SignalBalanced = V^{\tau}, \tag{6.5}$$

which like *datX* is an $m \times n$ dimensional matrix. Denote by D^- the (pseudo) inverse of the matrix of singular values. Then $datX.SignalBalanced = D^- U^\tau datX$ can be interpreted as the "balanced" projection of *datX* onto its left singular basis, where the balancing weights are inversely proportional to the singular values (Hibbs et al. 2007). Correlations between columns of *datX.SignalBalanced* equally weight each dimension of the orthonormal basis U and balance their contributions such that the least prominent patterns are amplified and more dominant patterns are dampened. Signal balancing can be a useful preprocessing method if dominant patterns (corresponding to the columns of U) represent noise variation (Hibbs et al. 2007). While signal balancing has been found useful in some gene expression applications, we generally advise against it since we have found that it often removes meaningful signal from the data.

6.1.2 Eigenvectors, Eigengenes, and Left Singular Vectors

In the following, we describe an important use of the left singular vectors of the singular value decomposition (6.1). The sign of the first left singular vector u_1 is fixed by requiring a positive average correlation with the columns of *datX*. When *datX* corresponds to the gene expression data underlying a co-expression network module, then u_1 (the first column of U) is referred to as the **module eigengene**. More generally, we refer to the first left singular vector

$$E = u_1 \tag{6.6}$$

as the **eigenvector** or the **eigennode** in the context of a correlation network.

Using (6.21), one can show that E is an eigenvector of the $m \times m$ dimensional matrix $datX(datX)^\tau/m$ corresponding to the largest eigenvalue. The eigenvector E is an optimal way of summarizing the scaled columns of *datX*. More precisely, it explains the maximum amount of the variation in the scaled columns. The **proportion of variance explained** by the eigenvector E is given by

$$propVarExpl = \frac{|d_1|^2}{\sum_{l=1}^{min(m,n)} |d_l|^2}. \tag{6.7}$$

Note that $0 \leq propVarExpl \leq 1$. The singular value decomposition of *datX* (whose columns are scaled) is related to the principal component analysis of the $n \times n$ dimensional correlation matrix $cor(datX) = (cor(x_i, x_j))$ whose entries correspond to the pairwise correlations between the columns of *datX*. Since the diagonal elements of $cor(datX)$ equal 1, the trace of $cor(datX)$ (i.e., the sum of the diagonal elements) is given by

$$trace(cor(datX)) = \sum_{i=1}^{n} 1 = n. \tag{6.8}$$

A well-known result from linear algebra is that the trace of a symmetric matrix equals the sum of its eigenvalues. Since the eigenvalues of $cor(datX)$ are given by $|d_l|^2/m$ (6.21), we find that

$$trace(cor(datX)) = \sum_{l=1}^{min(m,n)} |d_l|^2/m. \tag{6.9}$$

By combining (6.8) with (6.9), we find that $n = \sum_{l=1}^{min(m,n)} |d_l|^2/m$, which implies that $propVarExpl$ (6.7) is given by

$$propVarExpl = \frac{|d_1|^2}{mn}. \tag{6.10}$$

We find it useful to define the following measure of **eigenvector factorizability** (Horvath and Dong 2008):

$$EF(E) = \frac{|d_1|^4}{\sum_j |d_j|^4}. \tag{6.11}$$

Note that $0 \leq EF(E) \leq 1$ and the close resemblance to the proportion of variance explained by the module eigenvector (6.7). In Sect. 6.2, we argue that $EF(E) \approx 1$ implies that the correlation matrix factors as follows:

$$cor(x_i, x_j) \approx cor(x_i, E)\, cor(x_j, E).$$

6.2 Characterizing Approx. Factorizable Correlation Networks

We define the **eigenvector conformity**:

$$\text{kME}_i = cor(x_i, E) = v_{1,i}\frac{|d_1|}{\sqrt{m}}. \tag{6.12}$$

When E represents the eigengene of a co-expression module, then kME is also known as **module eigengene-based connectivity** or **module membership measure** (Ghazalpour et al. 2006; Fuller et al. 2007; Oldham et al. 2008). For a weighted correlation network, we define the **weighted eigenvector conformity** as:

$$A_{e,i} = |cor(x_i, E)|^\beta = |\text{kME}_i|^\beta, \tag{6.13}$$

where the power $\beta > 0$ is the soft threshold used in the weighted network construction. Since the eigenvector E summarizes the overall behavior of $datX$, the eigenvector conformity $|cor(x_i, E)|^\beta$ measures how well node i conforms to the overall module. This insight led us to coin the term "conformity" (Horvath and Dong 2008). If a sample trait $t = (t_1, \dots, t_m)$ is available, the weighted trait-based **eigenvector significance** is defined as follows:

$$A_{e,t} = |cor(t, E)|^{\beta}, \tag{6.14}$$

which is also denoted as *EigenvectorSignif*.

The i, jth element of the **eigenvector-based adjacency matrix** A_E is defined as:

$$A_{E,ij} = A_{e,i} A_{e,j}. \tag{6.15}$$

Note that ith diagonal element of A_E is given by $A_{E,ii} = (A_{e,i})^2$ and $A_E = A_e (A_e)^{\tau}$ is a rank 1 matrix.

Further, we define **eigenvector-based node significance measure**

$$GS_{E,i} = A_{e,i} A_{e,t}. \tag{6.16}$$

Thus, $GS_{E,i}$ equals the weighted eigenvector conformity times the eigenvector significance. We will derive the following:

Observation 6.1 (Characterizing approx. factorizable correlation networks). *If the eigenvector factorizability $EF(E)$ (6.11) is close to 1, the adjacencies of the weighted correlation network $A = |cor(datX)|^{\beta}$ and the trait-based node significance measure $GS_i = |cor(x_i, t)|^{\beta}$ can be factored as follows:*

$$\begin{aligned} A_{ij} &\approx A_{e,i} A_{e,j} = A_{E,ij}, \\ GS_i &\approx A_{e,i} A_{e,t} = GS_{E,i}, \end{aligned} \tag{6.17}$$

where $A_{e,i}$ (6.13) is the weighted eigenvector conformity, $A_{e,t}$ (6.14) is the eigenvector significance.

Observation 6.1 allows us to characterize weighted correlation networks for which the adjacency $A_{i,j}$ can be approximated by a product of the centroid conformities (1.26): $A_{i,j} \approx CentroidConformity_i CentroidConformity_j$. In our applications, we find that modules in co-expression networks are often factorizable (Dong and Horvath 2007; Horvath and Dong 2008). Observation 6.1 can be used to argue that modules comprised of correlated genes tend to be factorizable.

In the following, we outline a theoretical derivation for Observation 6.1. This technical part can be skipped at first reading. As described in (5.10), the correlation between two vectors x_i and x_j is given by

$$cor(x_i, x_j) = \frac{scale(x_i)^T scale(x_j)}{||scale(x_i)|| * ||scale(x_j)||}.$$

Recall that we assume that the vectors have been scaled as follows:

$$mean\,(x_i) = 0$$

$$\frac{||x_i||^2}{m} = \frac{sum\left((x_i)^2\right)}{m} = \frac{(m-1)}{m} * var\,(x_i) = 1. \tag{6.18}$$

This implies that $scale(x_i) = (x - mean(x_i))/\sqrt{var(x_i)}$ is given by $scale(x_i) = x_i * \sqrt{\frac{m-1}{m}}$. Under our scaling assumptions (6.18), one can show that $||scale(x_i)|| = \sqrt{m-1}$ and

$$cor\,(x_i, x_j) = \frac{x_i^T x_j}{m}. \tag{6.19}$$

The singular value decomposition of *datX* (6.1) implies that

$$x_i = \sum_{l=1}^{min(m,n)} u_l |d_l| v_{l,i}, \tag{6.20}$$

where $v_{l,i}$ denotes the ith component of right singular vector v_l. Equations (6.19) and (6.20) imply that

$$\begin{aligned} cor\,(x_i, x_j) &= \frac{x_i^T x_j}{m} = \sum_{l=1}^{min(m,n)} \sum_{h=1}^{min(m,n)} (u_l)^\tau u_h |d_l||d_h| v_{l,i} v_{h,j}/m \\ &= \sum_{l=1}^{min(m,n)} \frac{|d_l|^2}{m} v_{l,i} v_{l,j}, \end{aligned} \tag{6.21}$$

where use has been made of the fact that the left singular vectors u_l are orthonormal, i.e., $(u_l)^\tau u_h = 1$ if $l = h$ and $= 0$ otherwise.

Using the fact that the right singular vector v_j are orthonormal, (6.21) implies that

$$\sum_i \sum_j (cor\,(x_i, x_j))^2 = \frac{\Sigma_l |d_l|^4}{m^2}. \tag{6.22}$$

Similar to (6.21), one can derive that the correlation between x_i and u_l is given by

$$cor\,(x_i, u_l) = v_{l,i} |d_l| / \sqrt{m}. \tag{6.23}$$

For $l = 1$, (6.23) implies that the eigenvector $E = u_1$ is given by

$$cor\,(x_i, E) = v_{1,i} |d_1| / \sqrt{m}. \tag{6.24}$$

Since $cor(x_i,u_l)cor(x_j,u_l) = v_{l,i}|d_l|^2 v_{l,j}/m$, (6.21) implies

$$\begin{aligned} cor(x_i,x_j) &= \sum_{l=1}^{min(m,n)} cor(x_i,u_l)\,cor(x_j,u_l) \\ &= cor(x_i,E)\,cor(x_j,E) + \sum_{l=2}^{min(m,n)} cor(x_i,u_l)\,cor(x_j,u_l). \end{aligned} \quad (6.25)$$

Similar to (6.22), one can show that

$$\sum_i \sum_j (cor(x_i,x_j) - cor(x_i,E)\,cor(x_j,E))^2 = \frac{\sum_{l>1}|d_l|^4}{m^2}. \quad (6.26)$$

Equations (6.22) and (6.26) imply the following:

$$1 - \frac{\sum_i \sum_j (cor(x_i,x_j) - cor(x_i,E)\,cor(x_j,E))^2}{\sum_i \sum_j (cor(x_i,x_j))^2} = \frac{|d_1|^4}{\sum_l |d_l|^4} = EF(E), \quad (6.27)$$

where $EF(E)$ is the eigennode-based factorizability measure (6.11). Note that the left-hand side of (6.27) is close to 1, if the correlation matrix is approximately factorizable, i.e., when

$$cor(x_i,x_j) \approx cor(x_i,E)\,cor(x_j,E). \quad (6.28)$$

By raising both sides of (6.28) to a power β, we find

$$\begin{aligned} A_{ij} &= |cor(x_i,x_j)|^\beta \approx |cor(x_i,E)|^\beta\,|cor(x_j,E)|^\beta \\ &= A_{e,i}A_{e,j} \end{aligned}$$

if $EF(E)$ is close to 1. This completes our derivation of Observation 6.1. The last step of our derivation highlights an important theoretical advantage of the weighted correlation network construction method: it preserves the approximate factorizability of the underlying correlation matrix.

An alternative, possibly more direct way of motivating Observation 6.1 is based on the insight that the eigenvalues of the correlation matrix $cor(datX) = [cor(x_i,x_j)]$ are given by $|d_l|^2/m$. But our derivation allowed us to present important formulas that will be used in other context.

6.3 Eigenvector-Based Network Concepts

In the following, we will define eigenvector-based network concepts, which are very useful for theoretical and practical purposes. Recall that we define network concepts by evaluating a network concept function ((3.1) in Sect. 3.1) on different types of input matrices (Table 3.1). Here we provide the following.

Definition 6.1 (Eigenvector-based network concept). An **eigenvector-based network concept** is defined by evaluating a network concept function NCF (3.1) on A_E (6.15) and/or a node significance measure GS_E (6.16), i.e.,

$$NC_E = NCF(A_E, GS_E). \quad (6.29)$$

Below, we will use eigenvector-based network concepts (a) to reveal relationships between network concepts, (b) to provide a geometric interpretation of network concepts, and (c) to describe relationships between numeric vectors and correlation modules. *Note that eigenvector-based network concepts are analogous to approximate CF-based network concepts (3.49) where $CF_i = A_{e,i}$.* Thus, our expressions for approximate CF-based concepts (3.49) can be used for calculating eigenvector-based concepts, e.g.,

$$Connectivity_{E,i} = = A_{e,i} sum(A_{e,i}),$$
$$NetworkSignif_E = A_{e,t} \frac{sum(A_{e,})}{n}. \quad (6.30)$$

The relationships among approximate CF-based concepts (3.8), which were derived in Sect. 3.8, apply to eigenvector-based network concepts as well.

The bounded CF moment assumption (3.16) is typically satisfied in correlation networks. Therefore, (3.51) implies

$$ClusterCoef_E \approx \left(1 + Heterogeneity_E^2\right)^2 \times Density_E. \quad (6.31)$$

For low powers β, the maximum conformity assumption (3.9) is often satisfied since:

$$A_{e,max} = max_j \left(|cor(x_j, E)|^\beta\right) = max(|\text{kME}|)^\beta \approx 1. \quad (6.32)$$

Therefore, (3.10) implies

$$K_{E,i} = A_{e,i} = |cor(x_i, E)|^\beta = |\text{kME}_i|^\beta \quad (6.33)$$

$$Centralization_E \approx \sqrt{Density_E}\left(1 - \sqrt{Density_E}\right) \quad (6.34)$$

$$HubNodeSignif_E \approx A_{e,t}. \quad (6.35)$$

Under the assumptions of Observation 6.1, we find that $A \approx A_E$ and $GS_i \approx GS_{E,i}$. For a continuous network concept function $NCF(\cdot,\cdot)$, this implies $NCF(A, GS) \approx NCF(A_E, GS_E)$. We summarize this observation as follows.

Observation 6.2 (Eigenvector-based analogs). *If $A = |cor(datX)|^\beta$ and the eigenvector factorizability $EF(datX)$ is close to 1, fundamental network concepts can be approximated by their eigenvector-based analogs.*

In particular, this implies

$$K_i \approx |cor(x_i, E)|^{\beta} = |\text{kME}_i|^{\beta}. \tag{6.36}$$

Since the network eigenvector summarizes the overall behavior of the network, kME_i measures how well node i conforms to the overall network. Thus, a tongue-in-cheek social network interpretation of (6.36) is that group-conforming behavior leads to high popularity. An empirical illustration of this observation can be found in Table 5.1, which presents fundamental network concepts and their eigenvector-based analogs in the different mouse tissue networks. As predicted by Observation 6.2, we find a close relationship between the two types of network concepts if the eigenvector factorizability of the corresponding network is close to 1. For gender/tissue combinations with low factorizability, the correspondence may break down. This observation is also illustrated in the brain cancer network application (Fig. 6.3).

Observation 6.2 can be used to derive the following.

Observation 6.3 (Concepts in approx. factorizable correlation networks). *The relationships among eigenvector-based concepts approximately apply to their fundamental network analogs as well if* $A = |cor(datX)|^{\beta}$ *and the eigenvector factorizability is close to* 1.

6.3.1 *Relationships Among Density Concepts in Correlation Networks*

As described in Sect. 6.3, the simple relationships among approximately CF-based network concepts also apply to eigenvector-based network concepts. But additional relationships can be derived between eigenvector-based concepts and *propVarExpl* (6.10). For example, we find the following relationships for the mean squared kME value:

$$mean\left((\text{kME})^2\right) = \sum_i v_{1,i}^2 \frac{|d_1|^2}{mn} = \frac{|d_1|^2}{mn} = propVarExpl. \tag{6.37}$$

If the eigenvector factorizability $EF(E)$ (6.11) is close to 1, the mean correlation $meanCor = mean(cor(datX))$ can be approximated as follows:

$$\begin{aligned} meanCor &\approx (meanKME)^2 \\ &\approx \left(\frac{\sum_{i=1}^n v_{1,i}}{\sqrt{n}}\right)^2 propVarExpl = (cos(\theta))^2 propVarExpl, \end{aligned} \tag{6.38}$$

where θ is the angle between the vector v_1 and the vector $\mathbf{1} = (1, 1, \ldots, 1)$.

For a weighted correlation network, $A = |cor(datX)|^\beta$, approximate factorizability implies that

$$meanAdj \approx \left(\frac{\sum_i |\text{kME}_i|^\beta}{n} \right)^2. \tag{6.39}$$

For $\beta = 2$, (6.37) implies the following approximate relationship:

$$meanAdj \approx \left(\frac{\sum_i |\text{kME}_i|^2}{n} \right)^2 = propVarExpl^2. \tag{6.40}$$

6.4 Eigenvector-Based Approximations of Intermodular Concepts

Assume that we have two weighted correlation network modules q_1 and q_2 whose adjacencies are given by $A^{(q_1,q_1)} = |cor(datX^{(q_1)})|^\beta$ and $A^{(q_2,q_2)} = |cor(datX^{(q_2)})|^\beta$, respectively. Denote by $x_i^{(q_1)}$ and $x_j^{(q_2)}$ column vectors of $datX^{(q_1)}$ and $datX^{(q_2)}$, respectively. Further assume that both eigenvector factorizabilities $EF(E^{(q_1)})$ and $EF(E^{(q_2)})$ (6.11) are close to 1, i.e., the two correlation networks are approximately factorizable (Observation 6.1). Below, we derive the following.

Observation 6.4 (Factorizable intermodular correlation networks). *If the eigenvector factorizabilities $EF(E^{(q_1)})$ and $EF(E^{(q_2)})$ (6.11) are close to* 1, *then the intermodular adjacency matrix $A^{(q_1,q_2)} = |cor(datX^{(q_1)}, datX^{(q_2)})|^\beta$ can be factored as follows:*

$$A_{ij}^{(q_1,q_2)} \approx A_{e,i}^{q_1} A_{q_1 q_2} A_{e,j}^{q_2}, \tag{6.41}$$

where

$$A_{e,i}^{q_1} = \left|cor\left(x_i^{q_1}, E^{q_1}\right)\right|^\beta$$

$$A_{q_1 q_2} = \left|cor\left(E^{(q_1)}, E^{(q_2)}\right)\right|^\beta. \tag{6.42}$$

Abstractly speaking, Observation 6.4 allows us to characterize correlation networks for which the intermodular adjacency $A_{ij}^{(q_1,q_2)}$ can be approximated by a product of the centroid conformities (1.26): $A_{ij}^{(q_1,q_2)} \approx CF_i CF_j A_{q_1 q_2}$, where $A_{q_1 q_2}$ is an adjacency measure between modules q_1 and q_2 (as described in Sect. 6.5).

The eigenvector-based approximations on the right-hand side of (6.41) motivate us to define the **eigenvector-based intermodular adjacency matrix** $A_E^{(q_1,q_2)}$ as follows:

$$A_{E,ij}^{(q_1,q_2)} = A_{e,i}^{q_1} \left(A_{e,j}^{q_1}\right)^{\tau} A_{q_1 q_2}, \tag{6.43}$$

where $A_{q_1 q_2} = |cor(E^{(q_1)}, E^{(q_2)})|^{\beta}$. We are now ready to define.

Definition 6.2 (Eigenvector-based analog of an intermodular network concept). An **eigenvector-based intermodular network concept** $NCF(A_E^{(q_1,q_1)}, A_E^{(q_2,q_2)}, A_E^{(q_1,q_2)})$ is defined by evaluating a network concept function (3.1) on the eigenvector-based adjacency matrices (6.43).

Note that eigenvector-based intermodular network concepts are a special case of the approximate CF-based intermodular network concepts (where $CF_i^{q_1} = A_{e,i}^{q_1}$) described in Sect. 3.12 .

We define the **eigenvector-based module separability measure** as follows:

$$separability.ME\left(q_1,q_2\right) = 1 - \left|cor\left(E^{(q_1)}, E^{(q_2)}\right)\right|. \tag{6.44}$$

In the following, we will study the relationship between the fundamental network concept of average separability (1.47) and the eigenvector-based analog for the case that the adjacency matrix A corresponds to a unsigned weighted correlation network, i.e., $A_{ij} = |cor(x_i, x_j)|^{\beta}$. If modules q_1 and q_2 have high eigenvector factorizability (6.11), then

$$separability.ave_E\left(q_1,q_2\right) \approx 1 - \left|cor\left(E^{(q_1)}, E^{(q_2)}\right)\right|^{\beta}. \tag{6.45}$$

Equation (6.45) can be used to argue that by increasing the soft threshold β, correlation modules tend to become more separated. Since the eigenvector-based separability can be computed much more efficiently than the average separability (1.47), we prefer it in our correlation network applications.

In the following, we derive Observation 6.4. The reader may want to skip this technical part. Our scaling assumptions (6.18) and the orthonormality of the left-singular vectors U imply that the mean and variance of the eigenvector $E^{(q_1)} = u_1^{(q)}$ are given by $mean(u_1^{(q_1)}) = 0$ and $var(u_1^{(q_1)}) = 1/(m-1)$, respectively. These relations imply that

$$cor\left(E^{(q_1)}, E^{(q_2)}\right) = \left(u_1^{(q_1)}\right)^{\tau} u_1^{(q_2)}. \tag{6.46}$$

In the following, we will generalize (6.21) to the case of two column vectors $x_i^{(q_1)}$ and $x_j^{(q_2)}$ from $datX^{(q_1)}$ and $datX^{(q_2)}$, respectively.

$$\begin{aligned} cor\left(x_i^{(q_1)}, x_j^{(q_2)}\right) &= \frac{\left(x_i^{(q_1)}\right)^T x_j^{(q_2)}}{m} \\ &= \sum_{l=1}^{min\left(m,n^{(q_1)}\right)} \sum_{h=1}^{min\left(m,n^{(q_2)}\right)} \left(u_l^{(q_1)}\right)^\tau u_h^{(q_2)} |d_l^{(q_1)}|\,|d_h^{(q_2)}| v_{l,i}^{(q_1)} v_{h,j}^{(q_2)}/m \\ &\approx \left(u_1^{(q_1)}\right)^\tau u_1^{(q_2)} |d_1^{(q_1)}|\,|d_1^{(q_2)}| v_{1,i}^{(q_1)} v_{1,j}^{(q_2)}/m \end{aligned} \tag{6.47}$$

where the last approximation assumes that terms involving $|d_l^{(q_1)}||d_h^{(q_2)}|$ are negligible for $l > 1$ or $h > 1$. Equations (6.24) and (6.46) imply

$$cor\left(x_i^{(q_1)}, x_j^{(q_2)}\right) \approx cor\left(x_i^{(q_1)}, E^{(q_1)}\right) cor\left(E^{(q_1)}, E^{(q_2)}\right) cor\left(x_j^{(q_2)}, E^{(q_2)}\right). \tag{6.48}$$

Since (6.48) implies Observation 6.4, this concludes the derivation.

6.5 Networks Whose Nodes are Correlation Modules

Here we describe methods for constructing networks among modules that are defined for a correlation network. Assume that $datX^{(q_1)}$ and $datX^{(q_2)}$ are two subsets of columns of *datX*. Denote by $\mathscr{M}_{q_1}$ and $\mathscr{M}_{q_2}$ the sets of column (node) indices that correspond to modules q_1 and q_2, respectively. The adjacencies between nodes of the two modules can be represented by $A^{(q_1,q_2)} = |cor(datX^{(q_1)}, datX^{(q_2)})|^\beta$, which is an $n^{(q_1)} \times n^{(q_2)}$ dimensional sub-matrix of the full adjacency matrix A.

As described in (1.44), one can define measures of adjacency between the two modules q_1 and q_2 based on $A^{(q_1,q_2)}$, e.g., $A^{ave}_{q_1,q_2} = mean(A^{(q_1,q_2)})$. Alternatively, one can use eigenvectors $E^{(q_1)}$ and $E^{(q_2)}$ of the respective modules to define an **eigenvector-based measure of intermodular adjacency**:

$$A_{q_1q_2} = \left|cor\left(E^{(q_1)}, E^{(q_2)}\right)\right|^\beta \tag{6.49}$$

Using results from Sect. 6.4, one can easily prove that

$$A^{ave}_{q_1,q_2} \approx \left|cor\left(E^{(q_1)}, E^{(q_2)}\right)\right|^\beta = A_{q_1q_2}, \tag{6.50}$$

if the eigenvector factorizabilities (6.11) of the modules are close to 1. Note that the $A_{q_1q_2}$ can be interpreted as the off-diagonal elements of an adjacency matrix. In general, we define an **eigenvector network** as a correlation network between module eigenvectors. The module eigenvectors of different modules can be highly correlated (Fig. 6.2a). Detecting a high correlation between module eigenvectors may either be of biological interest (suggesting interactions between pathways) or it may be a methodological artifact (suggesting poorly defined modules that should be merged). In the context of gene co-expression networks, the correlations among eigengenes have been used to define **eigengene networks** in humans, chimpanzees, and mice (Langfelder and Horvath 2007). A sample trait $t = (t_1, \ldots, t_m)$ can be used to define a trait-based **eigenvector significance measure** using a correlation test.

Eigenvector network analysis can be viewed as a network reduction scheme that reduces a correlation network involving thousands of variables (vectors) to an orders of magnitude smaller meta-network involving module representatives (one eigenvector per module). Unlike traditional data reduction methods that impose orthogonality (e.g., principal component analysis) or independence (e.g., independent component analysis), an eigenvector network allows one to measure the dependencies among the modules, i.e., it can be considered a variant of oblique factor analysis. In biologic applications, it is unreasonable to assume that pathways (or modules) are "orthogonal" to each other. Thus, it is natural to use intermodular network methodology to analyze their relationships.

6.6 Dictionary for Fundamental-Based and Eigenvector-Based Concepts

To facilitate the communication between microarray data analysts and network theorists, we provide a short dictionary for translating between microarray data analysis- and network theory terminology. More specifically, for a subset (module) of nodes that have high expression factorizability, Table 6.1 describes the correspondence between general network terms and their eigenvector-based counterparts.

For example, the eigenvector significance and the eigenvector conformity are the eigenvector-based counterparts of the centroid significance (1.25) and centroid conformity (1.26), respectively.

While our theoretical derivations assume a weighted correlation network, our empirical robustness studies show that many of the findings apply to unweighted networks as well (Horvath and Dong 2008). In general networks, eigenvector-based network concepts are no substitute for fundamental network concepts. It is natural to use network concepts when describing the pairwise relationships among nodes and to use eigenvector-based network concepts when relating the vectors x_i to a module eigenvector. Since eigenvector-based network concepts tend to be relatively simple, they often simplify theoretical derivations. Further, many of them allow one to calculate a statistical significance level (p value) using a correlation- or regression-based test statistic.

Table 6.1 Dictionary for translating between general network terms and their eigenvector-based counterparts. A_e denotes a vector whose ith component is given by $A_{e,i}$

Term	General network adjacency matrix $A=[A_{ij}]$	Numeric data $datX$
Decomposition	factor analysis of A	SVD of $datX = UDV^{\tau}$
Centroid	intramodular hub node	module eigenvector E
Conformity(i)	CF_i defined as 1st factor of A	$A_{e,i} = \|cor(x_i, E)\|^{\beta}$
Approximately factorizable means	$A_{ij} \approx CF_i CF_j$	$x_i \approx E_i \|d_1\| v_{1j}$
Factorizability measure	$F(A) = 1 - \frac{\|\|(A-I)-(A_{CF}-I)\|\|_F^2}{\|\|A-I\|\|_F^2}$	$EF(X) = \frac{\|d_1\|^4}{\sum_j \|d_j\|^4}$
CentroidSignif(i)	$GS_{i.centroid}$	$A_{e,t} = \|cor(E,t)\|^{\beta}$
CentroidConformity(i)	$A_{i.centroid.i}$	$A_{e,i} = \|cor(E,x_i)\|^{\beta}$
Weighted correlation network and its eigenvector-based approximation if $EF(datX) \approx 1$		
	Correlation Network	eigenvector-based counterpart
network	$A = \|cor(datX)\|^{\beta}$	$A_E = A_e A_e^{\tau}$
Node significance(i)	$GS_i = \|cor(x_i, t)\|^{\beta}$	$GS_{E.i} = A_{e,i} A_{e,t}$
Connectivity(i)	$k_i = \sum_{j \neq i} A_{ij}$	$k_{E.i} = A_{e,i} \sum_j A_{e,j}$
Network concepts based on a network concept function $NCF(\cdot)$ if $EF(X) \approx 1$, $max_j(A_{e,j}) \approx 1$		
Concept	fundamental $NCF(A-I, GS)$	eigenvector-based $NCF(A_E, GS_E)$
Scaled Connectivity(i)	$K_i = \frac{k_i}{k_{max}}$	$K_{E.i} \approx A_{e,i} = \|cor(E, x_i)\|^{\beta}$
Density	$\frac{\sum_i \sum_{j \neq i} A_{ij}}{n(n-1)}$	$\left(\frac{sum(A_e)}{n}\right)^2$
Centralization	$\frac{n}{n-2}\left(\frac{k_{max}}{n-1} - Density\right)$	$\sqrt{Density_E}(1-\sqrt{Density_E})$
Heterogeneity	$\frac{\sqrt{variance(k)}}{mean(k)}$	$\frac{\sqrt{variance(A_e)}}{mean(A_e)}$
Clustering Coefficient(i)	$\frac{\sum_{l \neq i} \sum_{m \neq i,l} A_{il} A_{lm} A_{mi}}{\left(\sum_{l \neq i} A_{il}\right)^2 - \sum_{l \neq i} A_{il}^2}$	$\left(\frac{sum(A_e^2)}{sum(A_e)}\right)^2$
Hub node significance	$\frac{\sum_i GS_i K_i}{\sum_i (K_i)^2}$	$A_{e,t}$
Network significance	$\frac{\sum_{i \in I(\cdot)} GS_i}{\|I(\cdot)\|}$	$\sqrt{Density_E} \times A_{e,t}$
Module separability	$1 - \frac{mean(A^{(q_1, q_2)})}{\sqrt{Density^{(q_1)} Density^{(q_1)}}}$	$1 - \|cor(E^{(q_1)}, E^{(q_1)})\|^{\beta}$

6.7 Geometric Interpretation

The angular interpretation of correlation (5.11) facilitates a geometric interpretation of network concepts as will be illustrated in the following.

6.7.1 *Interpretation of Eigenvector-Based Concepts*

Observations 6.2 and 3.5 allow us to provide a geometric interpretation of network concepts in approximately factorizable correlation networks (e.g., co-expression network modules). If the maximum conformity assumption is satisfied, Equation

$K_i \approx A_{e,i} = |cor(x_i, E)|^\beta$ (6.36) facilitates a geometric interpretation of the scaled connectivity: the smaller the angle θ_i between the x_i and the eigenvector, the higher is the scaled connectivity value.

We provide two geometric interpretations of the density. The first makes use of the relationship $A_{ij} = |\cos(\theta_{ij})|^\beta$, where θ_{ij} denotes the angle between vectors i and j. By definition (1.18), the smaller the pairwise angles θ_{ij}, the higher is the network density. Equation $Density \approx mean(A_e)^2$ provides another interpretation: the smaller the angles θ_i between the network vectors and the eigenvector E, the higher is the density. Thus, the density can be interpreted as a measure of average closeness between the vectors x_i and E.

The eigenvector-based heterogeneity equals the coefficient of variation of the A_e, i.e., it is a measure of variability of the angles θ_i between the vectors and the network eigenvector. The heterogeneity equals 0 if the angles θ_i are all equal.

The ith node has high eigenvector-based significance $GS_{E,i}$ (6.16) if the eigenvector has a small angle with the sample trait and θ_i is small. Similarly, the geometric interpretation of the hub node significance (1.23) is straightforward: the smaller the angle between the network eigenvector and the sample trait, the higher is the hub node significance (3.28).

We provide two geometric interpretations of the network significance (1.24). The first interpretation is based on the definition of the network significance as average node significance; a network has high network significance if on average the angles between the vectors x_i and the sample trait t tend to be small. The second interpretation of the network significance is based on (3.29): a network has high significance if the network density is high and the angle between the network eigenvector and the sample trait is small.

6.7.2 Interpretation of a Correlation Network

In the following, we will provide a geometric interpretation of a correlation network adjacency. We assume a weighted correlation network $A_{ij} = |cor(x_i, x_j)|^\beta$ (5.21). Since the correlation is scale-invariant, i.e., $cor(ax_i + b, cx_j + d) = cor(x_i, x_j)$, we can assume without loss of generality that the vectors x_i have a mean 0 and the same variance. In an exercise, you are asked to prove that mean 0 and variance 1 imply that the vectors x_i have the same length, i.e., they can be visualized as points on the surface of a hyper-sphere. According to the angular interpretation of the Pearson correlation (5.11), $cor(x_i, x_j)$ equals the cosine of the angle θ_{ij} between the two scaled vectors $scale(x_i)$ and $scale(x_j)$, i.e., $A_{ij} = |cosine(\theta_{ij})|^\beta$. If $0 \le \theta_{ij} \le \pi/2$, then the cosine is a monotonically decreasing function of the angle θ_{ij}, which implies that A_{ij} is also a decreasing function of θ_{ij}. When the angle θ_{ij} equals 0 or $\pi/2$, the adjacency equals 1 or 0, respectively. The network adjacency is a monotonically decreasing function of the length of the shortest path (geodesic) between the two points on the hyper-sphere. The higher the soft threshold β, the more weight is assigned to short geodesic distances compared to large distances. Soft thresholding

(5.21) preserves the continuous nature of these distances, while hard thresholding (5.22) dichotomizes the distance information. Assume now that a quantitative sample trait $t = (t_1, \ldots, t_m)$ is used to define a trait-based node significance measure as follows: $GS_i = |cor(x_i, t)|^\beta$ (5.20). For this trait-based node significance measure, the sample trait t can also be considered a point on the hyper-sphere. Analogous to the network adjacency, the smaller the geodesic distance between the ith node x_i and the trait t, the higher the node significance of the ith node. In other words, the smaller the angle between $scale(t)$ and $scale(x_i)$, the more significant is the ith node.

6.7.3 Interpretation of the Factorizability

Here we provide an angular interpretation of a factorizable network. In the following, we argue that $EF(E) \approx 1$ if the vectors x_i lie in the same plane as the eigenvector $E = u_1$. Our scaling assumptions (6.18) and (5.10) imply that $cor(E, x_i)$ equals the cosine of the angle θ_i between E and x_i. Similarly, $cor(u_l, x_i) = cos(\theta_{l,i})$, i.e., the correlation equals the cosine of the angle $\theta_{l,i}$ between the lth left singular vector u_l (6.4) and x_i. Using $\sum_{i=1}^{n} cos(\theta_{l,i})^2 = d_l^2/m$, we are able to provide an **angular interpretation of the eigenvector factorizability**:

$$EF(E) = \frac{\left(\sum_{i=1}^{n} cos(\theta_{1,i})^2\right)^2}{\sum_{l=1}^{m} \left(\sum_i cos(\theta_{l,i})^2\right)^2}. \tag{6.51}$$

Thus, $EF(E) \approx 1$ if the vectors x_i are approximately orthogonal ($cos(\theta_{l,i}) \approx 0$) to the left singular vectors u_l for $l \geq 2$, i.e., if on average the vectors point in the direction of the network eigenvector $E = u_1$.

Figure 6.1 provides a geometric interpretation of an approximately factorizable correlation matrix. Under our scaling assumptions (6.18), $cor(x_i, x_j)$ equals the cosine of the angle θ_{ij} between x_i and x_j. In the following, we show that $\theta_{ij} \approx |\theta_i \pm \theta_j|$ and $sin(\theta_i)sin(\theta_j) \approx 0$ imply approximate factorizability of the correlation matrix, i.e., $cor(x_i, x_j) \approx cor(x_i, E)cor(x_j, E)$.

Using the assumptions described in Fig. 6.1a, $\theta_{ij} \approx |\theta_i \pm \theta_j|$, $sin(\theta_i)sin(\theta_j) \approx 0$, we find that

$$\begin{aligned} cor(x_i, x_j) &= cos(\theta_{ij}) \approx cos(|\theta_i \pm \theta_j|) \\ &= cos(\theta_i)cos(\theta_j) \mp sin(\theta_i)sin(\theta_j) \\ &\approx cos(\theta_i)cos(\theta_j) \\ &= cor(x_i, E)cor(x_j, E), \end{aligned} \tag{6.52}$$

i.e., the correlation matrix is approximately factorizable.

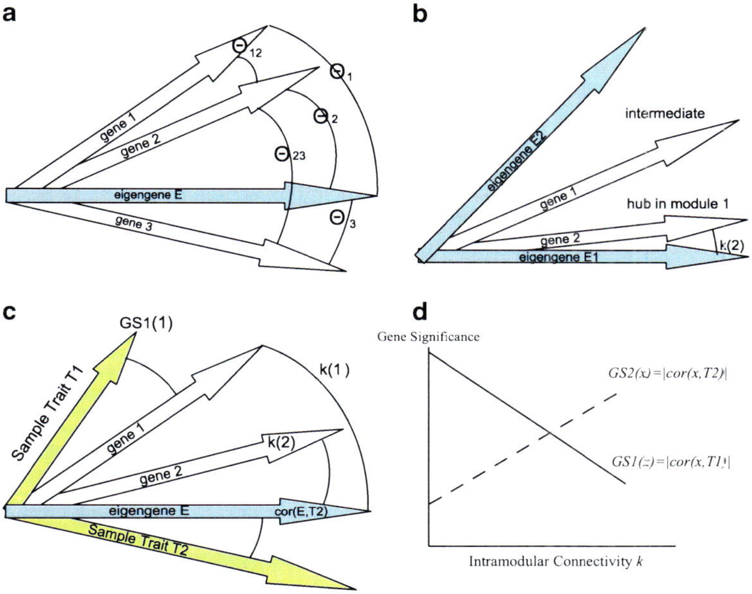

Fig. 6.1 Geometric interpretation of network theoretic results. Figure (**a**) presents a geometric interpretation of a factorizability in a co-expression network if the genes (i.e., numeric vectors x_i) and the network eigenvector E lie in a Euclidean plane. Then the angle θ_{12} between genes (nodes) x_1 and x_2 can be expressed in terms of angles with the network eigengene. Since $\theta_{12} = \theta_1 - \theta_2$ and $\theta_{23} = \theta_2 + \theta_3$, we find $\theta_{ij} = |\theta_i \pm \theta_j|$. Using a trigonometric formula (6.52), this implies that the correlation matrix is approximately factorizable. Figure (**b**) illustrates why intramodular hub nodes cannot be "intermediate" nodes between two distinct correlation modules. The large angle between module eigengenes $E1 = E^{(1)}$ and $E2 = E^{(2)}$ reflects that the corresponding modules are distinct. Since intermediate gene 1 does not have a small angle with either module eigengene, it is not an intramodular hub. By contrast, intramodular hub gene 2 has a small angle with eigengene $E1$ but is not close to module eigengene $E2$. Figures (**c,d**) illustrate that the hub node significance of a module depends on the relationship between the module eigenvector and the underlying sample trait (3.28). For sample traits $T2$ and $T1$, the hub node significance (and corresponding eigenvector significance $cor(E,T)$) are high and low, respectively. The geometry of Figure (**c**) implies relationships between the connectivity k of a node (determined by its angle with the eigenvector E) and node significance measure $GS1$ (its angle with trait $T1$) and $GS2$ (its angle with trait $T2$). As shown in Figure (**d**), the node significance measure $GS2$ increases with k since the small angle between E and $T2$ implies that nodes with high k (small angle with E) also have a small angle with $T2$. In contrast, high connectivity k implies a large angle with $T1$ and thus $GS1$ decreases as a function of k

6.8 Network Implications of the Geometric Interpretation

In the following, we provide several examples that illustrate potential uses of the geometric interpretation.

6.8.1 Statistical Significance of Network Concepts

While fundamental network concepts are defined as functions of the network adjacency matrix, their eigenvector-based analogs are often simple monotonic functions of correlation coefficients. This insight can be used to attach significance levels (p values) to several eigenvector-based network concepts. For example, the eigenvector-based hub node significance is a monotonic function of the correlation between the eigenvector and the sample trait (3.28). Thus, one can use a correlation test p value (described in Sect. 10.3) or a regression-based p value for assessing the statistical significance between E and the sample trait t. Analogously, one can attach a significance level to kME_i (6.53). We briefly mention that permutation tests can be used to determine significance for differential network concepts such as $Diff.K_i$ (1.51). By randomly permuting the samples that are used for constructing the reference and test network, one can easily establish permutation test p values (Fuller et al. 2007).

6.8.2 Intramodular Hubs Cannot be Intermediate Nodes

The geometric interpretation of correlation network analysis can be used to argue that a node that lies "intermediate" between two distinct modules cannot be a highly connected intramodular hub node in either module (Fig. 6.1b). More precisely, we refer to node i as hub node in module 1 if its scaled connectivity $K_i^{(1)}$ is very high (say larger than 0.9). Further, we refer to two modules as distinct if their respective eigenvectors have a low correlation, say $A_{(1)(2)} = |cor(E^{(1)}, E^{(2)})| < 0.3$. We refer to node i as intermediate between modules 1 and 2 if it has a moderately high connectivity with both modules, say $K_i^{(1)} > 0.5$ and $K_i^{(2)} > 0.5$.

Equation (3.27) allows us to translate statements about the scaled intramodular connectivity into statements about the angles between vectors (nodes) and module eigenvectors. A node is an intermediate node if it has a moderately small angle with both module eigenvectors. If the eigenvectors are distinct (i.e., the angle between them is large), the intermediate node cannot have a very small angle with either module eigenvector, i.e., it cannot be an intramodular hub node in either module. A geometric interpretation of this example can be found in Fig. 6.1b. As an important caveat, we mention that intermediate network nodes may well be highly connected "hub" nodes if the factorizability property does not hold such as in the entire network comprised of multiple distinct modules.

6.8.3 Characterizing Networks Where Hub Nodes Are Significant

For a trait-based node significance measure, the striking relationship between module significance and hub node significance (3.29) suggests a positive relationship between connectivity and node significance (high hub node significance) in modules that are enriched with significant nodes (high module significance). Further, (3.28) shows that the hub node significance of a network module is determined by the angle between the module eigenvector and the sample trait. This allows us to describe situations when a module has high hub node significance, i.e., when there is a strong positive relationship between node significance and intramodular connectivity. In the example provided in Fig. 6.1c and d, the angle between E and $T2$ is small which implies that the hub node significance with regard to $GS2_i = |cor(x_i, T2)|$ is high. By contrast, the angle between E and $T1$ is large, which implies that the hub node significance with regard to $GS1_i = |cor(x_i, T1)|$ is low.

6.9 Data Analysis Implications of the Geometric Interpretation

Here we illustrate how the geometric interpretation of correlation networks can be used to derive results, which may be interesting to data analysts.

In the following, we show that the geometric interpretation can be used to (a) derive **relationships between data reduction methods** and (b) argue that intramodular connectivity is a measure of module membership.

Two broad categories of data reduction techniques are often applied to high dimensional data (e.g., gene expression data). The first category, often used by network theorists, is to reduce the data into modules. Each module can be represented by a centroid, e.g., an intramodular hub node. The second category, often used by statisticians and data miners, reduces a large set of numeric variables to a small number of components that capture the essential behavior of the variables (Alter et al. 2000; Holter et al. 2000; West et al. 2001; Liao et al. 2003; Adrian et al. 2004; Shen et al. 2006; Tamayo et al. 2007). For example, to summarize the expression profiles of the genes inside a single module, one can apply the singular value decomposition to the expression data and summarize the module with the module eigenvector (referred to as module eigengene). Since $K_i \approx A_{e,i} = |cor(x_i, E)|^{\beta}$ (6.36) implies that hub nodes are highly correlated with the module eigenvector, we find that the two seemingly different approaches will lead to very similar results in practice (Fig. 6.2c).

We will now argue that **intramodular connectivity is a measure of module membership**. Analogous to the case of intramodular network concepts (Sect. 1.8), we define eigenvector-based intramodular network concepts by evaluating the network concept function $NCF(A_E^{(q)}, GS_E^{(q)})$ on the eigenvector-based module adjacency matrix $A_E^{(q)}$ (6.15) and the eigenvector-based node significance measure $GS_E^{(q)}$ (6.16).

Analogous to (6.12), we use the module eigenvector $E^{(q)}$ to define a quantitative measure of module membership (Horvath and Dong 2008):

$$\mathrm{kME}_i^{(q)} = cor(x_i, E^{(q)}). \tag{6.53}$$

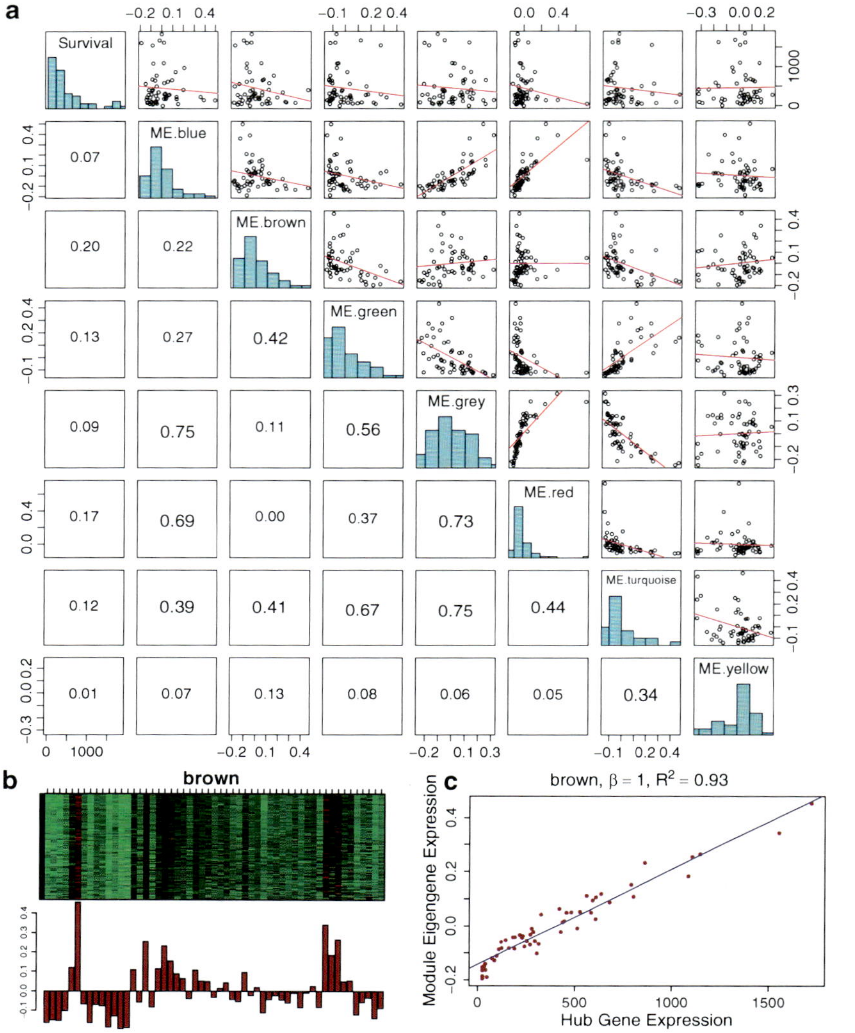

Fig. 6.2 Module eigengenes in the brain cancer gene co-expression network. Figure (**a**) depicts the pairwise scatter plots between the module eigengenes $E^{(q)}$ of different modules and cancer survival time t. Each dot represents a microarray sample. ME.blue denotes the module eigengene $E^{(blue)}$ of the *blue module*. Numbers below the diagonal are the absolute values of the corresponding correlations. Note that the module eigengenes of different modules can be highly correlated. The *brown module* eigengene has the highest absolute correlation ($r = 0.20$) with survival time. Frequency plots (histograms) of the variables are plotted along the diagonal. *Upper panel* of Figure (**b**): heat map plot of the *brown module* gene expression profiles (rows) across the microarray samples (columns). *Red* corresponds to high- and *green* to low-expression values. Since the genes of a module are highly correlated, one observes vertical bands. *Lower panel* of Figure (**b**): the values of the components of the module eigengene (y-axis) versus microarray sample number (x-axis). Note that vertical bands of red (*green*) in the upper panel correspond to high (*low*) values of the eigengene in the *lower panel*. Figure (**c**) shows that the expression profile of the module eigengene (y-axis) is highly correlated with that of the most highly connected hub gene (x-axis), which illustrates (6.36). A linear regression line has been added

The larger $\text{kME}_i^{(q)}$, the more correlated is node i with the summary profile of the qth module. In co-expression applications, $\text{kME}_i^{(q)}$ is sometimes referred to as signed module eigengene-based connectivity (Ghazalpour et al. 2006; Fuller et al. 2007). Observation 3.5 and (6.36) applied to a correlation module implies

$$\frac{\text{kIM}_i^{(q)}}{max(\text{kIM})} = K_i^{(q)} \approx \left|\text{kME}_i^{(q)}\right|^{\beta}.$$

Thus, both intramodular connectivity kIM and kME can be used as measures of module membership. While kIM is defined for all networks, kME is only defined for correlation networks. Theoretical advantages of kME include: (a) it is naturally scaled to take values between -1 and 1, (b) it maintains the sign information between x_i and the overall module behavior measured by E, and (c) a simple correlation test p value can be computed (as described in Sect. 10.3).

Module detection usually involves certain parameter choices. For some nodes, it may be difficult to decide whether they belong to a particular module or more than one module. Instead of reporting a binary indicator of module membership, it can be advantageous to report a fuzzy measure of module membership, which takes on values in the unit interval [0,1]. A natural choice for a fuzzy measure of module membership is $\text{kME}_i^{(q)}$ (6.53). The fuzzy module membership measures $\text{kME}_i^{(q)}$ specify how close node i is to modules $q = 1, \ldots, Q$. It is straightforward to use these measures for finding nodes that are close to two modules, i.e., intermediate nodes. In Fig. 6.4, we show the pairwise relationships among different $\text{kME}_i^{(q)}$ measures where nodes (genes) are colored by their original module assignment. Note that many of the non-module (gray) nodes lie intermediate between the proper module nodes.

6.10 Brain Cancer Network Application

Here we apply our methods to the brain cancer network application described in Sect. 5.7. Recall that we identified six proper modules (branches of a hierarchical clustering tree). The expression profiles of each module are represented by the eigengene. We defined the module eigenvector significance (6.14) as the absolute value of the correlation between the module eigenvector and patient survival time. The brown module eigenvector also had the highest eigenvector significance: $A_{e.t}^{brown} = |cor(E^{brown}, t)| = 0.202$. An advantage of the eigenvector-based hub gene significance (the eigenvector significance) is that it allows one to compute a corresponding p value. Using a correlation test, we find that the value of the eigenvector significance $A_{e.t}^{brown}$ is statistically insignificant ($p = 0.30$) in this data set. However, when we combined these data with an additional data set, we found that the brown module eigenvector is significantly related to survival time (Horvath et al. 2006).

We visualize the gene expression profiles of module genes with a heat map plot (Fig. 6.2b) where rows correspond to the genes, the columns to the samples, and the gene expression profiles have been standardized to a mean of 0 and a variance of 1. The heat map colors high and low expression values by red and green, respectively. For a given module, the heat map exhibits characteristic vertical bands that reflect the high correlation among module gene expression profiles. For the six proper modules of our brain cancer application, the proportion of variance explained by the first eigenvector ranges from 0.59 to 0.71 (Table 6.2). For the improper gray module genes (defined as genes outside of all proper modules), the proportion of variance explained by the first eigenvector is only 0.28. Similarly, when all network genes are used to define an improper module, the proportion of variance explained by the first eigenvector is only 0.32. As expected by module construction, we find that the gene expression data of proper modules have high eigenvector factorizabilities $EF(X) \geq 0.97$ (Table 6.2). By contrast, the factorizability of the gray genes (i.e., the genes outside of proper modules) is relatively low ($EF(X) = 0.66$).

For each module, Table 6.2 reports network properties including network size, density, centralization, heterogeneity, mean clustering coefficient, module

Table 6.2 Values of network concepts in weighted gene co-expression network modules (brain cancer data)

	Proper modules						Improper module
Concept	Blue	Brown	Green	Red	Turquoise	Yellow	Gray
Module size ($n^{(q)}$)	606	185	136	105	1112	143	1313
Eigenvector factorizab. ($EF(E^{(q)})$)	0.97	0.99	0.99	0.98	0.98	0.99	0.66
$PVE(E^{(q)})$	0.59	0.66	0.70	0.68	0.57	0.71	0.28
Max. conformity $max(A_{e,i})$	0.97	0.97	0.98	0.95	0.98	0.98	0.91
$Density$	0.58	0.65	0.69	0.67	0.55	0.70	0.29
$Density_E$	0.58	0.65	0.70	0.68	0.55	0.71	0.23
$Centralization$	0.16	0.13	0.12	0.11	0.17	0.12	0.15
$Centralization_E$	0.16	0.13	0.12	0.11	0.18	0.12	0.21
$Heterogeneity$	0.14	0.10	0.11	0.091	0.17	0.11	0.17
$Heterogeneity_E$	0.14	0.10	0.11	0.091	0.17	0.11	0.44
$mean(ClusterCoef)$	0.60	0.66	0.71	0.68	0.59	0.72	0.32
$ClusterCoef_E$	0.60	0.66	0.71	0.68	0.59	0.72	0.33
$ModuleNetworkSignif$	0.088	0.12	0.21	0.16	0.14	0.065	0.11
$ModuleNetworkSignif_E$	0.018	0.093	0.21	0.16	0.13	0.039	0.008
$HubNodeSignif$	0.11	0.14	0.25	0.18	0.19	0.074	0.15
$HubNodeSignif_E$	0.023	0.11	0.25	0.18	0.17	0.045	0.014
$EigenvectorSignif = A_{e,t}^{(q)}$	0.024	0.12	0.25	0.19	0.18	0.046	0.016

Here we report the results for soft thresholding with $\beta = 1$ (5.21). The results for higher powers β and for unweighted networks can be found in Horvath and Dong (2008). Gray color genes outside the six properly defined modules. The table shows that network concepts in the proper modules are close to their eigenvector-based analogs.

significance, hub gene significance, and eigenvector significance. For the proper (non-gray) modules, we find that the numerical values of the intramodular network concepts and their eigenvector-based analogs support our theoretical derivations.

Our empirical results illustrate Observation 6.2 regarding the relationship between intramodular network concepts and their eigenvector-based analogs. Figure 6.3 depicts the relationships among centralization, heterogeneity, clustering coefficient, module significance, hub gene significance, and their respective

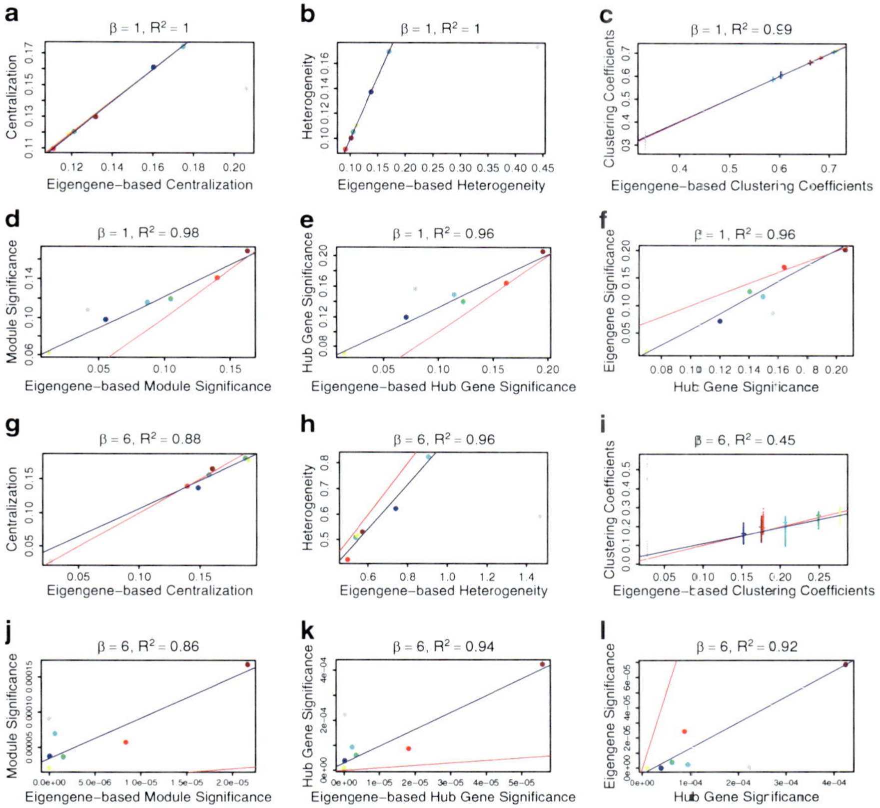

Fig. 6.3 Illustrating Observation 6.2 regarding the relationship among network concepts (y-axis) and their eigengene-based analogs (x-axis) in the brain cancer data. Each point corresponds to a module. Figures (**a**–**f**) and (**g**–**l**) correspond to a weighted network constructed with a soft threshold (5.21) of $\beta = 1$ and $\beta = 6$, respectively. (**a**,**g**) Centralization (y-axis) versus eigengene-based Centralization$_E$ (x-axis); analogous plot for (**b**,**h**) Heterogeneity (**c**,**i**) clustering coefficient; (**d**,**j**) module significance; and (**e**,**k**) hub gene significance; Figures (**f**,**l**) illustrate (1.23) regarding the relationship between eigengene significance and hub gene significance. The *blue line* is the regression line through the points representing proper modules (i.e., the gray, non-module genes are left out). While the *red reference line* (slope 1, intercept 0) does not always fit well, we observe high squared correlations R^2 between network concepts and their analogs. Since the gray point corresponds to the genes outside properly defined modules, we did not include it in calculations

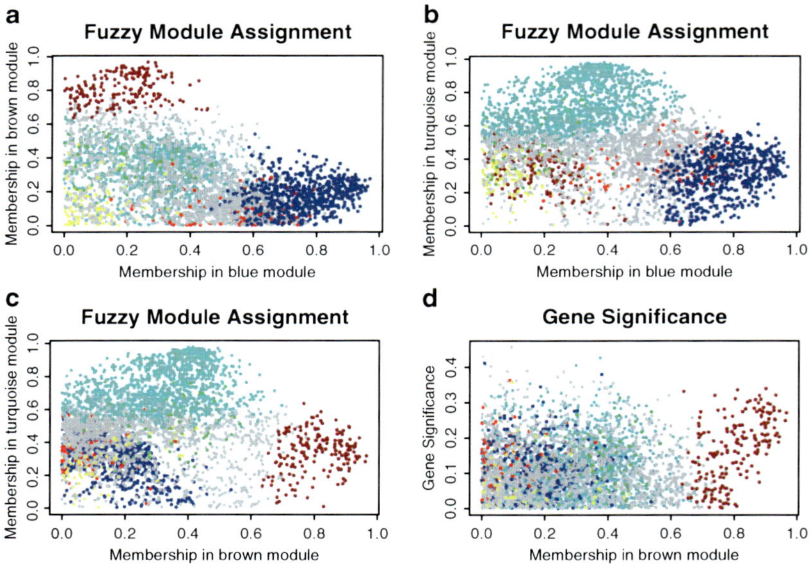

Fig. 6.4 Fuzzy module annotation of genes in the brain cancer network. A natural choice for a fuzzy measure of module membership is the generalized scaled connectivity measure $|\mathrm{kME}_i^{(q)}| = |cor(x_i, E^{(q)})|$ (6.53). Figure (**a**) shows the scatterplot of the *brown* module membership measure (y-axis) versus that of the *blue* module (x-axis). Note that *gray dots* corresponding to genes outside of properly defined modules may can be intermediate between module genes. Figure (**b**) shows the corresponding plot for blue versus turquoise module membership; Figure (**c**) shows brown versus turquoise module membership. Figure (**d**) shows the relationship between gene significance based on survival time (y-axis) and *brown* module membership (x-axis)

eigenvector-based analogs when a soft threshold of $\beta = 1$ is used for the weighted network construction (5.21). The analogous results for $\beta = 6$ are depicted in Fig. 6.3g–k. Fig. 6.3(f) and (l) depicts the relationship between hub gene significance (1.23) and module eigenvector significance (6.14) for $\beta = 1$ and $\beta = 6$, respectively. For completeness, we also report the results for the gray, non-module genes in the figures. But since our theoretical results assume proper modules, we exclude the gray genes from the calculation of the squared correlation coefficient R^2. The summary of a robustness analysis with regard to different soft thresholds β and hard thresholds τ is reported in Table 6.3. Overall, we find very high R^2 values (> 0.85), which confirm our theoretical results. Only the R^2 values for the relationship between clustering coefficient and its eigenvector-based analog is decreased if $\beta > 3$.

Figure 6.5 illustrates the implications of Observation 3.5 regarding the relationship among network concepts in the cancer co-expression network modules. Figure 6.5a shows that the scaled connectivity $K_i^{(q)}$ is highly correlated ($R^2 > 0.99$) with $A_{e,i}^{(q)}$, which illustrates (3.27). This relationship is highly robust with regard to high soft thresholds β as can be seen from Table 6.3.

Table 6.3 Robustness analysis of the brain cancer gene co-expression network results

Squared correlation R^2 across modules	Weighted networks soft threshold β						Unweighted threshold τ	
Relation	1	2	3	4	5	6	0.7	0.5
$Centralization \approx Centralization_E$	1.0	1.0	0.97	0.90	0.87	0.88	0.07	0.93
$Heterogeneity \approx Heterogeneity_E$	1.0	1.0	0.99	0.98	0.97	0.96	0.89	0.87
$ClusterCoef_i \approx ClusterCoef_E$	0.99	0.96	0.88	0.74	0.58	0.45	0.04	0.32
$ModuleSignif \approx ModuleSignif_E$	0.98	0.91	0.87	0.85	0.85	0.86	0.98	0.98
$HubGeneSignif \approx HubGeneSignif_E$	0.96	0.91	0.89	0.90	0.92	0.94	0.93	0.87
$EigenvectorSignif \approx HubGeneSignif$	0.96	0.89	0.87	0.88	0.90	0.92	0.93	0.87
$ClusterCoef_i \approx (1+(Heterog.)^2)^2 \times Density$	0.99	0.96	0.89	0.76	0.61	0.49	0.006	0.32
$ModuleSignif \approx \sqrt{Density} \times HubGeneSignif$	1.0	0.99	0.99	0.98	0.97	0.95	0.85	0.99
$Centralization \approx \sqrt{Density}(1-\sqrt{Density})$	0.90	0.68	0.058	0.016	0.16	0.35	0.20	1.0
$\frac{k_{max}}{n-1} \approx \sqrt{Density}$	0.94	0.94	0.94	0.94	0.93	0.92	0.95	0.98
$K_i \approx A_{e,i}$ (median R^2)	1.0	1.0	1.0	1.0	1.0	0.99	0.95	0.83

The table reports how the relation among network concepts changes as function of different soft threshold parameters β (5.21) or hard thresholds (5.22) used in the network construction. For each relationship and each network construction method, the table entry reports the squared correlation R^2 across the proper modules. For within module comparisons the table reports median R^2 values.

Figure 6.5b illustrates the relationship between the clustering coefficient (the mean corresponds to the short horizontal line) and $(1+Heterogeneity^2)^2 \times Density$ (3.30). This relationship is diminished for soft thresholds $\beta > 3$ as can be seen from Table 6.3. Figure 6.5c illustrates the relation $ModuleSignif^{(q)} \approx \sqrt{Density^{(q)}} \times HubGeneSignif^{(q)}$ (3.29), which is highly robust with regard to different choices of β (Table 6.3). $Centralization^{(q)} \approx \sqrt{Density^{(q)}}(1-\sqrt{Density^{(q)}})$ (3.31) is illustrated in Fig. 6.5d. This relationship is *not* robust with regard to β the R^2 value is only 0.058 for $\beta = 3$. Figure 6.5e illustrates $\frac{k^{(q)}_{max}}{n^{(q)}-1} \approx \sqrt{Density^{(q)}}$, which is highly robust with regard to β (Table 6.3).

Although our theoretical results were derived using relatively restrictive assumptions, we find that most results are robust in the weighted networks, see Fig. 6.6, Table 6.3 and (Horvath and Dong 2008). However, in unweighted networks, several relationships have lower R^2 values and show a strong dependence on the hard threshold τ (Table 6.3).

6.11 Module and Hub Significance in Men, Mice, and Yeast

Here we use several gene co-expression network applications to illustrate

$$NetworkSignif^{(q)} \approx \sqrt{Density^{(q)}} \times HubNodeSignif^{(q)}.$$

Consider an extended version of the mouse tissue gene expression data from Sect. 5.5. The data were already described above and in Fig. 5.2. Here we focus on the female mouse liver tissues of the above-mentioned F2 mouse cross. Specifically, 135 female mice were used to construct a weighted network comprised of 3,400 highly connected genes. The biological significance and gene ontology enrichment

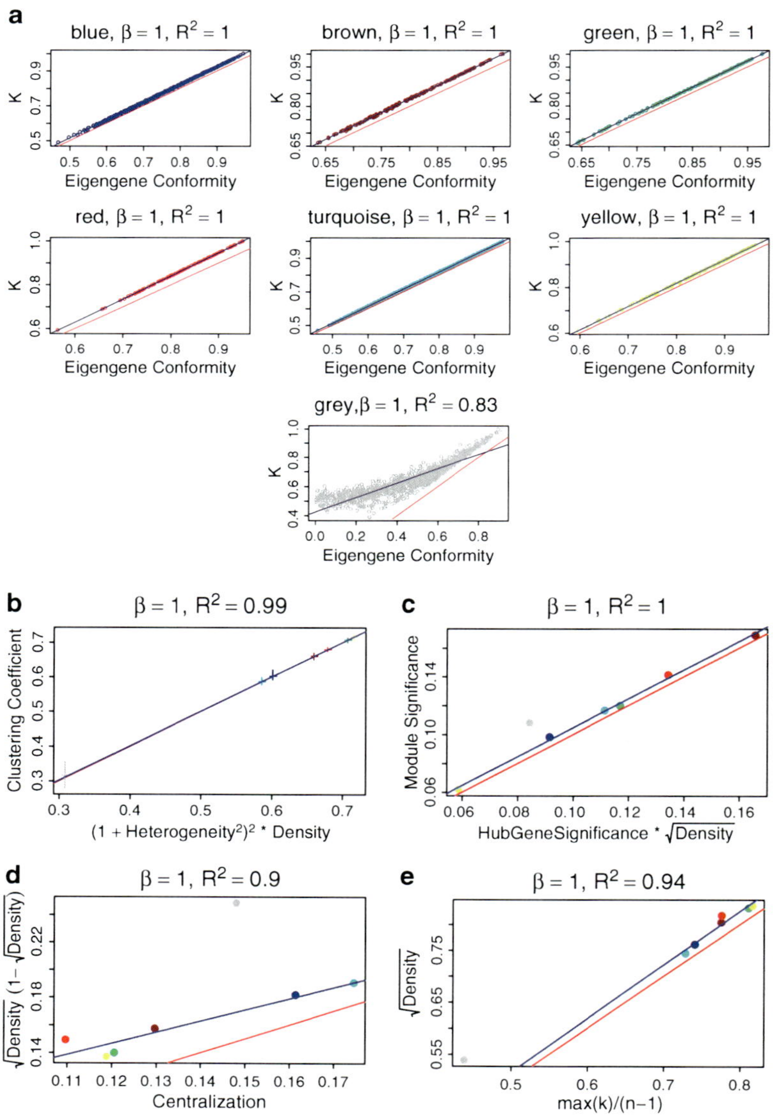

analysis of the 12 modules in this large network are described in Ghazalpour et al. (2006). In Fig. 6.6, we focus on the relationships among the network concepts. We find that many of our theoretical results hold approximately even if the expression factorizability is low. Overall, we find that our theoretical results are highly robust in weighted networks.

In the following, we illustrate our theoretical derivations using the yeast gene co-expression network. The yeast microarray data were derived from experiments designed to study the cell cycle (Spellman et al. 1998). A detailed biological description of the modules and the importance of intramodular connectivity can be found in previous work (Carlson et al. 2006). In Fig. 6.6, we use a gene significance measure that encodes knockout essentiality, i.e., $GS_i = 1$, if the ith gene is known to be essential and 0 otherwise. In contrast to the other applications, *this gene significance measure is not based on a sample trait.* Our theoretical derivations for relating module significance to hub gene significance (3.29) assumed a sample-trait-based gene significance measure. Although this important assumption is violated for knockout essentiality, it is striking that the relationship between hub gene significance and module significance can still be observed empirically (Fig. 6.6). To explain this relationship, recall the following relationship between approximate CF-based network concepts (3.8)

$$NetworkSignif_{CF,app} \approx \frac{\sqrt{Density_{CF,app}}}{CF_{max}} \times HubNodeSignif_{CF,app}.$$

Thus, approximate factorizability of the network module and the node significance measure can be used to justify the relationship.

Fig. 6.5 Using the brain cancer data to illustrate Observation 3.5 regarding the relationships among network concepts. Figure (**a**) illustrates (3.27) regarding the relationship between scaled intramodular connectivity $K_i^{(q)}$ (y-axis) and eigengene conformity $A_{e,i}$ (x-axis). Each dot corresponds to a gene colored by its module membership. We find a high squared correlation R^2 even for the gray genes outside properly defined modules. Figure (**b**) illustrates $mean(ClusterCoef^{(q)}) \approx (1 + (Heterogeneity^{(q)})^2)^2 \times Density^{(q)}$. Again each dot represents a gene. The clustering coefficients of gray genes vary more than those of genes in proper modules. The short *horizontal lines* correspond to the mean clustering coefficient of each module. Figure (**c**) illustrates $ModuleSignif^{(q)} \approx \sqrt{Density^{(q)}} \times HubGeneSignif^{(q)}$ (3.29); here each dot corresponds to a module. Since the *gray dot* corresponds to genes outside of properly defined modules, we have excluded it from the calculation of the squared correlation R^2. Figure (**d**) illustrates $Centralization^{(q)} \approx \sqrt{Density^{(q)}}(1 - \sqrt{Density^{(q)}})$ (3.31); Figure (**e**) illustrates $\frac{k_{max}^{(q)}}{n^{(q)}-1} \approx \sqrt{Density^{(q)}}$. A reference line (*red*) with intercept 0 and slope 1 has been added to each plot. The *blue line* is the *regression line* through the points representing proper modules (i.e., the gray, non-module genes are left out). A robustness analysis with regard to different network construction methods, e.g., $\beta > 1$, can be found (Horvath and Dong 2008)

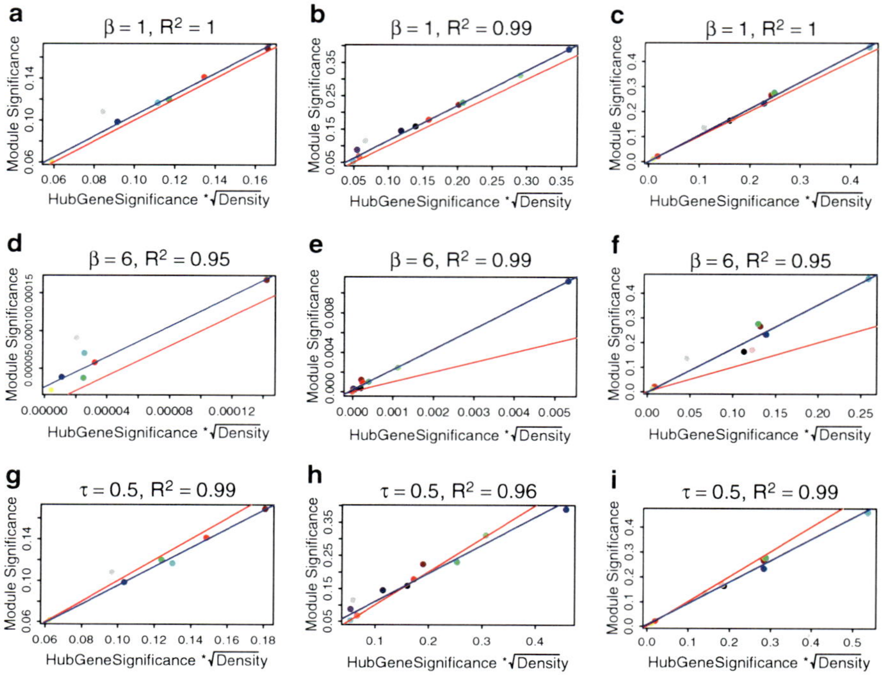

Fig. 6.6 Using three different data (brain cancer, mouse liver, yeast cell cycle) and three different network construction methods to illustrate (3.29) regarding the relationship between module significance (y-axis) and $\sqrt{Density^{(q)}} \times HubGeneSignif^{(q)}$ (x-axis). Points correspond to modules. The square of the correlation coefficient R^2 was computed without the gray, improper module. The figures in the first column (**a**,**d**,**g**), second column (**b**,**e**,**h**), and third column (**c**,**f**,**i**) correspond to the brain cancer, mouse liver, and yeast gene co-expression networks, respectively. The figures in the first (**a**–**c**) and second rows (**d**–**f**) correspond to weighted networks (5.21) constructed with soft thresholds $\beta = 1$ and $\beta = 6$, respectively. The figures in the third row (**g**–**i**) correspond to an unweighted network (5.22) that results from thresholding the correlation matrix at $\tau = 0.5$. Overall, we find that the reported relationship is quite robust with respect to our theoretical assumptions (e.g., factorizability). The *blue line* is the *regression line* through the points representing proper modules (i.e., the gray, non-module genes are left out). A *reference line* with slope 1 and intercept 0 is shown in *red*

6.12 Summary

The merging of network theory and data mining has spawned a new field: correlation network analysis. For example, gene expression data can be analyzed with gene co-expression network analysis. While the utility of network methods for analyzing high dimensional genomic data has been demonstrated in numerous publications, the utility of data mining methods for solving network theoretic problems has not yet been fully appreciated. One goal of this chapter is to show that simple geometric arguments can be used to derive network theoretic results if the networks are defined on the basis of a correlation matrix. The unification of correlation network methods with traditional data mining methods informs the application and development of systems level analysis methods. We exploit the relationship between network theory and data mining methods to clarify the meaning of and the relationship among network concepts in correlation networks. Conversely, data mining techniques (singular value decomposition) can also be used to address difficult problems in network theory.

Our theoretical results highlight another major advantage of correlation networks: the angular interpretation of correlations can be used to present a geometric interpretation of network concepts and results. For example, the sample trait-based node significance measure of the ith node is determined by the angle between x_i and the sample trait t (5.20); the scaled connectivity of the ith node (3.27) is determined by the angle between the x_i and the module eigenvector; the hub node significance (3.28) is determined by the angle between the eigenvector and the sample trait.

The geometric interpretation of correlation network analysis reveals a deep connection to other statistical methods. Since it projects the numeric vectors x_i onto the hyper-sphere in an m-dimensional Euclidean space, network analysis can be considered a special case of directional statistics. When focusing on the use of module eigenvectors, network analysis can be considered a variant of oblique factor analysis.

A high-level view of modules and their centroids (eigenvectors) can be used to define eigenvector networks (see Sect. 6.5) (Langfelder and Horvath 2007). High correlations (small angles) between module eigenvectors may suggest close relationships between the corresponding pathways. Eigenvector-based network concepts facilitate a low-level view of a correlation network module if this network is approximately factorizable (e.g., if the module is comprised of correlated vectors). This low-level view of a single module allows us to provide a geometric interpretation of intramodular network concepts. We use the singular value decomposition of the numeric data *datX* to characterize approximately factorizable correlation networks, i.e., adjacency matrices that satisfy $A_{ij} \approx CF_i CF_j$. We provide an intuitive formula of the conformity $CF_i \approx |cor(x_i,E)|^\beta$. Since the eigenvector E summarizes the overall behavior of *datX*, the eigenvector conformity $|cor(x_i,E)|^\beta$ measures how well node i conforms to the overall module. This insight led us to coin the term "conformity". Using the singular values, we define a measure of eigenvector factorizability (6.11) that is analogous to the proportion of variance explained by the module eigenvector (6.7). We provide a geometric interpretation of network factorizability in Fig. 6.1. The derivation of Observation 6.1 highlights a theoretical advantage of the soft-thresholding approach (5.21); the resulting weighted network maintains the approximate factorizability of the underlying correlation matrix:

$$A_{ij} = |cor(x_i,x_j)|^\beta \approx |cor(x_i,E)cor(x_j,E)|^\beta \approx |cor(x_i,E)|^\beta |cor(x_j,E)|^\beta .$$

While network concepts are functions of the adjacency matrix, eigenvector-based network concepts are analogous functions of the eigenvector conformities

$|cor(x_i, E)|^{\beta}$. Algebraically, eigenvector-based network concepts are closely related to "approximate conformity-based" network concepts (Sect. 3.3) but they allow for a geometric interpretation. Higher order generalizations of eigenvector-based network concepts are described in Sect. 3.6.

For approximately factorizable correlation networks, eigenvector-based network concepts can be used to provide a geometric interpretation of their corresponding fundamental network concepts. Observation 6.2 states that network concepts in weighted correlation networks are approximately equal to their eigenvector-based analogs. A major theoretical advantage of eigenvector-based network concepts is that they reveal simple relationships. Particularly simple relationships result if the maximum conformity assumption (6.32) is satisfied. Table 6.1 provides a rough dictionary for translating between correlation network analysis and the singular value decomposition if the underlying expression data have high eigenvector factorizability (say $EF(datX) > 0.95$) and if the maximum conformity assumption (6.32) is satisfied. However, even if the maximum conformity assumption does not hold, one can still find simple relationships among the network concepts (see (3.8)).

The geometric interpretation of correlation networks facilitates the derivation of several results that should be interesting to network theorists. For example, we argue that highly connected intramodular hub nodes cannot be intermediate between two distinct correlation modules (Fig. 6.1b). The geometric interpretation is particularly useful when studying node significance and module significance measures that are based on a sample trait (5.20). To study the relationship between connectivity and node significance, we use the measure of hub node significance (1.23). For a trait-based node significance measure, the hub node significance is determined by the angle between its eigenvector and the sample trait (3.28). Our geometric interpretation of co-expression networks allows us to describe situations when a factorizable network (e.g., a correlation module) has low hub node significance (Fig. 6.1).

The correspondence between factorizable correlation networks and the singular value decomposition (Table 6.1) can break down when a high soft threshold is used for constructing a weighted network or when dealing with an unweighted network. Thus, eigenvector-based concepts do not replace network concepts when describing interaction patterns between nodes. We will illustrate that the geometric interpretation of correlation networks has important theoretical and practical implications that may guide the development and application of network methods.

The gene expression data applications (e.g., the brain cancer data) illustrate our theoretical results. We provide empirical evidence that co-expression modules tend to have high eigenvector factorizability, and that the maximum conformity assumption (6.32) is satisfied for low values of β. A robustness analysis shows that many of our theoretical results apply even if the underlying assumptions are not satisfied (Figs. 6.3, 6.6, Table 6.3, and Horvath and Dong 2008). We find that the correspondence between network concepts and their eigenvector-based analogs is typically better in weighted networks than in unweighted networks. Further, we find that results in weighted networks tend to be more robust than those in unweighted networks with regard to changing the network construction thresholds β and τ, respectively. Thus, weighted correlation networks are preferable over unweighted networks when a geometric interpretation of network concepts is desirable.

We briefly mention that R code and data underlying the brain cancer and mouse gene co-expression network applications can be found at the following webpage: www.genetics.ucla.edu/labs/horvath/ModuleConformity/GeometricInterpretation.

6.13 R Code for Simulating Gene Expression Data

In the following, we illustrate functions in the `WGCNA` R library for simulating gene expression data. The gene expression data are stored in a $m \times n$ dimensional matrix *datExpr* (analogous to our earlier notation *datX* for a matrix of numeric vectors). The columns of *datExpr* correspond to gene expression profiles that follow a normal distribution. We will simulate $n = 3,000$ genes for which $m = 50$ sample measurements are available. Apart from the genes, we also simulate a binary sample trait y which results from thresholding a continuous sample trait *y.continuous* at its median value.

We assume that five modules (clusters) underly the expression data. Clusters are color coded (and labeled) by turquoise, blue, brown, green, and yellow. Only two of these modules (brown and green) are simulated to relate to y. Not all genes are simulated to be part of a module: a relatively large number of genes will be simulated outside of the proper modules and will be color-coded gray.

A co-expression module is simulated to be a matrix whose columns correspond to genes with varying correlations with a input vector, which is referred to as *seed module eigengene*. Thus, the resulting module genes have different levels of correlations with the seed module eigengene. The seed module eigengene is not part of the module.

The simulation proceeds along two steps. First, seed module eigengenes are simulated, and some of them are correlated with the sample trait y.continuous. Second, gene expression vectors are simulated around the seed module eigengenes. Toward this end, we use the function `simulateDatExpr5Modules`, which simulates five modules (corresponding to the colors turquoise, blue, brown, green, yellow). Module sizes (proportions) are specified using the function argument `simulateProportions`. The following contains relevant R code for step 1 of the simulations:

```
library(WGCNA)
#set the seed of the random number generator
set.seed(1)
# number of samples
m=50
#Step 1: simulate the seed module eigengenes
sMEturquoise= rnorm(m)
#expected cor(sMEblue,sMEturquoise)=0.6
sMEblue=0.6*sMEturquoise+sqrt(1-0.6^2)*rnorm(m)
sMEyellow=rnorm(m)
```

```
sMEgreen=rnorm(m)
#expected cor(y.continuous,seed.ME)=0.6
y.continuous=0.6*sMEgreen+sqrt(1-0.6^2)*rnorm(m)
#expected cor(y.continuous,seed.ME)=-0.6
sMEbrown=-0.6*y.continuous+sqrt(1-0.6^2)*rnorm(m)
#dichotomize y.continuous
y=ifelse(y.continuous>median(y.continuous),2,1)
datsME=
data.frame(y,sMEturquoise,sMEblue,sMEbrown,sMEgreen,sMEyellow)
```

This results in the following correlation structure between the binary outcome y and the seed module eigengenes.

```
>signif(cor(datsME,use="p"),1)
                 y  sMEturquoise  sMEblue  sMEbrown  sMEgreen  sMEyellow
y             1.00  -0.100        0.010    -0.500     0.3       0.10
sMEturquoise -0.10   1.000        0.500    -0.002    -0.3       0.03
sMEblue       0.01   0.500        1.000     0.003    -0.2       0.02
sMEbrown     -0.50  -0.002        0.003     1.000    -0.5      -0.05
sMEgreen      0.30  -0.300       -0.200    -0.500     1.0       0.20
sMEyellow     0.10   0.030        0.020    -0.050     0.2       1.00
```

The following contains relevant R code for step 2 of the simulations:

```
n=3000
# proportion in the turquoise,blue,brown,yellow,green module:
simulateProportions1=c(0.2,0.15,0.08,0.06,0.04)
# Note that the proportions don't add up to 1 since we allow for
#extramodular background genes, colored in gray.
dat1=simulateDatExpr5Modules(MEturquoise=sMEturquoise,
MEblue=sMEblue,MEbrown=sMEbrown,MEyellow=sMEyellow,
   MEgreen=sMEgreen,
nGenes=n,simulateProportions=simulateProportions1)
# simulated expression data
datExpr=dat1$datExpr
# next we assign names to the rows and columns of datExpr
datExpr=data.frame(datExpr)
ArrayName=paste("Sample",1:dim(datExpr)[[1]],sep="")
dimnames(datExpr)[[1]]=ArrayName
GeneName=paste("Gene",1:dim(datExpr)[[2]],sep="")
dimnames(datExpr)[[2]]=GeneName
#simulated true module color of each gene
truemodule = dat1$truemodule
```

The following R output reports the number of genes per module.

```
>table(truemodule)
# output
blue brown  green grey turquoise yellow
450  240    120   1410    600    180
```

Note that 1,410 genes are gray genes outside the proper modules.

The following R code shows how to carry out the singular value decomposition of the gene expression data corresponding to the brown module. Further, it shows how to create a heatmap which color-codes the values of the scaled expression data (red corresponds to high values and green to low values).

```
library(WGCNA)
# scale the columns to mean zero and variance one
datExpr=scale(datExpr)
colorh1=truemodule
# choose the module
q="brown"
datExpr.q=datExpr[,colorh1==q]
# R command for the singular value decomposition
svd.q=svd(datExpr.q)
VectorSingularValues=svd.q$d
U.q=svd.q$u
# here we define a diagonal matrix
D.q=diag(VectorSingularValues)
V.q=svd.q$v
dim(datExpr.q)
# check the dimensions of the matrices
dim(U.q); dim(D.q); dim(V.q)
# check that the columns are orthonormal
round(t(U.q)%*% U.q,10)
round(t(V.q)%*% V.q,10)
# right hand side of the SVD decomposition
datRHS=(U.q) %*% D.q %*% t(V.q)
#difference between left and right hand side is zero
summary(as.numeric(datExpr.q-datRHS))
# preliminary definition of the eigenvector (up to a sign)
E.q=U.q[,1]
# fix the sign of the eigenvector
signAverageCor=sign(sum(cor(E.q,datExpr.q,use="p"),na.rm=T))
E.q=signAverageCor*E.q
PVE=VectorSingularValues[1]/sum(VectorSingularValues^2)
# set the margins of the graphics window
par(mfrow=c(2,1),mar=c(0.3,5.5,3,2))
# create a heatmap whose columns correspond to the arrays
# and whose rows correspond to genes
plot.mat(t(scale(datExpr.q)),cex.axis=2,nrgcols=30,
rlabels=F,rcols=q,main=paste(q,"module"),cex.main=2)
par(mar=c(5,4.2,1,0.7))
barplot(E.q,col=q,main=paste("Eigenvector.",q,sep=""),
cex.main=2,cex.lab=2,ylab="eigengene expression",xlab="sample")
```

A heatmap of module expression values and the corresponding eigenvector can be found in Fig. 6.7. To define a data frame of module eigengenes corresponding to the module colors in the vector colorh1, one can use the following R code.

```
MEoutput=moduleEigengenes(datExpr,colorh1)
datME=MEoutput$eigengenes
#verify that the brown module eigengene equals E.q
summary(datME$MEbrown-E.q)
#This code can be used to get the correlation p values
#between the binary trait and the module eigengenes
ES.pvalue=corPvalueStudent(cor(y,datME,use="p"),nSamples=length(y))
```

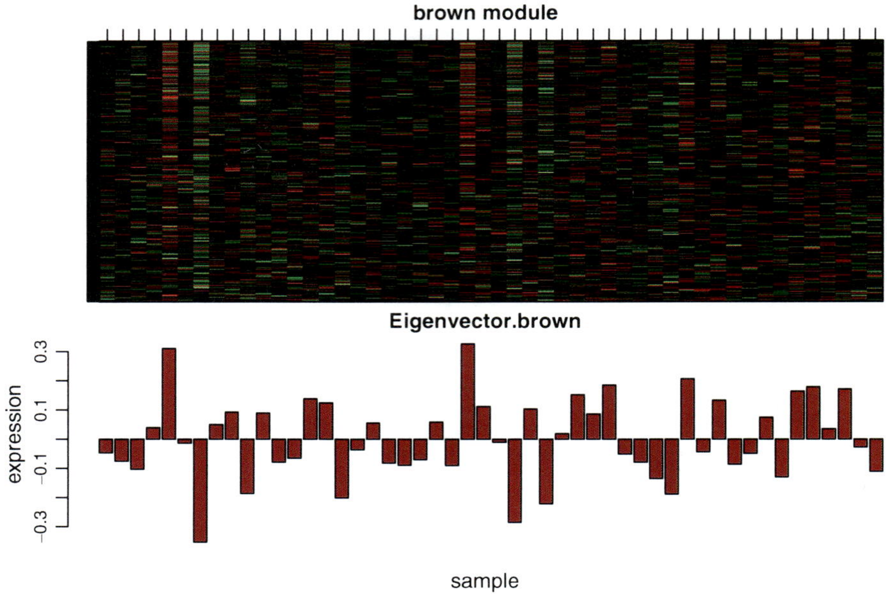

Fig. 6.7 Heatmap of the brown module gene expressions (rows correspond to genes, columns correspond to the 50 samples). The heatmap color-codes the scaled gene expression values: *red* corresponds to overexpression and green to underexpression. The barplots underneath the heatmap show the expression levels (y-axis) of the *brown* module eigenvector (eigengene) across the 50 samples (x-axis). Thus, each bar corresponds to a component of the eigenvector. Note that the eigenvector takes on a high value (high bar) in samples where the module genes tend to be overexpressed (*red*). Conversely, the eigenvector takes on a negative value in samples where the genes are underexpressed (*green*)

The correlation between an eigenvector and a sample trait is sometimes called eigenvector significance (ES). The following R code computes the p values for the eigengene significances and the correlations in our simulated data.

```
> signif(ES.pvalue,2)
   MEblue MEbrown MEgreen MEgrey MEturquoi MEyellow
[1,]     1 7.7e-05   0.042   0.38      0.55     0.35

# the variance explained by each module eigengene.
> varExplained1=signif(as.numeric(MEoutput$varExplained),2)
> rbind(names(datME),varExplained1)
  "MEblue" "MEbrown" "MEgreen" "MEgrey" "MEturquoise"
     "MEyellow"
varEx "0.29"   "0.29"    "0.28"    "0.077"      "0.28"  "0.28"
# Note the low value for the improper grey module
# correlate the seed MEs with the observed MEs
> signif(cor(data.frame(datsME,datME)),1)
           MEblue MEbrown MEgreen MEgrey MEturquoise MEyello
y           3e-04  -0.500     0.3 -0.100     -0.09    0.10
sMEturquo  5e-01  -0.009    -0.3  1.000      1.00    0.04
```

```
sMEblue     1e+00   0.020   -0.2  0.500    0.50    0.03
sMEbrown    2e-02   1.000   -0.5  0.004   -0.01   -0.04
sMEgreen   -2e-01  -0.500    1.0 -0.300   -0.30    0.20
sMEyellow   1e-02  -0.040    0.2  0.040    0.03    1.00
```

Note that the observed (inferred) module eigengenes have correlation 1 with the underlying simulated seed eigengenes.

6.14 Exercises

1. Exercise. Define a weighted correlation network with $\beta = 1$ for the simulated expression data described in Sect. 6.13. Use the binary vector y to define a trait-based node significance measure.

   (i) Use the WGCNA function `networkConcepts` to compute fundamental network concepts, conformity-based network concepts, approximate conformity based network concepts, and eigenvector-based network concepts. Hint:

   ```
   NC=networkConcepts(datExpr,power=1,trait=y)
   ```

   (ii) Do you expect that fundamental network concepts can be approximated by their eigenvector-based analogs when considering the entire network? Hint: What is the eigenvector factorizability of the entire network?

   (iii) Next construct a subnetwork corresponding the nodes in the turquoise module. Calculate the eigenvector factorizability of the turquoise network module. Does the maximum conformity assumption hold?

   (iv) For the turquoise network module, numerically verify the relationships mentioned in the dictionary for translating between general network terms and their eigenvector-based counterparts (Table 6.1). Hint:

   ```
   networkConcepts(datExpr[,truemodule=="turquoise"],power=1,
     trait=y)
   ```

2. Exercise regarding module separability. Assume that two modules (labeled q_1 and q_2) have been defined for an unsigned weighted correlation network $A_{ij} = |cor(x_i, x_j)|^{\beta}$. Denote by $M.q1$ and $M.q2$ the sets of node indices that correspond to modules q_1 and q_2. Show that if the two modules have eigenvector factorizability (6.11), then

$$separability^{average}(q_1, q_2) = 1 - \frac{\sum_{i \in M.q1} \sum_{j \in M.q1} A_{ij}}{n^{(q_1)} * n^{(q_2)} * \sqrt{density^{(q_1)} * density^{(q_1)}}}$$
$$\approx 1 - |cor(E^{(q_1)}, E^{(q_2)})|^{\beta}.$$

   Hint: Either show it mathematically or use the following R code to show it numerically.

```
library(WGCNA)
m=500
power1=3
n1=200
n2=300
# first eigennode
E1=rnorm(m)
# numeric data first module
datX1=simulateModule(M=E1,nGenes=n1)
ADJ1=adjacency(datX1,power=power1,type="unsigned")
# true correlation between the first and second eigenvector
r=0.5
# second eigennode
E2=r*E1+sqrt(1-r^2)*rnorm(m)
# numeric data first module
datX2=simulateModule(M=E2,nGenes=n2)
ADJ2=adjacency(datX2,power=power1,type="unsigned")
# combined numeric data
datX=data.frame(datX1,datX2)
# combined adjacency matrix
ADJ= abs(cor(datX))^power1
# adjacencies between the modules
BetweenADJ=ADJ[1:n1,c((n1+1):(n1+n2))]
Density1=mean(as.dist(ADJ1))
Density2=mean(as.dist(ADJ2))
separability=1-mean(BetweenADJ)/sqrt(Density1*Density2)
# fundamental network concept
separability
# approximation if the factorizability is high
1-abs(cor(E1,E2))^power1
```

3. Exercise regarding the definition of second-order eigenvector-based network concepts. Recall that eigenvector-based network concepts approximate fundamental network concepts if the eigenvector factorizability is close to 1 (Observation 6.2). But this approximation breaks down in networks with low factorizability.

   (i) Generalize eigenvector-based network concepts so that they provide a more accurate approximation of their fundamental analogs.
   Hint: A fundamental network concept in a correlation network can be written as a function $NCF(C)$ which depends on an $n \times n$ dimensional matrix $C = (C_{ij})$ whose i, jth component is given by

$$C_{ij} = \begin{cases} cor(x_i, x_j) & \text{if } i \neq j \\ 0 & \text{if } i = j \end{cases} \tag{6.54}$$

   Note that the diagonal elements of C equal 0. The eigenvector-based analog of this network concept can be written as $NCF(C_E)$, where the i, jth element of the matrix C_E is given by

$$C_{E,ij} = cor(x_i, E)cor(x_j, E). \tag{6.55}$$

If the eigenvector factorizability is low, then (6.25) suggests to approximate C with the matrix $C_{E,2nd\ order}$ whose ith element is given by

$$C_{E,2nd\ order} = cor(x_i,E)cor(x_j,E) + cor(x_i,u_2)cor(x_j,u_2). \quad (6.56)$$

If NCF is a continuous function, then the fundamental network concept $NCF(C)$ is better approximated by the **second-order eigenvector-based network concept**:

$$NCF(C_{E,2nd\ order}) \quad (6.57)$$

(ii) Use numerical studies in whole networks to show that $NCF(C_{E,2nd\ order})$ is more accurate than $NCF(C_E)$ when it comes to approximating the fundamental network concept $NCF(C)$. Hint: Illustrate this for the clustering coefficient, the density, and other concepts. For example, use the 3,600 genes from the mouse liver gene co-expression network described in Chap. 12.

References

Adrian D, Chris H, Beatrix J, Joseph R, Guang Y, West M (2004) Sparse graphical models for exploring gene expression data. J Multivar Anal 90(1):196–212

Alter O, Brown PO, Botstein D (2000) Singular value decomposition for genome-wide expression data processing and modelling. Proc Natl Acad Sci USA 97:10101–10106

Carlson M, Zhang B, Fang Z, Mischel P, Horvath S, Nelson SF (2006) Gene connectivity, function, and sequence conservation: Predictions from modular yeast co-expression networks. BMC Genomics 7(7):40

Dong J, Horvath S (2007) Understanding network concepts in modules. BMC Syst Biol 1(1):24

Fuller TF, Ghazalpour A, Aten JE, Drake T, Lusis AJ, Horvath S (2007) Weighted gene coexpression network analysis strategies applied to mouse weight. Mamm Genome 18(6–7):463–472

Ghazalpour A, Doss S, Zhang B, Plaisier C, Wang S, Schadt EE, Thomas A, Drake TA, Lusis AJ, Horvath S (2006) Integrating genetics and network analysis to characterize genes related to mouse weight. PloS Genet 2(2):8

Hibbs MA, Hess DC, Myers CL, Huttenhower C, Li K, Troyanskaya OG (2007) Exploring the functional landscape of gene expression: Directed search of large microarray compendia. Bioinformatics 23(20):2692–2699

Holter NS, Mitra M, Maritan A, Cieplak M, Banavar JR, Fedoroff NV (2000) Fundamental patterns underlying gene expression profiles: Simplicity from complexity. Proc Natl Acad Sci USA 97(15):8409–8414

Horvath S, Dong J (2008) Geometric interpretation of gene co-expression network analysis. PLoS Comput Biol 4(8):e1000117

Horvath S, Zhang B, Carlson M, Lu KV, Zhu S, Felciano RM, Laurance MF, Zhao W, Shu Q, Lee Y, Scheck AC, Liau LM, Wu H, Geschwind DH, Febbo PG, Kornblum HI, Cloughesy TF, Nelson SF, Mischel PS (2006) Analysis of oncogenic signaling networks in glioblastoma identifies ASPM as a novel molecular target. Proc Natl Acad Sci USA 103(46):17402–17407

Langfelder P, Horvath S (2007) Eigengene networks for studying the relationships between co-expression modules. BMC Syst Biol 1(1):54

Liao JC, Boscolo R, Yang YL, Tran LM, Sabatti C, Roychowdhury VP (2003) Network component analysis: Reconstruction of regulatory signals in biological systems. Proc Natl Acad Sci USA 100(26):15522–15527

Oldham MC, Konopka G, Iwamoto K, Langfelder P, Kato T, Horvath S, Geschwind DH (2008) Functional organization of the transcriptome in human brain. Nat Neurosci 11(11):1271–1282

Shen R, Ghosh D, Chinnaiyan A, Meng Z (2006) Eigengene-based linear discriminant model for tumor classification using gene expression microarray data. Bioinformatics 22(21):2635–2642

Spellman PT, Sherlock G, Zhang MQ, Iyer VR, Anders K, Eisen MB, Brown PO, Botstein D, Futcher B (1998) Comprehensive identification of cell cycle-regulated genes of the yeast Saccharomyces cerevisiae by microarray hybridization. Mol Biol Cell 9(12):3273–3297

Tamayo P, Scanfeld D, Ebert BL, Gillette MA, Roberts CW, Mesirov JP (2007) Metagene projection for cross-platform, cross-species characterization of global transcriptional states. Proc Natl Acad Sci USA 104(14):5959–5964

West M, Blanchette C, Dressman H, Huang E, Ishida S, Spang R, Zuzan H, Olson JA, Marks JR, Nevins JR (2001) Predicting the clinical status of human breast cancer by using gene expression profiles. Proc Natl Acad Sci USA 98(20):11462–11467

Chapter 7
Constructing Networks from Matrices

Abstract Methods for defining an adjacency matrix are also known as network construction-, inference-, or deconvolution methods. The network adjacency matrix can be defined by transforming a similarity or dissimilarity matrix, a symmetric matrix, or even a general square matrix. Multiple similarity matrices can be combined into a single "consensus" network, which allows one to define consensus modules. A signed correlation network turns out to be rank-equivalent to a Euclidean-distance-based network between scaled vectors. Sample networks, which are often defined as distance-based networks, are useful for identifying outlying samples or observations. A distance-based adjacency function yields an adjacency matrix A for which $dissA = 1 - A$ satisfies the triangle inequality and other distance properties. Distance-based adjacency functions are useful for generalizing the ARACNE algorithm to general networks. We describe how the Kullback–Leibler pre-dissimilarity can be used (a) for measuring the difference between matrices and (b) for network construction.

7.1 Turning a Similarity Matrix into a Network

A **similarity matrix** (also known as similarity measure) is defined by an $n \times n$ dimensional, symmetric matrix $S = (s_{ij})$ whose entries are nonnegative numbers $s_{ij} \geq 0$. Thus, the components of a similarity matrix satisfy the following conditions:

$$\begin{aligned} s_{ij} &\geq 0, \\ s_{ij} &= s_{ji}. \end{aligned} \tag{7.1}$$

A network adjacency matrix is a special case of a similarity matrix: it is a similarity matrix whose entries are smaller than or equal to 1 and whose diagonal elements equal 1. To construct a network adjacency matrix from a similarity matrix, one needs to transform the similarity matrix so that its off-diagonal elements lie between 0 and 1 and its diagonal elements equal 1. This can be achieved by finding upper bounds for the similarities. We call an $n \times n$ dimensional, symmetric matrix

S. Horvath, *Weighted Network Analysis: Applications in Genomics and Systems Biology*,
DOI 10.1007/978-1-4419-8819-5_7,

$Upperbounds(S) = (Upperbounds_{ij})$ a **matrix of upper bounds** for S if its elements satisfy the following conditions:

$$s_{ij} \leq Upperbounds_{ij} \text{ if } i \neq j \tag{7.2}$$

$$s_{ii} = Upperbounds_{ii} \tag{7.3}$$

$$Upperbounds_{ij} = Upperbounds_{ji}.$$

The **component-wise matrix division**

$$A = S/UpperBounds(S) \tag{7.4}$$

is defined as the matrix whose i,jth element is given by $s_{ij}/Upperbounds_{ij}$. Note that A (7.4) defines an adjacency matrix.

As an example, consider the similarity matrix $S{=}abs(cov(datX)){=}(|cov(x_i,x_j)|)$ of absolute values of covariances (5.12) between vectors x_i $(i = 1,\ldots,n)$. Equation (5.5) can be used to show that $Upperbounds(S) = \left(\sqrt{var(x_i)var(x_j)}\right)$ is a matrix of upper bounds for S. Then $A = S/UpperBounds(S) = (|cor(x_i,x_j)|)$ (7.4) defines an unsigned correlation network adjacency matrix.

7.2 Turning a Symmetric Matrix into a Network

In the following, we describe a very general way of defining an adjacency matrix based on a symmetric matrix whose elements are real (possibly negative) numbers. Denote the minimum and maximum entry of S by $min(S)$ and $max(S)$, respectively. Further, assume that $nonDecreasingF(s)$ is a nondecreasing function that maps the interval $[min(S), max(S)]$ into $[0,1]$. Then

$$A_{ij} = nonDecreasingF(s_{ij}) \text{ if } i \neq j,$$

$$A_{ii} = 1 \tag{7.5}$$

defines the elements of a weighted adjacency matrix. Note that any monotonically increasing function is also a nondecreasing function. Let us now describe several examples.

Example 1. For $\beta > 0$, the increasing function

$$increasingF^{power}(s) = \left(\frac{s - min(S)}{max(S) - min(S)}\right)^{\beta} \tag{7.6}$$

maps the interval $[min(S), max(S)]$ onto $[0,1]$.

Example 2. For any positive number $positiveNumber$,

$$increasingF(s) = 1 - exp\left(-positiveNumber * \left(\frac{s - min(S)}{max(S) - min(S)}\right)\right)$$

defines a monotonically increasing function that maps the interval $[min(S),\infty)$ into $[0,1)$.

Consider a symmetric matrix *Signed.S* that contains at least one negative entry. Further, assume that the matrix encodes "signed" relationships between n objects, i.e., its entries satisfy the following *intuitive requirements*:

$$\begin{aligned}
Signed.S_{ij} &> 0 \text{ Implies a positive, close relationship,}\\
Signed.S_{ij} &= 0 \text{ Implies a neutral, nonexistent or independent relationship,}\\
Signed.S_{ij} &< 0 \text{ Implies a negative, contrary, or opposite relationship.}
\end{aligned} \tag{7.7}$$

In a social network application, $Signed.S_{ij} = 0$ may encode that individuals i and j are neutral toward each other, while $Signed.S_{ij} > 0$ (or $Signed.S_{ij} < 0$) implies that they like (dislike) each other. For example, the matrix of pairwise correlations $cor(datX) = (cor(x_i, x_j))$ or the matrix of covariances $cov(datX) = (cov(x_i, x_j))$ is signed symmetric matrices.

To turn a signed similarity into an unsigned similarity, one can simply take the absolute values of the entries $S = abs(Signed.S) = (|Signed.s_{ij}|)$, which removes the sign information. Alternatively, one can preserve the sign information by defining a nondecreasing function *nonDecreasingF* that maps the interval $[min(signed.S), max(signed.S)]$ into $[0,1]$. Specifically, we define a **signed network adjacency matrix** $Signed.A = (Signed.A_{ij})$ as an adjacency matrix that arises from a nondecreasing function of the elements of *Signed.S*:

$$\begin{aligned}
Signed.A_{ij} &= nonDecreasingF(Signed.S_{ij}) \text{ if } i \neq j\\
Signed.A_{ii} &= 1.\\
Signed.A_{ij} &\geq 0.
\end{aligned} \tag{7.8}$$

While the signed similarity values $signed.s_{ij}$ may be negative, we require that the elements of *Signed.A* be **nonnegative** numbers smaller than 1. A signed correlation network can be defined as follows:

$$Signed.A_{ij} = (0.5 + 0.5cor(x_i, x_j))^{\beta},$$

where β is sometimes referred to as soft threshold. We typically choose a high soft threshold value, e.g., $\beta = 12$.

7.3 Turning a General Square Matrix into a Network

To turn a general $n \times n$ dimensional matrix M into a network, one can follow the following steps: First, M is symmetrized. Second, the symmetrized matrix is turned into a network using the methods outlined in Sect. 7.2. A matrix can be symmetrized in many different ways such as follows:

$$
\begin{aligned}
Symmetrize^{ave}(M)_{ij} &= \frac{M_{ij}+M_{ji}}{2},\\
Symmetrize^{max}(M)_{ij} &= max(M_{ij},M_{ji}),\\
Symmetrize^{min}(M)_{ij} &= min(M_{ij},M_{ji}).
\end{aligned}
\tag{7.9}
$$

The following R code can be used to symmetrize a matrix

```
Sym.Ave=(M+t(M))/2
Sym.Max=pmax(M,t(M),na.rm=T)
Sym.Min=pmin(M,t(M),na.rm=T)\, ,
```

where `pmax` and `pmin` are defined in (7.36).

7.4 Turning a Dissimilarity or Distance into a Network

A **dissimilarity** matrix is defined to be a symmetric matrix $D=(D_{ij})$ of nonnegative numbers with diagonal elements equal to 0, i.e., its elements satisfy the following conditions:

$$
\begin{aligned}
&(1)\ D_{ij} \geq 0 (\text{nonnegativity}),\\
&(2)\ D_{ij} = D_{ji} (\text{symmetry}),\\
&(3)\ D_{ii} = 0 (\text{zero diagonal}).
\end{aligned}
\tag{7.10}
$$

A **distance** measure (also known as metric) is a special type of dissimilarity, which satisfies the following additional conditions:

$$
\begin{aligned}
&(4)\ D_{ij} = 0 \text{ if, and only if, } i=j \text{ (identity of indiscernibles)},\\
&(5)\ D_{ij} \leq D_{ik}+D_{kj} \text{ (triangle inequality or sub-additivity)}.
\end{aligned}
\tag{7.11}
$$

The identity of indiscernibles (condition 4) and nonnegativity (condition 1) implies that a distance matrix is positive definite, i.e., its eigenvalues are positive.

A distance is called an **ultradistance** (or ultra-metric) if it satisfies the following stronger version of the triangle inequality where objects (e.g., vectors) can never fall "between" other objects:

$$
D_{ij} \leq max(D_{ik},D_{kj}). \tag{7.12}
$$

For example, the cophenetic distance (8.9) based on a hierarchical cluster tree is an ultradistance.

Any monotonically decreasing function can be used to turn a dissimilarity matrix into a similarity:

$$
S = decreasingFunction(D) = (decreasingFunction(D_{ij})).
$$

Next the resulting similarity matrix can be used to define a network adjacency matrix as described in Sect. 7.1. For example, if $UpperBounds(D)$ denotes a symmetric matrix of element-wise upper bounds for D then,

$$A = 1 - D/UpperBound(D) \tag{7.13}$$

defines an adjacency matrix. For any dissimilarity measure D, the monotonically decreasing function $decreasingFunction(s) = exp(-positiveNumber * s)$ yields a weighted adjacency matrix

$$A = exp(-positiveNumber * D), \tag{7.14}$$

where high values of the parameter $positiveNumber$ lead to smaller adjacencies.

7.5 Networks Based on Distances Between Vectors

In the following, we will review several distance measures between numeric vectors. For ease of notation, we will define the distances between two numeric vectors x and y. The p**-norm distance** is defined by

$$||x-y||_p = \left(\sum_{u=1}^{m}(x_u - y_u)^p\right)^{1/p}. \tag{7.15}$$

Choosing $p = 2$ results in the **Euclidean distance** (denoted simply by $||x-y||$), i.e.,

$$||x-y|| = ||x-y||_2 = \left(\sum_u (x_u - y_u)^2\right)^{1/2}.$$

The 1-norm distance is known as **Manhattan** or **taxicab distance**:

$$||x-y||_1 = \sum_u |x_u - y_u|.$$

The **infinity norm distance**, also known as **Chebyshev distance**, is given by

$$||x-y||_\infty = \lim_{p\to\infty}\left(\sum_u (x_u - y_u)^p\right)^{1/p} = max(|x_1 - y_1|, \ldots, |x_m - y_m|), \tag{7.16}$$

which is the maximum absolute value across the components of $|x-y|$. Like the Manhattan distance, the **Canberra distance**

$$||x-y||_{Canberra} = \sum_u \frac{|x_u - y_u|}{|x_u| + |y_u|} \tag{7.17}$$

is named after a location (capital of Australia), but it is *not* a special case of a p norm distance. Different from the p norm distances that measure absolute differences, the Canberra distance measures the relative differences between the components. If $x_u = 0$ and $y_u = 0$, then $\frac{|x_u - y_u|}{|x_u| + |y_u|} = \frac{0}{0}$ is set to 1.

Consider now a set of n numeric vectors represented as columns of the $m \times n$ dimensional matrix $datX$ and a distance measure $dist$ between them. Then the matrix of pairwise distances $D = dist(x_i, x_j)$ can be transformed into network (as described in Sect. 7.4). As an example network, consider the pairwise *squared* Euclidean distances $D = (||x_i - x_j||^2)$ between a set vectors x_i $(i = 1, \ldots, n)$. Denote by $maxDiss = max(D)$ the maximum dissimilarity (squared distance) between the vectors. Then the **canonical Euclidean distance-based network** (7.18) is given by

$$A_{ij} = 1 - ||x_i - x_j||^2 / maxDiss.$$

The connectivity of the ith node is given by (7.19)

$$k_i = (n-1) - \frac{n * ||x_i - \bar{x}||^2 + \sum_j ||x_j - \bar{x}||^2}{maxDiss}.$$

The following R code defines a function for computing the p norm distance function:

```
pNormDist=function(x,y,p){sum(abs(x-y)^p)^(1/p)}
```

To compute a matrix of Euclidean distances between the columns of a data frame `datX`, one can use the following R function in the `cluster` package (Kaufman and Rousseeuw 1990):

```
EuclideanDistBetweenRows=as.matrix(dist(t(datX),
  method="euclidean"))
```

Different choices of the `method` argument of `dist` result in other distances. For example, choosing `method="maximum"` results in the infinity norm distance (7.16), choosing `"manhattan"` leads to the Manhattan distance, and `"canberra"` leads to the Canberra distance (7.17).

7.6 Correlation Networks as Distance-Based Networks

Consider the matrix of pairwise squared Euclidean distances $D = (||x_i - x_j||^2)$ between a set vectors x_i $(i = 1, \ldots, n)$. Denote by $maxDiss = max(D)$ the maximum dissimilarity (squared distance) between the vectors. Then

$$A_{ij} = 1 - ||x_i - x_j||^2 / maxDiss \tag{7.18}$$

defines a network adjacency matrix, which we will call the **canonical Euclidean distance-based network**. In an exercise, you are asked to prove that the connectivity $k_i = \sum_{j \neq i} A_{ij}$ of the ith node is given by

$$k_i = (n-1) - \frac{n||x_i - \bar{x}||^2 + \sum_j ||x_j - \bar{x}||^2}{maxDiss}, \tag{7.19}$$

where $\bar{x} = \frac{1}{n} \sum_j x_j$. Thus, k_i is a monotonic function of the distance between x_i and the mean vector $\bar{x}$.

Consider the Euclidean distance-based network (7.18) between scaled vectors $scale(x_i),\ldots,scale(x_n)$. In an exercise, you are asked to show

$$A_{ij}^{Euclid} = 1 - ||scale(x_i) - scale(x_j)||^2/maxDiss = \frac{cor(x_i,x_j) - minCor}{1 - minCor}, \quad (7.20)$$

where $minCor = min(vectorizeMatrix(cor(datX)))$ denotes the minimum entry of the pairwise correlation matrix. Equation (7.20) can be derived using the following relationship:

$$||scale(x_i) - scale(x_j)||^2 = 2(m-1)(1 - cor(x_i,x_j)), \quad (7.21)$$

where m equals the number of components of x. Note that A_{ij}^{Euclid} is a monotonically increasing function of $cor(x_i,x_j)$. Thus, any signed correlation networks (defined as an increasing function of $cor(x_i,x_j)$) is rank-equivalent (Sect. 4.7) to the Euclidean distance-based network between scaled vectors $scale(x_i)$ (7.20).

Let us now consider two networks $A^{method1}$ and $A^{method2}$ which have been constructed between a set of n vectors $x_1,\ldots,x_n$. Define a Euclidean distance-based network

$$A_{ij}^{method1} = decreasingF(||scale(x_i) - scale(x_j)||)$$

as a monotonically decreasing function of the Euclidean distance of scaled numeric vectors. Further, define a signed correlation network

$$A_{ij}^{method2} = increasingF(cor(x_i,x_j)).$$

In an exercise, you are asked to prove that networks $A^{method1}$ and $A^{method2}$ are rank-equivalent.

7.7 Sample Networks for Outlier Detection

While networks methods are typically used to describe the relationships between genes and their gene products, we find that network methods and concepts are also very useful for describing relationships between samples (e.g., mice, patients, etc.). **Mike Oldham** has demonstrated that sample networks can greatly enhance unsupervised learning methods such as cluster analysis. In particular, sample networks are useful for identifying outliers (based on $Z.k$, (7.24)) and batch effects in genomic data (Oldham et al. 2011). In the following, we focus on sample networks for understanding the relationships between the *rows* of a numeric data matrix $datX$. In this notation, the uth sample S_u equals the uth row of $datX$ where $u = 1,\ldots,m$. The **Euclidean distance-based sample network** is simply the canonical Euclidean distance-based network (7.18) between the samples S_u, i.e.,

$$A_{uv} = 1 - ||S_u - S_v||^2/maxDiss, \quad (7.22)$$

where *maxDiss* is the maximum squared Euclidean distance between the samples. An advantage of the Euclidean distance-based sample network is that the resulting connectivity measure $k_u = \sum_{v \neq u} A_{uv}$ has the following nice interpretation: $k_u = (m-1) - \frac{m*||S_u - \bar{S}||^2 + \sum_v ||x_v - \bar{S}||^2}{maxDiss}$ (7.19), i.e., it is determined by the Euclidean distance between sample S_u and the average sample $\bar{S}$.

The **correlation-based sample network** is defined as follows:

$$A_{uv} = (0.5 + 0.5cor(S_u, S_v))^{\beta}, \tag{7.23}$$

where we typically choose $\beta = 2$. Note that (7.23) defines a signed weighted correlation network adjacency matrix (5.25). A major advantage of defining a network adjacency matrix (as opposed to a general similarity measure) between samples is that it allows specification of network concepts (e.g., in particular the connectivity and clustering coefficient). Signed correlation networks are superior to unsigned networks since it is essential to distinguish positive from negative sample correlations. Having said this, negative correlations between samples occur rarely in practice. Weighted networks are often preferable to unweighted networks since they preserve the continuous nature of the sample relationships. While any other power β could be used in (7.23), the choice of $\beta = 2$ has a major advantage: it results in an adjacency measure that is close to the correlation when the correlation is large (say larger than 0.6). Note that cor= $0.6, 0.7, 0.8, 0.9, 1.0$ leads to an adjacency of $0.64, 0.72, 0.81, 0.90, 1.0$, respectively. A major advantage of the correlation-based sample network ((7.23) with $\beta = 2$) is that the resulting connectivity measure k_u has the nice interpretation as sum of correlations:

$$k_u \approx \sum_{v \neq u} cor(S_u, S_v)$$

if $cor(S_u, S_v) \geq 0.6$,

While k_u has nice interpretations for our Euclidean distance-based and correlation-based sample networks, it can be challenging to determine what constitutes an extreme value in a given dataset. Therefore, we find it expedient to define a standardized measure ($Z.k_u$) as follows:

$$Z.k_u = scale(k)_u = \frac{k_u - mean(k)}{\sqrt{var(k)}}. \tag{7.24}$$

The resulting distributions of $Z.k$ values can be examined for a given sample network to determine whether any samples appear to be outliers (e.g., those samples that are two or three standard deviations below the mean for the dataset). This process can be performed iteratively until no samples meet this criterion, and/or a target sample network metric for the entire dataset has been attained (e.g., until the sample network density has reached a sufficiently high value). We find that sample network concepts (including the clustering coefficient) are useful for low-level analysis tasks (e.g., outlier detection) and for high-level tasks (e.g., group comparisons) (Oldham et al. 2011).

To calculate distance-based adjacency matrices, one can use the `adjacency` function in the `WGCNA` R package.For example, to calculate the Euclidean distance-based sample network (7.22) between the rows of *datX*, one can use the following R code:

```
# since the nodes correspond to rows, we transpose the data
A.Euclid=adjacency(t(datX),type = "distance")
```

To calculate a sample network based on the (squared) Manhattan distance (1-norm distance) (7.15), one can use the following code

```
A=adjacency(t(datExprFemale),type="distance",
distOptions="method='manhattan'")
```

Similarly, a network adjacency can be defined based on the square of other types of distances. For example, `method='maximum'` or `='canberra'` chooses the infinity norm distance or the Canberra distance (7.17) , respectively. These and other distances are described in Sect. 7.5. Additional R code and an application of sample networks can be found in Sect. 12.1.

7.8 KL Dissimilarity Between Positive Definite Matrices

Here we propose a dissimilarity measure between positive definite matrices. Let us briefly review basic facts about positive definite matrices. An $n \times n$ real symmetric matrix Σ is **positive definite** if $\mathbf{z}^{\tau}\Sigma\mathbf{z} > 0$ for each nonzero numeric vector $\mathbf{z}$ with n components, where τ denotes the transpose. A symmetric matrix is positive definite if, and only if, all of its eigenvalues are positive. Since the determinant $det(\Sigma)$ is the product of eigenvalues of Σ, this implies that the determinant is positive and that Σ is invertible. For example, the variance–covariance matrix Σ of a random vector is typically assumed to be positive definite (see Sect. 15.1.3). Or the sample variance covariance matrix $S = cov(datX)$ of pairwise covariances between the columns of *datX* is positive definite as long as it is invertible.

Assume two $n \times n$ dimensional positive definite matrices denoted by Σ_X and Σ_Y. The notation will make sense when we interpret these matrices as variance covariance matrices. Denote by $det(\Sigma_Y)$ the matrix determinant (which can be defined as the product of eigenvalues). As difference measure between two positive definite matrices, we propose to use

$$PreDissim_{KL}(\Sigma_X, \Sigma_Y) = \frac{1}{2}\left[log\left(\frac{det(\Sigma_Y)}{det(\Sigma_X)}\right) + trace(\Sigma_X \Sigma_Y^{-1}) - no.rows(\Sigma_X)\right], \tag{7.25}$$

which we refer to as the *KL pre-dissimilarity*. The letters "KL" stand for "Kullback–Leibler", which reflects that this measure is closely related to the Kullback–Leibler divergence between two multivariate normal probability densities (defined in Sect. 15.1.3). Specifically, denote by p_X and p_Y the multivariate normal densities $N(0, \Sigma_X)$ and $N(0, \Sigma_Y)$, respectively. Then the Kullback–Leibler divergence

$KL(p_X, p_Y)$ (15.13) between the distributions equals the KL pre-dissimilarity between the variance–covariance matrices, i.e.,

$$KL(p_X, p_Y) = PreDissim_{KL}(\Sigma_X, \Sigma_Y). \tag{7.26}$$

While the Kullback–Leibler divergence $KL(p_X, p_Y)$ measures the difference between two probability densities, the KL pre-dissimilarity measures the difference between two positive definite matrices. In Sect. 11.1.5, we further motivate the KL pre-dissimilarity measure by showing that it is equivalent to a likelihood ratio statistic for evaluating the fit of a causal model.

The reason why $PreDissim_{KL}$ is called a *pre*-dissimilarity is that it satisfies all properties of a dissimilarity measure (Sect. 7.4) except for symmetry. To construct a true dissimilarity measure, the pre-dissimilarity can be symmetrized using one of the following approaches (compare with Sect. 7.3):

$$\begin{aligned} D_{KL}^{average}(\Sigma_X, \Sigma_Y) &= \frac{PreDissim_{KL}(\Sigma_X, \Sigma_Y) + PreDissim_{KL}(\Sigma_Y, \Sigma_X)}{2} \\ D_{KL}^{min}(\Sigma_X, \Sigma_Y) &= min(PreDissim_{KL}(\Sigma_X, \Sigma_Y), PreDissim_{KL}(\Sigma_Y, \Sigma_X)). \end{aligned} \tag{7.27}$$

We now describe how to construct a network among n positive-definite matrices $\Sigma_{X_1}, \ldots, \Sigma_{X_n}$. The pairwise (symmetric) KL-based dissimilarities (7.27) has given rise to a symmetric $n \times n$ dimensional dissimilarity matrix. To turn the symmetric matrix into a network adjacency matrix, one can use any of the approaches described in Sect. 7.2. Note that the nodes in the resulting network correspond to positive definite matrices.

We briefly mention a computationally convenient way of calculating the trace (sum of diagonal elements) in (7.25). One can avoid the computationally intensive matrix multiplication inside the trace function using the following relationship:

$$trace(\Sigma_X \Sigma_Y^{-1}) = sum(\Sigma_X * (\Sigma_Y^{-1})^{\tau}),$$

where $*$ denotes the component-wise matrix multiplication and *sum*() denotes the sum over all matrix elements.

7.9 KL Pre-Dissimilarity for Parameter Estimation

Here we outline how to use the Kullback–Leibler pre-dissimilarity $PreDissim_{KL}$ (7.25) as objective function, which allows one to estimate network parameters by minimization. Assume that the positive definite matrix Σ_X represents the observed data and the positive definite matrix $\Sigma_Y(\theta)$ represents a model-based matrix which depends on parameter values denoted by θ. A parameter estimate $\hat{\theta}$ can be defined as the minimizing solution of the KL pre-dissimilarity-based objective function

$$f(\theta) = PreDissim_{KL}(\Sigma_X, \Sigma_Y(\theta)). \tag{7.28}$$

For example, assume that $\Sigma_X = cov(datX)$ is the matrix of pairwise covariances between the n columns of the numeric matrix $datX$. Thus, the i,jth element of the $n \times n$ dimensional matrix Σ_X is given by $\Sigma_{Xij} = cov(x_i, x_j)$. If $\Sigma_X = cov(datX)$ is invertible, then it is a positive definite matrix.

Further assume that the model-based matrix $\Sigma_Y(\theta)$ allows the factorization

$$\Sigma_Y(\theta) = L(\theta)L(\theta)^\tau, \tag{7.29}$$

where $L(\theta)$ denotes a lower triangular matrix with strictly positive diagonal elements. Assume that some of the lower diagonal elements of $L(\theta)$ are constrained to zero, while nonzero elements represent the parameters θ to be estimated. By minimizing the KL-based objective function (7.28) with respect to θ, we arrive at the model-based estimate of the matrices $L(\hat{\theta})$ and $\Sigma_Y(\hat{\theta})$. To turn the symmetric matrix $\Sigma_Y(\hat{\theta})$ into a network adjacency matrix, one can use any of the approaches described in Sect. 7.2.

Many approaches exist for defining a model-based positive definite matrix $\Sigma_Y(\theta)$. For example, in Chap. 11, we describe how to use path-analytic models and causal models to parameterize a positive definite matrix $\Sigma_Y(\hat{\theta})$.

In Sect. 11.1.6, we show that a maximum likelihood-based method for evaluating the fit of a causal model (structural equation model) is equivalent to the KL predissimilarity applied to multivariate normal distributions.

7.10 Adjacency Function Based on Distance Properties

Recall that a distance is called an **ultradistance** (7.12) (or ultra-metric) if it satisfies the following stronger version of the triangle inequality where objects (e.g., vectors) can never fall "between" other objects:

$$D_{ij} \le max(D_{ik}, D_{kj}). \tag{7.30}$$

In the following, define the adjacency-based dissimilarity matrix $D(A) = 1 - A$. One can easily verify that $D(A)$ satisfies the properties of a dissimilarity matrix (Sect. 7.4). If D is an ultradistance, it satisfies the following inequality (7.12)

$$D(A)_{ij} \le max(D(A)_{ik}, D(A)_{kj}) \tag{7.31}$$

for any $k \neq i, j$. In terms of the original adjacencies, this inequality is equivalent to requiring that

$$\begin{aligned} A_{ij} &\ge min(A_{ik}, A_{kj}) \text{ for all k} \\ &= max_k(min(A_{ik}, A_{kj})). \end{aligned} \tag{7.32}$$

We define an adjacency matrix that satisfies condition (7.32), as **ultra adjacency matrix** since the corresponding dissimilarity satisfies the properties of an ultradistance.

Any adjacency matrix $A^{original}$ can be transformed into an ultra adjacency matrix using the ultra adjacency function $AF^{ultra}(A^{original}, \varepsilon)$ which depends on the "tolerance" parameter ε. Specifically, the ultra adjacency function sets all adjacencies to zero that violate a more tolerant version of inequality (7.32), i.e.,

$$AF^{ultra}(A^{original}, \varepsilon)_{ij} = \begin{cases} 0 & \text{if } A_{ij}^{original} < max_k(min(A_{ik}^{original}, A_{kj}^{original})) - \varepsilon \\ A_{ij}^{original} & \text{otherwise} \end{cases} \tag{7.33}$$

If there is zero tolerance (i.e., $\varepsilon = 0$), then $AF^{ultra}(A^{original}, \varepsilon = 0)$ defines an ultra adjacency matrix.

Analogously, one can use the triangle inequality $D(A)_{ij} \leq D(A)_{ik} + D(A)_{kj}$ as a motivation for the triangle adjacency function:

$$AF^{triangle}(A^{original}, \varepsilon)_{ij} = \begin{cases} 0 & \text{if } A_{ij}^{original} < max_k(A_{ik}^{original} + A_{kj}^{original}) - \varepsilon \\ A_{ij}^{original} & \text{otherwise} \end{cases} \tag{7.34}$$

If there is zero tolerance (i.e., $\varepsilon = 0$), then $AF^{triangle}(A^{original}, \varepsilon = 0)$ defines an adjacency matrix whose corresponding dissimilarity matrix

$$D = 1 - AF^{triangle}(A^{original}, \varepsilon = 0)$$

satisfies the triangle inequality.

7.11 Constructing Networks from Multiple Similarity Matrices

Here we describe methods for constructing a single network adjacency matrix on the basis of multiple symmetric matrices. These techniques are useful for constructing "consensus" networks based on multiple disparate data sets. A related task is to define consensus modules that are present in multiple networks.

Let us introduce some notation that resembles the list notation from the R software. The d-th $n \times n$ dimensional matrix is denoted as $M[[d]]$, where $d = 1, \ldots, no.M$. Thus, $M[[d]]$ can be interpreted as the dth component of a list M with $no.M$ components.

The i, jth element $M[[d]]$ is denoted by $M[[d]]_{ij}$.

Let us now introduce some functions for combining multiple $n \times n$ dimensional matrices into a single matrix. The **parallel mean** is defined by

$$pmean(M) = pmean(M[[1]], \ldots, M[[no.M]]) = \frac{M[[1]] + \ldots + M[[no.M]]}{no.M}. \tag{7.35}$$

Denote by $pmax(M)$ and by $pmin(M)$ the parallel maximum and parallel minimum across the matrices whose i,jth components are given by

$$
\begin{aligned}
pmax(M)_{ij} &= pmax(M[[1]],\ldots,M[[d]])_{ij} = max(M[[1]]_{ij},\ldots,M[[d]]_{ij}) \\
pmin(M)_{ij} &= pmin(M[[1]],\ldots,M[[d]])_{ij} = min(M[[1]]_{ij},\ldots,M[[d]]_{ij}),
\end{aligned} \tag{7.36}
$$

respectively. The parallel **quantile transformation** $pquantile_q(M)$ yields a single matrix whose i,jth component is given by the qth quantile of the corresponding adjacencies, i.e.,

$$
\begin{aligned}
pquantile_q(M)_{ij} &= pquantile_q(M[[1]],\ldots,M[[d]])_{ij} \\
&= quantile_q(M[[1]]_{ij},\ldots,M[[no.networks]]_{ij}).
\end{aligned} \tag{7.37}
$$

The parallel minimum (7.36) and parallel maximum are special cases of the parallel quantile transformation: $pmin = pquantile_{q=0}$ and $pmax = pquantile_{q=1}$.

Let us now consider the case of multiple adjacency matrices $A[[d]]$. One can easily prove that the parallel mean $pmean(A)$ and parallel quantile $pquantile_q(A)$ satisfy the defining conditions of an adjacency matrix.

The `WGCNA` R package contains the above-mentioned functions, e.g., `pquantile`, `pmean`, `pmin`.

7.11.1 Consensus and Preservation Networks

We now introduce the notion of a **consensus network adjacency matrix** $Consensus(A)$ based on several input adjacency matrices $A[[1]],\ldots,A[[no.networks]]$. Roughly speaking, the adjacency between two nodes in the consensus network is high if the input networks "agree" on the strength of this connection. For the case of an unweighted network, a consensus network could be defined by linking two nodes if, and only if, a link exists in each of the input networks. Thus, a consensus adjacency matrix can be defined as minimum of the input adjacencies:

$$
\begin{aligned}
Consensus(A) &= Consensus(A[[1]],\ldots,A[[no.networks]]) \\
&= pmin\,(A[[1]],\ldots,A[[no.networks]]).
\end{aligned} \tag{7.38}
$$

Using the parallel minimum for the definition of the *Consensus* adjacency may be too stringent when dealing with more than a handful of networks. A less stringent approach uses the quantile transformation (7.38) to define the consensus network as follows:

$$
Consensus_q(A) = pquantile_q\,(A[[1]],\ldots,A[[no.networks]]). \tag{7.39}
$$

Note that our stringent *pmin*-based consensus network (7.38) corresponds to the choice of $q = 0$. Less stringent consensus network can be defined by raising the value of q, e.g., $q = 0.25$ corresponds to the first quartile and $q = 0.5$ corresponds to the median.

Since $Consensus(A)$ is an adjacency matrix, it allows one to define network concepts. For example, the density of the consensus network reflects the level of agreement between the input adjacency matrices.

We define **consensus modules** as modules in the consensus network (7.39). To define consensus modules, one can use one of the following consensus dissimilarity measures as input of a clustering procedure:

$$\begin{aligned} D^{Consensus,ADJ} &= 1 - Consensus(A) \\ D^{Consensus,TOM} &= 1 - Consensus(AF^{TOM}(A[[1]]), AF^{TOM}(A[[2]]), \ldots) \\ D^{Consensus,TOMversion2} &= 1 - AF^{TOM}(Consensus(A)), \end{aligned} \quad (7.40)$$

where AF^{TOM} (1.38) is the topological overlap matrix-based adjacency function. For example, if hierarchical clustering is used, consensus modules are defined as branches of the resulting clustering tree. If *pmin* is used to define the consensus network, then the resulting consensus modules can be detected in each of the input networks $A[[d]]$.

We will now mention another way of combining multiple adjacency matrices into a network whose connection strengths measure how well preserved the network adjacencies are across multiple networks. The **preservation network** $Preserv(A)$

$$Preserve(A) = 1 - [pmax(A) - pmin(A)] \quad (7.41)$$

can be used to determine whether adjacencies are preserved across the different input networks. When $no.networks = 2$, then $Preserve(A)$ can be a valuable starting point of differential network analysis (Langfelder and Horvath 2007). Often we use the abbreviation $Preserv[1, \ldots, no.networks] = Preserv(A[[1]], \ldots, A[[no.networks]])$. The adjacency between i and j is highly preserved across the networks if $Preserv[1, 2, \ldots]_{ij} \approx 1$. Note that $Preserv[1, \ldots, no.networks]$ can be interpreted as adjacency matrix (1.1) and we refer to it as the *preservation network*. The scaled preservation connectivity (see (1.9)) of the preservation network,

$$C_i(Preserv[1, 2, \ldots]) = 1 - \frac{\sum_{j \neq i} pmax(A)_{ij} - pmin(A)_{ij}}{n-1}, \quad (7.42)$$

is an aggregate measure of adjacency preservation for the ith gene. The density of the preservation network,

$$Density(Preserv[1, 2, \ldots]) = 1 - \frac{\sum_i \sum_{j \neq i} pmax(A)_{ij} - pmin(A)_{ij}}{n(n-1)}, \quad (7.43)$$

is an aggregate measure of adjacency preservation across the networks $A[[1]]$, $A[[2]], \ldots$. As an aside, we mention that the *Consensus* transformation can be related to the *Preserv* transformation (7.41): if $pmax(A)_{ij} \approx 1$ for all pairs of nodes i, j, we find $Preserv(A) \approx Consensus(A)$.

We describe R functions that implement the detection of consensus modules based on $D^{Consensus,TOM}$ (7.40) in Sect. 12.6. For example, the function `blockwiseConsensusModules` constructs consensus modules based on multiple correlation networks. Further, the `WGCNA` package implements the following related functions `pquantile`, `pmin`, `pmax`, `pmean`.

7.12 Exercises

1. Exercise regarding the construction of a signed network based on a symmetric matrix S.
   (i) Propose three monotonically increasing functions for turning the symmetric matrix $S_{ij} = cov(x_i, x_j)$ (based on the covariance between vectors) into a network adjacency matrix A^{signed}. (Note that the resulting adjacency matrix is a signed network since it is threshold implied by S.)
   (ii) Why is it advantageous to require that a signed adjacency matrix satisfy the properties of an adjacency matrix? Hint: Mention important network concepts that would be difficult to interpret if some of the adjacencies A_{ij} were negative.
2. Regarding the network construction from a symmetric matrix. Outline how the hyperbolic tangent function $tanh(s) = \frac{exp(2*s)-1}{exp(2*s)-1}$ can be used to construct a weighted adjacency matrix on the basis of a symmetric matrix whose elements are real (possibly negative) numbers. Hint: Use the first derivative of $tanh$ to determine where this function is increasing.
3. Assume that S is a symmetric matrix whose elements are real (possibly negative) numbers. Denote the minimum and maximum entry of S by $min(S)$ and $max(S)$, respectively. Further, assume that $nonDecreasingF(s)$ is a nondecreasing function that maps the interval $[min(S), max(S)]$ into $[0, 1]$. Prove that

$$\begin{aligned} A_{ij} &= nonDecreasingF(s_{ij}) \text{ if } i \neq j, \\ A_{ii} &= 1 \end{aligned} \tag{7.44}$$

   defines the elements of a weighted adjacency matrix.
4. Exercise. Define the elements of an $n \times n$ dimensional symmetric matrix $s_{ij} = r_i - r_j$ based on n real numbers $r_1, r_2, \ldots, r_n$. Use (7.6) with $\beta = 1$ to define an adjacency matrix A.
   (i) Express the connectivity $k_i = \sum_{j \neq i} A_{ij}$ as a function of the real numbers.
   (ii) Under what conditions on the real numbers does the resulting network satisfy approximate scale-free topology? Hints: Consider skipping this difficult

task. Alternatively, determine when $ScaleFreeFit(A - I, no.bins)$ (1.13) is close to 1. Recall that the discretize function and the correlation are invariant with regard to linear transformations.

5. Exercise regarding networks based on p norm distances. Consider a set of n vectors $x_1, \dots, x_n$ each of which with m components.

   (i) Define a weighted network adjacency matrix among the n vectors based on a p norm distance given by $||x_i - x_j||_p$ (7.15). Hint: For example, (7.13) or (7.14) can be used.
   (ii) In general, is the network based on the $p = 1$ norm distance (Manhattan distance) rank-equivalent (10.38) to that based on $p = \infty$ (Chebysheff distance)? Hints: See Sect. 4.7. Can you find a counter example?
   (iii) Define a weighted network adjacency matrix among the n vectors based on the Canberra distance (7.17).
   (iv) In general, is the network based on the $p = 1$ norm distance (Manhattan distance) rank-equivalent to that based on the Canberra distance? Hint: Prove rank-equivalence or disprove it with a counter example.

6. Exercise regarding a network based on squared Euclidean distances. For a set of n numeric vectors represented as columns of the $m \times n$ dimensional matrix $datX$, define the matrix of pairwise squared Euclidean distances $D = (||x_i - x_j||^2)$. and denote by $maxDiss = max(D)$ the maximum dissimilarity between the vectors.

   (i) Prove that

   $$A_{ij} = 1 - ||x_i - x_j||^2 / maxDiss$$

   defines an adjacency matrix.
   (ii) Prove that the connectivity $k_i = \sum_{j \neq i} A_{ij}$ is given by

   $$k_i = (n-1) - \frac{n * ||x_i - \bar{x}||^2 + \sum_j ||x_j - \bar{x}||^2}{maxDiss},$$

   where $\bar{x} = \frac{1}{n} \sum_j x_j$. Hint: Either derive it mathematically or use the following R code to verify it.

```
library(WGCNA)
datX=matrix(1:50,nrow=5)
n=dim(datX)[[2]]
# Squared Euclidean distance among columns
Diss=as.matrix(dist(t(datX)))^2
maxDiss=max(Diss)
# Euclidean distance based adjacency
ADJ=1-Diss/maxDiss
diag(ADJ)=0
```

```
# alternatively, calculate ADJ as follows
ADJ=adjacency(datX,type="distance")

k=as.numeric(apply(ADJ,1,sum))
x.bar=apply(datX,1,mean)
for (i in 1:dim(datX)[[2]] ) {
x.i=datX[,i]
k.i=(n-1)-1/maxDiss*(n*sum((x.i-x.bar)^2)+sum(
   (datX-x.bar)^2))
print(paste("i=",i,",k.i.formula=",k.i,", True k.i=",k[i]))
}
```

(iii) Argue that the closer x_i is to the mean $\bar{x}$, the higher is its network connectivity.

7. Exercise regarding the relationship between Euclidean distance-based networks (7.18) and correlation networks.

   (i) Prove relationship (7.21):

   $$||scale(x_i) - scale(x_j)||^2 = 2(m-1)(1 - cor(x_i, x_j)),$$

   where m equals the number of components of the scaled vector $scale(x)$.

   (ii) Derive (7.20), i.e.,

   $$A_{ij}^{Euclid} = 1 - \frac{||scale(x_i) - scale(x_j)||^2}{maxDiss} = \frac{cor(x_i, x_j) - minCor}{1 - minCor}$$

   where $minCor = min(vectorizeMatrix(cor(datX)))$ denotes the minimum entry of the pairwise correlation matrix.

   (iii) Argue that A_{ij}^{Euclid} is rank-equivalent to a signed weighted correlation network. Hint: Rank-equivalence of networks is defined in Sect. 4.7. Note that a signed weighted correlation network can be expressed as follows: $A_{ij}^{signed,weighted} = increasingF(cor(x_i, x_j))$.

8. Exercise regarding rank-equivalence between a signed correlation network and a distance-based network. Consider two networks $A^{method1}$ and $A^{method2}$ which have been constructed between a set of n vectors $x_1, \ldots, x_n$ each of which has length m. The first network is defined as a monotonically decreasing function of the Euclidean distance of scaled numeric vectors, i.e.,

   $$A_{ij}^{method1} = decreasingF(||scale(x_i) - scale(x_j)||).$$

   The second network is defined as a monotonically increasing function of the Pearson correlation, i.e.,

   $$A_{ij}^{method2} = increasingF(cor(x_i, x_j)).$$

In the following, you will prove that networks $A^{method1}$ and $A^{method2}$ are rank-equivalent (see Sect. 4.7). Denote by $decreasingG(s) = \sqrt{2(m-1)(1-s)}$ and by $decreasingF^{-1}$ the inverse of $decreasingF$.

(i) Prove that

$$AF^{1\to2}(A^{original}) = increasingF(decreasingG(decreasingF^{-1}(A^{original})))$$

defines a rank-preserving adjacency function. Hint: Show that $AF^{1\to2}$ maps an adjacency matrix to another adjacency matrix, and that it satisfies (4.19).

(ii) Prove that

$$A^{method2} = AF^{1\to2}(A^{method1}).$$

Hint: Use $||scale(x_i) - scale(x_j)|| = \sqrt{2(m-1)(1-cor(x_i,x_j))}$ (7.21).

References

Kaufman L, Rousseeuw PJ (1990) Finding groups in data: An introduction to cluster analysis. Wiley, New York

Langfelder P, Horvath S (2007) Eigengene networks for studying the relationships between co-expression modules. BMC Syst Biol 1(1):54

Oldham MC, Langfelder P, Horvath S (2011) Sample networks for enhancing cluster analysis of genomic data: Application to huntington's disease. Technical Report

Chapter 8
Clustering Procedures and Module Detection

Abstract Detecting clusters (also referred to as groups or modules) of closely related objects is an important problem in data mining in general. Network modules are often defined as clusters. Partitioning-around-medoids (PAM) clustering and hierarchical clustering are often used in network applications. Partitioning-around-medoids (aka. k-medoid clustering) leads to relatively robust clusters but requires that the user specify the number k of clusters. Hierarchical clustering is attractive in network applications since (a) it does not require the specification of the number of clusters and (b) it works well when there are many singleton clusters and when cluster sizes vary greatly. But hierarchical clustering requires the user to determine how to cut branches of the resulting cluster tree. Toward this end, one can use the dynamicTreeCut method and R library. The dynamic hybrid method combines the advantages of hierarchical clustering and partitioning-around-medoids clustering. Network concepts are useful for defining cluster quality statistics (e.g., to measure the density or separability of clusters). To determine whether the cluster structure is preserved in another data sets, one can use cross-tabulation-based preservation statistics. To measure the agreement between two clusterings, one can use the Rand index and other cross-tabulation-based statistics.

8.1 Cluster Object Scatters Versus Network Densities

Clustering procedures are described in many books (Kaufman and Rousseeuw 1990; Hastie et al. 2001). Many of these procedures take as input a dissimilarity matrix $D = (d_{ij})$, i.e., a symmetric matrix whose diagonal elements equal 0 (see Sect. 7.4). In the following, we adapt notation from Kaufman and Rousseeuw (1990) and Hastie et al. (2001). The *total object scatter* with regard to the dissimilarity $D = (d_{ij})$ is defined as:

$$TotalScatter(D) = \frac{1}{2}\sum_{i=1}^{n}\sum_{j=1}^{n} d_{ij} = \sum_{i=1}^{n}\sum_{j>i}^{n} d_{ij}. \tag{8.1}$$

S. Horvath, *Weighted Network Analysis: Applications in Genomics and Systems Biology*, DOI 10.1007/978-1-4419-8819-5_8,

Assume that $Cl(i)$ encodes a cluster assignment, i.e., $Cl(i) = q$ if the ith object is in the qth cluster (where $q = 1,\ldots,k$ indexes the k clusters). Then the *within cluster object scatter* is defined as follows:

$$WithinScatter(Cl,D) = \frac{1}{2}\sum_{q=1}^{k}\sum_{Cl(i)=q}\sum_{Cl(j)=q} d_{ij}. \tag{8.2}$$

and the *between cluster object scatter* is defined as follows:

$$BetweenScatter(Cl,D) = \frac{1}{2}\sum_{q=1}^{k}\sum_{Cl(i)=q}\sum_{Cl(j)\neq q} d_{ij}. \tag{8.3}$$

One can easily verify that

$$BetweenScatter(Cl,D) = TotalScatter(D) - WithinScatter(Cl,D) \tag{8.4}$$

In principle, the definition of a cluster assignment Cl is straightforward: simply find the cluster assignment Cl that minimizes the within-cluster object scatter $WithinScatter(Cl,D)$ over all possible assignments of the n objects to k clusters. Minimizing $WithinScatter(Cl)$ with respect to Cl is equivalent to maximizing $BetweenScatter(Cl)$ as can be seen from (8.4). Unfortunately, optimization by complete enumeration is feasible only if n is relatively small. A more limited goal is to identify at least a local minimum of $WithinScatter(Cl)$. Toward this end, one can use an iterative greedy descent algorithm for finding the local minimum of the within object scatter.

Let us briefly describe the relationships between the object scatters and the density (1.18) of a suitably defined network. Assume that the dissimilarity D arises from the dissimilarity transformation of an $n \times n$ adjacency matrix, i.e., $D = D(A) = 1 - A$ (1.39). Then the total object scatter

$$TotalScatter(D) = \frac{n(n-1)}{2}(1 - Density(A))$$

is inversely related to the network density (1.18).

For a cluster assignment Cl, we find the following relationship:

$$WithinScatter(Cl,D(A)) = \sum_{q=1}^{Q}\frac{n^{(q)}(n^{(q)}-1)}{2}(1 - Density(A^{(q)})), \tag{8.5}$$

where $A^{(q)}$ denotes the $n^{(q)} \times n^{(q)}$ dimensional adjacency matrix of the qth cluster and the cluster sizes $n^{(q)}$ add up to the total number of objects $\sum_{q=1}^{Q} n^{(q)} = n$.

Partitioning-around-medoid clustering (described in Sect. 8.2) can be used to find a clustering assignment Cl, which represents a (local) minimum of the function $WithinScatter(Cl,D(A))$. If the cluster sizes $n^{(q)}$ were fixed (which they are not!), then minimizing $WithinScatter(Cl, D(A))$ would amount to maximizing the module densities $Density(A^{(q)})$. Roughly speaking, clustering methods that aim to maximize *WithinScatter* tend to result in modules with high module network density. This suggests to use the density of the network $A^{(q)}$ as one of many measures of cluster quality.

8.2 Partitioning-Around-Medoids Clustering

Partitioning-around-medoids (aka. k-medoid clustering) is a clustering procedure that implements an iterative algorithm for minimizing the within-cluster scatter $WithinScatter(Cl)$. A medoid can be defined as that object of a cluster, whose average dissimilarity to all the objects in the cluster is minimal, i.e., it is the most centrally located object inside a given cluster. The most common realization of k-medoid clustering is the Partitioning Around Medoids (PAM) algorithm (Kaufman and Rousseeuw 1990). PAM is a classical partitioning technique of clustering that clusters the data set of n objects into k clusters where the integer k is a user-supplied parameter. For a suitably chosen dissimilarity measure, PAM can be more robust to noise and outliers as compared to k-means. In contrast to the k-means algorithm, PAM chooses actual data objects as centers (referred to as medoids or centroids). PAM clustering is generally applicable since it can input any dissimilarity measure. Instead, k-means is an iterative algorithm for minimizing the squared *Euclidean distance* between vectors. Since k-means clustering cannot deal with a general dissimilarity matrix, it is generally not applicable for finding modules in a network.

The **PAM algorithm** proceeds along the following steps:

Step 1: The algorithm begins with an initial set of k medoids. The number of medoids k is a subset of the n objects, which entails that $k \leq n$. The initial medoids could (a) be randomly chosen, (b) be user-supplied, or (c) represent the results of another clustering procedure (e.g., hierarchical clustering).

Step 2: Each of the remaining objects (i.e., the non-medoid objects) are assigned to the medoid which is least dissimilar (according to the user-supplied dissimilarity measure). This step results in a cluster assignment Cl. Compute the within-cluster scatter $WithinScatter(Cl)$.

Step 3: For each of the k clusters (encoded in Cl) determine the medoid, i.e., the object that minimizes the average dissimilarity to the other objects in the cluster. Unless this results in a new set of medoids, stop.

Step 4: Compute the within-cluster scatter $WithinScatter_{newmedoids}(Cl)$ after swapping the initial medoids with the new set of medoids.

Step 5: If $WithinScatter_{newmedoids}(Cl) < WithinScatter(Cl)$, then swap the initial set of k medoids with the new set. Repeat steps 2–5 until there is no change in the medoid assignments.

In the following, we will present R code and corresponding output that illustrates how to use the command `pam` in the R library `cluster` (Kaufman and Rousseeuw 1990) for analyzing the simulated data from Sect. 6.13. Further, we cross-tabulate the pam clustering results with the simulated truth.

```
library(cluster); library(WGCNA)
# here we define a similarity matrix as adjacency matrix
ADJ6=abs(cor(datExpr,use="p"))^6
#Turn the similarity measure into a dissimilarity measure
dissTOM=TOMdist(ADJ6)
pam6=pam(as.dist(dissTOM),k=6)
#cross-tabulate pam clusters versus simulated true clusters
table1=table(pam6$clustering,truemodule)
table1
# output
> truemodule
   blue brown green grey turquoise yellow
  1 5 8 8 566 299 2
  2 25 8 10 507 275 3
  3 413 8 7 185 16 7
  4 1 3 0 52 4 158
  5 1 1 87 35 2 6
  6 5 212 8 65 4 4

randIndex(table1,adjust=F)#0.7089
chisq.test(table1)#p-value<2.2e-16
```

Note that the pam cluster assignment is significantly related to the simulated true module structure as one can see from Pearson's chi-square test and the unadjusted Rand index (8.24) of 0.71. But overall, pam clustering does not work well in this example because it clusters the "gray" background objects into the six clusters instead of treating them as singleton clusters.

8.3 *k*-Means Clustering

k-means clustering is a very popular iterative descent clustering methods for the special case where the dissimilarities $d_{ij} = ||x_i - x_j||$ correspond to squared Euclidean distances between numeric vectors. The user-specified parameter is the number of clusters k. In this case, one can show that the within-object scatter can be expressed as follows:

$$\begin{aligned} WithinScatter(Cl, ||.||^2) &= \sum_{q=1}^{k} \sum_{\{i|Cl(i)=q\}} \sum_{\{j|Cl(j)=q\}} ||x_i - x_j||^2 \\ &= \sum_{q=1}^{k} n^{(q)} \sum_{\{i|Cl(i)=q\}} ||x_i - mean^{(q)}(x)||^2, \end{aligned} \tag{8.6}$$

where

$$mean^{(q)}(x) = \frac{\sum_{\{i|Cl(i)=q\}} x_i}{n^{(q)}} \tag{8.7}$$

is the average across the $n^{(q)}$ vectors in the qth cluster and $n^{(q)}$ denotes the number of objects in the qth cluster.

Thus, one can find the optimal cluster assignment by minimizing the enlarged within-scatter function

$$EnlargedWithinScatter(Cl, m^{(1)}, \ldots, m^{(k)}) = \sum_{q=1}^{k} n^{(q)} \sum_{\{i|Cl(i)=q\}} ||x_i - mean^{(q)}(x)||^2 \tag{8.8}$$

with respect to the cluster assignment Cl and the $vectorsm^{(q)}$.

This can be minimized using the following **steps of the k-means algorithm**.

Step 1: Given the current cluster assignment Cl, minimize $EnlargedWithinScatter$ $(Cl, m^{(1)}, \ldots, m^{(k)})$ (8.8) with respect to the vectors $m^{(q)}$. This yields the means of the currently assigned clusters (8.7).

Step 2: Given the current set of means, $EnlargedWithinScatter(Cl, m^{(1)}, \ldots, m^{(k)})$ is minimized with respect to Cl by assigning each object to the closest (current) cluster mean. That is the new cluster assignment

$$Cl.new(i) = argmin_q ||x_i - m^{(q)}||^2$$

is defined by assigning the ith object to the cluster whose mean is closest to the ith object.

Step 3: Steps 1 and 2 are iterated until the assignments do not change.

The algorithm will converge to a local minimum. One can either start with many different random choices of starting means or choose the solution having smallest value of the within object scatter function or use another clustering method for initializing the cluster assignment.

Assume `datX` is a matrix or data frame whose columns correspond to numeric variables.

To cluster the numeric variables (the columns) using k-means, one can use the following R code:

```
Cl.kmeans=kmeans(t(datX),3)$cluster
```

In the following, we show how to use k-means clustering to define clusters of positively correlated vectors. Note that the Euclidean distance between $scale(x)$ and $scale(y)$ is proportional to $1 - cor(x,y)$ (see (7.21)). In other words, if the Euclidean distance between two *scaled* vectors is small, the vectors are positively correlated. Assume that each column of the matrix `datX` has a positive sample variance so that its scaled version is defined. Alternatively, restrict attention to columns with positive variance using the following R code:

```
ColumnVariance=apply(datX,2,var,na.rm=T)
datX=datX[,ColumnVariance>0]
```

Then k-means can be used to cluster the columns of $datX$ into three clusters of positively correlated vectors using the following R code:

```
Cl=kmeans(t(scale(datX)),3)$cluster
```

8.4 Hierarchical Clustering

Hierarchical algorithms find successive clusters based on previously defined clusters. These algorithms can be either bottom-up (agglomerative) or top-down (divisive). Here we will focus on agglomerative hierarchical clustering.

Agglomerative hierarchical clustering begins objects as a separate cluster and merges them into successively larger clusters. Hierarchical clustering creates a hierarchy of clusters which may be represented in a tree structure called a dendrogram, dendogram, or cluster tree. The root of the tree consists of a single cluster containing all objects, and the leaves correspond to individual objects.

Agglomerative hierarchical clustering has two inputs: (a) a pairwise dissimilarity measure and (b) a method for constructing an **inter-cluster dissimilarity** (aka. inter-group distance) measure. The inter-cluster dissimilarity (also known as **linkage method** or agglomeration method) is based on the pairwise dissimilarities between objects inside the clusters. Here we review three widely used approaches for defining the inter-cluster dissimilarity between two clusters $clust.q1$ and $clust.q2$. The first approach defines it as the average dissimilarity between objects of each cluster (also called **average linkage hierarchical clustering**):

$$d_{average}(clust.q1,\, clust.q2) = \frac{\sum_{i \in clust.q1} \sum_{j \in clust.q2} d_{i,j}}{|clust.q1||clust.q2|},$$

where $d_{i,j}$ denotes the pairwise dissimilarity between objects i and j, and $|clust.q1|$ denotes the number of objects in $clust.q1$.

The second approach defines the inter-cluster dissimilarity as the maximum dissimilarity between objects of each cluster (also called **complete linkage clustering**):

$$d_{complete}(clust.q1,\, clust.q2) = max(\{d_{i,j}|i \in clust.q1,\; j \in clust.q2\}).$$

The third approach defines the inter-cluster dissimilarity as the minimum dissimilarity between objects of each cluster (also called single linkage clustering):

$$d_{single}(clust.q1,\, clust.q2) = min(\{d_{i,j}|i \in clust.q1,\; j \in clust.q2\}).$$

The complete linkage method finds small but similar clusters ('perls on a string'). The single linkage method represents a friends of friends clustering strategy. Average linkage clustering can be regarded as aiming for clusters with characteristics somewhere between the single and complete linkage methods. We typically use average linkage clustering since it leads to robust clusters but some authors prefer complete linkage.

Agglomerative hierarchical clustering produces a **dendrogram** (aka. **cluster tree**) which is a data structure containing information on which objects (singletons or clusters) were merged at each step and on their merging height, that is the dissimilarity of the two merged objects that were merged at the particular step. By construction, the sequence of merging heights is increasing. The number of merges (and merging heights) is always smaller than $n-1$, where n is the number of input objects to be merged. Dendrograms are typically plotted by arranging the input objects along the x-axis and joining them by lines at the height (y-axis) corresponding to their merging heights; the result looks like an (upside down) tree whose branches correspond to clusters and leaves correspond to the input objects. The input objects are permuted such that the lines in the resulting plot do not cross.

In general, a dendrogram resulting from single linkage is different from that resulting from complete linkage. But if the input dissimilarity is an ultradistance (7.12), then the two dendrogram are the same as will be shown in an exercise.

In the following, we present R code for studying the effect of different intergroup dissimilarity measures in hierarchical clustering. In this book, we will use two R functions that implement agglomerative hierarchical clustering. First, the (slow) function `hclust` from the `cluster` R package (Kaufman and Rousseeuw 1990) and the (fast) function `flashClust` from the `flashClust` R package (Langfelder and Horvath 2011). The hierarchical clustering algorithm implemented in the function `hclust` is an order n^3 (n is the number of clustered objects) version of a publicly available clustering algorithm (http://astro.u-strasbg.fr/ fmurtagh/mda-sw/). The R package `flashClust` implements the original algorithm by Murtagh, which is of order n^2, leading to substantial time savings when clustering large data sets.

Consider the following set of real numbers

$$RealNumbers = \{1,2,3,5,10,12,14,18,25,26,28.40\}$$

and the Manhattan distance measure between the numbers, i.e., $dist(r_i,r_j) = |r_i - r_j|$. The following R code shows how to create hierarchical cluster trees corresponding to average linkage, complete linkage, and single linkage.

```
RealNumbers=c(1,2,3,5,10,12,14,18,25,26,28,40)
distManhattan=dist(RealNumbers,method="manhattan")
hierAverage=hclust(distManhattan,method="a")
hierComplete=hclust(distManhattan,method="complete")
hierSingle=hclust(distManhattan,method="single")
par(mfrow=c(1,3))
plot(hierAverage,main="Average Linkage",
labels=RealNumbers,cex.lab=1.5,cex=1.5)
plot(hierComplete,main="Complete Linkage",
labels=RealNumbers,cex.lab=1.5,cex=1.5)
plot(hierSingle,main="Single Linkage",
labels=RealNumbers,cex.lab=1.5,cex=1.5)
```

Figure 8.1 shows the resulting three cluster trees.

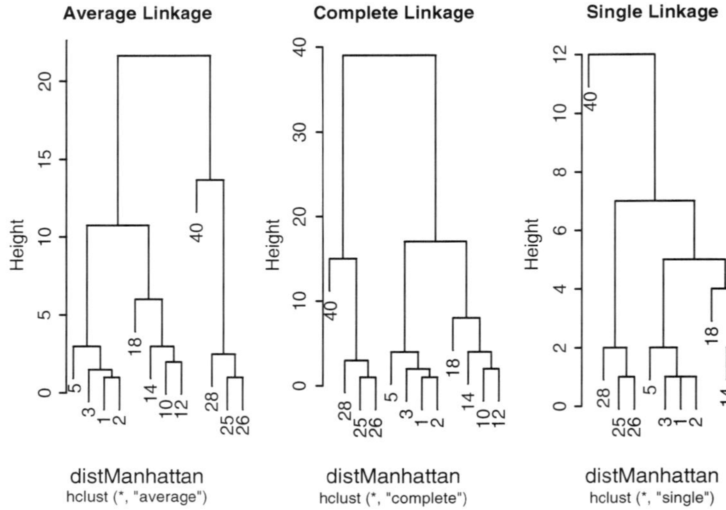

Fig. 8.1 Cluster trees corresponding to three intergroup dissimilarity measures (linkage methods). The leaves of the tree are labeled by the numbers. The y-axis shows the heights where leaves or clusters are merged. For example, the joining height between 1 and 2 corresponds to their Manhattan distance of $distManhattan(1,2) = 1$. Since the cluster signal is strong, all three linkage methods lead to branches that reflect the underlying three clusters. Note that the number 40 is an outlier that forms a singleton cluster

8.5 Cophenetic Distance Based on a Hierarchical Cluster Tree

Above we described how a dissimilarity measure can be used as input of hierarchical clustering to produce a cluster tree (dendrogram). Conversely, a dendrogram can be used to define a distance measure, which is referred to as **cophenetic distance**. The name "cophenetic" reflects its origin in taxonomy (Sokal and Rohlf 1962). The cophenetic distance between two objects is defined as the height value where the two objects are joined in the dendrogram, i.e.,

$$dist^{cophenetic}(i,j) = \text{joining height in the dendrogram.} \tag{8.9}$$

One can prove that the cophenetic distance forms an ultradistance, i.e., it satisfies (7.12).

In the following, assume that the dendrogram was constructed on the basis of a dissimilarity matrix $dissim^{original}$. The relationship between $dissim^{original}$ and $dissim^{cophenetic}$ depends on which intergroup dissimilarity method is used for constructing the dendrogram. Denote by $dist^{cophenetic}_{SingleLinkage}$, $dist^{cophenetic}_{CompleteLinkage}$, and $dist^{cophenetic}_{AverageLinkage}$ the cophenetic distances corresponding to dendrograms constructed using single linkage, complete linkage, and average linkage, respectively. Then one can show that

$$dist^{cophenetic}_{SingleLinkage}(i,j) \leq dissim^{original}(i,j) \leq dist^{cophenetic}_{CompleteLinkage}.$$

In an exercise, you are asked to show that the cophenetic distance based on single linkage is smaller than that based on average linkage:

$$dist^{cophenetic}_{SingleLinkage}(i,j) \leq dist^{cophenetic}_{AverageLinkage}(i,j) \tag{8.10}$$

Although, the cophenetic distance based on complete linkage tends to be larger than that based on average linkage, we do not find a strict inequality between them (when using the R function cophenetic).

To measure how faithfully the dendrogram preserves the pairwise distances between the original objects, one can define the cophenetic correlation coefficient (Sokal and Rohlf 1962)

$$\begin{aligned} copheneticCor = cor(&vectorizeMatrix(dissim^{original}), \\ &vectorizeMatrix(dist^{cophenetic})), \end{aligned} \tag{8.11}$$

which is defined as the Pearson correlation between the upper off-diagonal elements of $dissim^{original}$ and $dist^{cophenetic}$ (scatter plot in Fig. 8.2).

The following R code shows how to calculate the cophenetic distance and the cophenetic correlation coefficient.

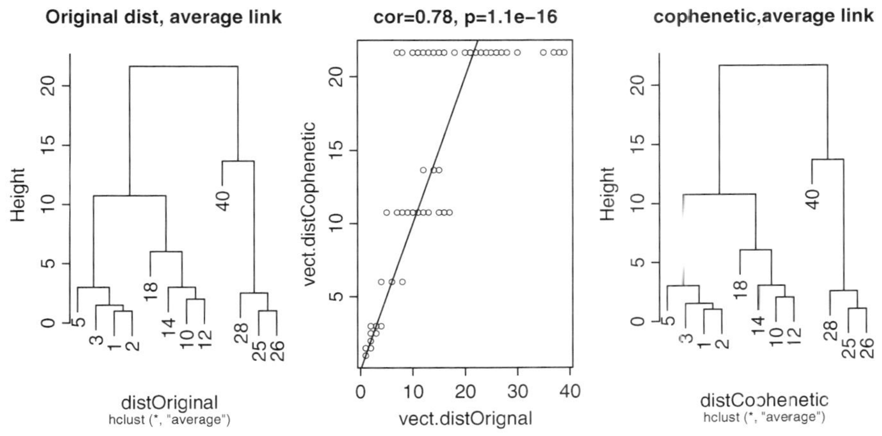

Fig. 8.2 Figure comparing the cophenetic distance (8.9) to the original distance. The scatter plot between the original and the cophenetic distance shows that the cophenetic correlation coefficient (8.11) equals 0.78. The average linkage hierarchical dendrogram based on the original distance equals that based on the cophenetic distance

```
library(WGCNA)
RealNumbers=c(1,2,3,5,10,12,14,18,25,26,28,40)
distOriginal=dist(RealNumbers,method="manhattan")
par(mfrow=c(1,3))
hierAverage=hclust(distOriginal,method="a")
plot(hierAverage,main="Original dist, average link",
labels=RealNumbers,cex.lab=1.5,cex=1.5)
distCophenetic=cophenetic(hierAverage)
vect.distOrignal=vectorizeMatrix(as.matrix(distOriginal))
vect.distCophenetic=vectorizeMatrix(as.matrix(distCophenetic))
copheneticCor=cor(vect.distOrignal,vect.distCophenetic)
#Compare the original distance to the cophenetic distance
verboseScatterplot(vect.distOrignal,vect.distCophenetic)
abline(0,1)
# cluster tree based on the cophenetic distance
hierCopheneticAverageLinkage=hclust(distCophenetic,method="a")
plot(hierCopheneticAverageLinkage,main="cophenetic,average link",
labels=RealNumbers,cex.lab=1.5,cex=1.5)
```

8.6 Defining Clusters from a Hierarchical Cluster Tree: The Dynamictreecut Library for R

We often define network modules as branches of a hierarchical clustering tree. The process of cluster detection based on a hierarchical cluster tree is sometimes referred to as tree cutting, branch cutting, or branch pruning. In the following, we review different branch (tree) cutting methods. The most method defines each contiguous branch below a fixed height cutoff as a separate cluster. We refer to this standard approach as the "static" tree cut method. The structure of cluster-joining heights often poses a challenge to cluster definition. Several distinct branches in a dendrogram can often not be cutoff using a fixed height cutoff. To address this challenge, **Peter Langfelder** and **Bin Zhang** developed a "dynamic" (as opposed to static) branch-cutting method based on analyzing the shape of the branches. The algorithm is implemented in the `dynamicTreeCut` R package (Langfelder et al. 2007).

As motivating example, consider Fig. 8.3a that shows a dendrogram for cluster detection in a protein–protein interaction network in *Drosophila* (fly). The Dynamic Tree Cut method succeeds at identifying branches that could not have been identified using the static cut method. The identified clusters (branches) are highly significantly enriched with known gene ontologies (Dong and Horvath 2007), which provides indirect evidence that the resulting clusters are biologically meaningful. In the following, we provide only a brief summary of the dynamic tree cut method. More technical details can be found in the Methods Supplement of Langfelder et al. (2007). To provide more flexibility, we present two variants of the method. The first variant, invoked using function `cutreeDynamicTree`, is a top-down algorithm that only inputs the cluster tree (dendrogram). This variant has been used to identify biologically meaningful gene clusters in a variety of gene co-expression networks from several species such as yeast (Carlson et al. 2006; Dong and Horvath 2007),

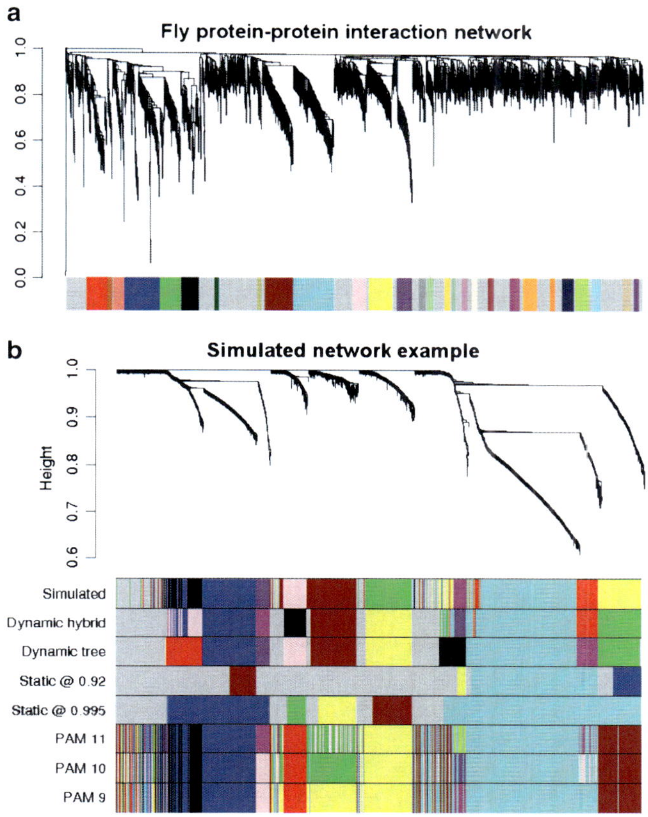

Fig. 8.3 (**a**) Average linkage hierarchical clustering using the Topological Overlap Matrix (Yip and Horvath 2007) and the Dynamic Tree cut applied to the protein–protein interaction network of *Drosophila* (PPI data from BioGRID, www.thebiogrid.org). Module assignment is depicted by the row of color immediately below the dendrogram, with gray representing unassigned proteins. A functional enrichment analysis has shown that the clusters are significantly enriched with known gene ontologies (Dong and Horvath 2007). Note that a fixed height cutoff would not be able to identify many of the shown clusters. (**b**) Hierarchical cluster tree and various cluster detection methods applied to a simulated gene expression data set. The color bands below the dendrogram show the cluster membership according to different clustering methods. The color *gray* is reserved for genes outside any proper cluster, i.e., the tree cut methods allow for unassigned objects. The first color band "Simulated" shows the simulated true cluster membership; color bands "Dynamic Hybrid" and "Dynamic Tree" show the results of the proposed tree-cutting methods; the color band "Static at 0.92" shows the results of the standard, constant height cutoff method at height 0.92. The height refers to the y-axis of the dendrogram. The color band "PAM 11" shows the results of $k = 11$ medoid clustering

mouse (Ghazalpour et al. 2006), and human cell lines (Gargalovic et al. 2006). The algorithm implements an adaptive, iterative process of cluster decomposition and combination and stops when the number of clusters becomes stable. It starts by obtaining a few large clusters by the static tree cut. The joining heights of each cluster are analyzed for a characteristic pattern of fluctuations indicating a sub-cluster structure; clusters exhibiting this pattern are recursively split. To avoid over-splitting, very small clusters are joined to their neighboring major clusters.

The second variant, invoked using function `cutreeHybrid`, is a bottom-up algorithm that inputs both a cluster tree and a dissimilarity measure. As a hybrid between hierarchical clustering and partitioning-around-medoids (PAM) clustering (described in Sect. 8.2), it may improve the detection of outlying members of each cluster. The hybrid cluster detection proceeds along two steps. First, branches that satisfy specific criteria for being clusters are detected. Next, all previously unassigned objects are tested for sufficient proximity to clusters detected in the first step; if the nearest cluster is close enough, the object is assigned to that cluster. Since partitioning-around-medoids also involves assigning objects to their closest medoids, the Dynamic Hybrid variant can be considered a hybrid of hierarchical clustering and modified PAM.

A major application of clustering is to find groups of highly correlated gene expression profiles that may correspond to biological pathways. In Fig. 8.3, we report a simulated gene expression data set (detailed description can be found in (Langfelder et al. 2007)). We simulated ten nested clusters labeled by different colors, shown in the first color band underneath the dendrogram. The results of the methods presented here are shown in correspondingly labeled color bands. Visual inspection shows that the Dynamic Hybrid method outperforms the Static height cutoff method whose clusters contain either too few genes or too many genes. PAM performs even worse in this example. A quantitative analysis of this example is presented in the Supplement.

The parameters of the dynamic tree-cutting methods afford great flexibility to the user to cut branches off a dendrogram. While our default values have worked well in several applications, we recommend to carry out a cluster stability/robustness analysis in practice.

The function `cutreeDynamic` in the `dynamicTreeCut` R package is a wrapper that provides a common access point for the two dynamic branch-cutting methods. In Sect. 12.2.4, we illustrate the function `cutreeDynamic` and related functions for identifying co-expression modules in female mouse liver samples. But here is some typical code:

```
# define a hierarchical tree with hclust or with flashClust
hierTOM=hclust(as.dist(dissTOM),method="average")
# Here we use static branch cutting with height 0.995
# and minimum cluster size of 30.
colorStaticTOM=cutreeStaticColor(hierTOM,cutHeight=0.995,
  minSize=30)
# We now use two Dynamic Tree Cut methods in which the height
# cut-off plays a minor role. The first method is called the tree
# method and only uses the dendrogram as input.
branch.number=cutreeDynamic(hierTOM,method="tree",cutHeight=0.995,
deepSplit=F,minClusterSize = 30)
# This function transforms the branch numbers into colors
colorDynamicTOM=labels2colors(branch.number)
```

The second branch-cutting method is called hybrid dynamic branch-cutting method. As input it requires both a dendrogram and the dissimilarity that was used to create the dendrogram.

```
labelDynamicHybrid=cutreeDynamic(hierTOM,distM=dissTOM,
cutHeight=0.995,deepSplit=1,pamRespectsDendro=FALSE,
   minClusterSize=30)
colorDynamicHybridTOM=labels2colors(labelDynamicHybrid)
# Plot results of all module detection methods together:
plotDendroAndColors(dendro =hierTOM,colors=data.frame(truemodule,
colorStaticTOM,colorDynamicTOM,colorDynamicHybridTOM),
dendroLabels = FALSE, marAll = c(0.2, 8, 2.7, 0.2),
main ="Cluster dendrogram and color-band")
```

Figure 8.4 and the following Rand index calculations

```
randIndex(table(colorStaticTOM,truemodule),adjust=F)
[1] 0.6683119
randIndex(table(colorDynamicTOM,truemodule),adjust=F)
[1] 0.6683119
randIndex(table(colorDynamicHybridTOM,truemodule),adjust=F)
[1] 0.7179655
```

show that hierarchical clustering and the three branch cutting work well at retrieving the true simulated cluster signal.

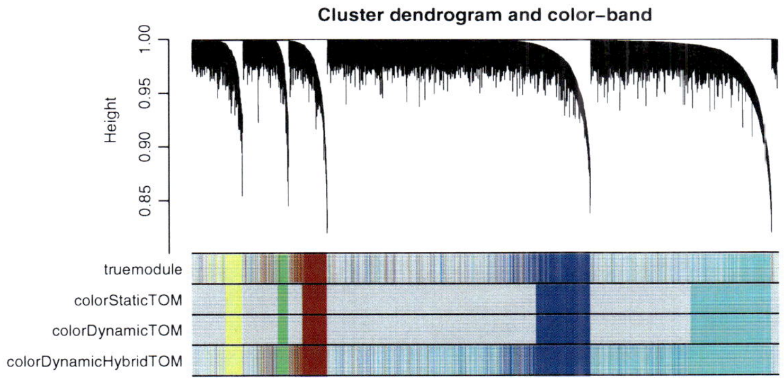

Fig. 8.4 Average linkage hierarchical clustering using the Topological Overlap Matrix (Yip and Horvath 2007) and different branch-cutting methods. The simulated true module assignment is depicted in the first color band underneath the tree with *gray* representing objects outside proper clusters. Note that the branches of the cluster tree correspond to the simulated true modules. The second color band shows the results of branch cutting with the static height cutoff method. The height refers to the y-axis of the dendrogram. The color *gray* is reserved for genes outside any proper cluster, i.e., the tree cut methods allow for unassigned objects. The third and fourth color bands show the results for the dynamic "tree" and the dynamic "hybrid" branch cutting method, respectively

8.7 Cluster Quality Statistics Based on Network Concepts

Assume that a cluster assignment is produced using a dissimilarity measure D. As described in Sect. 7.4, the dissimilarity can be turned into a network adjacency matrix, e.g., $A = 1 - D/max(D)$ (7.13).

Thus, for the qth cluster of the cluster assignment Cl, an $n^{(q)} \times n^{(q)}$ dimensional adjacency matrix can be defined by subsetting the rows and columns of A. Thus, the qth cluster can be interpreted as the qth module of A. This insight allows us to use intramodular density concepts and intermodular separability concepts to assess the quality of the cluster definition. Density statistics (concepts) can be used to determine how closely interconnected the nodes of a cluster are. Separability statistics can be used to determine how well a cluster is separated from other clusters. Particularly powerful cluster quality statistics can be defined if the dissimilarity measure D is related to a correlation measure between numeric vectors, which correspond to the columns of a matrix $datX$. In this case, correlation network concepts can be used. In Table 8.1, we provide an overview of the input required to calculate each of the cluster quality statistics presented here. All statistics require the cluster assignment (label) and output a quality statistic for each cluster (module). Specifically, the cluster quality statistics are defined by

$$meanAdj^{(q)} = mean\left(vectorizeMatrix(A^{(q)})\right) \tag{8.12}$$

$$meanCor^{(q)}_{unsigned} = mean\left\{vectorizeMatrix\left(|r^{(q)}_{ij}|\right)\right\} \tag{8.13}$$

$$meanCor^{(q)}_{signed} = mean\left\{vectorizeMatrix\left(r^{(q)}_{ij}\right)\right\} \tag{8.14}$$

$$meanKME^{(q)}_{unsigned} = mean_{i \in \mathcal{M}_q}\left\{|\mathrm{kME}^{(q)}_i|\right\} \tag{8.15}$$

$$meanKME^{(q)}_{signed} = mean_{i \in \mathcal{M}_q}\left\{\mathrm{kME}^{(q)}_i\right\} \tag{8.16}$$

$$propVarExpl^{(q)} = mean_{i \in \mathcal{M}_q}\left\{(\mathrm{kME}^{(q)}_i)^2\right\} \tag{8.17}$$

$$separability^{(q_1,q_2)} = 1 - cor(E^{(q_1)}, E^{(q_2)}). \tag{8.18}$$

Table 8.1 Overview of cluster quality statistics based on network concepts

				Network input		
No.	Statistic	Statistic type	Network type	Label	Adj	*datX*
1	*meanAdj*	Density	General	Yes	Yes	No
2	*separability*^{average}	Separability	General	Yes	Yes	No
3	*meanCor*	Density	Correlation	Yes	No	Yes
4	*propVarExpl*	Density	Correlation	Yes	No	Yes
5	*meanKME*	Density	Correlation	Yes	No	Yes
6	*separability*	Separability	Correlation	Yes	No	Yes

The table reports the input needed for module quality measures that measure either module density or separability in the reference network.

The mean adjacency (8.12) applies to a general network while the remaining statistics assume a correlation network. The *meanCor* and *meanKME* statistics each have slightly different versions for signed and unsigned correlation networks.

Thus, the correlation statistic of module quality is simply the mean of (the absolute values of) the pairwise correlations within the module. Low cluster separability may suggest that the two clusters q_1, q_2 are not really distinct and should be merged.

R code for calculating cluster quality statistics is presented in Sect. 12.5.

8.8 Cross-Tabulation-Based Cluster (Module) Preservation Statistics

In many clustering applications, one is interested to determine whether a cluster that was found in one data set can also be found in another data set, i.e., that the cluster is preserved. For example, to validate the existence of a cluster, it is desirable to show that it is reproducible (or preserved) in a second, independent data set. In Chap. 9, we will present network statistics for measuring module or cluster preservation. Here we present cluster preservation statistics (also known as **cluster validation statistics**) that do not require that a network has been defined. These three basic cross-tabulation-based statistics for determining whether clusters in a reference data set are preserved in a test data set. These statistics do not assume a network. Instead, cluster assignments in both the reference and the test data are needed. For each object i, $Cl_i^{[\text{ref}]}$ denotes its cluster label in the reference data. The clusters are labeled by $q = 1, 2, \ldots, Q^{[\text{ref}]}$, where $Q^{[\text{ref}]}$ is the number of clusters in the reference set. The number of objects in cluster q will be denoted by $n^{(q)}$. Assume that the cluster assignment of the test data ($Cl^{[\text{test}]}$) leads to $Q^{[\text{test}]}$ clusters labeled by $q' = 1, 2, \ldots, Q^{[\text{test}]}$. The goal is to determine whether a cluster q in the reference clustering $Cl^{[\text{ref}]}$ can be matched to a cluster q' in the test clustering $Cl^{[\text{test}]}$.

For each cluster specified in $Cl^{[\text{ref}]}$, a cross-tabulation-based preservation statistic defines a value. As convention, we assume that the higher the preservation value of cluster (module) q, the stronger the evidence that it corresponds to a cluster specified in $Cl^{[\text{test}]}$. Here we will describe two approaches that are based on cross-tabulating $Cl^{[\text{ref}]}$ and $Cl^{[\text{test}]}$, that is, creating a contingency table in which every row corresponds to a reference cluster, and each column to a test cluster.

We start with a cluster preservation statistic based on keeping track of the co-clustering of pairs of objects. For the qth cluster of $Cl^{[\text{ref}]}$, one can form

$$\binom{n^{(q)}}{2} = \frac{n^{(q)}(n^{(q)} - 1)}{2}$$

different pairs of objects. Let $n_{qq'}$ denote the number of objects which are in the qth cluster of $Cl^{[\text{ref}]}$ and in the q'th cluster of $Cl^{[\text{test}]}$. The number of *pairs* of objects in the qth cluster of $Cl^{[\text{ref}]}$ that are also part of the q'th cluster of $Cl^{[\text{test}]}$ is given by $\binom{n_{qq'}}{2}$. The proportion of pairs of objects in cluster q that also cluster in cluster q' is given by

$$propCoClustering(q,q') = \binom{n_{qq'}}{2} \Big/ \binom{n^{(q)}}{2}$$

Apart from pairs of objects ($tupletsize = 2$), one can also calculate the proportion of triplets ($tupletsize = 3$) or quadruplets of objects ($tupletsize = 4$) from the qth cluster of $Cl^{[\mathrm{ref}]}$ that co-cluster in the q'th cluster of $Cl^{[\mathrm{test}]}$:

$$propCoClustering(q,q',tupletsize) = \binom{n_{qq'}}{tupletsize} \Big/ \binom{n^{(q)}}{tupletsize}$$

Using the above notation, we define the co-clustering-based cluster preservation statistic (Langfelder et al. 2011) as

$$coClusteringPreservation(q,tupletsize) = \sum_{q'=1}^{Q^{[\mathrm{test}]}} propCoClustering(q,q',tupletsize) \tag{8.19}$$

which depends on the tuplet size. For the case $tupletsize = 2$ (pairs), the co-clustering preservation statistic is related to the prediction strength statistic in Kapp and Tibshirani (2007) and Tibshirani and Walther (2005). The coClusteringPreservation statistic is implemented in the WGCNA function coClusteringPreservation.

An alternative cross-tabulation statistic is the accuracy and the related Fisher's exact test p value. For each proper reference cluster q with n_q objects, we find the proper test cluster q' with the highest number of objects common to both the reference and the test cluster, $n_{qq'}$. We define the *accuracy* of the reference cluster q in the test clustering as

$$accuracy_q = \frac{n_{qq'}}{n_q} \tag{8.20}$$

By definition, the accuracy lies in $[0,1]$, and $accuracy_q = 1$ indicates that all objects that form the reference cluster q are part of the same cluster in the test data. We emphasize that this definition is only used for the proper clusters. For the improper cluster that contains all objects not assigned to any of the proper clusters (labeled by the label 0), we define its accuracy as

$$accuracy_0 = \frac{n_{00'}}{n_0} \tag{8.21}$$

where n_0 is the number of unassigned objects, and $n_{00'}$ is the number of objects unassigned both in the reference and in the test clusterings.

A potential disadvantage of the accuracy measure is that it does not take into account how likely it is to observe a particular maximum overlap by chance. As an example, consider a case in which there are say ten clusters in the reference clustering, and a test clustering in which the clusters are perfectly reproduced; so the accuracy of all ten reference clusters equals 1. Now consider another test clustering in which all objects belong to a single cluster. The accuracies of all clusters with respect to the second test clustering again equal 1, but intuitively the second test clustering is less interesting than the first clustering. To address this issue, we define a p value-based preservation statistic as follows. For each reference-test pair q, q' of proper clusters, we calculate the one-sided Fisher p value of the observed overlap of the two clusters, $p_{qq'}$. For each proper reference cluster q, we then report minus the logarithm of the lowest (most significant) p value:

$$mlfp_q = -log\left(min_{q'} p_{qq'}\right) . \tag{8.22}$$

For the improper cluster of unassigned variables that carry the label 0, we define

$$mlfp_0 = -log p_{00'} . \tag{8.23}$$

These cross-tabulation based preservation statistics are implemented in the WGCNA function overlapTable. Cross-tabulation-based cluster preservation measures are intuitive, which is why they are often used. For example, they work quite well when it comes to studying the preservation of clusters in human and chimpanzee brain networks (see Fig. 9.3 and Sect. 9.4). However, they suffer from one major limitation: they assume that a clustering algorithm produces a meaningful and robust assignment $Cl^{[test]}$ in the test data. Since many cluster algorithms are inherently non-robust and dependent on parameter choices, the resulting cross-tabulation statistics suffer from the same shortcomings. In Chap. 9, we describe network-based preservation statistics that do not require the specification of a test set clustering assignment $Cl^{[test]}$. Instead, these powerful and highly robust network-based statistics require the definition of an adjacency matrix in the reference and test data sets. Toward this end, one can transform the dissimilarity measure (used for defining the clusters) into weighted networks as described in Sect. 7.4.

8.9 Rand Index and Similarity Measures Between Two Clusterings

Assume that a set of n objects has been clustered in two different ways with resulting cluster assignments $Cl^{[ref]}$ and $Cl^{[test]}$. Here we will describe measures of similarity between two clusterings (or partitions). When comparing two clusterings, the cluster label is unimportant since the first cluster of $Cl^{[ref]}$ may correspond to the third cluster of $Cl^{[test]}$. Thus, meaningful measures of cluster similarity should be invariant with regard to changing the order of the cluster labels. We will now describe the Rand index that measures the co-clustering of pairs of objects. Specifically, let a be the number of pairs of objects that are in the same cluster according to $Cl^{[ref]}$ and also in the same cluster according to $Cl^{[test]}$ (but the cluster labels do not have to be identical). Let b be the number of pairs that are in different clusters according to $Cl^{[ref]}$ and in different clusters according to $Cl^{[test]}$. Let c be the number of pairs objects that are in the same cluster according to $Cl^{[ref]}$ but in different clusters according to

$Cl^{[\text{test}]}$. Let d be the number of pairs that are in different clusters according to $Cl^{[\text{ref}]}$ but in the same cluster according to $Cl^{[\text{test}]}$. Then the (unadjusted) Rand index (Rand 1971; Hubert and Arabie 1985) is defined as:

$$RandIndex(Cl^{[\text{ref}]}, Cl^{[\text{test}]}) = \frac{a+b}{a+b+c+d} = \frac{a+b}{n(n-1)/2}. \tag{8.24}$$

Note that $a+b$ can be thought of as the number of agreements, while $c+d$ can be thought of as the number of disagreements between $Cl^{[\text{ref}]}$ and $Cl^{[\text{test}]}$. The unadjusted Rand index takes on a value between 0 and 1, with 1 indicating that the cluster assignments are the same (up to permuting the labels) and 0 indicating that the two data clusters do not agree on any pair of objects. A major advantage of the unadjusted Rand index is that it has a very intuitive interpretation: it keeps track of the co-clustering of pairs of observations. But a disadvantage is that it does not correct for chance occurrence and that it does not lead to a p value. A modified version of the Rand index, which adjusts for chance occurrences, is given by the adjusted Rand index (Hubert and Arabie 1985):

$$AdjRandIndex(Cl^{[\text{ref}]}, Cl^{[\text{test}]}) = \frac{2(ab-cd)}{(a+d)(d+b)+(a+c)(c+b)}. \tag{8.25}$$

The adjusted and the unadjusted Rand index are symmetrical with regard to reversing the role of reference and test network, i.e.,

$$RandIndex(Cl^{[\text{ref}]}, Cl^{[\text{test}]}) = RandIndex(Cl^{[\text{ref}]}, Cl^{[\text{test}]}) \tag{8.26}$$

$$AdjRandIndex(Cl^{[\text{ref}]}, Cl^{[\text{test}]}) = AdjRandIndex(Cl^{[\text{test}]}, Cl^{[\text{ref}]}). \tag{8.27}$$

as one can easily show using the co-clustering formulation of the Rand index presented in the following.

8.9.1 Co-Clustering Formulation of the Rand Index

Here we will review an alternative definition of the Rand index using co-clustering notation. Denote the total number of pairs of observations by $N = \binom{n}{2}$. Denote the number of pairs of objects that co-cluster in $Cl^{[\text{ref}]}$ and also in $Cl^{[\text{test}]}$ by $A = \sum_r \sum_s \binom{n_{rs}}{2}$. Denote the number of pairs of objects that co-cluster in $Cl^{[\text{test}]}$, but not in $Cl^{[\text{ref}]}$ by

$$B = \sum_s \binom{n_{\cdot s}}{2} - \sum_r \sum_s \binom{n_{rs}}{2} = B' - A,$$

where $B' = \sum_s \binom{n_{\cdot s}}{2}$. Denote the number of pairs that co-cluster in $Cl^{[\text{ref}]}$, but not in $Cl^{[\text{test}]}$ by

$$C = \sum_r \binom{n_{r\cdot}}{2} - \sum_r \sum_s \binom{n_{rs}}{2} = C' - A,$$

where $C' = \sum_r \binom{n_{r\cdot}}{2}$.

Denote the number of pairs that co-cluster in neither $Cl^{[ref]}$ nor $Cl^{[test]}$ by $D = N - (A+B+C)$. Then the unadjusted Rand index (8.24) is given by

$$RandIndex(Cl^{[ref]}, Cl^{[test]}) = \frac{A+D}{N} = \frac{\binom{n}{2} - \sum_s \binom{n_{\cdot s}}{2} - \sum_r \binom{n_{r\cdot}}{2} + 2\sum_r \sum_s \binom{n_{rs}}{2}}{\binom{n}{2}},$$

and the adjusted Rand index (8.25) is given by

$$AdjRandIndex(Cl^{[ref]}, Cl^{[test]}) = \frac{A - \frac{B'C'}{N}}{\frac{B'+C'}{2} - \frac{B'C'}{N}}$$

$$= \frac{\sum_r \sum_s \binom{n_{rs}}{2} - \left(\sum_s \binom{n_{\cdot s}}{2} \sum_r \binom{n_{r\cdot}}{2}\right) / \binom{n}{2}}{\frac{1}{2}\left(\sum_s \binom{n_{\cdot s}}{2} + \sum_r \binom{n_{r\cdot}}{2}\right) - \left(\sum_s \binom{n_{\cdot s}}{2} \sum_r \binom{n_{r\cdot}}{2}\right) / \binom{n}{2}}.$$

The above definition of the Rand index shows that it keeps track of the co-clustering of *pairs* of objects. But it is straightforward to generalize it so that it keeps track of the co-clustering of tuplets by replacing 2 by *tupletsize* in the binomial coefficients.

Numerous alternative measures have been proposed for evaluating the agreement between two cluster assignments (reviewed in Hubert and Arabie 1985; Dudoit and Fridlyand 2002; Tibshirani and Walther 2005). Since a cluster assignment can be interpreted as a categorical variable, any similarity measure between categorical variables can be used for defining a measure of cluster agreement as long as the measure is invariant with regard to changing the cluster labels. For example, Pearson's chi-square test statistic and a likelihood ratio test statistic are attractive alternatives since they allow one to compute p values. Also, the mutual information can be used to relate the two clusterings.

8.9.2 R Code for Cross-Tabulation and Co-Clustering

Both the adjusted and the unadjusted Rand index can be calculated with the WGCNA R function `randIndex`. The WGCNA R function `matchLabels` matches each cluster label in $Cl^{[ref]}$ with a corresponding label of $Cl^{[test]}$ according to Fisher's exact test p value of pairwise overlap. More specifically, the R commands

```
library(WGCNA)
Cl2.new=matchLabels(Cl2,Cl1)
```

assign new labels to the clusters of $Cl^{[test]}$ according to their overlap with the reference clusters of $Cl^{[ref]}$.

For the qth cluster of $Cl^{[ref]}$, one can also create a 2×2 dimensional cross-tabulation table for relating cluster membership in $Cl^{[ref]}$ versus cluster membership in matchLabels(q) of $Cl^{[test]}$. Next, Fisher's exact test can be used to calculate a p value of overlap. We caution the reader that this Fisher's exact p value should not be interpreted as statistical significance measure of cluster preservation since it is highly biased due to the fact that an optimal matching cluster label has been found in the test data. Instead, we define the overlap p value-based cluster preservation measure

$$ClusterPreservation^{overlapP}(q) = -log_{10}(FisherPvalue_{q,matchLabels(q)}) \quad (8.28)$$

as minus log (base 10) of Fisher's exact p value as *descriptive* measure of cluster preservation. This cluster preservation measure can be calculated using the following R code:

```
Cl2.new=matchLabels(Cl2,Cl1)
OverlapTable.q=table(Cl1==q,Cl2.new==q)
FisherPvalue.q=fisher.test(OverlapTable.q)
ClusterPreservation.q=-log10(FisherPvalue.q))
```

Alternatively, one could use the WGCNA function `overlapTable` to calculate pairwise Fisher's exact p values. The co-clustering preservation statistics are implemented in the WGCNA function coClusteringPreservation.

8.10 Discussion of Clustering Methods

Clustering methods are often (but not always) used to define network modules. In most applications, we find that many network nodes (considered as objects) should not be grouped into a cluster, i.e., these "background" nodes should form their own singleton clusters. Partitioning methods (such as PAM or k-means clustering) do not work well when many objects form singleton clusters and/or when cluster sizes vary a lot. Further, partitioning methods require the user to specify the number of clusters k or at least they require the specification of a criterion for estimating k from the data. In contrast, hierarchical clustering methods work very well when dealing with singleton clusters, varying cluster sizes, and they do not require the specification of k. Further, hierarchical clustering leads to an intuitive depiction of the clustering results in a cluster tree. For these reasons, hierarchical clustering methods appear to be more widely used than partitioning methods in genomic applications. However, hierarchical clustering methods are typically less robust than partitioning methods. In particular, single linkage and complete linkage methods are not robust, which is why we prefer average linkage hierarchical clustering. But even the results of average linkage hierarchical clustering are not very robust. Ideally, one would like to have a "hybrid" clustering method that combines the advantages of hierarchical clustering with the advantage of partitioning methods (robustness). Such a method is the dynamic hybrid hierarchical clustering method (described in Sect. 8.6). This method first uses hierarchical clustering to produce a cluster tree, next defines clusters as branches, and finally uses a PAM-like algorithm to reshuffle objects into more robustly defined clusters. Because of its theoretical advantages and its successful use in numerous applications, we use the dynamic hybrid method as default method in many network analyses. Having said this, it is always useful to determine how the results of an analysis change when different clustering algorithms are being used.

Since clustering methods are non-robust, it is essential to assess the reproducibility and validity of clusters. In Sect. 8.8 and Chap. 9, we describe statistics for assessing the preservation of clusters in different data sets. Statistics for measuring the quality of clusters are presented in Sect. 8.7.

While we present major types of clustering procedures, we should point out that our review is incomplete and we refer readers to books that focus on clustering methods. We briefly mention biclustering (also known as two-way clustering or co-clustering) methods, which cluster both the rows and columns of a data matrix $datX$ simultaneously. Given a set of m columns (samples) and n rows (objects) (i.e., an $m \times n$ matrix $datX$), the biclustering algorithm generates biclusters defined as subsets of rows which exhibit similar behavior across a subset of columns, or vice versa. The biclustering problem can be formulated in different ways. In particular, the result depends on the merit function used to evaluate the quality of a given bicluster. Many variants of the biclustering problem are computationally challenging (NP-complete) and require the use of heuristics. Biclustering methods are attractive in the context of gene expression data since they may lead to the identification of co-regulated genes and simultaneously conditions when these genes are overexpressed. Numerous biclustering methods have been proposed for constructing gene co-expression modules. Since our interest lies in module detection methods that can be used for general association networks (based solely on an adjacency matrix), a review of different biclustering methods is beyond our scope.

Cluster analysis is and always will be somewhat of an artform. It requires good judgement by the data analyst since it suffers from the following shortcomings. Besides depending on the measure used to quantify similarities among samples, the results of cluster analysis can also depend on the specific clustering algorithm. For example, dendrograms produced by hierarchical clustering algorithms acting upon the same data may look quite different depending on whether single, average, or complete linkage is used to calculate distances between clusters. Other clustering procedures may involve additional parameter choices that can have a substantial effect on cluster assignments (e.g., the choice of k in k-means or partitioning-around-medoids clustering). In addition, cluster composition can be influenced by the presence of outliers. Finally, cluster analysis can be impractical for very large datasets, in which the sheer number of samples obscures the organization and characteristics of a dendrogram and produces ambiguous cluster boundaries. These shortcomings motivated us to use **sample networks** (described in Sect. 7.7) for characterizing sample relationships in genomic datasets. Most cluster analyses take as input a similarity (or dissimilarity) matrix. After transforming such a matrix to a weighted network, one is able to describe the properties of this matrix using network concepts. Techniques for turning a general similarity or dissimilarity matrix into a network are described in Sect. 7.1 and in Sect. 7.4. **Mike Oldham** has demonstrated that sample networks enhance the cluster analysis, e.g., by aiding the analysis of outliers (Oldham et al. 2011). R code and an application of sample networks can be found in Sect. 12.1.

8.11 Exercises

1. Exercise regarding object scatters and k-means clustering. Here we use the simulated data *datExpr* from Sect. 6.13 for which the simulated true cluster label (*truemodule*) is known.

   (i) Form the Euclidean distances between the $n = 3{,}000$ scaled columns (genes) of *datExpr* and calculate the corresponding total object scatter (8.1). Hint:

```
# The distance is defined by
EuclidDistScaled=dist(t(scale(datExpr)))
squaredEuclidDistMatrix=as.matrix(EuclidDistScaled)^2
totalscatterEuclid=1/2*sum(squaredEuclidDistMatrix)
```

   (ii) Use k-means clustering with $k = 6$ for clustering the scaled columns of *datExpr*. For the resulting cluster assignment *Cl.kmeansEuclid*, calculate the within-object scatter and the between-object scatter. Hint:

```
library(cluster)
k=6
kmeansEuclid=kmeans(t(scale(datExpr)),k)
Cl.kmeansEuclid=as.numeric(kmeansEuclid$cluster)
withinScatter=0
for (q in 1:k){
rest.q=Cl.kmeansEuclid==q
withinScatter.q=1/2*sum(squaredEuclidDistMatrix
   [rest.q,rest.q])
withinScatter=withinScatter.q+withinScatter
} #end of for loop
betweenScatter=totalscatterEuclid-withinScatter
```

   (iii) Numerically verify (8.6), which is given by

$$WithinScatter(Cl, ||.||^2) = \sum_{q=1}^{k} n^{(q)} \sum_{\{i|Cl(i)=q\}} ||x_i - mean^{(q)}(x)||^2$$

   Hint:

```
# Left hand side
withinScatter
# Right hand side
nq.vector=as.numeric(table(Cl.kmeansEuclid))
sum(kmeansEuclid$withinss*nq.vector)
```

   (iv) Derive (8.6) mathematically.

2. Exercise. Regarding the invariance of clustering results.

   (i) Assume that *diss1* and *diss2* are two $n \times n$ dimensional dissimilarity matrices for which $diss1 = a * diss2$ where a is a positive number. Prove that PAM clustering based on *diss1* leads to the same cluster assignment as that based on *diss2* (as long as the same set of initial medoids is chosen for the PAM algorithm).

(ii) For the simulated data *datExpr* from Sect. 6.13, determine the linear relationship between the correlation-based dissimilarity `dissCor=1-cor (datExpr)` and the squared Euclidean distance of the scaled columns

```
squaredEuclidDistMatrix=as.matrix(dist(t(scale
  (datExpr))))^2
```

Hint: According to (7.21)

$$||scale(x) - scale(y)||^2 = a(1 - cor(x,y)), \tag{8.29}$$

where $a = 2(m-1)$.

(iii) What is the relationship between the total scatter of dissSquaredEuclidScaled and the mean correlation between the columns of datExpr? Hint: (7.21).

(iv) Use dissCor and dissEuclidScaled as input of PAM clustering with $k = 6$. Verify numerically that the clustering results are the same (for the same initial medoids). Further, determine how the results compare with those from k-means clustering of the scaled columns. Hint:

```
set.seed(1);
Cl.pamCor=pam(as.dist(dissCor),k=6)$clustering
set.seed(1)
Cl.pamSqEuclidScaled=
pam(as.dist(squaredEuclidDistMatrix),k=6)$clustering
set.seed(1)
Cl.kmeans=kmeans(t(scale(datExpr)),6)$cluster
table(Cl.pamCor,Cl.pamSqEuclidScaled)
table(Cl.pamCor,Cl.kmeans)
```

(v) Discuss why the clustering results of k-means clustering could be different from those of PAM.

(vi) Cross-tabulate the resulting cluster labels with those corresponding to the true simulated clusters and report unadjusted Rand indices.

```
randIndex(table(Cl.pamSqEuclidScaled,truemodule),adjust=F)
randIndex(table(Cl.kmeans,truemodule),adjust=F)
```

Comment: There are two reasons for the relatively poor clustering performance. First, genes that are anti-correlated are very dissimilar with respect to the studied dissimilarity measures. But the simulated clusters contain both positively and negatively correlated genes. Second, these partitioning methods cannot deal well with background "gray" genes.

3. Exercise PAM and different network-based dissimilarity measures. Here we use the simulated data *datExpr* from Sect. 6.13 for which the simulated true cluster label (*truemodule*) is known. Define unsigned weighted correlation network adjacency matrices *ADJ1* and *ADJ6* corresponding to powers $\beta = 1$ and $\beta = 6$, respectively. Next define the adjacency-based dissimilarities $dissADJ1 = 1 - ADJ1$, $dissADJ6 = 1 - ADJ6$, and the topological overlap-based dissimilarity *dissTOM*. Hint:

```
library(WGCNA)
dissADJ1=1-adjacency(datExpr,power=1)
dissADJ6=1-adjacency(datExpr,power=6)
# the TOM calculation may take a few minutes...
dissTOM=TOMdist(adjacency(datExpr,power=6))
```

(i) For a random subset of genes, create pairwise scatter plots between the entries of dissADJ1, dissADJ6, and dissTOM. Hint:

```
set.seed(1)
subsetGenes=sample(1:ncol(datExpr),500,replace=F)
dissADJ1subset=dissADJ1[subsetGenes,subsetGenes]
dissADJ6subset=dissADJ6[subsetGenes,subsetGenes]
dissTOMsubset=dissTOM[subsetGenes,subsetGenes]
par(mfrow=c(2,2))
verboseScatterplot(vectorizeMatrix(dissADJ1subset),
vectorizeMatrix(dissTOMsubset),xlab="dissADJ1",
ylab="dissTOM",col=truemodule[subsetGenes])
verboseScatterplot(vectorizeMatrix(dissADJ6subset),
vectorizeMatrix(dissTOMsubset),xlab="dissADJ6",
ylab="dissTOM",col=truemodule[subsetGenes])
verboseScatterplot(vectorizeMatrix(dissADJ1subset),
vectorizeMatrix(dissADJ6subset),xlab="dissADJ6",
ylab="dissADJ6",col=truemodule[subsetGenes])
```

Note that these network dissimilarity measures are highly related.

(ii) Use dissADJ1, dissADJ6, and dissTOM as input of PAM clustering with $k = 6$ partitions. Measure the agreement between the cluster assignments using the unadjusted Rand index

```
Cl.pamADJ1=pam(as.dist(dissADJ1),k=6)$clustering
Cl.pamADJ6=pam(as.dist(dissADJ6),k=6)$clustering
Cl.pamTOM=pam(as.dist(dissTOM),k=6)$clustering
randIndex(table(Cl.pamADJ1,Cl.pamTOM),adjust=F)
randIndex(table(Cl.pamADJ6,Cl.pamTOM),adjust=F)
randIndex(table(Cl.pamTOM,Cl.pamADJ1),adjust=F)
```

(iii) Cross-tabulate the resulting cluster labels with those corresponding to the true simulated clusters and report unadjusted Rand indices.

```
randIndex(table(Cl.pamADJ1,truemodule),adjust=F)
randIndex(table(Cl.pamADJ6,truemodule),adjust=F)
randIndex(table(Cl.pamTOM,truemodule),adjust=F)
```

(iv) Discuss the performance of the TOM versus the ADJ-based dissimilarity. Hint: None of the different dissimilarities work well when used as input of PAM. Partioning methods do not work well when dealing with a lot of background (gray) genes as can be seen from `table(Cl.pamADJ1,truemodule)`. As can be seen from another exercise, hierarchical clustering works better in this situation.

4. Exercise regarding simulation studies for evaluating a clustering method.

   (i) Simulate a cluster structure for a network where PAM with the TOM-based dissimilarity measure will work best at retrieving the simulated cluster structure. Hint: For any adjacency matrix ADJ, define

   ```
   dissTOM=TOMdist(ADJ)
   TrueClusterTOMpam=pam(as.dist(dissTOM),k=3)$clustering
   ```

   (ii) Simulate a cluster structure for a network where PAM with the ADJ-based dissimilarity measure will work best. Hint:

   ```
   dissADJ=1-ADJ
   TrueClusterTOMpam=pam(as.dist(dissADJ),k=3)$clustering
   ```

   (iii) Simulate a cluster structure for a network where average linkage hierarchical clustering with TOM will work best.

   ```
   tree1=hclust(as.dist(dissTOM),method="average:)
   TrueClusterTOMhclust=
   cutreeDynamic(tree1,distM=dissTOM,minClusterSize=5)
   ```

   (iv) Discuss the following statements: (a) For any clustering procedure one can find a simulation model where it outperforms all other clustering procedures. (b) To assess the merits of a clustering procedure, it should perform well on multiple real data sets.

5. Exercise regarding hierarchical clustering with an ultradistance. Assume that an $n \times n$ dimensional matrix $D = (d_{ij})$, which satisfies the conditions of an ultradistance (7.12), i.e., a symmetric matrix with zero diagonal elements whose elements satisfy $d(i,j) \leq max(d(i,k),d(k,j))$. Further, assume that this ultra distance is used as an input of hierarchical clustering with different intergroup dissimilarity measures.

   (i) Show that single linkage and complete linkage lead to the same hierarchical clustering tree. Hint: Either prove your results mathematically or illustrate the finding using a numeric example.
   (ii) Discuss whether average linkage will produce the same cluster tree. Hint: (8.10).

6. Exercise regarding the cophenetic distance (8.9) based on the joining heights of a hierarchical cluster tree (dendrogram). Use the simulated data from Sect. 6.13 to define an adjacency matrix-based dissimilarity measure (called $dist^{original}$)

   ```
   no.nodes=800
   set.seed(1)
   randomSample=sample(1:dim(datExpr)[[2]],no.nodes,replace=F)
   ADJ=adjacency(datExpr[,randomSample])
   distOriginal=as.dist(1-ADJ)
   ```

   (i) Use $dist^{original}$ as input of single linkage, complete linkage, and average linkage hierarchical clustering. Calculate three cophenetic distances based on the joining heights of the three dendrogram. Hint:

```
hierSingle=hclust(distOriginal,method="single")
hierComplete=hclust(distOriginal,method="complete")
hierAverage=hclust(distOriginal,method="average")
distCopheneticSingle=cophenetic(hierSingle)
distCopheneticComplete=cophenetic(hierComplete)
distCopheneticAverage=cophenetic(hierAverage)
```

(ii) Verify that the cophenetic distances satisfy the properties of an ultradistance (symmetry, zero diagonal, and (7.12)). Hint:

```
distUltra= as.matrix(distCopheneticAverage)
i=1; j=2
table(distUltra[i,j]<=pmax(distUltra[i,],distUltra[j,]))
```

(iii) Show that a cophenetic distance based on single linkage satisfies

$$dist_{SingleLinkage}^{cophenetic}(i,j) \leq dissim^{original}(i,j),$$

while the cophenetic distance based on complete linkage satisfies

$$dist_{CompleteLinkage}^{cophenetic}(i,j) \geq dissim^{original}(i,j).$$

Hint:

```
table(distCopheneticSingle<=distOriginal)
table(distCopheneticComplete>=distOriginal)
```

(iv) Show that $dist_{SingleLinkage}^{cophenetic}(i,j) \leq dist_{AverageLinkage}^{cophenetic}(i,j)$

```
table(c(distCopheneticSingle)<=c(distCopheneticAverage))
```

(v) For the three cophenetic distances, compute the corresponding cophenetic correlations (8.11), i.e., the Pearson correlation coefficients with $dist^{original}$.

```
par(mfrow=c(2,2))
verboseScatterplot(c(distOriginal),c(distCopheneticSingle))
abline(0,1)
verboseScatterplot(c(distOriginal),
c(distCopheneticComplete)) abline(0,1)
verboseScatterplot(c(distOriginal),
c(distCopheneticAverage)) abline(0,1)
```

(vi) Which of the three dendrograms (corresponding to single, complete, and average linkage) provides the best representation of the original dissimilarity? Which linkage method would you choose as default method? Why?

7. Exercise regarding an adjacency function $AF^{cophenetic}()$ based on the cophenetic distance (8.9). Define the cophenetic adjacency function using the following steps. First, use $A^{original}$ to define the adjacency matrix-based dissimilarity measure $dist^{original} = 1 - A^{original}$. Second, use $dist^{original}$ as input of hierarchical clustering (any intergroup linkage method could be used, e.g., average linkage).

Third, use the resulting dendrogram to define the cophenetic distance $dist^{cophenetic}$ (8.9). Fourth, define

$$AF^{cophenetic}(A^{original}) = 1 - dist^{cophenetic}.$$

(i) Prove that the cophenetic distance satisfies the conditions of a distance matrix (see Sect. 7.4).
(ii) Prove that $A = AF^{cophenetic}(A^{original})$ is an adjacency matrix. Hint: Either prove it mathematically or verify it numerically on an example.
(iii) Determine whether $AF^{cophenetic}$ is threshold-preserving if $dist^{original}$ is an ultradistance (7.12).

8. Exercise regarding cross-tabulation-based measures of cluster preservation. Here we use R code to define two cluster assignments.

```
Cl1=c(1,1,1,1,1,2,2,2,2,3,3,3,3)
Cl2=c(2,2,2,1,3,3,3,3,3,1,1,1,2)
```

(i) Calculate the unadjusted Rand index (8.24) and the adjusted Rand index (8.25) of agreement. Hint: WGCNA R function `randIndex`.
(ii) Calculate Pearson's chi-square test statistic and p value for relating *Cl1* and *Cl2*. Hint:

```
chisq.test(table(Cl1,Cl2))
```

(iii) Relabel the clusters of Cl1 as follows: $1 \to 2$, $2 \to 3$, $3 \to 1$. Show that the above measures of agreement lead to the same values.
(iv) For each of the three clusters of Cl1, calculate co-clustering-based cluster preservation statistic $ClusterPreservation^{co-clustering}$ (8.19) for tuplet sizes 2, 3, and 10. Message: The measure becomes uninformative for large tuplet sizes.
(v) For each of the three clusters of $Cl^{[ref]}$ calculate the accuracy-based cluster preservation measure and the overlap p value-based cluster preservation measure (8.28).
(vi) Which of the clusters of Cl1 are most highly preserved in Cl2?

References

Carlson M, Zhang B, Fang Z, Mischel P, Horvath S, Nelson SF (2006) Gene connectivity, function, and sequence conservation: Predictions from modular yeast co-expression networks. BMC Genomics 7(7):40

Dong J, Horvath S (2007) Understanding network concepts in modules. BMC Syst Biol 1(1):24

Dudoit S, Fridlyand J (2002) A prediction-based resampling method for estimating the number of clusters in a dataset. Genome Biol 3(7):RESEARCH0036

Gargalovic PS, Imura M, Zhang B, Gharavi NM, Clark MJ, Pagnon J, Yang WP, He A, Truong A, Patel S, Nelson SF, Horvath S, Berliner JA, Kirchgessner TG, Lusis AJ (2006) Identification of inflammatory gene modules based on variations of human endothelial cell responses to oxidized lipids. Proc Natl Acad Sci USA 103(34):12741–12746

Ghazalpour A, Doss S, Zhang B, Plaisier C, Wang S, Schadt EE, Thomas A, Drake TA, Lusis AJ, Horvath S (2006) Integrating genetics and network analysis to characterize genes related to mouse weight. PloS Genet 2(2):8

Hastie T, Tibshirani R, Friedman J (2001) The elements of statistcal learning: Data mining, inference, and prediction. Springer, New York

Hubert L, Arabie P (1985) Comparing partitions. J Classif 2(1):193–218

Kapp AV, Tibshirani R (2007) Are clusters found in one dataset present in another dataset? Biostat 8(1):9–31

Kaufman L, Rousseeuw PJ (1990) Finding groups in data: An introduction to cluster analysis. Wiley, New York

Langfelder P, Horvath S (2011) Fast R functions for robust correlations and hierarchical clustering. J Stat Software. In press

Langfelder P, Zhang B, Horvath S (2007) Defining clusters from a hierarchical cluster tree: The Dynamic Tree Cut library for R. Bioinformatics 24(5):719–720

Langfelder P, Luo R, Oldham MC, Horvath S (2011) Is my network module preserved and reproducible? Plos Comput Biol 7(1):e1001057

Oldham MC, Langfelder P, Horvath S (2011) Sample networks for enhancing cluster analysis of genomic data: Application to huntington's disease. Technical Report

Rand WM (1971) Objective criteria for the evaluation of clustering methods. J Am Stat Assoc 66(336):846–850

Sokal RR, Rohlf FJ (1962) The comparison of dendrograms by objective methods. Taxon 11:33–40

Tibshirani R, Walther G (2005) Cluster validation by prediction strength. J Comput Graph Stat 14:511–528

Yip A, Horvath S (2007) Gene network interconnectedness and the generalized topological overlap measure. BMC Bioinform 8(8):22

Chapter 9
Evaluating Whether a Module is Preserved in Another Network

Abstract In network applications, one is often interested in studying whether modules are preserved across multiple networks. For example, to determine whether a pathway of genes is perturbed in a certain condition, one can study whether its connectivity pattern is no longer preserved. Non-preserved modules can be either biologically uninteresting (e.g., reflecting data outliers) or interesting (e.g., reflecting species-specific modules). An intuitive approach for studying module preservation is to cross-tabulate module membership. But this approach often cannot address questions about the preservation of connectivity patterns between nodes. Thus, cross-tabulation-based approaches often fail to recognize that important aspects of a network module are preserved. Cross-tabulation methods make it difficult to argue that a module is *not* preserved. The weak statement ("the reference module does not overlap with any of the identified test set modules") is less relevant in practice than the strong statement ("the module cannot be found in the test network irrespective of the parameter settings of the module detection procedure"). Network concepts allow one to determine whether a module is preserved and reproducible in another network. Module preservation statistics have important applications, e.g., the wiring of apoptosis genes in a human cortical network differs from that in chimpanzees. It is advantageous to aggregate multiple preservation statistics into summary preservation statistics, e.g., $Z_{summary}$ and *medianRank*. Our applications show that the correlation structure underlying correlation networks facilitates the definition of particularly powerful module preservation statistics. Evaluating module preservation is in general different from evaluating cluster preservation. However, when modules are defined as clusters, then close relationships exist with cluster preservation statistics. This chapter describes results from a collaboration with Peter Langfelder et al. (Plos Comput Biol 7(1):e1001057, 2011).

9.1 Introduction

In many network applications, one is interested in studying the properties of network modules and their change across conditions (Fuller et al. 2007; van Nas et al. 2009; Wang et al. 2009; Plaisier et al. 2009). For example, Cai et al. (2010) studied which

S. Horvath, *Weighted Network Analysis: Applications in Genomics and Systems Biology*,
DOI 10.1007/978-1-4419-8819-5_9, © Springer Science+Business Media, LLC 2011

human brain co-expression modules can also be found in blood, Miller et al. (2010) studied module preservation between human and mouse brains, Keller et al. (2008) studied module preservation across multiple mouse tissues, and van Nas et al. (2009) studied network differences between male and female mice.

We describe several module preservation statistics for determining which properties of a network module are preserved in a second (test) network. The module preservation statistics allow one to quantify which aspects of within-module topology are preserved between a reference network and a test network. For brevity, we will refer to these aspects as connectivity patterns, but we note that our statistics are not based on network motifs. We use the term "module" in a broad sense: a network module is a subset of nodes that forms a subnetwork inside a larger network. Any subset of nodes inside a larger network can be considered a module. This subset may or may not correspond to a cluster of nodes.

Many cluster validation statistics proposed in the literature can be turned into module preservation statistics. In the following, we briefly review cluster validation statistics. Traditional cluster validation (or quality) statistics can be split into four broad categories: cross-tabulation-, density-, separability-, and stability statistics (Chen et al. 2002; McShane et al. 2002; Chen et al. 2007b; Kapp and Tibshirani 2007). Since cross-tabulation statistics compare cluster assignments in the reference and test clusterings, they require that a clustering procedure is also applied to the test data. On the other hand, density and density/separability statistics do not require a clustering in the test data set. These statistics typically evaluate clusters by how similar objects are within each cluster, and/or how dissimilar objects are between different clusters (Rousseeuw 1987; Shamir and Sharan 2001). Stability statistics (that typically study cluster stability when a controlled amount of artificial noise is added to the data) can also be used to evaluate clusters, but stability comparisons are more relevant to evaluating clustering procedures rather than cluster preservation and hence we do not consider them here.

Although many cluster validation statistics are based on within- and/or between cluster variance, several articles used prediction error to evaluate the reproducibility (or validity) of clusters in gene expression data (Dudoit and Fridlyand 2002; Tibshirani and Walther 2005; Kapp and Tibshirani 2007). These papers argued that the use of a measure of test set clusters defined by a classifier made from the reference data is a useful approach to cluster validation when the aim is to identify reproducible clusters of genes or samples with similar gene expression profile (Kapp and Tibshirani 2007).

The in-group proportion (IGP) approach for cluster validation takes advantage of the connection between cluster reproducibility and prediction accuracy (Kapp and Tibshirani 2007). The IGP measure is similar to the cluster cohesion measure (Bailey and Dubes 1982). For each IGP measure, one can also calculate a significance level (p value). We use the IGP measure as benchmark statistic for assessing the use of module preservation statistics in case that modules are defined as clusters. Below, we use real data applications and simulation studies to compare our module preservation statistics to the IGP. Our applications show that one of our module preservation statistics is sometimes closely correlated with the IGP statistic if the modules are defined as clusters. But cluster validation statistics (such as the IGP) may not be appropriate when modules are not defined as clusters. **In general, assessing module preservation is a different task from assessing cluster preservation**. In our simulations, we demonstrate that module preservation statistics can detect aspects of module preservation that are missed by existing cluster validation statistics.

In the following, we study several types of network preservation statistics that do not require a module assignment in the test network. We distinguish network preservation statistics by the type of the underlying network. Some preservation statistics are defined for a general network (defined by an adjacency matrix), while others are only defined for a correlation network (constructed on the basis of pairwise correlations between numeric variables).

9.2 Module Preservation Statistics

Table 9.1 presents an overview of the module preservation statistics studied in this chapter. We distinguish between cross-tabulation-based and network-based preservation statistics. **Cross-tabulation-based preservation statistics** require independent module detection in the test network and take the module assignments in both reference and test networks as input. Several cross-tabulation-based statistics are described in Sect. 8.8.

In our application, we describe situations when cross-tabulation-based preservation statistics are not applicable. But we should point out that cross-tabulation statistics also have the following advantages. First, they are often intuitive. Second, they can be applied when no network structure is present. Third, they work well when module assignments are strongly preserved and the modules remain separate in the test network. Cross-tabulation-based statistics have several disadvantages. To begin with, they are only applicable if the module assignment in the test data results from applying a module detection procedure to the test data. For example, a cross-tabulation based module preservation statistic would be meaningless when modules are defined as gene ontology categories since both reference and test networks contain the same sets of genes. But a nontrivial question is whether the network connections of a module (gene ontology category) in the reference network resemble those of the same module in the test network. To measure the resemblance of network connectivity, we propose several measures based on network concepts.

Even when modules are defined using a module detection procedure, cross-tabulation-based approaches face potential pitfalls. A module found in the reference data set will be deemed non-reproducible in the test data set if no matching module can be identified by the module detection approach in the test data set. Such non-preservation may be called the **weak non-preservation:** *"the module cannot be found using the current parameter settings of the module detection procedure"*. On the other hand, one is often interested in **strong non-preservation:** *"the module cannot be found irrespective of the parameter settings of the module detection*

Table 9.1 Overview of module preservation statistics

	Preservation statistic			Network	Ref. input			Test input			Used in composite		
No.	Name	Eq.	Type	Adj.	Lbl	Adj	*datX*	Lbl	Adj	*datX*	*Zsum.*	*medR.*	*Zsum.A*
1	coClustering	Supp.	Cross-tab	Not used	Yes	No	No	Yes	No	No	No	No	No
2	*accuracy*	Supp.	Cross-tab	Not used	Yes	No	No	Yes	No	No	No	No	No
3	-log(*p* value)	Supp.	Cross-tab	Not used	Yes	No	No	Yes	No	No	No	No	No
4	*meanAdj*	(9.2)	Density	General	Yes	No	No	No	Yes	No	No	No	Yes
5	*meanClCoef*	(9.3)	Density	General	Yes	No	No	No	Yes	No	No	No	No
6	*meanMAR*	(9.4)	Density	General	Yes	No	No	No	Yes	No	No	No	No
7	*cor.Adj*	(1.56)	Connect.	General	Yes	Yes	No	No	Yes	No	Yes	Yes	Yes
8	*cor.kIM*	(1.57)	Connect.	General	Yes	Yes	No	No	Yes	No	Yes	Yes	Yes
9	*cor.ClCoef*	(9.5)	Connect.	General	Yes	Yes	No	No	Yes	No	No	No	No
10	*cor.MAR*	(9.6)	Connect.	General	Yes	Yes	No	No	Yes	No	No	No	No
11	$separability^{ave}$	(1.47)	Separab.	General	Yes	Yes	No	No	Yes	No	No	No	No
12	*meanCor*	(9.8)	Den.+Con.	Cor	Yes	No	Yes	No	No	Yes	Yes	Yes	No
13	*cor.cor*	(9.9)	Connect.	Cor	Yes	No	Yes	No	No	Yes	Yes	Yes	No
14	*propVarExpl*	(9.10)	Density	Cor	Yes	No	Yes	No	No	Yes	Yes	Yes	No
15	*meanKME*	(9.11)	Den.+Con.	Cor	Yes	No	Yes	No	No	Yes	Yes	Yes	No
16	*cor.kME*	(9.12)	Connect.	Cor	Yes	No	Yes	No	No	Yes	Yes	Yes	No
17	*cor.kMEall*	(9.13)	Connect.	Cor	Yes	No	Yes	No	No	Yes	No	No	No
18	*separability*	(9.14)	Separab.	Cor	Yes	No	Yes	No	No	Yes	No	No	No
19	$Z_{summary}$	(9.1)	Compos.	Cor	Yes	Yes	Yes	No	Yes	Yes			
20	$p_{summary}$		Compos.	Cor	Yes	Yes	Yes	No	Yes	Yes			
21	*medianRank*	(9.20)	Compos.	Cor	Yes	Yes	Yes	No	Yes	Yes			
22	$Z_{summaryADJ}$	(9.21)	Compos.	General	Yes	Yes	No	No	Yes	No			

The columns report the names, types, and input of individual preservation statistics (Lbl, module label; Adj, general network adjacency; *datX*, numeric data from which a correlation network is constructed). The last three columns indicate which of the individual statistics are used in the composite summary statistics $Z_{summary}$, *medianRank*, and $Z_{summaryADJ}$, respectively. The definition of cross-tabulation-based statistics can be found in Sect. 8.8.

procedure". Strong non-preservation is difficult to establish using cross-tabulation approaches that rely on module assignment in the test data set. A second disadvantage of a cross-tabulation-based approach is that it requires that for each reference module one finds a matching test module. This may be difficult when a reference module overlaps with several test modules or when the overlaps are small. A third disadvantage is that it may fail to detect preservation of connectivity patterns. For example, it may fail to detect that highly connected hub nodes in the reference module remain hub nodes in the text module.

Network-based statistics do not require the module assignment in the test network but require the user to input network adjacency matrices. We distinguish the following three types of network-based module preservation statistics: (1) density-based, (2) separability-based, and (3) connectivity-based preservation statistics. **Density-based** preservation statistics can be used to determine whether module nodes remain highly connected in the test network. **Separability-based** statistics can be used to determine whether network modules remain distinct (separated) from one another in the test network. While numerous measures proposed in the literature combine aspects of density and separability, we keep them separate and provide evidence that density-based approaches can be more useful than separability-based approaches in determining whether a module is preserved.

The preservation statistics are based on network properties that can be measured using network concepts, e.g., fundamental network concepts (Chap. 1) or eigenvector-based network concepts (Chap. 3). Here we will show how to use both *intra*modular network concepts (e.g., $kME_i^{(q)}$ (6.53) and $kIM_i^{(q)}$ (1.41)) and *inter*modular network concepts (e.g., module separability (1.47)) to study the preservation of network modules. By measuring how these network concepts are preserved from a reference network to a test network, one can define network module preservation statistics as described below.

In the following, we describe *network*-based preservation statistics.

Connectivity-based preservation statistics can be used to determine whether the connectivity pattern between nodes in the reference network is similar to that in the test network.

Table 9.1 reports the **required input** for each preservation statistic. Since each preservation statistic is used to evaluate the preservation of modules defined in a reference network, it is clear that each statistic requires the module assignment from the reference data. But the statistics differ with regard to the module assignment in the test data. Only cross-tabulation-based statistics require a module assignment in the test data. Network-based preservation statistics do not require a test set module assignment. Instead, they require the test set network adjacency matrix (for a general network) or the test data set *datX* of numeric variables (for a correlation network).

We distinguish network statistics by the underlying network. Some preservation statistics are defined for a general network (defined by an adjacency matrix), while others are only defined for a correlation network (constructed on the basis of pairwise correlations between numeric variables). Our applications show that the correlation structure facilitates the definition of particularly powerful module preservation statistics. Preservation statistics 4–11 (Table 9.1) can be used for general networks while statistics 12–19 assume correlation networks. Network density and module separability statistics only need the test set adjacency matrix, while the connectivity preservation statistics also require the adjacency matrix in the reference data.

It is often not clear whether an observed value of a preservation statistic is higher than expected by chance. We attach a significance level (permutation test p value) to observed preservation statistics, using a permutation test procedure which randomly permutes the module assignment in the test data. Based on the permutation test we are also able to estimate the mean and variance of the preservation statistic under the null hypothesis of no relationship between the module assignments between reference and test data. By standardizing each observed preservation with regard to the mean and variance, we define a Z statistic for each preservation statistic (9.15). Under certain assumptions, each Z statistic follows (approximately) a standard normal distribution if the module is not preserved. The higher the value of a Z statistic, the stronger the evidence that the observed value of the preservation statistic is significantly higher than expected by chance.

9.2.1 Summarizing Preservation Statistics and Threshold Values

Because preservation statistics measure different aspects of module preservation, their results may not always agree. We find it useful to aggregate different module preservation statistics into **composite preservation statistics**. Composite preservation statistics also facilitate the fast evaluation of many modules in multiple networks. We define several composite statistics.

For correlation networks based on quantitative variables, the four density preservation statistics are summarized by $Z_{density}$ (9.16), the three connectivity-based statistics are summarized by $Z_{connectivity}$ (9.17), and all individual Z statistics are summarized by $Z_{summary}$ defined as follows:

$$Z_{summary} = \frac{Z_{density} + Z_{connectivity}}{2}. \tag{9.1}$$

Our simulations suggest the following thresholds for $Z_{summary}$: if $Z_{summary} > 10$, there is strong evidence that the module is preserved (Langfelder et al. 2011). If $2 < Z_{summary} < 10$, there is weak to moderate evidence of preservation. If $Z_{summary} < 2$, there is no evidence that the module preserved. For general networks defined by an adjacency matrix, we find it expedient to summarize the preservation statistics into a summary statistic denoted $Z_{summaryADJ}$ (9.21).

Since biologists are often more familiar with p values as opposed to Z statistics, our R implementation in function `modulePreservation` also calculates empirical p values. Analogous to the case of the Z statistics, the p values of individual preservation statistic are summarized into a descriptive measure called $p_{summary}$. The smaller the $p_{summary}$, the stronger the evidence that the module is preserved. In practice, we observe an almost perfect inverse relationship (Spearman correlation $r \approx -0.97$) between $Z_{summary}$ and $p_{summary}$.

The Z statistics and permutation test p values often depend on the module size (i.e., the number of nodes in a module). This fact reflects the intuition that it is more significant to observe that the connectivity patterns among hundreds of nodes are preserved than to observe the same among say only five nodes. Having said this, there will be many situations when the dependence on module size is not desirable, e.g., when preservation statistics of modules of different sizes are to be compared. In this case, we recommend to either focus on the observed values of the individual statistics or alternatively to summarize them using the composite module preservation statistic *medianRank* (9.20). The *medianRank* is useful for comparing relative preservation among multiple modules: a module with lower median rank tends to exhibit stronger observed preservation statistics than a module with a higher median rank. Since *medianRank* is based on the observed preservation statistics (as opposed to Z statistics or p values), we find that it is much less dependent on module size.

9.2.2 Module Preservation Statistics for General Networks

We define a module as a subset of nodes that forms a subnetwork in the original network. Modules are labeled by integer labels $q = 1, 2, \ldots, Q$, and sometimes by color labels. Color labels can be convenient for visualizing modules in network plots. For module q with $n^{(q)}$ nodes, the $n^{(q)} \times n^{(q)}$ dimensional adjacency matrix between the module nodes is denoted by $A^{(q)}$. Denote by $\mathscr{M}_q$ the set of node indices of the $n^{(q)}$ nodes in module q. Network concepts (such as the connectivity, clustering coefficient, MAR, etc.) defined for $A^{(q)}$ are defined as intramodular network concepts. For example, the density $Density^{(q)}$ (1.42) of module q is defined as the mean adjacency of the off-diagonal elements of $A^{(q)}$.

We will now describe module preservation statistics that can be used to determine whether a module that is present in a reference network (with adjacency $A^{[\mathrm{ref}]}$) can also be found in an independent test network (with adjacency $A^{[\mathrm{test}]}$). Specifically, assume that the vector $Cl^{[\mathrm{ref}]}$ encodes the module assignments in the reference network. Thus, $Cl_i^{[\mathrm{ref}]} = q$ ($q \in \{1, \ldots, Q^{[\mathrm{ref}]}\}$) if node i is assigned to module q. We reserve the label $Cl = 0$ (and color gray) for nodes that are not assigned to any module. For a given module q with n_q nodes, the $n_q \times n_q$ module adjacency matrices are denoted by $A^{[\mathrm{ref}](q)}$ and $A^{[\mathrm{test}](q)}$ in the reference and test networks, respectively. We propose network concepts that can be useful for determining whether a module q (found in the reference network) is preserved in the test network.

Intuitively, one may call a module q preserved if it has a high-density $Density^{[\mathrm{test}](q)}$ (1.18) in the test network. We define the *mean adjacency* for module q as the module density in the test network,

$$\begin{aligned} meanAdj^{[\mathrm{test}](q)} &= Density^{[\mathrm{test}](q)} \\ &= mean\left(vectorizeMatrix\left(A^{[\mathrm{test}](q)}\right)\right). \end{aligned} \tag{9.2}$$

Some of the **density preservation statistics** such as the mean adjacency are similar to previously described methods based on within-cluster and between-cluster dissimilarities (Chen et al. 2002). For example, the mean intramodular adjacency $meanAdj^{[\mathrm{test}](q)}$ (9.2) is oppositely related to the within-cluster as described in Sect. 8.1. The network density measure can be considered a generalization of the cluster cohesiveness measure (Bailey and Dubes 1982) to (possibly weighted) networks.

Other network concepts may be used to obtain a summary statistic of a module. For example, the `WGCNA` function `modulePreservation` also calculate preservation statistics based on the mean clustering coefficient (1.22):

$$meanClCoef^{[\text{test}](q)} = mean\left(ClusterCoef_i^{[\text{test}]}\right) \tag{9.3}$$

and mean MAR (1.16):

$$meanMAR^{[\text{test}](q)} = mean\left(MAR_i^{[\text{test}]}\right) \tag{9.4}$$

in the test network. **Connectivity preservation statistics** quantify how similar connectivity of a given module is between a reference and a test network. For example, module connectivity preservation can mean that, within a given module q, nodes with a high connection strength in the reference network also exhibit a high connection strength in the test network. This property can be quantified by the correlation of intramodular adjacencies in reference and test networks. Specifically, if the entries of the first adjacency matrix $A^{[\text{ref}](q)}$ are correlated with those of the second adjacency matrix $A^{[\text{test}](q)}$, then the adjacency pattern of the module is preserved in the second network. To measure preservation, we use the *adjacency correlation* of the qth module (1.56), (1.55) defined as follows:

$$cor.Adj^{(q)} = \text{cor}\left(vectorizeMatrix\left(A^{[\text{ref}](q)}\right), vectorizeMatrix\left(A^{[\text{test}](q)}\right)\right).$$

If module q is preserved in the second network, the highly connected hub nodes in the reference network will often be highly connected hub nodes in the test network. In other words, the intramodular connectivity $\text{kIM}^{[\text{ref}](q)}$ in the reference network should be highly correlated with the corresponding intramodular connectivity $\text{kIM}^{[\text{test}](q)}$ in the test network. To measure the preservation of intramodular connectivity, we use the correlation of intramodular connectivities $cor.kIM^{(q)}$ (1.57). Analogously, we define the correlation of clustering coefficients and maximum adjacency ratios,

$$cor.ClCoef^{(q)} = \text{cor}\left(ClusterCoef^{[\text{ref}](q)}, ClusterCoef^{[\text{test}](q)}\right), \tag{9.5}$$

$$cor.MAR^{(q)} = \text{cor}\left(MAR^{[\text{ref}](q)}, MAR^{[\text{test}](q)}\right). \tag{9.6}$$

9.2.3 Module Preservation Statistics for Correlation Networks

The specific nature of correlation networks allows us to define additional module preservation statistics. The underlying information carried by the sign of the correlation can be used to further refine the statistics irrespective of whether a signed or unsigned similarity is used in network construction. To simplify notation, we define

$$r_{ij}^{[\text{ref}]} = \text{cor}\left(x_i^{[\text{ref}]}, x_j^{[\text{ref}]}\right),$$
$$r_{ij}^{[\text{test}]} = \text{cor}\left(x_i^{[\text{test}]}, x_j^{[\text{test}]}\right). \tag{9.7}$$

We will use the notation $r_{ij}^{[\text{ref}](q)}$ for the correlation matrix restricted to the nodes in module q. We define the mean correlation density of module q as:

$$meanCor^{[\text{test}](q)} = mean\left\{vectorizeMatrix\left(sign\left(r_{ij}^{[\text{ref}](q)}\right) r_{ij}^{[\text{test}](q)}\right)\right\}. \tag{9.8}$$

Thus, the correlation measure of module preservation is the mean correlation in the test network multiplied by the sign of the corresponding correlations in the reference network. We note that a correlation that has the same sign in the reference and test networks increases the mean, while a correlation that changes sign decreases the mean. Because the preservation statistic keeps track of the sign of the corresponding correlation in the reference network, we call it the mean *sign-aware* correlation.

To measure the preservation of connectivity patterns within module q between the reference and test networks, we define a correlation-based measure $cor.cor$ similar to the $cor.Adj$ statistic (1.56):

$$cor.cor^{(q)} = \text{cor}\left(vectorizeMatrix\left(r^{[\text{ref}](q)}\right), vectorizeMatrix\left(r^{[\text{test}](q)}\right)\right). \tag{9.9}$$

In applications, we have found that the correlation-based preservation statistic $cor.cor$ is preferable to its general network counterpart $cor.Adj$; therefore, we only report $cor.cor$ (Langfelder et al. 2011).

9.2.3.1 Eigennode-Based Density Preservation Statistics

Many module construction methods lead to correlation network modules comprised of highly correlated variables. In this case, one can summarize the variables with the module eigennode (i.e., the singular vector). As explained in Sect. 6.1.2, the module eigennode $E^{(q)}$ of the q-th module can be interpreted as a weighted average of the scaled variables. The module eigennode $E^{(q)}$ (6.6) can be used to define the module eigengene-based connectivity measure $\text{kME}_i^{(q)} = \text{cor}(x_i, E^{(q)})$ (5.53).

The module eigennode also gives rise to several preservation statistics that in effect measure module density, or, from a different point of view, how well the eigennode represents the whole module. For example, one can use the proportion of variance explained *propVarExpl* by the module eigennode to arrive at a density measure.

Both *intra*modular network concepts (e.g., $\text{kME}_i^{(q)}$) and *inter*modular network concepts (e.g., module separability (1.47)) can be used to study the preservation of network modules. By measuring how these network concepts are preserved from a reference network to a test network, one can define network module preservation statistics as described below.

In Sect. 6.3.1, we prove that the *proportion of variance explained* (*propVarExpl*) (defined in (6.7)) can also be calculated as mean squared kME value:

$$propVarExpl^{[\text{test}](q)} = mean_{i \in \mathcal{M}_q} \left\{ \left(\text{kME}_i^{[\text{test}](q)} \right)^2 \right\}, \tag{9.10}$$

where $E^{[\text{test}](q)}$ is the eigennode of module q in the test network.

The *mean sign-aware module membership* is defined as:

$$meanKME^{[\text{test}](q)} = mean_{i \in \mathcal{M}_q} \left\{ sign\left(\text{kME}_i^{[\text{ref}](q)} \right) \text{kME}_i^{[\text{test}](q)} \right\}. \tag{9.11}$$

It measures the mean module membership, (6.53), in which nodes whose module memberships in the reference and test networks have the same sign contribute positively, and nodes whose module memberships in the reference and test networks have opposite signs contribute negatively. Our statistic *meanKME* is conceptually related to the homogeneity score (Chen et al. 2002) which is defined as the average correlation between a cluster's centroid and the members of the cluster.

9.2.3.2 Eigennode-Based Connectivity Preservation Statistics

Intuitively, if the internal structure of a module is preserved between a reference and a test network, we expect that a variable with a high module membership in the reference network will have a high module membership in the test network as well; conversely, variables with relatively low module membership in the reference network should also have a relatively low module membership in the test network. In other words, intramodular hubs in the reference network should also be intramodular hubs in the test network. For a given module q, we define the $cor.kME^{(q)}$ statistic as

$$cor.kME^{(q)} = \text{cor}_{i \in \mathcal{M}_q} \left(\text{kME}_i^{[\text{ref}](q)}, \text{kME}_i^{[\text{test}](q)} \right), \tag{9.12}$$

where the correlation runs only over variables that belong to module q. We also define an analogous statistic by correlating the module membership of *all* network variables in the reference and test networks:

$$cor.kMEall_i^{(q)} = \text{cor} \left(\text{kME}_i^{[\text{ref}](q)}, \text{kME}_i^{[\text{test}](q)} \right). \tag{9.13}$$

The advantage of using all nodes is that the statistic is less dependent on cutoffs (e.g., branch cut parameters) of the method used to define modules. On the other hand, for relatively small modules (compared to the size of the full network) the signal of the few nodes with high module membership may be overwhelmed by the noise contribution of the many nodes that have very low module membership.

9.2.3.3 Module Separability Statistics

A network module is distinct if it is well separated from the other modules in the network. A distinct module in a reference network may be considered well preserved in a test network if it remains well separated from the other modules in the test network. Recall the definition of the separability measure: (1.47)

$$separability.ave(q_1,q_2) = 1 - \frac{A^{ave}(q_1,q_2)}{Density(q_1,q_2)}.$$

The separability statistics take on (possibly negative) values smaller than 1. The closer a separability statistic value is to 1, the more separated (distinct) are the two modules.

In clustering applications based on Euclidean distance, it is customary to measure module distinctiveness, or separability, by the between-cluster distance. For correlation networks, we propose to measure module separability by 1 minus the correlation of their respective eigennodes. Specifically, for two modules q_1,q_2, their test separability is defined as:

$$separability.ME^{[\text{test}]}(q_1,q_2) = 1 - \text{cor}\left(E^{[\text{test}](q_1)}, E^{[\text{test}](q_2)}\right). \tag{9.14}$$

Low test separability suggests the modules are not preserved as separate clusters. Differences in separability between networks may also reflect biologically interesting differences in correlation relationships between whole modules (Langfelder and Horvath 2007).

In the Sect. 9.7, we outline when close relationships exist between *separability.ave* and eigennode-based separability *separability.ME*. Since the eigennode-based separability can be computed much more efficiently, we typically use the eigennode based separability in our applications.

The separability statistic is conceptually related to the separability score used in Chen et al. (2002) which for cluster q is the weighted average of the correlation between the centroid of cluster q and every other centroid q',

$$SS(q) = \frac{\sum_{q' \neq q} n^{q'} \text{cor}(E^{q'}, E^{q})}{\sum_{q'} n^{q'}}.$$

Since we wanted to put all modules on the same footing irrespective of module size, we do not use module size in the definition of the separability statistics. Having said this, it is straightforward to adapt the definition to include module size.

9.2.4 Assessing Significance of Observed Module Preservation Statistics by Permutation Tests

Typical values of module preservation statistics depend on many factors, for example, on network size, module size, number of observations, etc. It is advantageous to use a permutation test to assess the significance of an observed module preservation statistic. Specifically, we randomly permute the module labels in the test network and calculate corresponding preservation statistics. This procedure is repeated n_{perm} times. For each statistic labeled by index a, we then calculate the mean μ_a and the standard deviation σ_a of the permuted values. We define the corresponding Z_a statistic as:

$$Z_a = \frac{obs_a - \mu_a}{\sigma_a}, \tag{9.15}$$

where obs_a is the observed value for the statistic a. Under certain conditions, one can prove that under the null hypothesis of no preservation, the statistic Z_a asymptotically follows the standard normal distribution $N(0,1)$. Thus, under the assumption that the number of permutations is large enough to approximate the asymptotic regime, one can convert the Z statistics to p values using the standard normal distribution. The `WGCNA` R function `modulePreservation` outputs the asymptotic p values for each statistic. But we should point out that it would be preferable to use a full permutation test to calculate permutation test p values. We often report Z statistics (instead of p values) for the following two reasons: First, permutation p values of preserved modules are often astronomically significant (say $p < 10^{-30}$) and it is more convenient to report the results on a Z scale. The second reason is computational speed. The calculation of a Z statistic only requires one to estimate the mean and variance under the null hypothesis, for which fewer permutations are needed. To estimate a permutation test p value accurately would require computational time far beyond practical limits.

9.2.5 Composite Preservation Statistic $Z_{summary}$

In the Sect. 9.7, we describe when close relationships exist between many of the preservation statistics presented above. This suggests that one can combine the individual preservation statistics into a composite preservation statistic. We propose two composite preservation statistics. The first composite statistic $Z_{summary}$ (9.1) summarizes the individual Z statistic values that result from the permutation test. The second composite statistic *medianRank* (9.20) summarizes the ranks of the observed preservation statistics.

The relationships among preservation statistics derived in Sect. 9.7 suggest to summarize the density-based preservation statistics as follows:

$$Z_{density} = median\left(Z_{meanCor}, Z_{meanAdj}, Z_{propVarExpl}, Z_{meanKME}\right). \tag{9.16}$$

Similarly, the connectivity-based preservation statistics can be summarized as follows:

$$Z_{connectivity} = median\left(Z_{cor.kIM}, Z_{cor.Adj}, Z_{cor.kME}, Z_{cor.kMEall}, Z_{cor.cor}\right). \quad (9.17)$$

When density- and connectivity-based preservation statistics are equally important for judging the preservation of a network module, one can consider the composite Z summary statistic (9.1)

$$Z_{summary} = \frac{Z_{density} + Z_{connectivity}}{2}.$$

Alternatively, a weighted average between $Z_{density}$ and $Z_{connectivity}$ can be formed to emphasize different aspects of module preservation. Future research could investigate alternative ways of aggregating preservation statistics. In practice, we recommend to consider all individual preservation statistics.

Our simulated as well as empirical data show that the separability tends to have low agreement (as measured by correlation) with the other preservation statistics (Fig. 9.9). Since the *separability* statistic often performs poorly, we did not include it into the composite statistics.

Since $Z_{summary}$ is not a permutation statistic but rather the median of other Z statistics, we do not use it to calculate a p value. The function `modulePreservation` calculates a summary p value ($p_{summary}$) as the geometric mean of the p values corresponding to the calculated permutation Z statistics. Specifically, for each permutation Z statistic, it calculates the corresponding p value assuming that, under the null, the Z statistic has a normal distribution $N(0,1)$. The normal distribution can be justified using relatively weak assumptions described in statistics textbooks. As a caveat, we mention that we use preservation p values as descriptive (and not inferential) measures. On the other hand, we cannot assume normality for $Z_{summary}$. Hence, instead of calculating a p value corresponding to $Z_{summary}$, we calculate a summary log-p value instead, given as the median of the log-p values of the corresponding permutation Z statistics. Because of the often extremely significant p values associated with the permutation Z statistics, it is desirable to use logarithms (here base 10). We emphasize that the summary log-p value is not directly associated with $Z_{summary}$; rather, it is a separate descriptive summary statistic that summarizes the p values of the individual permutation Z statistics.

9.2.5.1 Thresholds for Module Preservation Statistics

It seems intuitive to call a module with $Z_{summary} > 2$ preserved. but our simulation studies argue for a more stringent threshold (Langfelder et al. 2011). We recommend the following threshold guidelines: if $Z_{summary} > 10$, there is strong evidence that the module is preserved. If $2 < Z_{summary} < 10$, there is weak to moderate evidence of preservation. If $Z_{summary} < 2$, there is no evidence that the module preserved. As discussed below, these threshold values should be considered rough guidelines since more (or less) stringent thresholds may be required depending on the application.

The `modulePreservation` R function calculates multiple preservation Z statistics and corresponding asymptotic p values. Similar to the case of p values, a threshold that is appropriate in one context may not be appropriate in another. Given the strong relationships among some preservation statistics, we have found it useful to aggregate the Z statistics (and optionally the empirical p values) in a statistically robust fashion using the median, but many alternative procedures are possible.

9.2.6 Composite Preservation Statistic *medianRank*

In some applications such as the human–chimpanzee example described above, one is interested in ranking modules by their overall preservation in the test set, i.e., one is interested in a relative measure of module preservation. Since our simulations and applications reveal that $Z_{summary}$ (9.1) strongly depends on module size, this statistic may not be appropriate when comparing modules of very different sizes. Here we define an alternative rank-based measure that relies on observed preservation statistics rather than the permutation Z statistics. For each statistic a, we rank the modules based on the observed values $obs_a^{(q)}$. Thus, each module is assigned a rank $rank_a^{(q)}$ for each observed statistic. We then define the median density and connectivity ranks

$$medianRank.density^{(q)} = median_{a \in \text{Density statistics}}(rank_a^{(q)}), \tag{9.18}$$

$$medianRank.connectivity^{(q)} = median_{a \in \text{Connectivity statistics}}(rank_a^{(q)}). \tag{9.19}$$

Analogously to the definition of $Z_{summary}$, we then define the statistic *medianRank* as the mean of *medianRank.density* and *medianRank.connectivity*,

$$medianRank = \frac{medianRank.density + medianRank.connectivity}{2}. \tag{9.20}$$

Alternatively, a weighted average of the ranks could be formed to emphasize different aspects of module preservation. It is worth repeating that a composite rank preservation statistic is only useful for studying the relative preservation of modules, e.g., we use *medianRank* for studying which human brain co-expression modules are least preserved in chimpanzee brain networks.

9.2.6.1 Composite Preservation Statistic $Z_{summaryADJ}$ for General Networks

While our examples deal with correlation (in particular, co-expression) networks, we have also implemented preservation statistics that can be applied to general networks specified only by an adjacency matrix. For example, this function could be used to study module preservation in protein–protein interaction networks.

In particular, we define a composite statistic $Z_{summaryADJ}$, which is defined for a general network specified by an adjacency matrix (9.21).

$$Z_{summaryADJ} = \frac{Z_{densityADJ} + Z_{ConnectivityADJ}}{2}, \tag{9.21}$$

where $Z_{densityADJ} = Z_{meanAdj}$ and $Z_{ConnectivityADJ} = median(Z_{cor.kIM}, Z_{cor.Adj})$. Note that $Z_{summaryADJ}$ is only computed with regard to a subset of the individual statistics. Set `dataIsExpr=FALSE` in the `modulePreservation` R function to calculate this statistic.

9.3 Cholesterol Biosynthesis Module Between Mouse Tissues

Several studies have explored how co-expression modules change between mouse tissues (Keller et al. 2008) and/or sexes (van Nas et al. 2009). Here we reanalyze gene expression data from the liver, adipose, muscle, and brain tissues of an F2 mouse intercross described in Sect. 5.5, Ghazalpour et al. (2006) and Fuller et al. (2007). The expression data contain measurements of 17,104 genes across the following numbers of microarray samples: 137 (female (F) adipose), 146 (male (M) adipose), 146 (F liver), 145 (M liver), 125 (F muscle), 115 (M muscle), 148 (F brain), and 141 (M brain).

We consider a single module defined by the genes of the gene ontology (GO) term "Cholesterol biosynthetic process" (CBP, GO id GO:0006695 and its GO offspring). Of the 29 genes in the CBP, 24 could be found among the 17,104 genes. Cholesterol is synthesized in liver and we used the female liver network as the reference network module. As test networks, we considered the CBP co-expression networks in other tissue/sex combinations.

Each circle plot in Fig. 9.1 visualizes the connection strengths (adjacencies) between CBP genes in different mouse tissue/sex combination. The color and width of the lines between pairs of genes reflect the correlations of their gene expression profiles across a set of microarray samples. Before delving into a quantitative analysis, we invite the reader to visually compare the patterns of connections. Clearly, the male and female liver networks look very similar. Because of the ordering of the nodes, the hubs are concentrated on the upper right size of the circle and the right side of the network is more dense. The adipose tissues also show this pattern, albeit much more weakly. On the other hand, the figures for the brain and muscle tissues do not show these patterns. Thus, the figure suggests that the CBP module is more strongly preserved between liver and adipose tissues than between liver and brain or muscle.

We now turn to a quantitative assessment of this example. We start out by noting that a **cross-tabulation-based approach** of module preservation is meaningless in this example since the module is a GO category whose genes can trivially be found in each network. However, it is a very meaningful exercise to measure

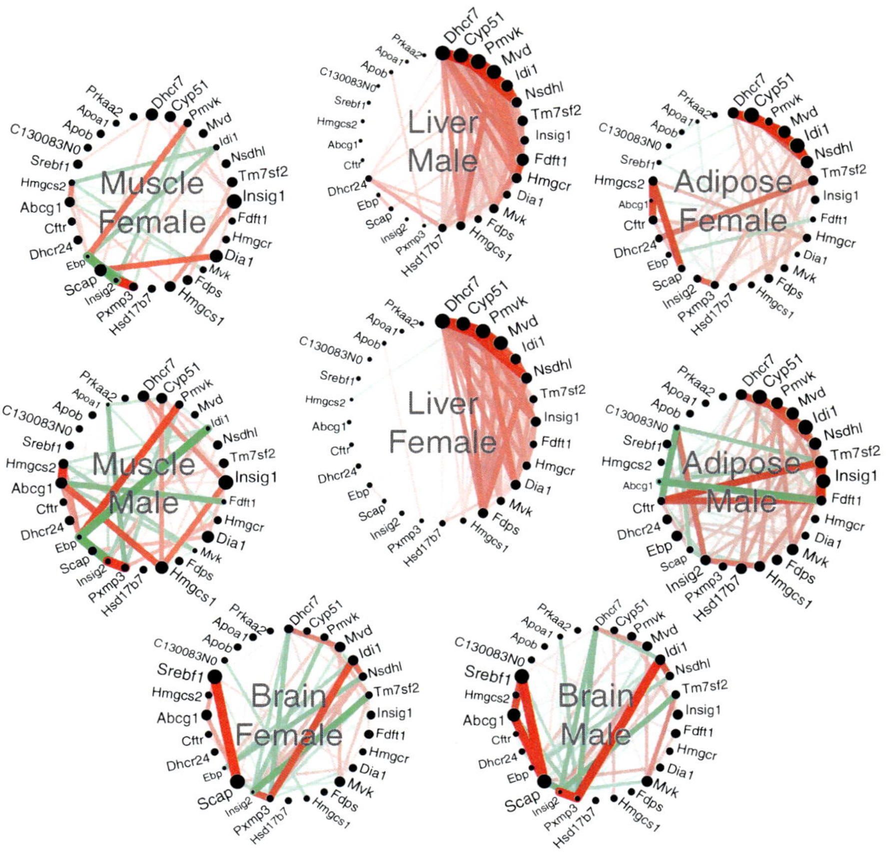

Fig. 9.1 Network plot of the module of cholesterol biosynthesis genes in different mouse tissues. Here a module is defined as a signed weighted correlation network (see Sect. 5.3) among genes from the GO category Cholesterol Biosynthetic Process. Module preservation statistics allow one to quantify similarities between the depicted networks. The figure depicts the connectivity patterns (correlation network adjacencies) between cholesterol biosynthesis genes in four different mouse tissues from male and female mice of an F2 mouse cross (described in application 4). The *thickness* of the line reflects the absolute correlation. The *line* is colored in *red* if the correlation is positive and *green* if it is negative. The size of each *black circle* indicates the connectivity of the corresponding gene; hubs (i.e., highly connected) genes are represented by *larger circles*. Visual inspection suggests that the male and female liver networks are rather similar and show some resemblance to those of the adipose tissue. Module preservation statistics can be used to measure the similarity of connectivity patterns between pairs of networks

the similarity of the connectivity patterns of the module genes across networks. To provide a quantitative assessment of the connectivity preservation, it is useful to adapt network concepts described in previous chapters. Figure 9.2 provides a quantitative assessment of the preservation of the connectivity patterns of the cholesterol biosynthesis module between the female liver network and networks

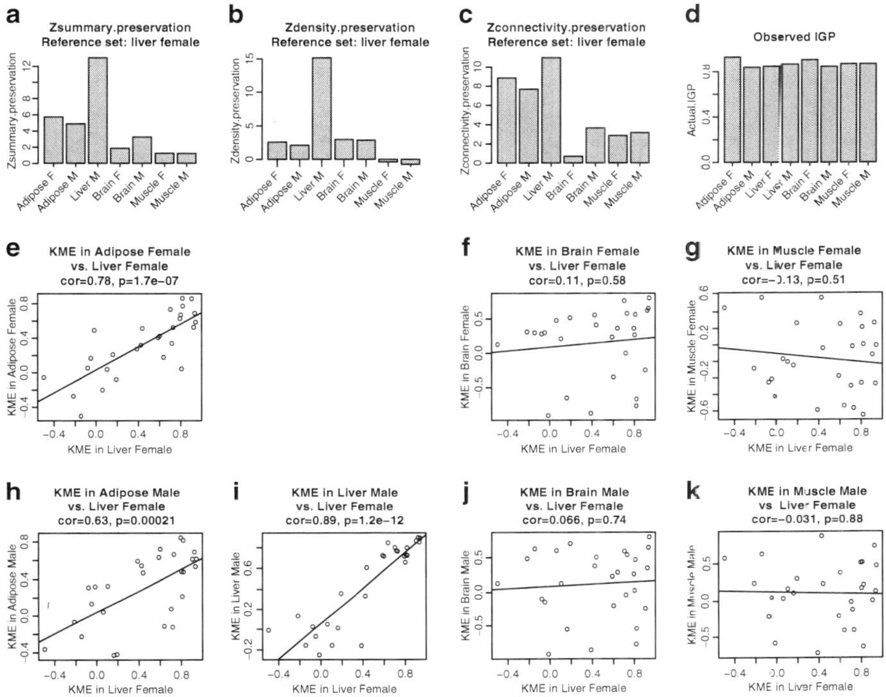

Fig. 9.2 Preservation of GO term cholesterol biosynthetic process across mouse tissues. This figure presents quantitative evaluation of the similarities among the networks depicted in Fig. 9.1. As reference module, we define a correlation network among the genes of the GO term "Cholesterol biosynthetic process" (CBP) in the female mouse liver network. Panels (**a**–**c**) show summary preservation statistics in other tissue and sex combinations. Panel (**a**) shows the composite preservation statistic $Z_{summary}$. The CBP module in the female liver network is highly preserved in the male liver network ($Z_{summary} > 10$) and moderately preserved in adipose networks. There is no evidence of preservation in brain or muscle tissue networks. Panels (**b**) and (**c**) show the density and connectivity statistics, respectively. Panel (**d**) shows the results of the in-group proportion analysis (Kapp and Tibshirani 2007). According to the IGP analysis, the CBP module is equally preserved in all networks which reflects that it ignores connectivity preservation. (**e**–**k**) show the scatter plots of kME (6.53) in one test data set (indicated in the title) versus the liver female reference set. Each point corresponds to a gene; Pearson correlations and the corresponding p values are displayed in the title of each scatter plot. The eigengene-based connectivity kME is strongly preserved between adipose and liver tissues; it is not preserved between female liver and the muscle and brain tissues

from other sex/tissue combinations. Figure 9.2a presents the composite summary statistic ($Z_{summary}$, (9.1)) in each test network. Overall, we find strong evidence of preservation ($Z_{summary} > 10$, (9.1)) in the male liver network but no evidence ($Z_{summary} < 2$) of preservation in the female brain and muscle networks. We find that the connectivity of the female liver CBP is most strongly preserved in the male liver network. It is also weakly preserved in adipose tissue but we find no evidence for its preservation in muscle and brain tissues. The summary preservation statistic $Z_{summary}$ measures both aspects of density and of connectivity preservation. We now evaluate which of these aspects are preserved. Figure 9.2b shows that the module shows strong evidence of density preservation ($Z_{density} > 10$) (9.16) in the male liver network but negligible density preservation in the other networks. Interestingly, Fig. 9.2c shows that the module has moderate connectivity preservation $Z_{connectivity} > 5$ (9.17) in the adipose networks.

The $Z_{connectivity}$ summarizes the statistical significance of several connectivity-based preservation statistics. Several of our connectivity measures evaluate whether highly connected intramodular hub nodes in the reference network remain hub nodes in the test network. Preservation of intramodular connectivity reflects the preservation of hub gene status between the reference and test network. One measure of intramodular connectivity is the module eigengene-based connectivity measures kME_i (6.53). A high correlation of kME between reference and test network can be visualized using a scatter plot and quantified using the correlation coefficient $cor.kME$. For example, Fig. 9.2i shows that kME in the female liver module is highly correlated with that of the male liver network ($r = cor.kME = 0.89$, $p = 1.2 \times 10^{-12}$). Further, the scatter plots in Fig. 9.2 show that the kME measures between liver and adipose networks show strong correlation (preservation): $cor.kME^{F\ liver,F\ adipose} = 0.78$ ($p = 1.7 \times 10^{-7}$), $cor.kME^{F\ liver,M\ adipose} = 0.63$ ($p = 2.1 \times 10^{-4}$), $cor.kME^{F\ liver,M\ liver} = 0.89$ ($p = 1.2 \times 10^{-12}$), while the correlation between kME in female liver and the brain and muscle data sets is not significant. This example demonstrates that connectivity preservation measures can uncover a link between CBP in liver and adipose tissues that density preservation statistics fail to identify.

We briefly compare the performance of our network-based statistics with those from the IGP method (Kapp and Tibshirani 2007). The R implementation of the IGP statistic requires that at least two modules are being evaluated. To get it to work for this application that involves only a single module, we defined a second module by randomly sampling half of the genes from the rest of the entire network. Figure 9.2d shows high, nearly constant values of the IGP statistic across networks, which indicates that the CBP module is present in all data sets. Note that the IGP statistic does not allow us to argue that the CBP module in the female liver network is more similar to the CBP module in the male liver than in other networks. This reflects that the IGP statistic, which is a cluster validation statistic, does not measure connectivity preservation.

9.4 Human Brain Module Preservation in Chimpanzees

Here we study the preservation of co-expression between human and chimpanzee brain gene expression data. The data set consists of 18 human brain and 18 chimpanzee brain microarray samples (Khaitovich et al. 2004). The samples were taken from six regions in the brain; each region is represented by three microarray samples. Since we used the same weighted gene co-expression network construction and module identification settings as in the original publication, our human modules are identical to those in Oldham et al. (2006). Because of the relatively small sample size, only few relatively large modules could be detected in the human data. The resulting modules were labeled by colors: turquoise, blue, brown, yellow, green, black, red (see Fig. 9.3a). Oldham et al. (2006) determined the biological meaning of the modules by determining in which brain regions the module genes were over-expressed. For example, heat maps of module expression profiles revealed that the turquoise module contains genes highly expressed in cerebellum, the yellow module contains genes highly expressed in caudate nucleus, the red module contains genes highly expressed in anterior cingulate cortex (ACC) and caudate nucleus, and the black module contains mainly genes expressed in white matter. The blue, brown, and green modules contained genes highly expressed in cortex, which is why we refer to these modules as cortical modules. Visual inspection of the module color band below the dendrograms in Fig. 9.3a and b suggests that most modules show fairly strong preservation. Oldham et al. argued that modules corresponding to evolutionarily older brain regions (turquoise, yellow, red, black) show stronger preservation than the blue and green cortical modules (Oldham et al. 2006). Here we reanalyze these data using module preservation statistics.

The most common **cross-tabulation approach** starts with a contingency table that reports the number of genes that fall into modules of the human network (corresponding to rows) versus modules of the chimpanzee network (corresponding to columns). The contingency table in Fig. 9.3c shows that there is high agreement between the human and chimpanzee module assignments. The human modules black, brown, red, turquoise, and yellow have well-defined chimpanzee counterparts (labeled by the corresponding colors). On the other hand, the human green cortical module appears not to be preserved in chimpanzee since most of its genes are classified as unassigned (gray color) in the chimpanzee network. Further, the human blue cortical module (360 genes) appears to split into several parts in the chimpanzee network: 27 genes are part of the chimpanzee blue module, 85 genes are part of the chimpanzee brown module, 52 fall in the chimpanzee turquoise module, 155 genes are gray in the chimpanzee network, etc. As described in Sect. 8.8 one can use the cross-tabulation based cluster preservation statistics (8.22) to attach a significance level (Fisher's exact p value) to each module overlap. The contingency table in Fig. 9.3c shows that every human module has significant overlap with a chimpanzee module. However, even if the resulting p value of preservation were not significant, it would be difficult to argue that a module is truly a human-specific module since an alternative module detection strategy in chimpanzee may arrive at a module with more significant overlap. In order to quantify the preservation of human modules in chimpanzee samples more objectively, one needs to consider statistics that do not rely on a particular module assignment in the chimpanzee data.

We now turn to approaches for measuring module preservation that do not require that module detection has been carried out in the test data set. Figure 9.4a,b show composite module preservation statistics of human modules in chimpanzee samples. The overall significance of the observed preservation statistics can be assessed using $Z_{summary}$ (9.1) that combines multiple preservation Z statistics into a

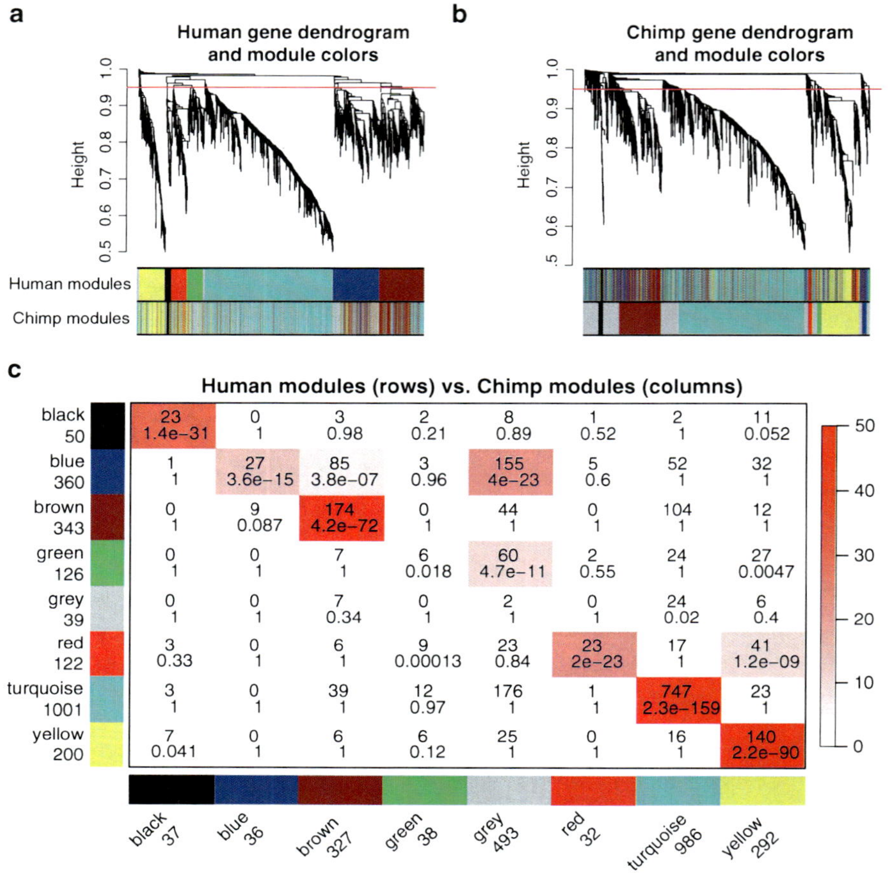

Fig. 9.3 Comparing modules (defined as clusters) in human and chimpanzee brain networks using cross-tabulation-based approaches. (**a**) Hierarchical clustering tree (dendrogram) of genes based on human brain co-expression network. Each "leaf" (*short vertical line*) corresponds to one gene. The color rows below the dendrogram indicate module membership in the human modules (defined by cutting branches of this dendrogram at the *red line*) and in the chimpanzee network (defined by branch cutting the dendrogram in panel (**b**).) The color rows show that most human and chimpanzee modules overlap (e.g., the turquoise module). (**b**) Hierarchical clustering tree of genes based on the chimpanzee co-expression network. The color rows below the dendrogram indicate module membership in the human modules (defined by cutting branches of dendrogram in panel (**a**)) and in the chimpanzee network (defined by branch cutting the dendrogram in this panel). (**c**) Cross-tabulation of human modules (rows) and chimpanzee modules (columns). Each row and column are labeled by the corresponding module color and the total number of genes in the module. In the table, numbers give counts of genes in the intersection of the corresponding row and column module. The table is color-coded by $-log(p)$, the Fisher's exact test p value, according to the color legend on the right. Note that the human yellow network is highly preserved, while the human blue network is only weakly preserved in the chimpanzee network

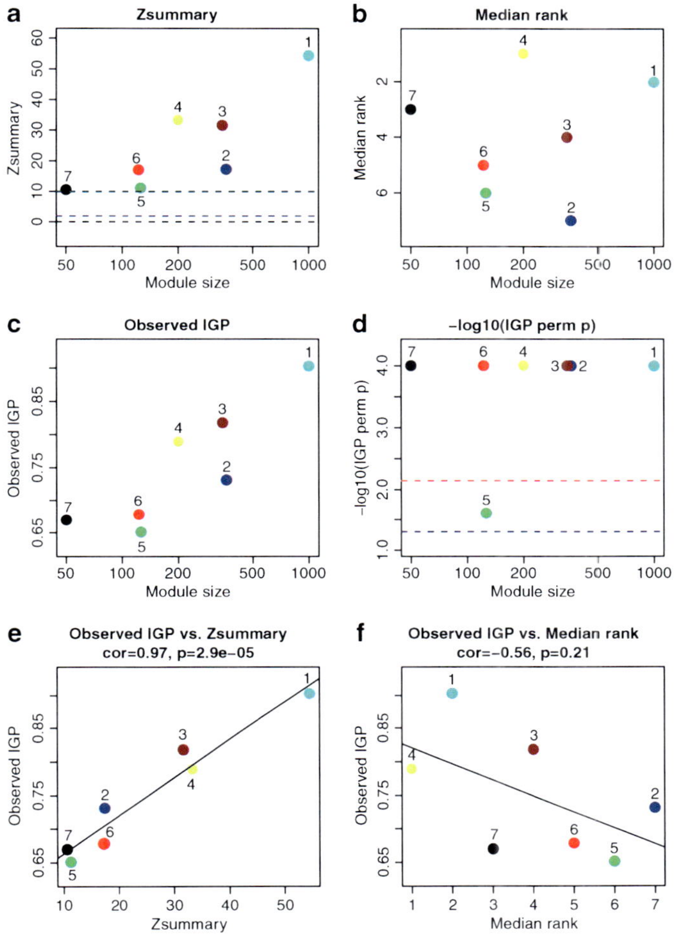

Fig. 9.4 Composite preservation statistics of human modules in chimpanzee samples. (**a**) The summary statistic $Z_{summary}$ (y-axis), (9.1), as a function of the module size. Each point represents a module, labeled by color and a secondary numeric label (*1* turquoise, *2* blue, *3* brown, *4* yellow, *5* green, *6* red, *7* black). The *dashed blue* and *green lines* indicate the thresholds $Z = 2$ and $Z = 10$, respectively. (**b**) The composite statistic *medianRank* (y-axis), (9.20), as a function of the module size. Each point represents a module, labeled by color and a secondary numeric label as in panel (**a**). As can be seen from the labels on the y-axis, the order on the y-axis is reversed. Low numbers on the y-axis indicate a high preservation. (**c**) Observed IGP statistic versus module size. (**d**) P value of the IGP statistic versus module size. (**e**) and (**f**) show scatter plots between the observed IGP statistic and $Z_{summary}$ and *medianRank*, respectively. In this example, where modules are defined as clusters, the IGP statistic has a high positive correlation ($r = 0.97$) with $Z_{summary}$ and a moderately large negative correlation ($r = -0.56$) with *medianRank*. The negative correlation is expected since low median ranks indicate high preservation

single overall measure of preservation (Fig. 9.4a). Note that $Z_{summary}$ shows a strong dependence on module size, which reflects the fact that observing module preservation of a large module is statistically more significant than observing the same for a small module. However, here we want to consider all modules on an equal footing irrespective of module size. Therefore, we focus on the composite statistic *medianRank* which shows no dependence on module size (Fig. 9.4b). The median rank is useful for comparing relative preservation among modules: a module with lower median rank tends to exhibit stronger observed preservation statistics than a module with a higher median rank. In Fig. 9.4b, we show the median ranks of the human brain modules. The median rank of the yellow module is 1, while the median ranks of the blue module is 6, indicating that the yellow module is more strongly preserved than the blue module. Our quantitative results show that modules expressed mainly in evolutionarily more conserved brain areas such as cerebellum (turquoise) and caudate nucleus (yellow and partly red) are more strongly preserved than modules expressed primarily in the cortex that is very different between humans and chimpanzees (green and blue modules). Thus, the module preservation results of *medianRank* corroborate **Mike Oldham**'s original finding regarding the relative lack of preservation of cortical modules.

Since the modules of this application are defined as clusters, it makes sense to evaluate their preservation using cluster validation statistics. Figure 9.4c shows that the IGP statistic also shows a strong dependence on module size in this application. The IGP values of all modules are relatively high. However, the permutation p values (panels c and d) identify the green module as less preserved than the other modules ($p = 0.06$, Bonferroni corrected p value 0.43). Figure 9.4e,f shows scatter plots between the observed IGP statistic and $Z_{summary}$ and *medianRank*, respectively. In this example, where modules are defined as clusters, the IGP statistic has a high positive correlation ($r = 0.97$) with $Z_{summary}$ and a moderately large negative correlation ($r = -0.56$) with *medianRank*. The negative correlation is expected since low median ranks indicate high preservation.

While composite statistics summarize the results, it is advisable to understand which properties of a module are preserved (or not preserved). For example, **module density-based** statistics allow us to determine whether the genes of a module (defined in the reference network) remain densely connected in the test network. As an illustration, we will compare the module preservation statistics for the human yellow module whose genes are primarily expressed in caudate nucleus (an evolutionarily old brain area), and the human blue module whose genes are expressed mostly in the cortex which underwent large evolutionary changes between humans and chimpanzees. In chimpanzees, the mean adjacency of the genes comprising the human yellow module is significantly higher than expected by chance, with a high permutation statistic $Z_{meanAdj}(\text{yellow}) = 62$, $p_{meanAdj}(\text{yellow}) < 10^{-200}$. But the corresponding permutation Z statistic for the human blue module is only weakly significant, $Z_{meanAdj}(\text{blue}) = 2.2$, $p_{meanAdj}(\text{blue}) = 0.014$. Thus, the mean adjacency permutation statistic suggests that the blue module is less preserved than the yellow module.

For co-expression modules, one can define an alternative density measure based on the module eigengene (Fig. 9.5a, e). The higher the proportion of variance explained by the module eigengene in the test set data, the tighter is the module in the test set. The human yellow module exhibits a high proportion of variance explained, $propVarExpl = 0.65$, and the corresponding permutation Z statistic is $Z_{propVarExpl}(\text{yellow}) = 22$, $p_{propVarExpl}(\text{yellow}) = 7.6 \times 10^{-105}$. In contrast, for the human blue module we find $propVarExpl = 0.33$ and the corresponding permutation Z statistic is $Z_{propVarExpl}(\text{blue}) = 4.6$, $p_{propVarExpl}(\text{blue}) = 2.6 \times 10^{-6}$. The permutation statistics again suggest that the yellow module is more preserved than the blue module.

Although density-based approaches are intuitive, they may fail to detect another form of module preservation, namely the **preservation of connectivity patterns** among module genes. For example, network module connectivity preservation can mean that, within a given module q, a pair of genes with a high connection strength (adjacency) in the reference network also exhibits a high connection strength in the test network. This property can be quantified by correlating the pairwise adjacencies or correlations between reference and test networks. For the genes in the human yellow module, the scatter plot in Fig. 9.5b shows pairwise correlations in the human network (x-axis) versus the corresponding correlations in the chimpanzee network (y-axis). The correlation between pairwise correlations (denoted by $cor.cor$) equals 0.91 and is highly significant, $p < 10^{-200}$. The analogous correlation for the blue module, Fig. 9.5f is lower, 0.56, but still highly significant, $p < 10^{-200}$, in part because of the higher number of genes in the blue module.

A related but distinct connectivity preservation statistic quantifies whether intramodular hub genes in the reference network remain intramodular hub genes in the test network. Preservation of intramodular connectivity kIM (1.41) reflects the preservation of hub gene status between the reference and test network . For example, the intramodular connectivity of the human yellow module is preserved between the human and chimpanzee samples, $cor.kIM = 0.66, p = 9.2 \times 10^{-29}$ (Fig. 9.5c). In contrast, the human blue (cortical) module exhibits a lower correlation (preservation) $cor.kIM = 0.56, p = 6 \times 10^{-33}$ (Fig. 9.5g). The p value is more significant because of the higher number of genes in the blue module.

Another intramodular connectivity measure is kME_i, which turns out to be highly related to kIM as described in Sect. 6.9 and Sect. 9.7. Figure 9.5d shows that kME for the human yellow module is highly preserved in the chimpanzee network ($r = 0.95, p = 9 \times 10^{-146}$). The corresponding correlation in the human blue module is lower, $r = 0.74, p = 4 \times 10^{-72}$ (Fig. 9.5h). In summary, the observed preservation statistics show that the human yellow module (related to the caudate nucleus) is more strongly preserved in the chimpanzee samples than the human blue module (related to the cortex).

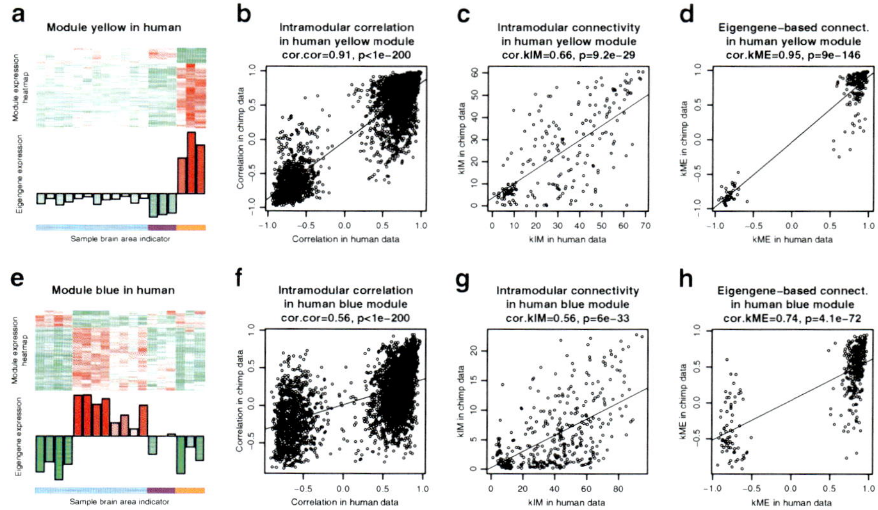

Fig. 9.5 Connectivity-based statistics for evaluating the preservation of the human *yellow and blue modules* in the chimpanzee network. (**a**) Heatmaps and eigengene plots for visualizing the gene expression profiles of the yellow module genes (rows) versus human brain microarray samples (columns). In the heat map, *green* indicates underexpression, *red* overexpression, and *white* mean expression. The module eigengene expression depicted underneath the heat map shows how the eigengene expression (y-axis) changes across the samples (x-axis) which correspond to the columns of the heat map. The eigengene can be interpreted as a weighted average gene expression profile. The color bar below the eigengene indicates the region from which the sample was taken: *light blue* color indicates cortical samples, magenta indicates cerebellum samples, and orange indicates caudate nucleus samples. Scatter plots (**b**)–(**d**) show that the connectivity patterns of the yellow module genes tends to be highly preserved between the two species. (**b**) Scatter plot of gene–gene correlations in chimpanzee samples (y-axis) versus human samples (x-axis) within the human yellow module. Each point corresponds to a gene–gene pair. The scatter plot exhibits a significant correlation (cor.cor and p value displayed in the title), indicating that the correlation pattern among the genes is preserved between the human and chimpanzee data. (**c**) Scatter plot of intramodular connectivities, (1.41), of genes in the human yellow module in chimpanzee samples (y-axis) versus human samples (x-axis). Each point corresponds to one gene. The scatter plot exhibits a significant correlation (cor.kIM and p value displayed in the title), indicating that the hub gene status in the human *yellow module* is preserved in the chimpanzee samples. (**d**) Scatter plot of eigengene-based connectivities, (6.53), of genes in chimpanzee samples (y-axis) versus human samples (x-axis). Each point corresponds to one gene. The scatter plot exhibits a significant correlation (cor.kME and p value displayed in the title), indicating that fuzzy module membership in the human *yellow module* is preserved in the chimpanzee samples. Scatter plots (**e**)–(**h**) show that the human blue module is less preserved in the chimpanzee network. Note that the correlations in scatter plots (**f**)–(**h**) are lower than the corresponding correlations in the yellow module plots (**b**)–(**d**) indicating weaker preservation of the human *blue module* in the chimpanzee samples. Overall, these results agree with those from the cross-tabulation-based analysis reported in Fig. 9.3

9.5 KEGG Pathways Between Human and Chimpanzee Brains

To further illustrate that modules do not have to be clusters, we now describe an application where modules correspond to Kyoto Encyclopedia of Genes and Genomes (KEGG) pathways. KEGG is a knowledge base for systematic analysis of gene functions, linking genomic information with higher order functional information (Kanehisa and Goto 2000). KEGG also provides graphical representations of cellular processes, such as signal transduction, metabolism, and membrane transport. To illustrate the use of the module preservation approach, we studied the preservation of selected KEGG pathway networks across human and chimpanzee brain correlation networks. While pathways in the KEGG database typically describe networks of proteins, our analysis describes the correlation patterns between mRNA expression levels of the corresponding genes. As before, we define a signed weighted correlation network adjacency matrix between the genes. For the sake of brevity, we focused the analysis on the following eight signaling pathways: Hedgehog signaling pathway (12 genes in our data sets), apoptosis (24 genes in our data sets), TGF-beta signaling pathway (26 genes), phosphatidylinositol signaling system (39 genes), Wnt signaling pathway (55 genes), endocytosis (59 genes), calcium signaling pathway (78 genes), and MAPK signaling pathway (93 genes). All of these pathways have been shown to play critical roles in normal brain development and function (Cerpa et al. 2009; Traiffort et al. 2010; Greer and Greenberg 2008; Berridge 2009; Samuels et al. 2009).

Figure 9.6a,b shows composite preservation statistics $Z_{summary}$ and *medianRank*. Both statistics show that the apoptosis module is the least preserved module. To visualize the lack of preservation, consider the circle plots of apoptosis genes in Fig. 9.7 l, m that shows pronounced differences in the connectivity patterns among apoptosis genes. While we caution the reader that additional data are needed to replicate these differences, prior literature points to an evolutionary difference for apoptosis genes. For example, a scan for positively selected genes in the genomes of humans and chimpanzees found that a large number of genes involved in apoptosis show strong evidence for positive selection (Nielsen et al. 2005). Further, it has also been hypothesized that natural selection for increased cognitive ability in humans led to a reduced level of neuron apoptosis in the human brain (Arora et al. 2009).

Figure 9.6a shows that $Z_{summary}$ exhibits some dependence on module size. Since we want to compare module preservation irrespective of module size, we focus on the results for the *medianRank* statistic (Fig. 9.6b).

Since KEGG pathways are not defined via a clustering procedure, it is not clear whether cluster preservation statistics are appropriate for analyzing this example. But to afford a comparison, we also report the findings for the IGP statistic (Kapp and Tibshirani 2007). Figure 9.6c, d shows that IGP identifies phosphatidilinositol and TGF-beta as the least preserved modules, while apoptosis genes are highly preserved. We find no significant relationship between the IGP statistic and our module preservation statistics $Z_{summary}$ and *medianRank* (Fig. 9.6e, f). This example highlights that module preservation statistics can lead to very different results from cluster preservation statistics.

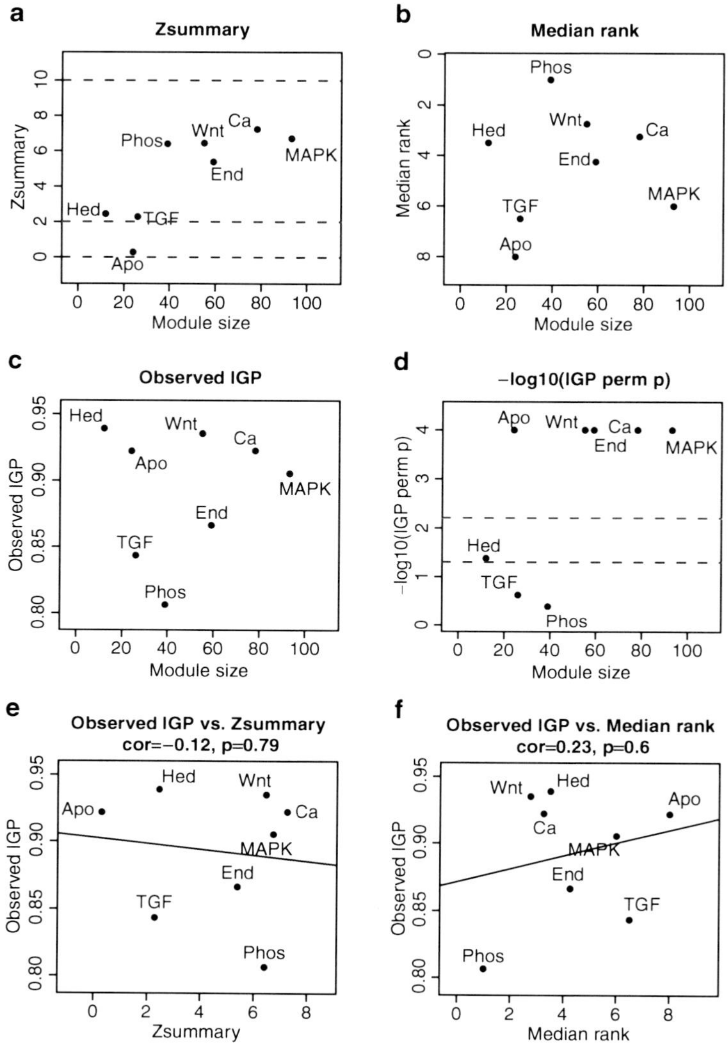

Fig. 9.6 Composite preservation statistics for KEGG pathways between human and chimpanzee brain co-expression networks. Here we compare the results of $Z_{summary}$ (panel **a**) to those of *medianRank* (panel **b**) and the IGP statistic (panels **c** and **d**). In panel (**b**) the order on the y-axis is reversed as can be seen from the labels on the y-axis. (**e**) and (**f**) show scatter plots between the observed IGP statistic and $Z_{summary}$ and *medianRank*, respectively. Here we find no significant relationship between the IGP statistic and the composite module preservation statistic. Since KEGG modules do not correspond to clusters, it is not clear whether cluster preservation statistics are useful in this example

To understand which aspects of the pathways are preserved, one can study the preservation of density statistics (Fig. 9.7b) and of connectivity statistics (Fig. 9.7c). According to $Z_{summary}$, the co-expresssion network formed by apoptosis genes is not preserved. It shows evidence of neither connectivity preservation ($Z_{connectivity} = 1.9$) nor density preservation ($Z_{density} = -1.4$, $p = 0.91$). The Hedgehog pathway also shows no evidence of density preservation ($Z_{density} = 1.7$, $p = 0.02$) but it shows weak evidence of connectivity preservation ($Z_{connectivity} = 3.2$, $p = 7.0 \times 10^{-4}$). The relatively low preservation Z statistics of the Hedgehog pathway may reflect a higher variability due to a small module size (it contains only 12 genes while the other pathways contain at least 22 genes). To explore this further, we studied the observed preservation statistics, which are less susceptible to network size effects than the corresponding Z statistics. The scatter plots in Fig. 9.7d–h show the correlations *cor.kME* between eigengene-based connectivity measures kME between the two species. For the Hedgehog pathway, we find that $cor.kME = 0.69$ ($p = 0.01$) which turns out to be higher than that of the TGF-β pathway.

The lack of preservation of the apoptosis pathway cannot be explained in terms of low module size. Figure 9.7e shows that it has the lowest observed *cor.kME* statistic, $r = 0.07, p = 0.76$.

9.6 Simulation Studies of Module Preservation

To illustrate the utility and performance of the proposed methods, we have considered seven different simulation scenarios that were designed to reflect various correlation network applications (Langfelder et al. 2011). An overview of these simulations can be found in Fig. 9.8. A more detailed description of the simulation scenarios is provided below.

Table 9.2 shows the performance grades of module preservation statistics in the different simulation scenarios. The highest grade of 4 indicates excellent performance.

We find that the proposed composite statistics $Z_{summary}$ (mean grade 3.4) and *medianRank* (mean grade 3.7) perform very well in distinguishing preserved from non-preserved modules. In contrast, cross-tabulation-based statistics only obtain a mean grade of 2.6. Since many simulation scenarios test the ability to detect connectivity preservation (as opposed to density preservation), it is no surprise that on average cluster validation statistics do not perform well in these simulations. For example, the IGP cluster validation statistic obtains a mean grade of 2.3 across the scenarios. But the IGP performs very well (grade 4) when studying the preservation of strongly preserved clusters (scenario 2).

The table also shows the performance of individual preservation statistics. Note that density-based preservation statistics perform well in scenarios 1–5 but fail in scenarios 6 and 7. But all connectivity-based statistics perform well in scenarios 6 and 7. The relatively poor performance of *separability* is one of the reasons why we did not include it into our composite statistics.

In the following, we describe the different simulation scenarios in more detail.

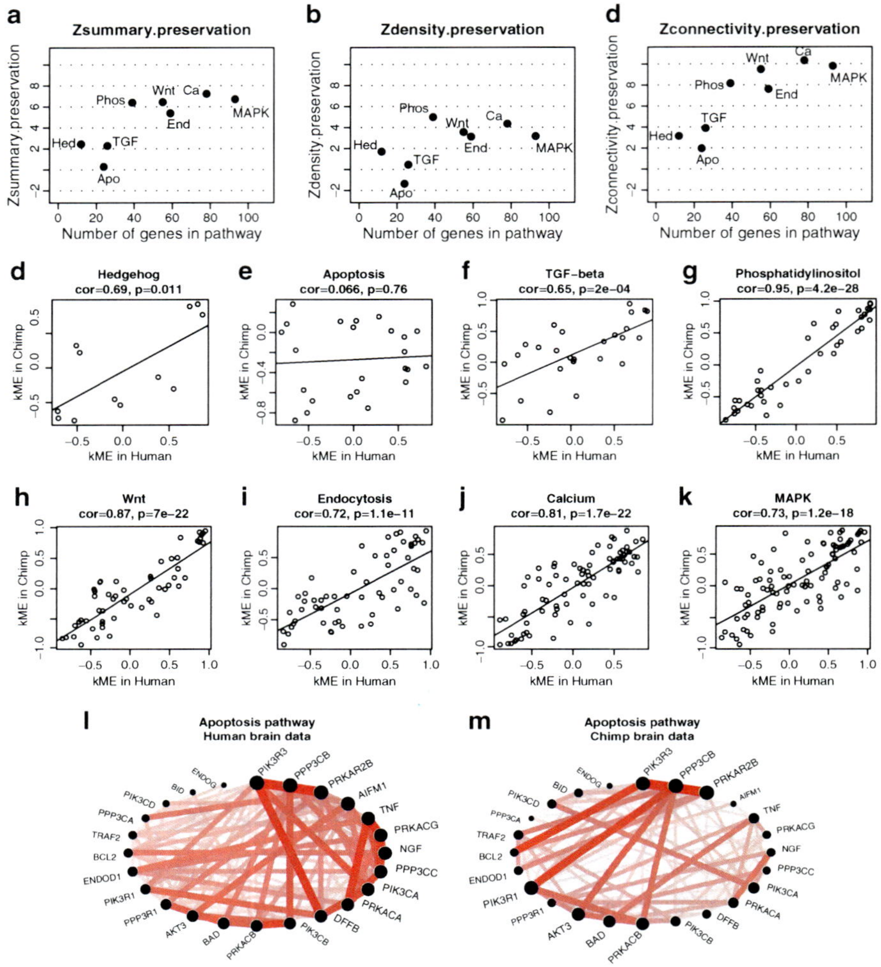

Fig. 9.7 Detailed preservation analysis of selected KEGG pathways between human (reference) and chimpanzee (test) brain co-expression networks. In the upper row, we present summary preservation Z statistics (y-axis) for selected KEGG pathways (interpreted as modules) versus the number of genes in the pathway (x-axis). Panel (**a**) shows $Z_{summary}$ (9.1), panel (**b**) shows the density summary statistic $Z_{density}$ (9.16), and panel (**c**) shows the connectivity summary statistic $Z_{connectivity}$ (9.17). Pathway names are shortened for readability. Panel (**a**) shows that MAPK, calcium, endocytosis, Wnt, and phosphatidylinositol show strong evidence of preservation ($Z_{summary} > 10$), while the apoptosis module is not preserved. Panel (**c**) shows that this preservation signal mainly reflects connectivity preservation $Z_{connectivity}$ (9.17), while panel (**b**) reveals that most modules have weak to moderate density preservation ($Z_{density} < 10$) (9.16). Note that the apoptosis pathway shows no evidence of preservations. Panels (**d**–**h**) display scatter plots of eigengene-based connectivities in the chimpanzee data (y-axis) versus in the human data (x-axis). Each point represents a gene in the pathway. Higher correlation means that the internal co-expression structure of the pathway is more strongly preserved. The apoptosis pathway has the lowest *cor.kME* statistic, while the phosphatidylinositol pathway has the highest. The *circle plots* in panels (**l**) and (**m**) show connection strengths among apoptosis genes in humans and chimpanzees, respectively

1. In the **weak preservation simulation** scenario, we simulate a total of 20 module in the reference data. Each of the 20 reference modules contains 200 nodes. But only 10 of the 20 modules are simulated to be preserved in the test network. We call it the weak preservation simulation since the intramodular correlations of preserved modules are relatively low. The intramodular correlations of non-preserved modules are expected to be zero. Note that the summary statistic $Z_{summary}$ successfully distinguishes preserved from non-preserved modules (second column of Fig. 9.8), with $Z_{summary} > 10$ for 8 of the 10 preserved modules. Similarly, the *medianRank* statistic distinguishes preserved from non-preserved modules (third column of Fig. 9.8). In comparison, the IGP permutation p value (fourth column of Fig. 9.8) is less successful: only one of the ten preserved modules pass the Bonferroni-corrected threshold; of the six modules that pass the $p = 0.05$ threshold, four are preserved and two are non-preserved. In this simulation, we observe a moderate relationship between the observed IGP and $Z_{summary}$, with Pearson correlation $r = 0.53$ ($p = 0.02$).
2. In the **half-preserved simulation** scenario, we simulate ten modules of varying sizes (between 50 and 1000 nodes), labeled 1–10. Modules 1–5 are preserved in the test set, while modules 6–10 are not preserved. All five preserved modules have $Z_{summary} > 10$, and all non-preserved modules have $Z_{summary} < 2$. Likewise, *medianRank* separates preserved and non-preserved modules. Permutation p values of IGP are also successful with respect to the Bonferroni-corrected threshold. In this simulation, we observe a strong correlation between IGP and $Z_{summary}$: $r = 0.97$ ($p = 3 \times 10^{-8}$).
3. In the **permuted simulation** scenario, none of the ten modules are preserved. Specifically, we simulate ten modules of varying sizes in the reference set and ten modules of the same sizes in the test set, but there is no relationship between the modules: the module membership is randomly permuted between the networks. The low value of the summary preservation statistic $Z_{summary} < 2$ accurately reflects that none of the modules are preserved. In contrast, the IGP permutation p value for two of the ten modules is lower than the Bonferroni threshold 0.005. In this simulation, the correlation between IGP and $Z_{summary}$ is not significant.
4. In the **half-permuted simulation** scenario, we simulate ten modules labeled 1–10 in the reference set. Modules 1–5 are preserved in the test set, while modules 6–10 are not. The test set contains modules 6'–10' of the same sizes as modules 6–10, but their module membership is randomly permuted with respect to the modules 6–10. The summary preservation statistic $Z_{summary}$ is quite accurate: all five preserved modules have $Z_{summary} > 10$ and three non-preserved modules have $Z_{summary} < 2$. The observed values of the IGP statistic are highly correlated ($r = 0.96$, $p = 1 \times 10^{-6}$) with $Z_{summary}$, but the IGP permutation p values do not work well: two preserved modules have an IGP p value above 0.05.
5. In the **intramodular permuted** scenario, we simulate modules whose density is preserved but whose intramodular connectivity is not preserved. Specifically, we simulate a total of ten modules labeled 1–10 in the reference set. The

density of modules 1–5 is preserved in the test set but the node labels inside each module are permuted, which entails that their intramodular connectivity patterns is no longer preserved in the test network. For modules 6–10, neither the density nor the connectivity is preserved. The $Z_{summary}$ and *medianRank* work well although not as good as in the previous studies. Both composite statistics successfully detect the preservation of the density preservation. IGP performs quite well: it misclassifies only one non-preserved module as preserved. In this simulation, we observe a strong correlation between IGP and $Z_{summary}$: $r = 0.91$ $(p = 5 \times 10^{-5})$.

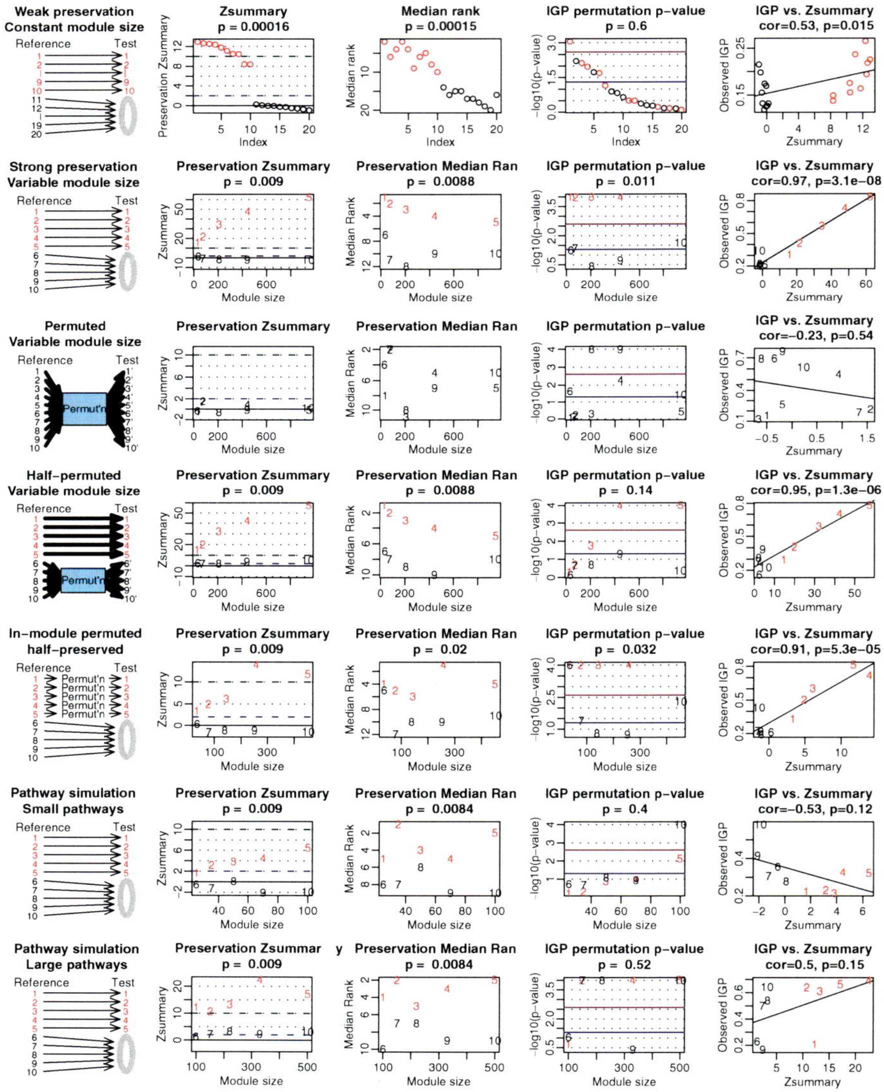

6–7. In the **pathway simulations** scenario, we simulate five (preserved) modules whose connectivity patterns are preserved but whose density is not. Further, we simulate five modules for which neither connectivity nor density is preserved. In the following description, we refer to the clusters from scenario 4 as clusters. The five preserved (non-preserved) modules of the pathway scenario are created by randomly selecting nodes (objects) from the preserved (non-preserved) clusters in scenario 4. Thus, the preserved modules contain nodes from multiple preserved clusters of scenario 4. Since the pairwise correlations between and within the preserved clusters (of scenario 4) are preserved, the intramodular connectivity patterns of the resulting pathway modules are preserved in the test network. But since nodes from different clusters may have low correlations, the density of the pathway modules tends to be low. The two pathway simulations differ by the module sizes: in the small scenario, modules range from 25 to 100 nodes; in the large scenario, modules range from 100 to 500 nodes. Because module membership is trivially preserved between reference and test networks, cross-tabulation statistics are not applicable. The composite statistics $Z_{summary}$ and *medianRank* distinguish preserved from non-preserved modules (Fig. 9.8) since they also measure aspects of connectivity preservation. By considering individual preservation statistics, we find that all connectivity preservation statistics successfully distinguish preserved from non-preserved modules. As expected, density-based statistics and the IGP statistic fail to detect the preservation of the connectivity patterns of the five preserved modules (Fig. 9.8), but these statistics correctly note that the density is not preserved.

Additional descriptions of the simulations can be found in Langfelder et al. (2011). As caveat, we mention that we only considered seven scenarios that aim to emulate selected situations encountered in co-expression networks. The performance of these preservation statistics may change in other scenarios. A comprehensive evaluation in other scenarios is beyond our scope. R software tutorials describing the results of our simulation studies can be found on our web page and will allow the reader to compare different methods using our simulated data.

Fig. 9.8 Design and main results of simulation studies of module preservation. The first column outlines the seven simulation studies (corresponding to the seven rows). Each preserved modules is marked in *red*. The gray 0 represents genes whose profiles are simulated to be independent (i.e., without any correlation structure). The second and third columns report representative values for composite statistics $Z_{summary}$ and *medianRank*, respectively. Preserved and non-preserved modules are marked in *red* and *black*, respectively. The *blue* and *green horizontal lines* show the thresholds of $Z_{summary} = 2$ and $Z_{summary} = 10$, respectively. Each figure title reports the Kruskal–Wallis test p value for testing whether the preservation statistics differ between preserved and non-preserved modules. Note that the proposed thresholds ($Z_{summary} > 10$ for preserved and $Z_{summary} < 2$ for non-preserved modules) work quite well. The fourth column shows the permutation p values of IGP obtained by the R package `clusterRepro`. The *blue* and *brown lines* show p value thresholds of 0.05 and its Bonferroni correction, respectively. The IGP permutation p value is less successful than $Z_{summary}$ at distinguishing preserved from non-preserved modules. The fifth and last column shows scatter plots of observed IGP versus $Z_{summary}$. We observe that IGP and $Z_{summary}$ tend to be highly correlated when modules correspond to clusters with varying extents of preservation

Table 9.2 Overview of the performance of various module preservation statistics in our simulation studies

				Simulation scenario							
				1	2	3	4	5	6	7	
No.	Statistic	Type	Network	Weak pres.	Strong pres.	Perm.	Half perm.	In-mod. perm.	Small pathw.	Large pathw.	Mean grade
1	coClustering	Cross-tab	Not used	1	4	3	1	4	NA[a]	NA[a]	2.6
2	*accuracy*	Cross-tab	Not used	1	4	4	1	3	NA[a]	NA[a]	2.6
3	- log(pvalue)	Cross-tab	Not used	1	4	3	1	4	NA[a]	NA[a]	2.6
4	*meanAdj*	Density	General	4	4	3	3	4	1	1	2.9
6	*cor.kIM*	Connectiv.	General	2	3	4	3	3	3	4	3.1
12	*meanCor*	Den.+Con.	Cor	4	4	4	4	1	1	2	2.9
13	*cor.cor*	Connectiv.	Cor	3	4	4	4	1	3	4	3.3
14	*propVarExpl*	Density	Cor	3	4	3	2	4	1	1	2.6
15	*meanKME*	Den.+Con.	Cor	4	3	4	4	3	1	2	3
16	*cor.kME*	Connectiv.	Cor	2	4	3	3	1	3	4	2.9
17	*cor.kMEall*	Connectiv.	Cor	4	2	4	3	3	1	1	2.6
18	*separability*	Separabil.	Cor	1	1	3	3	1	1	1	1.6
19	$Z_{summary}$	Composite	Cor	3	4	4	3	3	3	4	3.4
21	*medianRank*	Composite	Cor	4[b]	4	NA[b]	4	2	4	4	3.7
Comparison to the IGP cluster validation statistic											
	IGP	Dens+Sep	General	1	4	3	3	3	1	1	2.3
	IGP perm. p	Dens+Sep	General	1	4	3	2	2	1	1	2.0

Evaluating preservation statistics in different simulation scenarios. Best and worst performance correspond to grade 4 and 1, respectively. The columns report the name, type, and input of individual preservation statistics, and their performance in the simulation studies: the weak module preservation (Weak pres.) study, strong preservation study (Strong pres.), permuted preservation study (Perm.), half-permuted preservation study (Half perm.), in-module permuted (In-mod. perm.), and pathway (Path.) simulations summarized in Fig. 9.8. We graded each statistic as follows: Statistics that accurately distinguish preserved and non-preserved modules within the thresholds of $Z \lesssim 2$ for non-preserved modules and $Z \gtrsim 10$ for preserved modules get grade 4; statistics that accurately distinguish preserved and non-preserved modules but whose values may lie on the wrong side of the thresholds are graded 3; statistics that distinguish preserved and non-preserved modules with high accuracy (allowing 20%, that is 1 of 5, misclassifications) are graded 2, and statistics that perform worse are graded 1. The grading of *medianRank* is explained in footnote b. We only report selected statistics defined for correlation networks. Further, we also include the results for the IGP cluster validation statistic.

[a]value not available since the pathway membership is determined beforehand and the same in both data sets.

[b]No thresholds can be defined for the statistic *medianRank*. A top grade of 4 was assigned if the *medianRank* perfectly distinguished preserved from non-preserved modules, i.e., if the *medianRank* of each preserved module was smaller than that of all non-preserved modules. For simulated scenario 3 (Perm.), a grade is not available (NA) since none of the modules were preserved.

9.7 Relationships Among Module Preservation Statistics

In Table 9.1, we categorize the statistics according to which aspects of module preservation they measure. For example, we present several seemingly different versions of density- and connectivity-based preservation statistics. But for correlation network modules, close relationships exist between them as illustrated in Fig. 9.9. The hierarchical clustering tree in Fig. 9.9 shows the correlations between the observed preservation statistics in our real data applications. As input of hierarchical clustering, we used a dissimilarity between the observed preservation statistics, which was defined as one minus the correlation across all studied reference and test data sets. Overall, we observe that statistics within one category tend to cluster together. We also observe that separability appears to be weakly related to the density and connectivity preservation statistics. Cross-tabulation statistics correlate strongly with density and connectivity statistics in the study of human and chimpanzee brain data, but the correlation is weak in the study of sex differences in human brain data.

In the following, we derive relationships between module preservations statistics. In particular, the geometric interpretation of correlation networks (Sect. 6 3 and Sect. 6.4) can be used to describe situations when close relationship exist-among

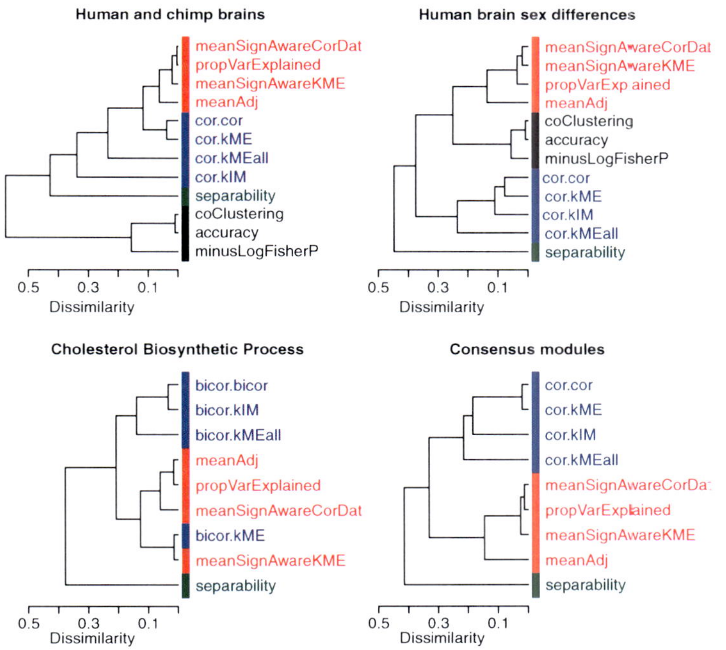

Fig. 9.9 Relationships between module preservation statistics in our real-data applications. The (average linkage) hierarchical cluster trees visualize the correlations between the preservation statistics. The preservation statistics are colored according to their type: density statistics are colored in *red*, connectivity preservation statistics are colored in *blue*, separability is colored in *green*, and cross-tabulation statistics are colored in *black*

the density-based preservation statistics (*meanCor*, *meanAdj*, *propVarExpl*, *meanKME*), among the connectivity-based preservation statistics (*cor.kIM*, *cor.kME*, *cor.kMEall*, *cor.cor*), and between the separability statistics (*separability.ave*, *separability.ME*). These relationships justify aggregating the module preservation statistics into composite preservation statistics such as $Z_{summary}$ (9.1) and *medianRank* (9.20).

Relationships Among Preservation Statistics in Correlation Networks

One can easily derive theoretical relationships between preservations statistics in case of approximately factorizable adjacency matrices ($A_{ij} \approx CF_i CF_j$) and correlation modules ($cor(x_i, x_j) \approx \mathrm{kME}_i \mathrm{kME}_j$). Eigenvector-based network concepts (Chap. 6) reveal that close relationship exists among density-based preservation statistics (*meanCor*, *meanAdj*, *propVarExpl*, *meanKME*), among the connectivity-based preservation statistics (*cor.kIM*, *cor.kME*, *cor.kMEall*, *cor.cor*), and between the separability statistics (*separability.ave*, *separability.ME*) as described in the following.

Relationships Among Density Preservation Statistics

A theoretical advantage of an (unsigned) weighted correlation network with adjacency $A_{ij} = |cor(x_i, x_j)|^{\beta}$ is that it allows one to derive simple relationships between network concepts. Recall that a correlation module with high eigennode factorizability $EF(E^{(q)})$ (6.11) leads to an approximately factorizable correlation matrix

$$cor\left(x_i^{(q)}, x_j^{(q)}\right) \approx \mathrm{kME}_i \mathrm{kME}_j. \tag{9.22}$$

For a factorizable correlation module comprised of positively correlated vectors, we have found the following relationship (6.38):

$$meanCor^{[\text{test}](q)} = \left(cos\left(\theta^{[\text{test}](q)}\right)\right)^2 propVarExpl^{[\text{test}](q)}.$$

Further, (6.38) implies that

$$meanCor^{[\text{test}](q)} \approx \left(meanKME^{[\text{test}](q)}\right)^2,$$

Equation (6.39) implies that for unsigned weighted correlation modules

$$meanAdj^{[\text{test}](q)} \approx \left(\frac{\sum_i |\mathrm{kME}_i^{[\text{test}](q)}|^{\beta}}{n^{(q)}}\right)^2.$$

and (6.40) implies that for $\beta = 2$

$$meanAdj^{[\text{test}](q)} = propVarExpl^{[\text{test}](q)}.$$

Relationships Among Connectivity Preservation Statistics in Correlation Networks

If the eigennode factorizability of a module is high, then (3.8) (with $CF_i = |\text{kME}|_i^\beta$) implies that

$$|\text{kME}|_i^\beta \approx \frac{\text{kIM}_i}{\sqrt{\sum_j \text{kIM}_j}} \tag{9.23}$$

for an unsigned weighted correlation network constructed with the soft threshold β. For the correlation between intramodular connectivities $cor.kIM^{(q)}$ (.57), this implies

$$cor.kIM^{(q)} \approx \text{cor}\left(|\text{kME}^{[\text{ref}](q)}|^\beta, |\text{kME}^{[\text{test}](q)}|^\beta\right). \tag{9.24}$$

Thus, for the special case of a weighted network module with $\beta = 1$ and *positive* values of kME

$$cor.kIM \approx cor.kME. \tag{9.25}$$

The relationships between *cor.kME* and *cor.kMEall* depends on the extramodular nodes, i.e., the nodes outside the network module under consideration. Only if their contribution to the correlation *cor.kMEall* is negligible, then

$$cor.kMEall \approx cor.kME. \tag{9.26}$$

By adapting the results from Sect. 3.13, one can easily show that $cor.cor^{(q)}$ (9.9) is approximately equal to $cor.kME^{(q)}$ (9.12), i.e.,

$$cor.cor^{(q)} \approx cor.kME^{(q)}. \tag{9.27}$$

Similarly, results from Sect. 3.13 can be used to show that approximate factorizability implies that

$$cor.Adj \approx \text{cor}\left(|\text{kME}_i|^\beta, |\text{kME}_j|^\beta\right). \tag{9.28}$$

Thus, in the special case where $\beta = 1$ and the module is comprised of genes with positive values of kME, *cor.Adj* is approximately equal to *cor.kME*. Similarly, one can show that

$$cor.Adj \approx cor.kIM \tag{9.29}$$

in approximately factorizable network modules, where $A_{ij} \approx \frac{\text{kIM}_i \, \text{kIM}_j}{\sum_l \text{kIM}_l}$.

Relationships Among Separability Statistics in Correlation Networks

Results from Sect. 3.12 can be used to derive a relationship between the average separability $separability.ave(q_1,q_2)$ (1.47) and an eigennode-based analog for an unsigned weighted correlation network whose adjacency is given by $A_{ij} = |cor(x_i,x_j)|^{\beta}$. In Sect. 3.12, we show that high eigennode factorizability (6.11) implies that

$$separability^{average}(q_1,q_2) \approx 1 - |\mathrm{cor}\left(E^{(q_1)},E^{(q_2)}\right)|^{\beta}. \tag{9.30}$$

Equation (9.30) can be used to argue that by increasing the soft threshold β, correlation modules tend to become more separated. Further, note that for a weighted network with $\beta = 1$, we find that

$$separability.ave \approx separability.ME.$$

9.8 Discussion of Module Preservation Statistics

The chapter describes powerful module preservation statistics that capture different aspects of module preservation. The network-based preservation statistics only assume that each module forms a subnetwork of the original network. Thus, we define a module as a subset of nodes with their corresponding adjacencies. In particular, our connectivity preservation statistics (*cor.Adj*, *cor.cor*, *cor.kIM*, and *cor.kME*) do not assume that modules are defined as clusters. We demonstrate that it is advantageous to aggregate multiple preservation statistics into composite statistics $Z_{summary}$ and *medianRank*. While we propose module preservation statistics for general networks (e.g., $Z_{summaryADJ}$), all of our applications involve gene co-expression networks.

An advantage of an (unsigned) weighted correlation network is that it allows one to derive simple relationships between network concepts. Eigenvector based network concepts (Chap. 6) allow us to characterize correlation modules where simple relationships exist between (a) density-based preservation statistics, (b) connectivity-based preservation statistics, and (c) separability-based preservation statistics (see the Sect. 9.7). Using conformity based concepts (Sect. 3.13) we are able to describe relationships between preservation statistics in general networks.

Our applications provide a glimpse of the types of research questions that can be addressed with the module preservation statistics. In general, methods for quantifying module preservation have several uses. First and foremost, they can be used to determine which properties of a network module are preserved in another network. Thus, module preservation statistics are a valuable tool for validation as well as differential network analysis. Second, they can be used to define a global measure of module structure preservation by averaging the preservation statistic across multiple modules or by determining the proportion of modules that are preserved.

A third use of module preservation statistics is to define measures of module quality (or robustness), which may inform the module definition. For example, to measure how robustly a module is defined in a given correlation network, one can use resampling techniques to create reference and test sets from the original data and evaluate module preservation across the resulting networks. Thus, any module preservation statistic naturally gives rise to a module quality statistic by applying it to repeated random splits (interpreted as reference and test set) of the data. By averaging the module preservation statistic across multiple random splits of the original data, one arrives at a module quality statistic.

Table 9.2 shows the performance grades of module preservation statistics in different simulation scenarios. We find that composite statistics $Z_{summary}$ and *medianRank* perform very well in distinguishing preserved from non-preserved modules. While the dependence of $Z_{summary}$ on the module size is often attractive, our applications show situations when it is unattractive. In this case, we recommend to use the composite statistic *medianRank*, which has an added bonus: its computation is much faster than that of $Z_{summary}$ since it does not involve a permutation test procedure. Our applications provide evidence that the *medianRank* statistic can lead to biologically meaningful preservation rankings among modules.

While cluster analysis and network analysis are different approaches for studying high-dimensional data, there are some commonalities. For example, it is straightforward to turn a network adjacency matrix (which is a similarity measure) into a dissimilarity measure which can be used as input of a clustering procedure (e.g., hierarchical clustering or partitioning-around-medoids) (Rousseeuw 1987). If a module is defined using a clustering procedure, one can use cluster preservation statistics as module preservation statistics. Conversely, our adjacency-based module preservation statistics give rise to cluster preservation statistics since a dissimilarity measure (used for the cluster definition) can also be transformed into a network adjacency matrix. In some of our applications where modules are defined as clusters, we find that $Z_{summary}$ is highly correlated with the IGP cluster validation statistic (Kapp and Tibshirani 2007) across modules. In our simulations, we observe that IGP and $Z_{summary}$ tend to be highly correlated when modules correspond to clusters with varying extents of preservation. When modules are not defined via a clustering procedure (e.g., in our KEGG pathway application), we find pronounced differences between $Z_{summary}$ and the IGP statistic.

Module preservation statistics have several limitations including the following. First, the preservation statistics only apply to undirected networks. In an exercise, you are asked to generalize the module preservation statistics to directed networks.

A second limitation concerns statistics of connectivity preservation that are based on correlating network adjacencies, intramodular connectivities, etc., between the reference and the test networks. Because Pearson correlation is sensitive to outliers, it may be advantageous to use an outlier-resistant correlation measure, e.g., the Spearman correlation or the biweight midcorrelation (5.17) (Wilcox 1997; Hardin et al. 2007) implemented in the WGCNA package (Langfelder and Horvath 2008). Robust correlation options are available in our implementation in R function `modulePreservation`.

A third limitation is that a high value of a preservation statistic does not necessarily imply that the module could be found by a de novo module detection analysis in the test data set. For example, if a module is defined using cluster analysis, then the resulting test set modules may not have significant overlap with the original reference module according to a module membership cross-tabulation table. As explained before, this potential limitation is a small price to pay for making a module preservation analysis independent from the vagaries of module detection.

A fourth limitation is that the different preservation statistics may disagree with regard to the preservation of a given module. While certain aspects of a module may be preserved, others may not be. In our simulation studies, we present scenarios where connectivity statistics show high preservation but density measures do not and vice versa. Since both types of preservation statistics will be of interest in practice, our R function `modulePreservation` outputs all preservation statistics. Although we aggregate several preservation statistics into composite statistics, we recommend to consider all of the underlying preservation statistics to determine which aspects of a module are preserved.

Given the above-mentioned limitations, it is reassuring that the module preservation statistics have performed in many applications (Langfelder et al. 2011; Cai et al. 2010; Miller et al. 2010). Although it would be convenient to have a single statistic and a corresponding threshold value for deciding whether a module is preserved, this simplistic view fails to realize that module preservation should be judged according to multiple criteria (e.g., density preservation, connectivity preservation, etc.). The preservation statistics provide a more nuanced and detailed view of module preservation. Before deciding on module preservation, the data analyst needs to decide which aspects of a module preservation are relevant for the particular application.

A major application of density-based statistics is to measure module *quality* in the reference data (e.g., to compare various module detection procedures). Module quality measures can be defined using density-based and separability-based module preservation measures: the density and separability of a module in the *reference* network measures its homogeneity and separateness, respectively. In Sect. 8.7, we describe cluster quality statistics that can sometimes be used as module quality statistics. In contrast, connectivity-based measures (which contrast the reference adjacency matrix with the test network adjacency matrix) are not directly related to module quality measures (unless a data splitting approach is used in the reference data). Module quality measures based on density and separability measures can be used to confirm that the reference modules are well defined. These module quality statistics are also implemented in the `modulePreservation` R function.

9.9 R Code for Studying the Preservation of Modules

Preservation statistics have been implemented in the R function `modulePreservation`, which is part of the WGCNA R package. For each user-defined reference network both preservation and quality statistics are calculated considering each of the remaining networks as test network. Data, R code, and tutorials can be downloaded from www.genetics.ucla.edu/labs/horvath/CoexpressionNetwork/modulePreservation.

In Sect. 12.5, we present R code for studying the preservation of gene co-expression modules between female and male mouse liver networks.

9.10 Exercises

1. Exercise regarding module preservation statistics for directed networks. Recall that a directed network can be represented by a (typically) asymmetrical adjacency matrix.
   (i) Adapt connectivity-based preservation statistics to directed networks.
   (ii) Adapt density-based preservation statistics to directed networks.
   (iii) Adapt composite preservation statistics $Z_{summary}$ and *medianRank* to directed networks.
2. Exercise. Simulate modules whose densities are preserved in a test network but whose intramodular connectivities are not.
   Apply the `modulePreservation` function to the simulated data and verify that $Z_{density}$ is high while $Z_{connectivity}$ is low.
3. Exercise. Simulate modules whose densities are not preserved in a test network but whose intramodular connectivities are preserved.
   Apply the `modulePreservation` function to the simulated data and verify that $Z_{density}$ is low while $Z_{connectivity}$ is high.
4. Exercise regarding co-expression module preservation between male and female mouse liver samples. Hint: This exercise requires that the user is familiar with the material from Chap. 12. Adapt the R code in Sect. 12.5 to study whether male mouse modules are preserved in female mouse liver. Before that, you need to use a clustering procedure to define co-expression modules based on the male mouse liver samples. In Sect. 12.2, we present R code for defining modules as branches of a hierarchical clustering tree.

References

Arora G, Polavarapu N, McDonald JF (2009) Did natural selection for increased cognitive ability in humans lead to an elevated risk of cancer? Med Hypotheses 73(3):453–456

Bailey TA, Dubes R (1982) Cluster validity profiles. Pattern Recognit 15(2):61–83

Berridge MJ (2009) Inositol trisphosphate and calcium signalling mechanisms. Biochim Biophys Acta – Molecular Cell Research 1793(6):933–940

Cai C, Langfelder P, Fuller TF, Oldham M, Luo R, van den Berg L, Ophoff R, Horvath S (2010) Is human blood a good surrogate for brain tissue in transcriptional studies? BMC Genomics 11(1):589

Cerpa W, Toledo EM, VarelaNallar L, Inestrosa NC (2009) The role of Wnt signaling in neuroprotection. Drug News Perspect 22(10):579–591

Chen G, Jaradat SA, Banerjee N, Tanaka TS, Ko MS, Zhang MQ (2002) Evaluation and comparison of clustering algorithms in analyzing ES cell gene expression data. Stat Sin 12:241–262

Chen LS, EmmertStreib F, Storey JD (2007b) Harnessing naturally randomized transcription to infer regulatory relationships among genes. Genome Biol 8:R219

Dong J, Horvath S (2007) Understanding network concepts in modules. BMC Syst Biol 1(1):24

Dudoit S, Fridlyand J (2002) A prediction-based resampling method for estimating the number of clusters in a dataset. Genome Biol 3(7):RESEARCH0036

Fuller TF, Ghazalpour A, Aten JE, Drake T, Lusis AJ, Horvath S (2007) Weighted gene coexpression network analysis strategies applied to mouse weight. Mamm Genome 18(6–7):463–472

Ghazalpour A, Doss S, Zhang B, Plaisier C, Wang S, Schadt EE, Thomas A, Drake TA, Lusis AJ, Horvath S (2006) Integrating genetics and network analysis to characterize genes related to mouse weight. PloS Genet 2(2):8

Greer PL, Greenberg ME (2008) From synapse to nucleus: Calcium-dependent gene transcription in the control of synapse development and function. Neuron 59(6):846–860

Hardin J, Mitani A, Hicks L, VanKoten B (2007) A robust measure of correlation between two genes on a microarray. BMC Bioinformat 8(1):220

Horvath S, Dong J (2008) Geometric interpretation of gene co-expression network analysis. PLoS Comput Biol 4(8):e1000117

Kanehisa M, Goto S (2000) KEGG: Kyoto encyclopedia of genes and genomes. Nucleic Acids Res 28(1):27–30

Kapp AV, Tibshirani R (2007) Are clusters found in one dataset present in another dataset? Biostat 8(1):9–31

Kaufman L, Rousseeuw PJ (1990) Finding groups in data: An introduction to cluster analysis. Wiley, New York

Keller MP, Choi YJ, Wang P, Belt Davis D, Rabaglia ME, Oler AT, Stapleton DS, Argmann C, Schueler KL, Edwards S, Steinberg HA, Chaibub Neto E, Kleinhanz R, Turner S, Hellerstein MK, Schadt EE, Yandell BS, Kendziorski C, Attie AD (2008) A gene expression network model of type 2 diabetes links cell cycle regulation in islets with diabetes susceptibility. Genome Res 18(5):706–716

Khaitovich P, Muetzel B, She X, Lachmann M, Hellmann I, Dietzsch J, Steigele S, Do H, Weiss G, Enard W, Heissig F, Arendt T, NieseltStruwe K, Eichler EE, Paabo S (2004) Regional patterns of gene expression in human and chimpanzee brains. Genome Res 14(8):1462–1473

Langfelder P, Horvath S (2007) Eigengene networks for studying the relationships between co-expression modules. BMC Syst Biol 1(1):54

Langfelder P, Horvath S (2008) WGCNA: An R package for weighted correlation network analysis. BMC Bioinformatics 9(1):559

Langfelder P, Luo R, Oldham MC, Horvath S (2011) Is my network module preserved and reproducible? Plos Comput Biol 7(1):e1001057

McShane LM, Radmacher MD, Freidlin B, Yu R, Li MC, Simon R (2002) Methods for assessing reproducibility of clustering patterns observed in analyses of microarray data. Bioinformatics 18(11):1462–1469

Miller JA, Horvath S, Geschwind DH (2010) Divergence of human and mouse brain transcriptome highlights Alzheimer disease pathways. Proc Natl Acad Sci USA 107(28):12698–12703

van Nas A, GuhaThakurta D, Wang SS, Yehya N, Horvath S, Zhang B, Ingram-Drake L, Chaudhuri G, Schadt EE, Drake TA, Arnold AP, Lusis AJ (2009) Elucidating the role of gonadal hormones in sexually dimorphic gene coexpression networks. Endocrinology 150(3):1235–1249

Nielsen R, Bustamante C, Clark AG, Glanowski S, Sackton TB, Hubisz MJ, FledelAlon A, Tanenbaum DM, Civello D, White TJ, Sninsky J, Adams MD, Cargill M (2005) A scan for positively selected genes in the genomes of humans and chimpanzees. PLoS Biol 3(6):e170

Oldham MC, Horvath S, Geschwind DH (2006) Conservation and evolution of gene coexpression networks in human and chimpanzee brains. Proc Natl Acad Sci USA 103(47):17973–17978

Plaisier CL, Horvath S, Huertas-Vazquez A, Cruz-Bautista I, Herrera MF, Tusie-Luna T, Aguilar-Salinas C, Pajukanta P (2009) A systems genetics approach implicates USF1, FADS3, and other causal candidate genes for familial combined hyperlipidemia. PLoS Genet 5(9):e1000642

Rousseeuw PJ (1987) Silhouettes: A graphical aid to the interpretation and validation of cluster analysis. J Comput Appl Math 20:53–56

Samuels IS, Saitta SC, Landreth GE (2009) MAPing CNS development and cognition: An ERK-some process. Neuron 61(2):160–167

Shamir R, Sharan R (2001) Algorithmic approaches to clustering gene expression data. In: Current topics in computational biology. The MIT, Cabridge, MA, pp 269–300

Tibshirani R, Walther G (2005) Cluster validation by prediction strength. J Comput Graph Stat 14:511–528

Traiffort E, Angot E, Ruat M (2010) Sonic Hedgehog signaling in the mammalian brain. J Neurochem 113(3):576–590

Wang J, Zhang S, Wang Y, Chen L, Zhang XS (2009) Disease-aging network reveals significant roles of aging genes in connecting genetic diseases. PLoS Comput Biol 5(9):e1000521

Wilcox RR (1997) Introduction to robust estimation and hypothesis testing. Academic, San Diego, CA

Chapter 10
Association Measures and Statistical Significance Measures

Abstract An association measure can be used to measure the relationships between two random variables. These variables may be numeric, categorical, or binary. Statistical test statistics can often be defined for deriving association measures. For example, several statistical test statistics (Fisher Z, Student t-test, Hotelling) can be used to calculate a statistical significance level (p value) for a correlation coefficient. Multiple comparison correction (MCC) procedures are needed to protect against false positives due to multiple comparisons. The Bonferroni- and Sidak correction are very conservative MCC procedures. The q-value (local false discovery rate) MCC is often advantageous since it allows one to detect more significant variables. To calculate the false discovery rate (FDR), one considers the shape of the histogram of p values. MCC procedures can often be interpreted as transformations that increase the p value to account for the fact that multiple comparisons have been carried out. For example, the Bonferroni correlation multiplies each p value by the number of comparisons. The q-value transformation is sometimes improper, i.e., it decreases significant p values. p values and q-values can be used to screen for significant variables. The WGCNA library contains several R functions that implement standard screening criteria for finding variables (e.g., gene expression profiles) associated with a sample trait y. In practice, many seemingly different gene screening methods turn out to be significant. p values (or q-values) can be used to formulate a statistical criterion for choosing the (hard) threshold τ when defining an unweighted correlation network. Many methods for defining unweighted networks on the basis of pairwise linear relationships between variables turn out to be equivalent.

10.1 Different Types of Random Variables

Variables are often referred to as "features" (in data mining books), "covariates" (in regression analysis books), and "characteristics" or "properties" in other contexts. Variables refer to measurable attributes, as these typically vary (across conditions, between individuals, over time, etc.). Variables can be of different types: they can be categorical, numeric, or neither. A **categorical** variable (also known as a **nominal** variable) takes on values that belong to two or more categories, but there is no intrinsic ordering to the categories. For example, hair color is a categorical variable

S. Horvath, *Weighted Network Analysis: Applications in Genomics and Systems Biology*, DOI 10.1007/978-1-4419-8819-5_10,

having a number of categories (brown, black, blonde, red, etc.). Note that there is no way to order these hair colors from highest to lowest. In contrast, if a clear ordering exists between the values of a variable, then that variable is called an **ordinal variable**. For example, grades or quantitative measurements are ordinal.

A **numeric variable** (also referred to as **quantitative variable**) is a variable whose possible values are numbers (e.g., real numbers or integers). There are two types of numeric variables: discrete and continuous. A **discrete numeric variable** is one which may take on only a countable number of distinct values such as $0, 1, 2, 3, \ldots$. Discrete random variables are often (but not always) counts (e.g., number of sunny days per year). If a numeric variable can take only a finite number of distinct values, then it must be discrete. A **continuous numeric variable** is one which takes an infinite number of possible values. Continuous random variables are usually measurements (e.g., gene expression values).

A random sample of m measurements of a general random variable V can be represented by a vector $v = (v_1, \ldots, v_m)$ with m components. Often we denote categorical random variables by DX, DY, and DZ, and numeric random variables by X, Y, and Z. Correspondingly, we denote random samples from categorical vectors by dx, dy, and dz, and random samples from numeric vectors by x, y, and z etc.

Assume that we want to quantify the relationship (or association) between two (possibly non-numeric) random variables V and W based on joint random samples $(v_1, w_1), \ldots, (v_m, w_m)$. Abstractly speaking, we define an **association measure** $AssocMeasure(v, w)$ as a real-valued function of the sample vectors $v = (v_1, \ldots, v_m)$ and $w = (w_1, \ldots, w_m)$. For example, if x and y are two numeric vectors then $Assoc(x, y) = |cor(x, y)|$ defines an association measure.

10.2 Permutation Tests for Calculating p Values

Statistical significance is often quantified by a p value which results from the statistical test of a null hypothesis. For example, a p value for a correlation coefficient can be calculated under the null hypothesis that the true correlation is zero. The p value is defined as the probability of falsely rejecting the null hypothesis when the null hypothesis is actually true. If the p value is smaller than a threshold (called the significance level), then null hypothesis is improbable and should be rejected. Statistics offers many approaches and test statistics for calculating p values. A general approach for testing whether two variables are related is based on the permutation test approach.

In the following, we describe a permutation test procedure for assessing the statistical significance (p value) of an association measure $AssocMeasure(v, w)$ between two (possibly non-numeric) vectors v and w. The permutation test computes a p value for the null hypothesis that v and w represent samples from *independent* random variables V and W, i.e., random variables that have no relationship with each other. Denote the observed value of the association measure by

$$M^{observed} = AssocMeasure(v, w).$$

To test the null hypothesis of independence, one randomly permutes the entries of each vector and arrives at $v.permuted, w.permuted$. Specifically, consider the **sample function** $sample(v)$ (and R command `sample(v)`) which randomly permutes the components of the input vector. Permuted versions of the original vectors are defined as $v.permuted = sample(v)$ and $w.permuted = sample(w)$. The permuted vectors are used to calculate the value of the association measure $M^{permuted} = AssocMeasure(v.permuted, w.permuted)$ under the null hypothesis of no relationship between the two variables. This step is repeated $no.perms$ times and the results are stored as the components of the vector

$$M^{perm} = (M_1^{permutation}, \ldots, M_{no.perms}^{permutation}).$$

Note that M^{perm} is a random sample of the association measure under the null hypothesis of independence between the variables. Next, one can calculate a permutation Z statistic:

$$Z^{perm} = \frac{M^{observed} - mean(M^{permutation})}{\sqrt{var(M^{perm})}}. \tag{10.1}$$

The larger that Z^{perm}, the stronger the evidence that the observed association measure is larger than expected by chance. Under certain conditions (including large sample size $m = length(v)$), one can show that Z^{perm} follows an asymptotic standard normal distribution $N(0,1)$. Then $Z^{perm} > 1.645$ allows us to reject the 1-sided null hypothesis of independence at a significance level of $\alpha = 0.05$. To compute a Z statistic-based (1-sided) p value, one can use the following R command:

```
p.Zbased.perm=pnorm(-Z.permutation)
```

The above p value has the advantage of computational convenience since relatively few permutations are required to estimate the mean and variance of $M^{permutation}$ under the null hypothesis. But there is a danger that Z^{perm} does not follow a standard normal distribution under the null hypothesis. A statistically superior but computationally more challenging approach for computing a permutation p value is based on the extended vector

$$M^{extended} = (M^{observed}, M_1^{perm}, \ldots, M_{no.perms}^{perm}),$$

which has $no.perms + 1$ components. The (1-sided) permutation p value $p^{permutation}$ is defined as the proportion of components of $M^{extended}$ which are larger than or equal to the observed value $M^{observed}$, i.e.,

$$permutation.p.value = \frac{count\{M^{extended} >= M^{observed}\}}{no.perms + 1}. \tag{10.2}$$

Relevant R code can be found in the exercises.

The permutation test p value (10.2) depends on the number of permutations $no.perms$: if only $no.perms = 1$ permutation is used, the permutation p value ≥ 0.5 irrespective of the validity of the null hypothesis. The more permutations $no.perms$ are used, the better. To reject the null hypothesis at significance level α, at least

$$no.perms = \frac{1}{\alpha} \tag{10.3}$$

permutations are needed. This shows that permutation tests can be computationally demanding when very stringent significance levels are needed. For example, to protect against false positives due to one million independent statistical tests, the Bonferroni method suggests a significance level of $\alpha = 0.05/10^6 = 5 * 10^{-8}$. In this case, at least 20 million ($=1/\alpha$) permutations are needed to estimate the permutation test p value. In the following, we describe fast, model-based approaches for calculating p values between numeric variables.

10.3 Computing p Values for Correlations

In the following, we will outline several approaches for calculating p value for the null hypothesis that the correlation $\rho = cor(X,Y)$ between two numeric variables X and Y is zero. The correlation coefficient between the respective random samples x and y is denoted by $r = cor(x,y)$. If either X or Y has a normal distribution, one can use a Student t-test statistic for calculating a p value. Specifically, the **Student t statistic** $Z^{Student}$:

$$Z^{Student} = \sqrt{m-2}\frac{r}{\sqrt{1-r^2}} \tag{10.4}$$

follows a Student T distribution with $m-2$ degrees of freedom under the null hypothesis of zero correlation. Based on the Student distribution, one can calculate a p value, e.g., a two-sided p value can be computed with the following R code:

```
pvalueStudent=2 pt(-abs(Z.Student),df=m-2)
```

It is worth repeating that the Student t-test p value calculation only assumes that one of the variables follows a normal distribution. For example, it is applicable for a binary vector y as long as x is drawn from a normal distribution. Since the case of a **binary** vector y is very important in practice, we will discuss it in more detail. Denote the number of observations with $y = 1$ and $y = 2$ by $m1$ and $m2$, respectively. Let $\bar{x}_1$ be the mean value of x across the $m1$ observations with value $y = 1$. In R notation, $\bar{x}_1 = mean(x[y == 1])$. Similarly, define $\bar{x}_2 = mean(x[y == 2])$. Further, assume that the $m1$ observations with $y = 1$ have the same variance in x as the $m2$ observations corresponding to $y = 2$. Then the variance of x can be estimated using the following pooled variance estimator:

$$var.pooled = \frac{(m1-1)var1 + (m2-1)var2}{m1 + m2 - 2}, \tag{10.5}$$

where $var1$ denotes the variance of x across observations with value $y = 1$. In R notation, $var1 = var(x[y == 1])$. Then the well-known Student t statistic for testing the equality of the two means is given by

$$t.different.means = \frac{\bar{x}_1 - \bar{x}_2}{\sqrt{var.pooled\left(\frac{1}{m1} + \frac{1}{m2}\right)}}. \quad (10.6)$$

Under the null hypothesis of equal means, $t.different.means$ follows a Student t-distribution with $df = m1 + m2 - 2 = m - 2$ degrees of freedom. Note that two Student t-test statistics can be used for relating a binary vector y with x: $t.different.means$ (10.6) is based on comparing mean values and $Z^{Student}$ (10.4) is based on the Pearson correlation. Fortunately, both of them lead to the same 2-sided p value since the absolute values of the two t-statistics are equal:

$$|Z^{Student}| = |t.different.means|. \quad (10.7)$$

Let us now return to the general case where y may not be binary. If r is a partial correlation coefficient (defined in (13.30)), then a Student t-test statistic can be calculated by replacing $m-2$ in (10.4) with $m-2-N$, where N is the number of conditioning vectors.

For large sample sizes (say $m > 50$), one can use the **Fisher's Z statistic** (also known as Z transformation) to calculate p value for a correlation coefficient r. The Fisher Z statistic is given by

$$Z^{Fisher} = 0.5\sqrt{m-3}\,log\left(\frac{1+r}{1-r}\right) = \sqrt{m-3}\,atanh(r), \quad (10.8)$$

where

$$atanh(r) = \frac{1}{2}log\left(\frac{1+r}{1-r}\right)$$

denotes the **area hyperbolic tangent** (also known as **inverse hyperbolic tangent function** $tanh^{-1}$). Note that $atanh(r)$ maps the interval $(-1,1)$ to $(-\infty,+\infty)$ in a monotonically increasing fashion. Z^{Fisher} follows asymptotically a normal distribution with mean $\mu = \sqrt{m-3}\,atanh(\rho)$ and variance $\sigma^2 = 1$ under weak assumptions (Hawkins 1989). Thus, to calculate a p value for the null hypothesis of zero correlation $\rho = 0$, one can assume that Z^{Fisher} follows a standard normal distribution.

The **Hotelling statistic**

$$Z^{Hotelling} = \sqrt{m-1}\left(0.5log\left(\frac{1+r}{1-r}\right) - \frac{1.5log\left(\frac{1+r}{1-r}\right) + r}{4(m-1)}\right) \quad (10.9)$$

is a more accurate (but more complicated) way for calculating an asymptotic p value for the correlation coefficient.

Importantly, the above statistics can also be used to calculate a p value for the Spearman correlation as long as m is relatively large (Sokal and Rohlf 1981). We describe a method for calculating a **correlation test p value for two binary variables** in Sect. 14.3.3.

Clearly, all of the above test statistics depend on the sample size m and the effect size (correlation r). It is worth pointing out that negligible correlations may lead to statistically significant p values if the sample size is sufficiently large. For example, in most biological applications a correlation of $cor(x,y) = 0.01$ has the same interpretation ("no relationship") as a correlation of 0. But $r = 0.01$ leads to a statistically significant p value (< 0.05) if the sample size is very large ($m > 40,000$). Statistical significance is not the same as practical significance. This is why one should always report effect sizes (e.g., correlation coefficients) along with p values.

Apart from the above-mentioned approaches for calculating p values, one can also use a likelihood ratio test (see Sect. 13.1) or a permutation test approach (described in Sect. 10.2).

10.4 R Code for Calculating Correlation Test p Values

For any test statistic Z that follows a standard normal distribution under the null hypothesis, one can use the following R code for computing a 2-sided p value

```
p.value=2*pnorm(-abs(Z)).
```

The R function `corPvalueStudent` in the `WGCNA` package inputs a matrix of correlations and a corresponding sample size m and outputs a matrix of corresponding p values (based on (10.4)). For example, assume that `datX` is a data frame with m rows. The `WGCNA` function `corFast` implements a fast way of calculating pairwise correlations. The R code

```
corPvalueStudent(cor=corFast(datX),nSamples=m)
```

will produce a matrix of Student t-test p values whose i,jth component corresponds to the correlation between the ith and jth column of `datX`. Similarly,

```
corPvalueFisher(cor=corFast(datX),nSamples=dim(datX)[[1]])
```

calculates a matrix of pairwise correlation p values based on the Fisher's Z transformation (10.8).

Assume now that y is a binary vector. The following R code can be used to prove that the Student t-test $t.different.means$ (10.6) based on comparing mean values leads to the same p value as the correlation-based Student t-test $Z^{Student}$ (10.4):

```
x=rnorm(10)
y.binary=sample(c(1,0),10,replace=T)
t.test(x~factor(y.binary),var.equal=T)
cor.test(x,y.binary)
```

10.5 Multiple Comparison Correction Procedures for p Values

The choice of the threshold t for a p value depends on the desired significance level (also known as alpha level) α. Typically, one chooses $t = \alpha = 0.05$ but this significance level is not stringent enough if many hypotheses are being tested

In genomic applications, one often relates tens of thousands of variables x_i (also known as features, covariates, characteristics) to a special variable y. For example, a gene expression study may relate each of the 25,000 gene expression profiles x_i to a sample trait y (e.g., body weight) using a correlation test or a regression model. For each variable, a null hypothesis is tested against an alternative hypothesis. For example, a correlation test can be used to test the null hypothesis that the correlation is zero.

Statistical tests of the null hypothesis allow one to calculate a p value for each variable. A variable is called significant if its p value lies below a pre-defined significance threshold. How to choose a significance threshold for a p value has been the subject of considerable research. A major challenge is the tendency for multiple comparisons to yield spurious significant p values even when the multiple null hypotheses hold true. To make these statements more precise, we will use the following notation/terminology:

$pvalueVector = (p_1, \ldots, p_n)$ is a vector of n p values corresponding to the "family" of n tests.

TrueNullHypothesis is a vector whose ith component equals 1 if the ith null hypothesis is true and 0 otherwise. Of course, this vector is not available in real data applications. A variable is called truly null if the null hypothesis is true, and a variable is truly alternative if the alternative hypothesis is true.

n_0 denotes the number of truly null variables in the study. In R notation, $n0 = sum(TrueNullHypothesis)$. π_0 is the proportion of truly null hypotheses among the family of n hypotheses:

$$\pi_0 = \frac{n_0}{n} = mean(TrueNullHypothesis). \tag{10.10}$$

In real data applications, the vector *TrueNullHypothesis* is unavailable. But surprisingly, statistical methods exist for estimating π_0 from the data (Storey 2002; Storey and Tibshirani 2003). The estimation methods consider the shape of the p value distribution (see Fig. 10.1). Specifically, they exploit the fact that p values follow a uniform distribution if all of the underlying null hypotheses are true. Thus, in this truly null case, a histogram (frequency plot) of p values will look flat. While p values of truly null variables follow a uniform distribution, the p values of truly significant (alternative) variables will tend to be close to zero. Thus, if a p value distribution is skewed, then it is unlikely that all null hypotheses hold true. Consider Fig. 10.1 that shows a histogram of p values. Most of the p values near 1 are likely to correspond to truly null hypotheses. The horizontal dashed line corresponds to the expected frequency distribution if all null hypotheses (of zero correlation) hold true. The increased height of the left-most bar in the histogram suggests that more

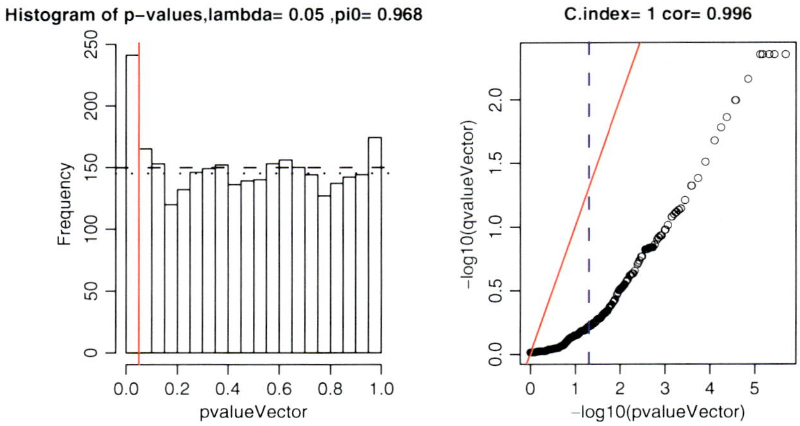

Fig. 10.1 False discovery rate estimation and q-values for the simulated gene expression data described in Sect. 10.7. The left panel shows a histogram (frequency plot) of $n = 3{,}000$ correlation test p values. The *dashed horizontal line* shows the histogram we would expect if all genes were null (not correlated with y). In this case, the number of genes per bin is expected to be $n/no.bins$ where $no.bins = 20$ is the number of bins. The *red vertical line* corresponds to the choice of $\lambda = 0.05$ for calculating the q-values and estimating the proportion of null p values $\hat{\pi}_0(\lambda) = 0.968$. The *dotted horizontal line* corresponds to the estimated proportion of null p values in each bin $n * \hat{\pi}_0(\lambda)/no.bins$. The right-hand panel shows the relationship between $-log10(pvalue)$ (logarithm with base 10) and $-log10(q - value)$. Note that the q-value is a nondecreasing function of the p value which is reflected in a C-index (5.18) of 1. The Spearman correlation equals 0.996. As long as the *black line* is below the *red straight line* (corresponding to $y = x$), the p value is smaller than the corresponding q-value. In particular, $p_i < q_i$ as long as p lies below the threshold $\alpha = 0.05$ (corresponding to the *blue dashed vertical line*)

variables are significant than expected by chance. Figure 10.1 also shows that the histogram of p values to the right of 0.05 looks fairly flat, which indicates that there are mostly null p values in the subinterval $[\lambda, 1]$ where $\lambda = 0.05$. It turns out that the height of this flat portion (dotted horizontal line) can be used to arrive at the following estimate

$$\hat{\pi}_0(\lambda) = \frac{\text{number of p values } > \lambda}{n(1-\lambda)}, \tag{10.11}$$

which involves the tuning parameter $\lambda = 0.05$. This estimate is conservative in the sense that it tends to be higher than the true value. To see this, note that erroneous inclusion of p values (corresponding to truly significant variables) leads to an overestimate of π_0. If we take $\lambda = 0$, then $\hat{\pi}_0(\lambda) = 1$ which implies that there is no signal, i.e., none of the variables are significant. In genomic applications, a large number of variables is expected to be truly significant which suggests to choose a relatively large value of λ. However, the closer λ is to 1, the higher the variance of the estimate $\hat{\pi}_0(\lambda)$, i.e., the less reliable is the estimate. Visual inspection of the histogram in Fig. 10.1 suggests that $\lambda = 0.05$ is a good choice. The `qvalue` R function provides an automatic method for estimating π_0 which obviates the need of specifying λ (Storey 2002; Storey and Tibshirani 2003). In the following, we will show how to compute the probability of rejecting the null hypothesis at least once in a family of n tests when the null hypothesis is true for all of the tests. Toward this end, we will use the following notation:

1. $F(t)$ is the number of false positives, i.e., the number of p values that lie below the significance threshold t even though they correspond to truly null hypotheses. In R notation, `F.t=sum(pvalueVector[TrueNullHypothesis==1]<=t)`.
2. $T(t)$ is the number of true positives, i.e., the number of null hypotheses that are correctly rejected since the corresponding p values lie below t. In R notation, `T.t=sum(pvalueVector[TrueNullHypothesis==0]<=t)`.
3. $S(t) = F(t) + T(t)$ is the total number of p values below the threshold t. In R notation, `S.t=sum(pvalueVector<=t)`.

The most widely used approach for dealing with multiple comparisons is to "correct" the original p values for the fact that multiple comparisons have been carried out. The **Bonferroni correction** is derived by observing Boole's inequality which implies that for a finite set of statistical tests (referred to as family of tests), the probability that at least one of them is significant is no greater than the sum of the probabilities of the individual tests. For a p value threshold of t, the resulting **family-wise error rate** is bounded as follows:

$$\begin{aligned} FamilyWiseErrorRate(t) &= P(F(t) \geq 1) \\ &= P(\text{for at least one } i{:}p_i{<}t|\text{Null Hypothesis holds}) \end{aligned} \tag{10.12}$$

With Boole's inequality, we find

$$\begin{aligned} &\leq \sum_i P(p_i < t|\text{Null Hypothesis holds}) \\ &= n * t, \end{aligned}$$

where we made use of the fact that $P(p_i \leq t|\text{Null Hypothesis holds}) = t$ since p values follow a uniform distribution under the null hypothesis. Note that we can guarantee that the family-wise error rate $P(F \geq 1) \leq \alpha$ by calling all variables significant if their p values are smaller than the **testwise significance level** α/n. Note that $p_i \leq \alpha/n$ is equivalent to requiring that $p_i^{Bonferroni} \leq \alpha$, where the **Bonferroni corrected** p **value** is defined as follows:

$$p^{Bonferroni} = p * n, \tag{10.13}$$

and n is the number of tests. Note that the Bonferroni corrected p value could be larger than 1.

Another multiple comparison correction method is the **Sidak correction**, which assumes that the individual tests are independent. Under this independence assumption, the family-wise error rate satisfies

$$
\begin{aligned}
FamilyWiseErrorRate(t) &= P(F(t) \geq 1) \\
&= P(\text{for at least one } i\text{: } p_i < t|\text{Null Hypothesis holds}) \\
&= 1 - P(\text{for each } i\text{: } p_i \geq t|\text{Null Hypothesis holds}).
\end{aligned}
$$

Independence implies

$$
\begin{aligned}
&= 1 - \prod_{i=1}^{n} P(p_i \geq t|\text{Null Hypothesis holds}) \\
&= 1 - \prod_{i=1}^{n} (1-t) = 1-(1-t)^n. \qquad (10.14)
\end{aligned}
$$

Note that $FamilyWiseErrorRate(t) \leq \alpha$ if each individual p value is smaller than the testwise error rate $1-(1-\alpha)^{1/n}$. Note that $p_i \leq 1-(1-\alpha)^{1/n}$ is equivalent to requiring that $p_i^{Sidak} \leq \alpha$, where the **Sidak corrected** p **value Sidak corrected** p **value** is defined as follows:

$$
p^{Sidak} = 1-(1-p)^n. \qquad (10.15)
$$

The Sidak MCC correction is slightly less conservative than the Bonferroni correction:

$$
p^{Sidak} \leq p^{Bonferroni}.
$$

A standard approach for defining a set of significant variables is to impose a threshold of $\alpha = 0.05$ to the Bonferroni or Sidak corrected p values. While the Bonferroni correction is the most conservative and credible method for choosing a significance threshold, its excessive rigueur is unpopular with researchers who cannot collect large data sets. Controlling the family-wise error rate $P(F(t) \geq 1)$ (10.12) is much too conservative for many genomic studies where many variables are expected to be truly alternative.

10.6 False Discovery Rates and q-values

We will now describe a "liberal" multiple comparison correction technique, which is widely used in genomics. If the sample size m is low, guarding against any single false-positive occurring (i.e., setting a low p value threshold) may be too strict and could lead to many missed findings. More liberal approaches have a more moderate goal: they aim to identify as many significant variables as possible, while incurring a relatively low proportion of false positives. To address this goal, methods for controlling the **false discovery rate (FDR)** have been developed (Benjamini and Hochberg 1995; Storey and Tibshirani 2003). Although the term "false positive rate" and "false discovery rate" sound similar, they are very different concepts. Assume a rule (or selection criterion) for calling variables significant (i.e., for creating a set of significant variables), the false-*positive* rate is the rate that at least one of the truly null variables is called significant. In contrast, the false *discovery* rate is the proportion of variables that are truly null among the variables deemed significant.

The false discovery rate calculation exploits the shape of the distribution of p values. Recall that for a given threshold t, $F(t)$ denotes the number of false-positive variables, and $S(t)$ denotes the number of p values that lie below the threshold. The proportion of false-positive variables among all significant variables is given by:

$$\frac{F(t)}{S(t)} = \frac{\text{number of false positive variables}}{\text{number of significant variables}}. \tag{10.16}$$

The false discovery rate $FDR(t)$ is defined as the expected value of this quantity:

$$FDR(t) = E\left(\frac{F(t)}{S(t)}\right). \tag{10.17}$$

We briefly mention that $\frac{F(t)}{S(t)}$ is not defined if $S(t) = 0$. One can require that $S(t) > 0$ and define $FDR(t)$ as a conditional expectation:

$$FDR(t) = E(F(t)/S(t)|S(t) > 0)P(S(t) > 0).$$

For example, assume a set of 100 significant variables (e.g., variables whose p values lie below a threshold t). If $FDR = 0.05$ for this set of significant variables, then only five of these significant variables are expected to be false positives. This would be a negligible error rate in most applications. Of course, in real applications $FDR(t)$ is unknown and must be estimated from the data. In the following, we describe how to estimate the false discovery rate of a list of significant variables (Storey 2002; Storey and Tibshirani 2003). If the number of variables n is large, then it can be shown that the false discovery rate (10.17) can be approximated as follows:

$$FDR(t) \approx \frac{E(F(t))}{E(S(t))}, \tag{10.18}$$

where $E(F(t))$ denotes the expected value of $F(t)$. A simple estimate of the denominator $E(S(t))$ is the observed number of significant variables

$$S(t) = (\text{number of } p \text{ values } \leq t);$$

that is, the number of observed p values $\leq t$. To estimate $E(F(t))$, recall that p values corresponding to truly null hypotheses should be uniformly distributed. Thus, the probability a null p value is $\leq t$ is simply t, and it follows that

$$E(F(t)) = n_0 t = n\pi_0 t,$$

where $\pi_0 = \frac{n_0}{n}$ (10.10) is the proportion of truly null p values, which can be estimated by $\hat{\pi}_0$ (10.11). By plugging the estimates into the right-hand side of (10.18), we arrive at the following estimate:

$$F\hat{D}R(t) = \frac{\hat{\pi}_0 nt}{S(t)} = \frac{(\text{number of } p \text{ values } > \lambda)}{(\text{number of } p \text{ values } \leq t)} \frac{t}{(1-\lambda)}, \tag{10.19}$$

where λ is the tuning parameter used for estimating $\hat{\pi}_0$. The estimated FDR is a very useful measure for error rate of a *set* of significant variables. But it is also desirable to have a false discovery rate-based measure of significance for each individual variable. Toward this end, we will review the definition of the **q-value** which is the FDR-based analogue of the p value (Storey 2002; Storey and Tibshirani 2003). The q-value of the ith hypothesis test is the minimum FDR at which the test would be called significant. To clarify the definition of the q-value corresponding to p_i, we define it in three steps. First, for each threshold $t \geq p_i$ identify the set of significant variables. Second, for each of these significant sets determine the false discovery rate. Third, define the q-value as the minimum of the false discovery rates. These three steps can be summarized in the following definition of the q-value:

$$q(p_i) = min_{t \geq p_i} FDR(t). \tag{10.20}$$

We can estimate the q-value of the ith variable by simply plugging $F\hat{D}R(t)$ into the definition (10.20) of the q-value:

$$\hat{q}(p_i) = min_{t \geq p_i} F\hat{D}R(t). \tag{10.21}$$

Note that the definition (using a minimum) guarantees that the estimated q-values are increasing in the same order as the p values. Similar to the p value, the q-value gives each variable its own individual measure of significance. While the p value is a measure of significance in terms of the false-positive rate, the q-value is a measure in terms of the FDR. By thresholding the q-value, one can define sets of significant variables which have a low false discovery rate (Storey 2002; Storey and Tibshirani 2003). Limitations of the q-value are described in Sect. 10.8.

10.7 R Code for Calculating q-values

Given a vector $pvalueVector = (p_1, \ldots, p_n)$ of p values, the R package `qvalue` allows one to calculate a corresponding vector $qvalueVector = (q_1, \ldots, q_n)$ of q-values. Here we use the simulated gene expression data from Sect. 6.13 to show how to calculate q-values. Recall that we had simulated a binary sample trait y (of length $m = 50$) and 3,000 numeric variables (gene expression profiles) that corresponded to columns of a matrix $datExpr$. Each of the 3,000 variables is correlated with y, and Fisher's transformation (10.8) is used to calculate a vector of p values. Next we use (10.11) to estimate the proportion of truly null hypotheses, i.e., the proportion of variables that have zero correlation with y.

```
library(qvalue)
# sample trait based node significance measure
# defined as correlation
GS=as.numeric(cor(y,datExpr,use="p"))
pvalueVector=corPvalueFisher(GS,nSamples=length(y))
n=length(pvalueVector)
# now we estimate the proportion of truly null p-values
lambda=0.05
pi0.estimated=sum(pvalueVector>lambda)/(n*(1-lambda)) #=0.968
```

In this simulated example, we can define a vector whose ith component equals 1 if the ith null hypothesis is truly null. Next, this vector can be used to calculate the exact value π_0.

```
TrueNullHypothesis
=is.element(truemodule,c("turquoise","blue","yellow","grey"))
# number of truly null variables
n0=sum(TrueNullHypothesis)
# proportion of truly null variables
pi0=n0/n
# alternatively, define
pi0=mean(TrueNullHypothesis) #=0.88
```

Note that the estimated value $\hat{\pi}_0 = 0.968$ is overly conservative compared to the true value $\pi_0 = 0.88$. Let us now consider a p value threshold of $t = 0.01$. The following code can be used to estimate the false discovery rate for the resulting set of significant variables.

```
# p value threshold
t=0.01
# total number of p values below the threshold t
S.t=sum(pvalueVector<=t) #=94
# Estimated false discovery rate
estimatedFDR=pi0.estimated*n*t/S.t #=0.309
```

Let us now calculate the true FDR for this list of genes:

```
# number of false positives
F.t=sum(pvalueVector[TrueNullHypothesis==1]<=t) #=30
# number of true positives
T.t=sum(pvalueVector[TrueNullHypothesis==0]<=t) #=64
# False discovery rate for the set of significant variables
trueFDR=F.t/S.t #=0.319
```

Thus for the set of 94 significant genes (whose p values lie below $t = 0.01$), the estimated FDR of 30.9% is an amazingly accurate estimate of the true FDR of 31.9%. We should mention that such a high FDR would be unacceptable in many applications and a more stringent value for t should be chosen. In our example, $t = 0.001$ leads to a set of 27 significant genes and a corresponding estimated FDR of 11%.

Next we will provide R code for computing the q-values and for creating Fig. 10.1.

```
lambda=0.05
qq=qvalue(pvalueVector,lambda=.05)
qvalueVector=qq$qvalues
```

```
pi0.estimated=signif(qq$pi0,3) #=0.968
par(mfrow=c(1,2))
no.bins=20
m1=paste("Histogram of p,lambda=",lambda,",pi0=",pi0.estimated)
hist(pvalueVector,no.bins,main=m1)
ExpectedNumberPerBin=length(pvalueVector)/no.bins
abline(h=ExpectedNumberPerBin,lwd=2, lty=2)
ExpectedNumberNullPvaluesPerBin=
        length(pvalueVector)*pi0.estimated/no.bins
abline(h=ExpectedNumberNullPvaluesPerBin,lwd=2, lty=3)
abline(v=lambda,col="red",lwd=2)
# now we calculate the C-index between the vectors
library(Hmisc)
C.index=signif(rcorr.cens(pvalueVector,qvalueVector,outx=T)
                [[1]],3)
SpearmanCor=signif(cor(pvalueVector,qvalueVector,method="s"),3)
plot(-log10(pvalueVector),-log10(qvalueVector),
main=paste("C.index=",C.index, "cor=",SpearmanCor))
abline(0,1,col="red",lwd=2)
abline(v=-log10(.05),col="blue",lwd=2,lty=2)
```

Note that the left panel of Fig. 10.1 shows a histogram of the p values and the expected histogram if all genes were null (not correlated with y). The right panel shows the relationship between the p value and the q-value. Note that the q-values are a nondecreasing function of the p values. Clearly, the estimates of the q-values depend on λ. After inspecting the histogram of p-values, we chose $\lambda = 0.05$. Let us now investigate how the q-value estimates change if the default method is used for estimating π_0.

```
qq.default=qvalue(pvalueVector)
qvalue.default=qq.default$qvalues
pi0.default=qq.default$pi0 #=1
cor(qvalue.default,qvalueVector) #=1
table(qvalue.default<=0.05,qvalueVector<=0.05)#perfect agreement
```

We find that the default method for estimating the q-value leads to virtually the same estimates as that based on $\lambda = 0.05$. In particular, both q-values lead to the same set of 16 significant genes if a q-value threshold of 0.05 is chosen. Also the two q-value vectors have a perfect correlation of 1.

10.8 Multiple Comparison Correction as p Value Transformation

Many approaches for choosing a p value threshold have been developed in the literature. Psychologically speaking, these procedures can be interpreted as "displacement" strategies, which allow one to avoid the unpleasant reality of choosing a seemingly ad hoc threshold by shifting the problem to that of choosing a conventional 0.05 threshold for a "transformed" version of the p value. Joking aside, multiple correction procedures are useful if a convincing statistical rationale can be provided for transforming the p value. Most p value transformations are nondecreasing functions that map the original p values to new values, which are referred to as "corrected" p values, "adjusted" p values, or in case of a false discovery rate approach as q-values or local false discovery rates. Abstractly speaking, we define a **multiple comparison transformation** $MCCtrafo()$ as a nondecreasing function that maps the original p values to real numbers (referred to as corrected p values), i.e.,

$$p_i \leq p_j \text{ implies } MCCtrafo(p_i) \leq MCCtrafo(p_j). \tag{10.22}$$

Further, we will call $MCCtrafo()$ **proper** with regard to the significance threshold α if it satisfies the following criterion:

$$MCCtrafo(p) >= p \text{ if } p \leq \alpha. \tag{10.23}$$

In other words, a proper MCC transformation satisfies the basic intuition that an MCC transformation should increase significant p values (i.e., decrease the statistical significance) to account for the fact that multiple comparisons have been carried out. In the following, we will describe several MCC transformations.

The Bonferroni-based MCC transformation maps each p value to its corresponding Bonferroni corrected version (10.13):

$$MCCtrafo^{Bonferroni}(p_i) = p_i^{Bonferroni} = p_i * n. \tag{10.24}$$

Similarly, the Sidak-based MCC transformation maps each p value to its Sidak corrected version (10.15):

$$MCCtrafo^{Sidak}(p_i) = p_i^{Sidak} = 1 - (1 - p_i)^n. \tag{10.25}$$

One can easily prove that $MCCtrafo^{Bonferroni}(p_i) \geq p_i$ and $MCCtrafo^{Sidak}(p_i) \geq p_i$, i.e., both MCC transformations are proper (10.23) for any choice of α.

The q-value-based transformation MCC^{qvalue} maps each p value to the corresponding q-value:

$$MCCtrafo^{qvalue}(p_i) = q_i. \tag{10.26}$$

For the simulated example in Fig. 10.1, we find that $MCCtrafo^{qvalue}$ is proper ($MCCtrafo^{qvalue}(p_i) > p_i$) if $p_i \leq \alpha = 0.05$. But it is easy to construct situations when $MCCtrafo^{qvalue}$ is improper. To illustrate this point, we find it useful to digress and to introduce the **crazy power correction method** $MCCtrafo^{crazy}$

$$MCCtrafo^{crazy}(p_i, \gamma) = p_i^{crazy} = \begin{cases} 1 & \text{if } \gamma <= 1 \\ \frac{1}{\gamma}(p_i)^{(\gamma-1)/\gamma} & \text{if } \gamma > 1 \end{cases}, \tag{10.27}$$

which depends on the parameter γ. $MCCtrafo^{crazy}$ is a crazy multiple comparison correction procedure for two reasons. First of all, no good statistical rationale is given for it. Second, it is highly improper for large values of γ (when $\gamma-1)/\gamma \approx 1$). In this case, $p_i^{crazy} \approx \frac{p_i}{\gamma}$ which implies that $MCCtrafo^{crazy}(p_i) < p_i$, i.e., it is improper. Now we are ready to show a situation when the q-value-based MCC transformation $MCCtrafo^{qvalue}$ is improper. Assume a large (say $n = 50,000$) sample of p values $pvalueVector = (p_1, \dots, p_n)$ which follow a Beta distribution with shape parameters $shape1 = 1/\gamma$ and $shape2 = 1$ (Fig. 10.2). Next we use the default settings of the R function `qvalue` to compute a corresponding vector of q-values $qvalueVector = (q_1, \dots, q_n)$. In an exercise, you are asked to show that in this special case

$$qvalueVector = \frac{pvalueVector^{(\gamma-1)/\gamma}}{\gamma} \tag{10.28}$$

$$= MCCtrafo^{crazy}(pvalueVector, \gamma). \tag{10.29}$$

Thus, in this case the q-value-based MCC transformation turns out to be identical to the crazy p value transformation (10.27). Thus, if the p values follow a Beta distribution with shape parameters $shape1 = 1/\gamma \le 0.35$ and $shape2 = 1$, then $MCCtrafo^{qvalue}(0.05) < 0.05$, i.e., $MCCtrafo^{qvalue}$ is improper at level $\alpha = 0.05$. This example demonstrates that a naive application of a false discovery rate approach can lead to improper results. Figure 10.2 and the following underlying R code illustrate this example.

```
library(qvalue)
library(WGCNA)
n=50000
gamma=5
# shape parameters of the Beta distribution
shape1=1/gamma; shape2=1
#either use pvalueVector=rbeta(n,shape1,shape2) or use
unif.vector=runif(50000)
pvalueVector=unif.vector^(1/shape1)
par(mfrow=c(1,3))
hist(pvalueVector,
main=paste("p-value distribution: Beta(",shape1,",",shape2,")"))
qvalueVector=qvalue(pvalueVector)$qvalues
qvalueVectorEstimated=1/gamma*pvalueVector^((gamma-1)/gamma)
xlab1="1/gamma*p^((gamma-1)/gamma)"
verboseScatterplot(qvalueVectorEstimated,qvalueVector,
xlab=xlab1,ylab="q-value")
abline(0,1,col="red",lwd=2)
# now we calculate the C-index between the vectors
library(Hmisc)
C.index=
  signif(rcorr.cens(pvalueVector,qvalueVector,outx=T)[[1]],3)
SpearmanCor=signif(cor(pvalueVector,qvalueVector,method="s"),3)
plot(-log10(pvalueVector),-log10(qvalueVector),
main=paste("C.index=",C.index, "cor=",SpearmanCor),
```

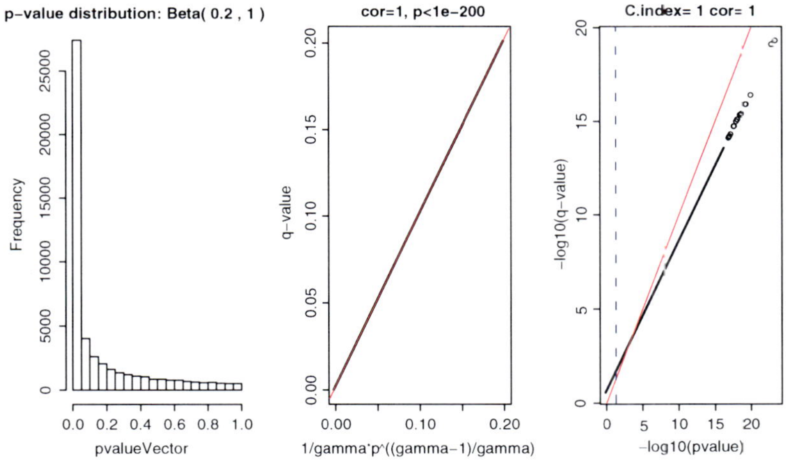

Fig. 10.2 Q-value calculation assuming that the p values follow a Beta distribution shape parameters $1/\gamma = 0.2$ and 1. The left-most panel shows a histogram (frequency plot) of the $n = 50,000$ simulated p values. The default values of the R function q-value are used to calculate corresponding q-values. The middle panel shows a scatter plot of the observed q-value versus our predicted q-value based on (10.28). The line $y = x$ is colored in *red*. Note that the predicted q-value provides an excellent fit to the observed q-value. The right-most panel shows a scatter plot between $-log10(pvalue)$ and $-log10(q-value)$. Note that the q-value is a nondecreasing function of the p value which is reflected in a C-index (5.18) of 1. For a p value of 0.05 (corresponding to the *dashed vertical line*), the *black line* lies above the *red line*, i.e., the q-values are larger than the p values. In summary, the q-value-based multiple comparison correction transformation is not proper (10.23) at level 0.05

```
ylab="-log10(q-value)",xlab="-log10(pvalue)")
abline(0,1,col="red",lwd=2)
abline(v=-log10(.05),col="blue",lwd=2,lty=2)
```

10.9 Alternative Approaches for Dealing with Many p Values

Multiple comparison correction methods allow one to avoid having to choose a significance level (threshold) α for the original p values by reframing the problem. Statistical rationales are given that a significance level α is appropriate for corresponding corrected p values. Unfortunately, we have to face the reality that threshold choices will always be somewhat ad hoc. For example, we wonder why many authors choose a threshold of 0.05 for q-values. Since multiple comparison correction transformations are threshold-implied by the original p values (see Sect. 10.11), any threshold for corrected p values corresponds to a threshold on the original p values. In practice, it is highly recommended to determine how the findings of a particular study depend on the choice of the threshold.

Another method for defining a set of significant variables is to (a) choose a fairly liberal threshold (say $\alpha = \frac{0.05}{\sqrt{n}}$, and (b) to impose additional selection criteria according to relevant complementary information (e.g., an additional object significance measure). For example, assume that we want to find genes that are significant predictors of brain cancer survival. For each of 25,000 genes, a p value is available for relating the expression profile to cancer survival time. Further assume that a second variable is available, which indicates for each gene whether it is already known to play a role in cancer. Due to the accumulation of data sets, there is substantial prior knowledge on which pathways relate to cancer survival. For example, genes that are members of the cell proliferation pathway are often associated with poor prognosis. In this case, most reviewers would find the following gene selection criteria credible: (a) the p value is smaller than 0.001, (b) the gene is part of the cell proliferation pathway.

Of course, the best way of dealing with the multiple comparison problem is to avoid it in the first place. One way to avoid it is to first "reduce" the original set of variables to a smaller set. For example, subsets of the original variables can be combined if they are highly correlated. Clusters of highly correlated variables can be summarized in multiple ways, e.g., one can choose a representative variable (e.g., a hub node) or one can define an aggregate representative (e.g., the first singular vector or eigenvector). Correlation network methods are a valuable data reduction method. Focusing the analysis on modules (clusters of variables) and their representatives (eigenvectors) involves far fewer comparisons than relating each of the individual variables to a sample trait.

10.10 R Code for Standard Screening

A standard variable screening analysis with respect to the variable y involves calculating an association measure between y and each column of an input data frame $datX$. Next columns (variables) with significant associations with y are being selected. The word "standard" implies that marginal (or univariate) associations are being computed.

Here we describe several `WGCNA` R functions that can be used to find variables associated with a numeric variable, a binary variable, or a censored time outcome. The function `standardScreeningBinaryTrait` computes widely used statistics for relating the columns of the input data frame (argument `datE`) to a binary sample trait (argument `y`). The binary trait can contain numbers or character strings. The statistics include Student t-test p value, the corresponding local false discovery rate (q-value), the fold change, the area under the ROC curve (i.e., the C-index), mean values, etc. The Kruskal–Wallis test is a nonparametric, rank-based group comparison test, which should be used if the numeric data do not follow a normal distribution. Kruskal–Wallis test p value and corresponding q-values are calculated if the user sets `kruskalTest=T`. The function `standardScreeningNumericTrait` does the analogous calculations for a numeric outcome variable y.

The function `standardScreeningCensoredTime` computes association measures between the columns of an input data frame `datE` and a censored time variable (e.g., survival time). The censored time is specified using two input variables: `time` and `event`. The event variable is binary where 1 indicates that the event took place (e.g., the person died) and 0 indicates censored (i.e., lost to follow-up). The function fits univariate Cox regression models (one for each column of `datE`) and outputs a Wald test p value, a log-rank p value, corresponding local false discovery rates (q-values), hazard ratios. Further, it reports the concordance (C) index (also known as area under the ROC curve).

We briefly mention the `WGCNA` function **rankPvalue**, which is useful for combining screening results from multiple independent data sets. Combining the results from multiple independent studies is sometimes referred to as meta analysis.

As an example, assume that you applied the function `standardScreeningBinaryTrait` to five different gene expression data involving 20,000 genes. For each of the five data sets, one can calculate a Z statistic (or Student t-test statistic) of differential expression. The five vectors of Z statistics form the columns of the $20,000 \times 5$ dimensional matrix `datS`. Assume that one gene has ranks 1, 1, 1, 1, and 20 in the five lists. The function `rankPvalue` can be used to calculate a p value for this very rare occurrence. In general, it estimates the p value for ranking consistently high (or low) on multiple lists. `rankPvalue` implements methods for calculating p values for observing that an object (corresponding to a row of the input data frame `datS`) has a consistently high ranking (or low ranking) according to multiple ordinal scores (corresponding to the columns of the input data frame `datS`). The function calculates asymptotic p values (and optionally q-values) for testing the null hypothesis that the values in the columns of `datS` are independent. More details can be found in the help file of `rankPvalue`.

10.11 When Are Two Variable Screening Methods Equivalent?

Statistical significance measures such as p values or q-values are often used to define sets of significant variables. In genomic applications, the identification of significant gene expression profiles x_i is also known as gene screening, gene filtering, or simply gene ranking. Here we will study the relationships of different variable screening methods, i.e., methods for defining a set of "significant" variables. As an example, consider the task of finding variables x_i that are related to a trait y. To address this task, several approaches are possible. One approach is to define a variable significance measure $OS_i^{abs.cor} = |cor(x_i, y)|$ as the absolute correlation coefficient. To define a set of significant variables, one can threshold this significance measure (with a threshold τ). Alternatively, assume that we have a vector of Student t-test p values $pvalueVector = (p_1^{Student}, \ldots, p_n^{Student})$. To turn this into a measure of object significance, we define

$$OS_i^{pvalue} = -log10(p_i^{Student}),$$

where $log10$ denotes the logarithm with base 10. Note that OS_i^{pvalue} is proportional to the number of zeros past the decimal point of the ith p value. A natural question is whether a set of significant variables based on thresholding OS^{pvalue} can be arrived at by thresholding $OS^{abs.cor}$? This turns out to be the case. A variable ranking based on OS^{pvalue} turns out to be identical to that based on $OS^{abs.cor}$ since the two object significance measures are related by a monotonically increasing function. Let us now consider the q-value-based object significance measure $OS_i^{qvalue} = -log10(q_i)$. Rankings based on OS^{qvalue} can be different from those of OS^{pvalue} due to ties in the q-value: two variables with different p values may have the same q-value. Many researchers use p values or q-values to identify sets of significant variables. This fundamental data analysis task is also known as variable selection, variable screening, or variable ranking. To study these variable screening methods in more detail, we find it useful to use a slightly more abstract terminology. Instead of talking about variables, we will talk about n "objects". For example, objects may correspond to variables, network nodes, or network edges. Further, we define a **measure of object significance** OS as a vector $OS = (OS_1, \ldots, OS_n)$ of real numbers whose n components correspond to the n objects. The *higher* the value of OS_i, the more significant is the ith object with regard to a certain application. For example, when considering a set of network nodes (interpreted as objects), then a node significance measure is a special case of an object significance measure. Our definition of an object significance measure is more general than that of a statistical significance measure. For example, to screen for wealthy individuals, one can use "network-worth" or "annual salary" as object significance measures.

In many applications, several object significance measures are possible and one wants to study the relationships of the resulting sets of significant variables. For example, assume that two (object) significance measures $OS^{(1)}$ and $OS^{(2)}$ are available. We define that $OS^{(1)}$ **threshold-implies** $OS^{(2)}$ if, and only if, a nondecreasing function $nonDecreasingF$ can be defined such that

$$OS^{(2)} = nonDecreasingF(OS^{(1)}).$$

Let us now provide another (equivalent) definition of threshold-implication, which explains its name. Assume that we use a threshold $\tau^{(2)}$ for $OS^{(2)}$ to define the **set of significant objects**

$$SignificantSet(OS^{(2)}, \tau^{(2)}) = \{i | OS_i^{(2)} > \tau^{(2)}\}.$$

$OS^{(1)}$ **threshold-implies** $OS^{(2)}$ if, and only if, for any threshold $\tau^{(2)}$ for $OS^{(2)}$, one can find a threshold $\tau^{(1)}$ for $OS^{(1)}$ such that

$$SignificantSet(OS^{(1)}, \tau^{(1)}) = SignificantSet(OS^{(2)}, \tau^{(2)}). \tag{10.30}$$

As a special case, consider that a monotonically increasing function F can be defined such that $OS^{(2)} = F(OS^{(1)})$. In this case, (10.30) is satisfied for $\tau^{(1)} = F^{-1}(\tau^{(2)}$.

A computationally easy way for studying threshold-implications is based on the C-index (5.18): one can show:

$$OS^{(1)} \text{ threshold-implies } OS^{(2)} \text{ if, and only if, } C.index(OS^{(1)}, OS^{(2)}) = 1. \quad (10.31)$$

Recall that the C-index is not symmetrical, i.e., $C.index(OS^{(1)}, OS^{(2)}) = 1$ does not necessarily imply $C.index(OS^{(2)}, OS^{(1)}) = 1$. The C-index can be used to measure the extent to which one significance measure threshold-implies another. Recall that $C.index = 0.5$ implies random concordance (no better than flipping a coin), and $C.index < 0.5$ implies opposite concordance (worse than random, but if you flip the direction it becomes a good concordance).

We call two object significance measures $OS^{(1)}$ and $OS^{(2)}$ **threshold-equivalent** if, and only if, $OS^{(1)}$ threshold-implies $OS^{(2)}$ and $OS^{(2)}$ threshold-implies $OS^{(1)}$. Thus, the two measures are threshold-equivalent if, and only, if

$$C.index(OS^{(1)}, OS^{(2)}) = C.index(OS^{(2)}, OS^{(1)}) = 1.$$

As an example, consider a vector of p values $pvalueVector = (p_1, \ldots, p_n)$ and a corresponding multiple comparison transformation $MCC(pvalueVector)$ (10.22). Then the significance measure $OS^{pvalue} = -log10(pvalueVector)$ threshold-implies $OS^{corrected.pvalue} = -log10(MCCtrafo(pvalueVector))$ since $MCCtrafo()$ is, by definition (10.22) a nondecreasing function. If $MCCtrafo()$ is a monotonically increasing function (e.g., the Bonferroni correction or the Sidak correction), then the two measures are threshold-equivalent since $OS^{corrected.pvalue}$ also threshold-implies OS^{pvalue}.

Recall that the rank function $rank(x)$ is defined in Sect. 5.1.2. Two (object) significance measures $OS^{(1)}$ and $OS^{(2)}$ are called **rank-equivalent** if

$$rank(OS^{(1)}) = rank(OS^{(2)}). \quad (10.32)$$

In an exercise, you are asked to show that rank-equivalence implies threshold-equivalence.

10.12 Threshold-Equivalence of Linear Significance Measures

Assume that we want to measure the linear relationship between n vectors $\{x_1, \ldots, x_n\}$ and another vector y. All vectors have length m. The absolute value of the Pearson correlation

$$OS_i^{abs.cor} = |cor(x_i, y)| = |r_i|$$

defines a "linear" object significance measure for relating these vectors. Many alternatively **linear object significance measures** can be defined based on correlation test p values, q-values, and model fitting indices. Many of these measures are threshold-implied (10.31) by $OS_i^{abs.cor} = |cor(x_i, y)|$. Since the sample size m and the total number of vectors n are fixed for the set of vectors, $OS^{abs.cor}$ threshold implies the following other linear object significance measures:

1. The absolute value of the Student t-test statistic (10.4):

$$OS_i^{abs.Tstatistic} = |Z_i^{Student}| = \sqrt{m-2}\frac{|r_i|}{\sqrt{1-(r_i)^2}}.$$

2. Minus log of the Student t-test p value:

$$OS_i^{pvalue} = -log10(p_i^{Student}).$$

3. Minus log of the Bonferroni-corrected Student t-test p value:

$$OS_i^{Bonferroni} = -log10(p_i^{Student,Bonferroni}) = -log10(n * p_i^{Student}). \quad (10.33)$$

4. Minus log of the q-value (10.26) which corresponds to the Student p value:

$$OS_i^{qvalue} = -log10(q_i^{Student}) = -log10(MCCtrafo^{qvalue}(p_i^{Student})). \quad (10.34)$$

5. The model fitting index $R_i^2 = cor(y, \hat{y})^2$ (13.26) of the simple univariate linear regression model $E(y) = \gamma_0 + \gamma_1 x_i$

$$OS_i^{lm} = R_i^2 = cor(x_i, y)^2,$$

   where we used the fact that $cor(y, \hat{y})^2 = cor(x_i, y)^2$ (see an Exercise).
6. The absolute value of the Fisher's Z transformation statistic of the correlation (10.8):

$$OS_i^{abs.Fisher} = |Z_i^{Fisher}| = \sqrt{m-3} * atanh(|r_i|). \quad (10.35)$$

7. Minus log of the Fisher's Z transform p value:

$$OS_i^{p.Fisher} = -log10(p_i^{Fisher}).$$

8. Minus log of the corresponding q-value:

$$OS_i^{qvalueFisher} = -log10(q_i^{Fisher}) = -log10(MCCtrafo^{qvalue}(p_i^{Fisher})). \quad (10.36)$$

Thus, any set of "significant" vectors defined by thresholding one of the above-mentioned significance measures can be arrived at by thresholding $OS^{abs.cor}$ with a suitably chosen threshold $\tau^{(1)}$. Except for the q-value-based measure, the above-mentioned linear measures are rank equivalent.

10.13 Network Screening

Here we consider network-based variable selection approaches (referred to as network screening). Roughly speaking, network screening makes use of network concepts and/or module membership to identify variables x_i that relate to a sample trait y. In contrast, standard screening only considers the marginal association between x_i and y. Many network screening are conceivable and this remains an area of active research. Here we briefly present two possible approaches. The first approach is based on network neighborhood analysis. Toward this end, y is included as additional node in an association network (e.g., a correlation network). This allows on to define the adjacency A_{yi} between y and x_i. Next, neighborhood analysis is applied to identify nodes (i.e., vectors x_i) that are close to node y (Li and Horvath 2007). The second approach considers module membership. It starts out identifying modules that relate to y and next selects variables according to their module membership (kME) or intramodular connectivity kIM (Horvath et al. 2006). This module-based screening approach is particularly promising if it is likely that the significant module is preserved in other (validation) data sets. For example, we used module membership with regard to a proliferation module to identify brain cancer-related genes (Horvath et al. 2006).

Network-screening approaches often avoid the pitfalls of multiple comparisons since they can effectively harness prior biological knowledge resulting from studying the biological meaning of modules. A problem with current network-screening methods is that they typically require substantial domain knowledge, i.e., they are not amenable to push-button queries. For example, genes inside a module that arises from a technical artifact should obviously be ignored in module-based screening approaches. An experienced data analyst will look at module heatmaps to check for artifacts and possibly outliers. How to optimally blend standard screening and network-based screening approaches remains a topic of active research. Authors of network screening approaches often report simulations and one successful application. However, it is very difficult to define *automatic* network screening approaches that consistently work well across many applications. We briefly mention the experimental function `networkScreening` in the `WGCNA` library which works quite well in several simulation studies (e.g., simulated data from Sect. 6.13), but it requires further refinements. In an exercise, you are asked to explore how well `networkScreening` performs in simulated data. Let us briefly describe the input and output of the function. It inputs a numeric sample trait y, module eigenvectors $datME$, and a data frame of numeric variables `datExpr`. It outputs the following three association measures for relating the columns of `datExpr` with y. First, the association measure $p.Weighted$ is referred to as **weighted p value** but it should only be interpreted as a descriptive measure (and not as inferential measure). Second, the association measure $q.Weighted$ is the q-value calculated based on $p.Weighted$. Third, $cor.Weighted$ is referred to as weighted correlation with y. These three association measures can be used to rank variables. $p.Weighted$ threshold-implies $q.Weighted$. The absolute value $|cor.Weighted|$ is rank-equivalent with $p.Weighted$. More details can be found in the help file of the function.

10.14 General Definition of an Association Network

We define an association network as a network (adjacency matrix) whose nodes correspond to vectors and whose adjacency matrix is based on an association measure between pairs of vectors. Each vector (node) may correspond to a random sample (of size m) of a corresponding random variable. The n vectors form the columns of an $m \times n$ dimensional matrix

$$datV = [v_1, \dots, v_n]. \tag{10.37}$$

For example, a correlation network is a special case of an association network where the ith node corresponds to a numeric vector x_i. How to define the association measure depends on the research question. For example, association measure could be defined with respect to a p values or other measures of dependency between pairs of variables. We define an association measure $AssocMeasure(,)$ as a real valued function on the set of $N = n^2$ vector pairs $\{Pair_1 = (v_1, v_1), Pair_2 = (v_1, v_2), \dots, Pair_N = (v_n, v_n)\}$. Note that an association measure defines an $n \times n$ dimensional matrix

$$S = (AssocMeasure(v_i, v_j)).$$

To complete the definition of an association network, one needs to specify how the association matrix S is transformed into an adjacency matrix. Methods for transforming a general (possibly nonsymmetric) matrix into an adjacency matrix are described in Sect. 7.3.

10.15 Rank-Equivalence and Threshold-Equivalence

It turns out that many seemingly different association measures are essentially equivalent. Given that seemingly countless approaches exist for defining association measures between pairs of vectors, it is useful to introduce terminology for describing their relationship. It useful to interpret an association measure *AssocMeasure* as an object significance measure (see Sect. 10.15) where the objects correspond to pairs of vectors (v_i, v_j). Then the object significance of (v_i, v_j) equals the i, jth element of the matrix $S = (AssocMeasure(v_i, v_j))$. Let us now study the case of association measures which are represented by two $n \times n$ dimensional matrices $S^{(1)}$

and $S^{(2)}$, respectively. Denote the vectorized version of two matrices by $OS^{(1)} = vectorizeMatrix(S^{(1)})$ and $OS^{(2)}$. We say that **matrix** $S^{(1)}$ **threshold-implies matrix** $S^{(2)}$ if, and only if, $OS^{(1)}$ threshold-implies $OS^{(2)}$. Thus, $S^{(1)}$ threshold-implies $S^{(2)}$ if, and only if, a nondecreasing function *nonDecreasingF* can be defined such that

$$vectorizeMatrix(S^{(2)}) = nonDecreasingF(vectorizeMatrix(S^{(1)})).$$

Similarly, we adapt the notions of threshold-equivalence and rank-equivalence, e.g., the two matrices are called rank-equivalent if, and only if,

$$rank(vectorizeMatrix(S^{(1)})) = rank(vectorizeMatrix(S^{(2)})). \qquad (10.38)$$

10.16 Threshold-Equivalence of Linear Association Networks

Assume that we have a set of n numeric vectors $x_1, \ldots, x_n$ which correspond to the columns of the $m \times n$ dimensional matrix $datX$. Any of the linear significance measures in Sect. 10.12 can be used to relate $x = x_i$ to $y = x_j$. Thus, each of the linear significance measures gives rise to an $n \times n$ dimensional symmetric matrix S^{linear} whose i, jth element measures the extent of a linear relationship between x_i and x_j. We refer to the resulting networks as **linear association networks**. For example, consider the matrix $S^{abs.cor}$ whose i, jth element equals the absolute value of the Pearson correlation

$$S_{ij}^{abs.cor} = |cor(x_i, x_j)|.$$

In Sect. 7.2, we describe several approaches for turning a symmetric matrix into an adjacency matrix. Here we describe focus on the construction of an unweighted network, which can be accomplished using a (hard) threshold τ for thresholding the linear association measure. For example, an unweighted correlation network adjacency matrix A (5.22) is defined by thresholding $|cor(x_i, x_j)|$.

Recall that we say that a matrix $S^{(1)}$ threshold-implies another matrix $S^{(2)}$ if, and only if, a nondecreasing function *nonDecreasingF* can be defined such that

$$vectorizeMatrix(S^{(2)}) = nonDecreasingF(vectorizeMatrix(S^{(1)})).$$

Analogous to Sect. 10.12, one can show that the matrix $S^{(abs.cor)}$ threshold-implies the following matrices $S_{ij}^{abs.Tstatistic} = |Z_{ij}^{Student}|$, $S_{ij}^{pvalue} = -log10(p_{ij}^{Student})$, $S_{ij}^{Bonferroni} = -log10(p_{ij}^{Bonferroni})$, $S_{ij}^{qvalue} = -log10(q_{ij})$, S_{ij}^{lm}, etc. In other words, an unweighted network defined by thresholding any of the above-mentioned linear significance measures with a hard threshold $\tau^{(2)}$ is identical to a network that results from thresholding $S^{abs.cor}$ with a suitably chosen threshold $\tau^{(1)}$. Thus, these unweighted linear networks are threshold-implied by the correlation network. How to choose the hard threshold τ will be discussed in the following.

10.17 Statistical Criteria for Choosing the Threshold τ

Consider the matrix $|cor(datX)| = (|cor(x_i,x_j)|)$ of absolute pairwise correlations between a set of vectors. In Sect. 10.3, we describe several methods for calculating a matrix of p values $MatrixPvalues = (p_{ij})$ corresponding to the correlation matrix. If the sample size m is fixed, choosing a threshold for a two-sided p value is equivalent to choosing a threshold for the absolute value of a correlation coefficient. An obvious approach is to choose the correlation threshold τ based on a statistical significance level. Since a correlation network of n nodes involves calculating $n(n-1)/2$ pairwise correlations, even moderately sized networks require that correlation test p values be corrected for multiple comparisons. Since a Bonferroni correction approach may be too conservative in this setting, it could be advisable to use a false discovery rate approach. This can be accomplished by transforming the matrix of p values into a matrix of q-values $MatrixQvalues = (q_{ij})$. Then a hard threshold τ for $|cor(datX)|$ can be determined so that $|cor(x_i,x_j) \geq \tau$ is equivalent $q_{ij} \leq 0.05$. The larger the threshold τ, the smaller is the corresponding false discovery rate. Although this approach for thresholding a correlation matrix is reasonable from a statistical point of view, it is worth noting that approach could easily lead to networks with unreasonable topological properties. As extreme case consider the situation when all (or none) of the correlations pass the threshold τ. In this case, the network adjacency matrix would be constant and uninteresting. In Sect. 4.3, we describe topological criteria for choosing thresholds. An advantage of topological criteria is that they can be used for constructing both unweighted and weighted networks.

10.18 Exercises

1. Exercise regarding the calculation of correlation p values. Use R to simulate two normally distributed vectors x and y of length $m = 20$ with underlying correlation $\rho = 0.1$. Hint:

   ```
   set.seed(1);x=rnorm(m);y=rho*scale(x)+sqrt(1-rho^2)*rnorm(m)
   ```

   (i) Create a scatter plot using the WGCNA R command `verboseScatterplot(x,y)`. Repeat the exercise for the following values of the underlying correlation $\rho = -0.3, -0.7, 1$. How do the observed correlations $r = cor(x,y)$ compare to their underlying true correlations ρ?
   (ii) Simulate two vectors x and y using a true correlation of $\rho = 0.5$. Use the Student t-test, the Fisher's transformation, and the Hotelling transformation

to calculate p values. Which of these equals the output from the R command `cor.test(x,y)`?

(iii) Denote the value of the Student t statistic by $Tstatistic$ and define an F statistic $Fstatistic = Tstatistic^2$ as square of the Student t statistic. Use R code to verify that the (two-sided) p value of the Student t test:

```
pvalueStudent=2*(1-pt(abs(Tstatistic,df=m-2)))
```

equals that of the F statistic with 1 and $m-2$ degrees of freedom: `1-pf(Tstatistic^2,df1=1,df2=m-2)`.

(iv) Find the Student t-test p value in the following R output:

```
summary(lm(y~x))
```

Hint: see the previous result.

(v) Use a linear model to regress $scale(y)$ on $scale(x)$. How do the estimates of the slope and the intercept compare to their expected exact values? Hint: Numeric inaccuracies may lead to values slightly different from $cor(x,y)$ and 0.

2. Exercise regarding the Sidak correction.

(i) Prove that the Sidak multiple comparison correction method is less conservative than the Bonferroni correction, i.e., prove that

$$p^{Sidak} \leq p^{Bonferroni}.$$

(ii) Which assumption regarding the statistical tests underlies the derivation of the Sidak correction?

3. Exercise regarding q-values for a relatively small set of variables. Use the R code of Sect. 6.13 to simulate $n = 200$ gene expression profiles, i.e., modify only one line of R code in this section. Next rerun the entire R code presented in Sect. 10.7 to estimate false discovery rates and q-values. Use the output to address the following.

(i) How well does the estimated value $\hat{\pi}_0$ approximate the true value π_0?
(ii) For a p value threshold of $t = 0.01$, how well does the estimated FDR $\hat{FDR}$ approximate the true FDR?
(iii) Create a histogram of p values and discuss its implications for the estimation of the FDR. Hint: Does the p value distribution look like a random sample from a uniform distribution? What do you think of the choice of $\lambda = 0.05$?
(iv) Create a scatter plot between -log10(p value) and -log10(q-value).
(v) Show that for any threshold for the q-value, one can find a threshold for the p value so that the resulting lists of significant genes are identical. Hint: Compute a C-index to argue that one measure threshold implies the other.

4. Exercise. Discuss whether it makes sense to estimate the false discovery rate of a set of significant variables if a total of $n = 10$ hypothesis tests have been carried out. Hint: Would the estimate of the FDR be accurate?
5. Exercise regarding q-values. Assume a large (say $n = 50,000$) vector of p values $pvalueVector = (p_1, \ldots, p_n)$ which follow a Beta distribution with shape parameters $shape1 = 1/\gamma$ and $shape2 = 1$. Next we use the default settings of the R function `qvalue` to compute a corresponding vector of q-values $qvalueVector = (q_1, \ldots, q_n)$. Show numerically that $qvalueVector = \frac{pvalueVector^{(\gamma-1)/\gamma}}{\gamma}$ (10.28). Hint: For different choices of $\gamma > 1$, apply the R code in Sect. 10.8.
6. Exercise regarding the dependence of q-values on the tuning parameter λ. For the numeric example presented in Sect. 10.7, determine how the q-value estimates change if λ is set to 0.2 and 0.5. Hint: Determine the minimum values and the medium values of the resulting q-values. Further, calculate correlations between the vectors of q-values.
7. Exercise regarding the threshold equivalence of linear significance measures.

   (i) Prove that $OS_i^{abs.cor} = |cor(x_i, y)|$ threshold-implies all other linear significance measures mentioned in Sect. 10.12. Hint: Since m and n are fixed, it is straightforward to show that each linear significance measure is a nondecreasing function of $OS^{abs.cor}$.
   (ii) Which measures are threshold-equivalent to $OS^{abs.cor}$? Hint: Which measures are an increasing function of $OS^{abs.cor}$?

8. Exercise regarding unweighted linear networks (Sect. 10.16). Assume n numeric vectors that form the columns of the matrix $datX$. An unweighted linear network adjacency matrix A (5.22) is defined by (hard) thresholding a linear significance measure. In Sect. 10.12, we describe several measures for measuring the linear relationships between a pair of vectors $x = x_i$ and $y = x_j$. For example, consider the matrix $M^{(1)}$ whose i, jth element equals the absolute value of the Pearson correlation
$$M_{ij}^{abs.cor} = |cor(x_i, x_j)|.$$
Show that the matrix $M^{(abs.cor)}$ threshold-implies the following matrices: $M_{ij}^{abs.Tstatistic} = |Z_{ij}^{Student}|$, $M_{ij}^{pvalue} = -log10(p_{ij}^{Student})$, $M_{ij}^{Bonferroni} = -log10(p_{ij}^{Bonferroni})$, $M_{ij}^{qvalue} = -log10(q_{ij})$, M_{ij}^{lm}, etc. Hint: Study vectorized versions of the matrices. This exercise is analogous to the previous exercise. Message: This shows that an unweighted network based on thresholding any of the linear significance measures can be arrived at by thresholding the absolute Pearson correlation matrix $M_{ij}^{abs.cor} = |cor(x_i, x_j)|$.
9. Exercise regarding the threshold-equivalence of two significance measures $OS^{(1)}$ and $OS^{(2)}$.

   (i) Show that rank equivalence implies threshold equivalence (as defined in (10.30)). Hint: Show that $rank(OS^{(1)}) = rank(OS^{(2)})$ (where ties are broken by averaging their ranks) implies

$$C.index(OS^{(1)}, OS^{(2)} = C.index(OS^{(2)}, OS^{(1)} = 1.$$

This follows from the fact that the C-index is invariant with regard to monotonically increasing transformations, i.e.,

$$C.index(OS^{(1)}, OS^{(2)} = C.index(rank(OS^{(1)}), rank(OS^{(2)}).$$

References

Benjamini Y, Hochberg Y (1995) Controlling the false discovery rate: A practical and powerful approach to multiple testing. J R Stat Soc Ser B 57:289–300

Hawkins DL (1989) Using U statistics to derive the asymptotic distribution of Fisher's Z statistic. Am Stat 43(4):235–237

Horvath S, Zhang B, Carlson M, Lu KV, Zhu S, Felciano RM, Laurance MF, Zhao W, Shu Q, Lee Y, Scheck AC, Liau LM, Wu H, Geschwind DH, Febbo PG, Kornblum HI, Cloughesy TF, Nelson SF, Mischel PS (2006) Analysis of oncogenic signaling networks in glioblastoma identifies ASPM as a novel molecular target. Proc Natl Acad Sci USA 103(46):17402–17407

Li A, Horvath S (2007) Network neighborhood analysis with the multi-node topological overlap measure. Bioinformatics 23(2):222–231

Sokal RR, Rohlf FJ (1981) Biometry: The principles and practice of statistics in biological research, 3rd edn. WH Freeman, New York

Storey JD (2002) A direct approach to false discovery rates. J R Stat Soc Ser B 64:479–498

Storey JD, Tibshirani R (2003) Statistical significance for genomewide studies. Proc Natl Acad Sci USA 100(16):9440–9445

Chapter 11
Structural Equation Models and Directed Networks

Abstract Undirected networks (encoded in symmetric adjacency matrices) cannot be used to describe causal relationships between random variables. Instead, causal information is encoded by directed networks where the arrow $A \to B$ indicates that variable A causally influences variable B. We refer to the process of assigning a causal direction to edges in an association network as "edge orienting". We review structural equation model (SEM)-based approaches for constructing directed networks between random variables. SEMs lead to predictions about the variance–covariance matrices of the observed variables, which is why they are also known as covariance structure models. We review likelihood-based approaches for evaluating the fit of a structural equation model. We provide a short review of SEMs and show how these techniques can be used for defining directed networks. In particular, we describe how local structural equations based on causal anchors can be used to infer causal networks among variables. Causal networks have been used in systems genetics applications for inferring causal relationships based on genetic markers. The network edge orienting (NEO) R software and method can be used to orient the edges of correlation networks (aka. quantitative trait networks) if the edges can be anchored to causal anchors (e.g., genetic polymorphisms). This section reviews and extends work with **Jason Aten** and **Jake Lusis**.

11.1 Testing Causal Models Using Likelihood Ratio Tests

Structural equation modeling descends from **Sewall Wright**'s **path analysis** and is a generalization of multivariate linear regression analysis. SEMs have become a widely used tool to explore the causal relationship between multiple variables (Kline 2005; Loehlin 2004; Pearl 2000; Fox 2006, 1984; Shipley 2000a). Excellent text books and literature exist on SEMs (Shipley 2000a). **John Fox** created the sem R library (Fox 2006, 1984), which will be described below.

SEMs are multiple-equation regression models representing putative causal (and hence structural) relationships among a number of variables, some of which may affect one another mutually. The response variable in one structural equation can appear as an explanatory variable in another structural equation. Two variables in an

S. Horvath, *Weighted Network Analysis: Applications in Genomics and Systems Biology*, DOI 10.1007/978-1-4419-8819-5_11, © Springer Science+Business Media, LLC 2011

SEM can even affect one-another reciprocally, either directly or indirectly through a feedback loop. SEMs can include variables that are not measured directly. Unmeasured variables are sometimes called **latent variables** or **factors**.

SEMs focus on the pattern of covariation among variables, which is why they are also known as covariance structure models. An SEM model involves several parameters, e.g., regression coefficients and variances. The set of parameters of an SEM model is denoted by θ.

SEM analysis typically starts with variables centered on their means and focuses on the covariance relationships. Traits or nodes are connected by arrows denoting causal relationships. The causal relationships define a system of linear regression models where the parents of a node are used to predict the child node's response. The system of resulting linear equations imposes constraints on the structure of the expected covariance matrix. Given the observed variables, we denote the observed sample covariance matrix by S and the expected covariance matrix under the causal model by $\Sigma(\theta)$, where θ denotes the set of model parameters.

The model-based estimates of the covariance parameters between observed variables is calculated by minimizing the difference between observed and predicted covariances. More precisely, maximum likelihood estimates of the parameters can be obtained by maximizing a multivariate normal likelihood.

The fit of two nested models can be compared using likelihood ratio tests. In the following sections, we describe the following steps for evaluating the fit of a causal model:

1. Depicting causal relationships in a path diagram
2. Translating the path diagram into a set of structural equations
3. Deriving model-based predictions of covariances between each pair of observed variables
4. Calculating maximum likelihood estimates of the model parameters
5. Evaluating the causal fit using likelihood ratio tests and model fitting statistics

We provide more details on each of these steps in the following sections.

11.1.1 Depicting Causal Relationships in a Path Diagram

Path diagrams are used as intuitive representations of causal relationships among variables. The nodes in a path diagram correspond to random variables and arrows encode causal relationships. For example, the arrow in $X \rightarrow Y$ indicates that random variable X has a direct causal effect on variable Y. A causal diagram contains two types of arrows. A unidirectional arrow from a variable X to another variable Y indicates that X has a direct causal effect on Y. A unidirectional arrow corresponds to a coefficient parameter in the corresponding structural equation, e.g., the arrow in $X \rightarrow Y$ corresponds to the coefficient parameter β_{YX}. In contrast, a bidirectional (double-headed) arrow corresponds to a free covariance parameter, e.g., the bidirectional arrow in $X \leftrightarrow Y$ corresponds to the covariance parameter σ_{XY}. Bidirectional arrow indicates unknown causal relationships between the two variables.

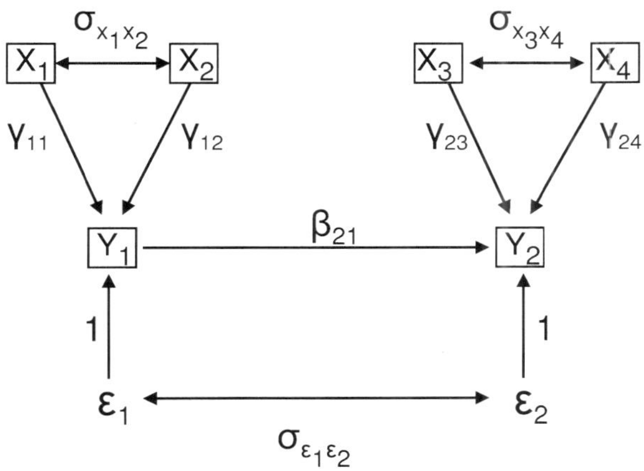

Fig. 11.1 Path diagram of causal model. There are four exogenous variables $(X_1,\ldots,X_4)$, two endogenous variables (Y_1,Y_2), and two error variables $\varepsilon_1,\varepsilon_2$. Observed variables are enclosed in squares. A unidirectional arrow between an exogenous and an endogenous variable corresponds to a regression coefficient γ. A unidirectional arrow between two endogenous variables corresponds to a coefficient denoted by β. The arrows between error variables and endogenous variables correspond to coefficients that have been set to 1

For example, Fig. 11.1 represents the causal relationships among different types of variables. Path diagrams can contain the following three types of variables:

- **Observed random variables** are enclosed in squares. We typically denote observed variables by $X_1,X_2,\ldots$ and $Y_1,Y_2,\ldots$.
- **Latent variables** are enclosed in circles. Latent variables are unobserved variables that are hypothesized to have a causal effect on some of the other variables. Sometimes we denote latent (hidden) variables by $H_1,H_2,\ldots$.
- **Error variables** (also known as disturbances) are not enclosed at all. Error variables represent measurement error and unmodeled causes affecting the variable into which it points. We usually denote error variables by $\varepsilon_1,\varepsilon_2,\ldots$.

Another classification of the random variables results from the causal model encoded in the path diagram:

- **Endogenous variables** have at least one causal parent. Each endogenous variable is a response variable in a corresponding structural equation. We assume that the SEM has p endogenous (i.e., response) variables denoted by $Y_1,\ldots,Y_p$. Each endogenous variable forms the outcome variable of a regression equation, which is referred to as **structural equation**. But an endogenous variable may also appear as an explanatory variable in other structural equations.
- **Exogenous variables** have no causal parent(s). Exogenous variables appear only as explanatory variables in the structural equations. The values of exogenous variables are therefore determined outside of the model (hence the term). Like the explanatory variables in a linear model, exogenous variables are assumed to be measured without error (but measurement error can be modeled using

latent-variable models). Exogenous variables can be categorical or ordinal. Assume that there are q exogenous variables denoted by X_1, X_q and respective random samples given by $x_1, \ldots x_q$.

- **Structural error variables** (aka. disturbances) represent the aggregated omitted causes affecting the endogenous variables, along with measurement error (and possibly intrinsic randomness) in the endogenous variables. There is one structural error variable for each endogenous variable (and hence for each structural equation). As in linear regression models, we will usually assume that the errors are normally distributed. Since we assume a model with p endogenous variables, there will be p error variables (and vectors) denoted by $\varepsilon_1, \ldots, \varepsilon_p$.

We assume that a random sample of m observations is available for each random variable. For example, random variable Y_1 leads to a vector y_1 (note the small cap) whose uth component is given by y_{1u}. In the following, we will often be sloppy and not distinguish a random variable from its corresponding random sample, e.g., we will not distinguish Y_1 from y_1.

11.1.2 Path Diagram as Set of Structural Equations

Each path diagram can be translated into a set of linear regression equations (referred to as structural equations). Each endogenous variable corresponds to exactly one structural equation, i.e., we assume p structural equations. Structural equations contain two types of regression coefficients (aka. **structural coefficients**), which will be referred to as **gamma coefficients** and **beta coefficients**. Gamma coefficients represent the effects of an exogenous variable on an endogenous variable, e.g., γ_{ij} represents the effect of exogenous variable x_j on endogenous variable y_i. Beta coefficients represent the effects of an endogenous variable on another endogenous variable, e.g., β_{ij} represents the effect of endogenous variable y_j on endogenous variable y_i. For example, Fig. 11.1 translates to

$$\begin{aligned} y_{1u} &= \gamma_{10} + \gamma_{11} x_{1u} + \gamma_{12} x_{2u} + 1\varepsilon_{1u} \\ y_{2u} &= \gamma_{20} + \beta_{21} y_{1u} + \gamma_{23} x_{3u} + \gamma_{24} x_{4u} + 1\varepsilon_{2u}, \end{aligned} \tag{11.1}$$

where the u indexes the uth observation. Note that the coefficient for the error terms have been set to 1 so that the parameters in the model are identifiable. It is convenient to simplify the structural equations by expressing all x's and y's as deviations from the population means (or from their sample means). This gets rid of the constant (intercept) terms γ_{10} and γ_{20}. Thus, (11.1) can be simplified to

$$\begin{aligned} y_{1u} &= \gamma_{11} x_{1u} + \gamma_{12} x_{2u} + \varepsilon_{1u} \\ y_{2u} &= \beta_{21} y_{1u} + \gamma_{23} x_{3u} + \gamma_{24} x_{4u} + \varepsilon_{2u}. \end{aligned} \tag{11.2}$$

It can be convenient to cast structural equations into matrix form. To illustrate this, (11.2) is rewritten as follows:

$$y_{1u} = 0y_{1u} + 0y_{2u} + \gamma_{11}x_{1u} + \gamma_{12}x_{2u} + 0x_{3u} + 0x_{4u} + \varepsilon_{1u}$$
$$y_{2u} = \beta_{21}y_1 + 0y_{2u} + 0x_{1u} + 0x_{2u} + \gamma_{23}x_{3u} + \gamma_{24}x_{4u} + \varepsilon_{2u}. \quad (11.3)$$

After defining the vectors ($\mathbf{y}_u = (y_{1u}, y_{2u})$, $\mathbf{x}_u = (x_{1u}, x_{2u}, x_{3u}, x_{4u})$, $\varepsilon_u = (\varepsilon_{1u}, \varepsilon_{2u})$ note the boldface for vectors), (11.3) can be expressed as follows:

$$\mathbf{y}_u = B\mathbf{y}_u + \Gamma\mathbf{x}_u + \varepsilon_u, \quad (11.4)$$

where

$$B = \begin{pmatrix} 0 & 0 \\ \beta_{21} & 0 \end{pmatrix}$$
$$\Gamma = \begin{pmatrix} \gamma_{11} & \gamma_{12} & 0 & 0 \\ 0 & 0 & \gamma_{23} & \gamma_{24} \end{pmatrix}. \quad (11.5)$$

Equation (11.4) can be recast as follows:

$$\mathbf{y}_u = (I - B)^{-1}\Gamma\mathbf{x}_u + (I - B)^{-1}\varepsilon_u. \quad (11.6)$$

The equation assumes that $(I - B)$ is invertible. If not, one could use a pseudo-inverse.

11.1.3 Deriving Model-Based Predictions of Covariances

The structural equations allow one to derive model-based predictions of the covariances between each pair of variables. The sem package in R automatically derives these predicted values. But in the following, we review the definition and properties of the covariances between vectors of random variables. Consider the column vector

$$\mathbf{X} = \begin{bmatrix} X_1 \\ . \\ . \\ . \\ X_q \end{bmatrix}$$

whose q components correspond to random variables. Then the **variance–covariance matrix** (aka. variance matrix) Σ_x is the matrix whose i,jth entry is defined as

$$\Sigma_{xij} = cov(X_i, X_j) = E((X_i - \mu_i)(X_j - \mu_j)),$$

where $\mu_i = E(X_i)$ denotes the expected value of the ith random variable. We typically assume centered variables (i.e., $\mu_i = 0$). Σ_x can be expressed in matrix form as follows:

$$\Sigma_x = cov(\mathbf{X}, \mathbf{X}) = E((\mathbf{X} - E(\mathbf{X}))(\mathbf{X} - E(\mathbf{X}))^{\tau}),$$

where $^\tau$ denotes the transpose. The covariance matrix between random variables satisfies the following properties:

1. $cov(X,Y) = E(\mathbf{X}\mathbf{Y}^\tau) - E(\mathbf{X})E(\mathbf{Y})^\tau$
2. $var(X) = E(\mathbf{X}\mathbf{X}^\tau) - E(\mathbf{X})E(\mathbf{X})^\tau$
3. $var(M\mathbf{X}+\mathbf{c}) = Mvar(\mathbf{X})M^\tau$ if M is a numeric matrix with q columns and c is a numeric vector of length q
4. $cov(\mathbf{X},\mathbf{Y}) = cov(\mathbf{Y},\mathbf{X})^\tau$ if $\mathbf{Y}$ denotes a vector whose p entries correspond to random variables
5. If $q = p$ then $var(\mathbf{X}+\mathbf{Y}) = cov(\mathbf{X}+\mathbf{Y},\mathbf{X}+\mathbf{Y}) = var(\mathbf{X}) + cov(\mathbf{X},\mathbf{Y}) + cov(\mathbf{Y},\mathbf{X}) + var(\mathbf{Y})$
6. If X and Y are independent, then $cov(\mathbf{X},\mathbf{Y}) = 0$
7. $cov(M\mathbf{X}+\mathbf{c}, N\mathbf{Y}+\mathbf{d}) = Mcov(\mathbf{X},\mathbf{Y})N^\tau$ if N denotes a matrix with p columns and d denotes a numeric vector

These properties can be used to derive SEM-based predictions of the covariance matrix between the variables.

Two exogenous variables may be correlated, e.g., the covariance $\sigma_{X_1X_2} = Cov(X_1,X_2)$ may represent a model parameter that needs to be estimated from the data. The variance of X_1 is denoted as $\sigma^2_{X_1}$ or as $\sigma_{X_1X_1} = Cov(X_1,X_1)$.

The error variables are assumed to have zero expectations and to be independent of (or at least to have zero covariance with) the exogenous variables X_i, i.e., we assume that the covariance between exogenous variables and structural errors is zero:

$$cov(X_i,\varepsilon_j) = 0. \tag{11.7}$$

The errors for different observations are assumed to be independent of one another. For example, we assume that the error variable ε follows a normal distribution $N(0,\sigma^2_\varepsilon)$ whose variance σ^2_ε is a model parameter that needs to be estimated from the data. Depending on the causal model, different error variables may be correlated, e.g., the covariance $\sigma_{\varepsilon_1\varepsilon_2} = cov(\varepsilon_1,\varepsilon_2)$ between arrow variables ε_1 and ε_2 could represent a model parameter that needs to be estimated from the data.

The path diagram in Fig. 11.1 implies that $cov(\varepsilon_{1u},\varepsilon_{2u}) = \sigma_{\varepsilon_1,\varepsilon_2}$. The variance–covariance matrix $\Sigma_{\varepsilon\varepsilon}$ of the error vector ε_u can be expressed as follows:

$$\Sigma_{\varepsilon\varepsilon} = var(\varepsilon_u) = cov(\varepsilon_u,\varepsilon_u^\tau) = \begin{pmatrix} \sigma^2_{\varepsilon_1} & \sigma_{\varepsilon_1\varepsilon_2} \\ \sigma_{\varepsilon_1\varepsilon_2} & \sigma^2_{\varepsilon_2} \end{pmatrix}. \tag{11.8}$$

Further, the path diagram in Fig. 11.1 implies $cov(x_{1u},x_{2u}) = \sigma_{x_1x_2}$, $cov(x_{3u},x_{4u}) = \sigma_{x_3x_4}$, and that all other covariances between exogenous variables are zero. Thus, the variance–covariance matrix of $\mathbf{x}_u$ is given by

$$\Sigma_{xx} = var(\mathbf{X}) = cov(\mathbf{X},\mathbf{X}) = \begin{pmatrix} \sigma^2_{x_1} & \sigma_{x_1x_2} & 0 & 0 \\ \sigma_{x_1x_2} & \sigma^2_{x_2} & 0 & 0 \\ 0 & 0 & \sigma^2_{x_3} & \sigma_{x_3x_4} \\ 0 & 0 & \sigma_{x_3x_4} & \sigma^2_{x_4} \end{pmatrix}. \tag{11.9}$$

In an exercise, you are asked to use covariance algebra to derive the covariance Σ_{yx} between the vector of endogenous variables $\mathbf{Y}$ and the vector of exogenous variables $\mathbf{X} = (X_1, \ldots, X_q)$:

$$\Sigma_{yx} = cov(\mathbf{Y}, \mathbf{X}) = (I - B)^{-1} \Gamma \Sigma_{xx} \tag{11.10}$$

Since the exogenous variables X_i are assumed to have vanishing covariances with the error variables ε_j (11.7), we find that

$$cov(\mathbf{X}, \varepsilon) = 0. \tag{11.11}$$

Covariance algebra applied to (11.6) implies the following model-based predictions of the covariance matrix:

$$\Sigma_{yy} = var(\mathbf{Y}) = (I - B)^{-1} \Gamma \Sigma_{xx} \Gamma^{\tau} (I - B)^{-1\tau} + (I - B)^{-1} \Sigma_{\varepsilon\varepsilon} (I - B)^{-1\tau}. \tag{11.12}$$

Note that the variance matrices Σ_{xx}, Σ_{xy}, and Σ_{yy} depend on parameters. The SEM model depicted in Fig. 11.1 contains the following $t = 14$ parameters: β_{21}, γ_{11}, γ_{12}, γ_{23}, γ_{24}, $\sigma^2_{x_1}, \ldots, \sigma^2_{x_4}$, $\sigma_{x_1 x_2}$, $\sigma_{x_3 x_4}$, $\sigma^2_{\varepsilon_1}, \sigma^2_{\varepsilon_2}, \sigma_{\varepsilon_1, \varepsilon_2}$. We follow the usual convention of denoting the set of free model parameters by θ. To highlight the dependence of the variance matrices on the model parameters, we sometimes write $\Sigma_{xx}(\theta)$, $\Sigma_{xy}(\theta)$, $\Sigma_{yy}(\theta)$, etc.

Consider the vector $\mathbf{V}_{obs} = (\mathbf{X}, \mathbf{Y})$ of length $q + p$ whose components correspond to the observed variables. Note that the first q components of $\mathbf{V}_{obs}$ contain the exogenous variables and the last p components contain the endogenous variables. The SEM predicts the following structure for the variance–covariance matrix:

$$\Sigma(\theta) = var(\mathbf{V}_{obs}) = \begin{pmatrix} \Sigma_{xx}(\theta) & \Sigma_{yx}(\theta)^{\tau} \\ \Sigma_{yx}(\theta) & \Sigma_{yy}(\theta) \end{pmatrix}. \tag{11.13}$$

Note that $\Sigma(\theta)$ is symmetrical since $\Sigma_{yx}(\theta)^{\tau} = \Sigma_{yx}$.

11.1.4 Maximum Likelihood Estimates of Model Parameters

Here we describe two approaches for estimating variance–covariance matrices based on random samples of the variables. First, the sample estimates ignore the SEM. Second, the maximum likelihood approach uses the SEM.

Given a random sample $\mathbf{x}_1, \ldots, \mathbf{x}_m$ from the q-dimensional random variable X, the sample mean is a vector

$$\hat{\mu}_x = \frac{1}{m} \sum_{u=1}^{m} \mathbf{x}_u,$$

whose ith component is given $\hat{\mu}_{xi} = \frac{1}{m}\sum_{u=1}^{m} x_{iu}$. The sample covariance matrix S_x is a $q \times q$ dimensional symmetric matrix whose i,jth element is given by $S_{xij} = \frac{1}{m-1}\sum_u (x_{iu} - \hat{\mu}_{xi})(x_{ju} - \hat{\mu}_{xj})$. In matrix form,

$$S_{xx} = \frac{1}{m-1}\sum_u (\mathbf{x}_y - \hat{\mu}_x)(\mathbf{x}_u - \hat{\mu}_x)^{\tau}.$$

The sample mean and the sample covariance matrix are unbiased estimates of the mean and the covariance matrix of the random variable X and ignore any relationship specified by the SEM. Analogously, the sample estimate of the variance–covariance matrix between random vectors $\mathbf{X}$ and $\mathbf{Y}$ is defined as follows: $S_{yx} = \frac{1}{m-1}\sum_u (\mathbf{y}_u - \hat{\mu}_y)(\mathbf{x}_u - \hat{\mu}_x)^{\tau}$. Random samples of the observed variables $\mathbf{V}_{obs} = (\mathbf{X}, \mathbf{Y})$ can be used to define the following sample-based estimate of the variance–covariance matrix

$$S = S_{v_{obs},v_{obs}} = \begin{pmatrix} S_{xx} & S_{yx}^{\tau} \\ S_{yx} & S_{yy} \end{pmatrix}. \tag{11.14}$$

Each SEM makes predictions about the variance–covariance matrix $\Sigma(\theta)$ (11.13) among the observed exogenous and endogenous variables. SEMs typically constrain many of the elements of the variance–covariance matrix, e.g., constraints may specify that certain parameters are 0. If the constraints are sufficient for identifying all parameters, then maximum likelihood estimation (MLE) method can be used to estimate the parameters from the data. Since we assume normally distributed variables, the log-likelihood of the SEM is given by

$$log(L(\theta)) = -\frac{m}{2}\left[(q+p)log(2\pi) + log(det(\Sigma(\theta))) + trace(S\Sigma(\theta)^{-1})\right], \tag{11.15}$$

where log denotes the logarithm with base e, $\Sigma(\theta)$ (11.13) denotes the model-based variance matrix, S (11.14) denotes the sample variance matrix, and $det()$ denotes the matrix determinant. Since the determinant function satisfies the following property $det(2\pi\Sigma) = (2\pi)^{no.rows(\Sigma)}\Sigma$, we can simplify the log-likelihood to

$$log(L(\theta)) = -\frac{m}{2}\left[log(det(2\pi\Sigma(\theta))) + trace(S\Sigma(\theta)^{-1})\right]. \tag{11.16}$$

The log-likelihood can be interpreted as a similarity measure between the sample covariance matrix S and the model-based matrix $\Sigma(\theta)$. The MLE estimates $\hat{\theta}$ are the values of the parameters that maximize the log-likelihood function under the constraints placed on the model, e.g., that certain entries of B, Γ, and possibly $\Sigma_{\varepsilon\varepsilon}$ are 0. The entire MLE methodology is available, e.g., alternative nested models can be compared by a likelihood ratio test (LRT) as will be described in the following.

11.1.5 Model Fitting *p* Value and Likelihood Ratio Tests

A structural equation with fewer than q, equal to q, or more than q structural parameters is called over-identified, just-identified, or under-identified, respectively. If the structural equation (and the corresponding SEM) is over-identified then we have more than enough information to estimate the parameters. The overidentifying constraints of an overidentified SEM model can be tested by contrasting the maximized log-likelihood with the log-likelihood of a **just-identified model** whose model matrix is identical to the observed sample variance–covariance matrix S. The (maximized) log-likelihood of a just-identified is given by

$$log(L_{justidenfied}) = -\frac{m}{2}\left[log(det(2\pi S)) + no.rows(S)\right], \tag{11.17}$$

where $no.rows(S) = q + p$ denotes the number of rows of S. Denoting the maximized log-likelihood for the over-identified model by

$$log(L_{overidentified}) = log(L(\hat{\theta})),$$

the LRT statistic is twice the difference between the log-likelihoods of the two models, i.e.,

$$\begin{aligned} LRT(\hat{\theta}) &= 2\left[log(L_{justidenfied}) - log(L_{overidenfied})\right] \\ &= m\left[log\left(\frac{det(\Sigma(\hat{\theta}))}{det(S)}\right) + trace(S\Sigma(\hat{\theta})^{-1}) - no.rows(S)\right]. \end{aligned} \tag{11.18}$$

A computationally convenient way of calculating the trace results from the following equality is

$$trace(S\Sigma(\hat{\theta})^{-1}) = sum(S * (\Sigma(\hat{\theta})^{-1})^{\tau}),$$

where $*$ denotes the component-wise matrix product and *sum* denotes the sum over all matrix components. Under the null hypothesis that the over-identified model is correct, the LRT statistic χ^2 is distributed as a chi-square statistic with degrees of freedom, df, given by the difference between the number of parameters of the just-identified model $\left(\frac{(q+p)(q+p+1)}{2}\right)$ and the number of parameters in the overidentified model (t), i.e.,

$$df = \frac{(q+p)(q+p+1)}{2} - t. \tag{11.19}$$

For the overidentified SEM depicted in Fig. 11.1, we find $t = 14$, $p = 2$, and $q = 4$ imply that $df = 7$.

11.1.6 Model Fitting Chi-Square Statistics and LRT

Many books (and the `sem` R package) make use of a slightly different notation for calculating model fitting p values. To facilitate reader-friendliness, we briefly review this notation.

To arrive at a maximum likelihood estimate of the model covariance matrix $\hat{\Sigma} = \Sigma(\hat{\theta})$, the following fitting criterion is minimized:

$$FittingCriterion(\theta) = log\left(\frac{det(\Sigma(\theta))}{det(S)}\right) + trace(S\Sigma(\theta)^{-1}) - no.rows(S), \quad (11.20)$$

where the *trace* function denotes the sum of the diagonal elements of a square matrix, and $no.rows(S) = q + p$ denotes the number of rows of S. Note that the fitting criterion is twice the KL pre-dissimilarity $PreDissim_{KL}(S, \Sigma(\theta))$ (7.25) between S and $\Sigma(\theta)$. As a matter of fact, this relationship motivated us to propose the KL pre-dissimilarity as a difference measure between positive definite matrices.

The parameter estimate $\hat{\theta}$ is defined as the minimizing solution of the objective function $FittingCriterion(\theta)$. The (minimized) fitting criterion is closely related to the LRT statistic (11.18) as can be seen from the following relationship:

$$FittingCriterion(\hat{\theta}) = \frac{LRT.statistic(\hat{\theta})}{m}.$$

The SEM model chi-square statistic is defined as follows:

$$FittingChisquare(\hat{\theta}) = (m-1)FittingCriterion(\hat{\theta}). \quad (11.21)$$

Note that the following relationship exists:

$$FittingChisquare = LRT.statistic\frac{m-1}{m}. \quad (11.22)$$

The null hypothesis states that the expected covariance matrix equals that of the underlying causal model. In large samples and assuming multivariate normality, *FittingChisquare* is distributed as a Pearson chi-square statistic. This statistic is known as the model chi-square or generalized likelihood ratio statistic. If $FittingChisquare = 0$, the causal model perfectly fits the data. If the causal model is correct, then *FittingChisquare* asymptotically follows a central chi-square distribution

$$FittingChisquare \sim \chi^2(df), \quad (11.23)$$

where $df = \frac{(q+p)(q+p+1)}{2} - t$ (11.19). Note that *FittingChisquare* (11.21 and 11.23) can be used to compute a model p value for each causal model.

The model chi-square statistic *FittingChisquare* (11.21 and 11.23) can be used to assess how well the data fit the hypothesized causal model. For example, $P(X \to Y1 \to Y2) = P(data|X \to Y1 \to Y2)$ denotes the p value for the model in which M causally affects variable A which in turn affects variable B. *FittingChisquare* tests the null hypothesis that the model is correct. A small model p value (say $p < 0.05$) indicates that the causal model does not fit well. Following the logic of an 'accept-support' context (Steiger and Fouladi 1997; Kline 2005) where the null hypothesis represents the researchers belief, it is the failure to reject the null hypothesis that supports the causal model. A *high p* value indicates that the causal model fits the data.

11.2 R Code for Evaluating an SEM Model

Here we present R code for simulating data corresponding to the model specified by the path diagram in Fig. 11.1.

```
# specification of the model parameters
beta21=3
gamma11=1;gamma12=1.5;gamma23=2.5;gamma24=3
sigma.x1x1=1;sigma.x2x2=2;sigma.x3x3=3;sigma.x4x4=4
sigma.x1x2=.5;sigma.x3x4=1
sigma.eps1eps1=1; sigma.eps2eps2=2
sigma.eps1eps2=.5
#let us check the validity of the parameter values:
if (sigma.x2x2-(sigma.x1x2/sigma.x1x1)^2<0){
print("sigma.x1x2 is too large")}
if (sigma.x4x4-(sigma.x3x4/sigma.x3x3)^2<0){
print("sigma.x3x4 is too large")}
if (sigma.eps2eps2-(sigma.eps1eps2/sigma.eps1eps1)^2<0){
print("sigma.eps1eps2 too large")}

m=150 # number of observations
set.seed(1) # seed for the random number generator
x1=rnorm(m,sd=sqrt(sigma.x1x1))
sig.ratio=sigma.x1x2/sigma.x1x1
x2=sig.ratio*x1+rnorm(m,sd=sqrt(sigma.x2x2-sig.ratio^2))
x3=rnorm(m,sd=sqrt(sigma.x3x3))
sig.ratio=sigma.x3x4/sigma.x3x3
x4=sig.ratio*x3+rnorm(m,sd=sqrt(sigma.x4x4-sig.ratio^2))
eps1=rnorm(m,sd=sqrt(sigma.eps1eps1))
sig.ratio=sigma.eps1eps2/sigma.eps1eps1
eps2=sig.ratio*eps1+rnorm(m,sd=sqrt(sigma.eps2eps2-sig.ratio^2))
y1=gamma11*x1+gamma12*x2+eps1
y2=beta21*y1+gamma23*x3+gamma24*x4+eps2
# data frame with exogenous variables
datX=data.frame(x1,x2,x3,x4)
datY=data.frame(y1,y2) # endogenous variables
p=dim(datY)[[2]] # number of endogeous variables
q=dim(datX)[[2]] # number of exogenous variables
dat.observedVariables=data.frame(datX,datY)
#sample var-covariance matrix S
S=cov(dat.observedVariables)
```

The `sem` function in the `sem` R package (created by **John Fox**) may be used to fit a wide variety of causal models. The function has the following three required arguments:

- `ram`, which is the reticular-action model (below) formulation of the causal diagram
- `S`, which is the sample variance–covariance matrix among the observed variables
- `N` which corresponds to our parameter m, i.e., the number of observation on which the variance matrix is based

Since the R code relies on the **reticular-action model** (RAM) formulation of a causal model, we will briefly describe this formulation. The RAM approach is

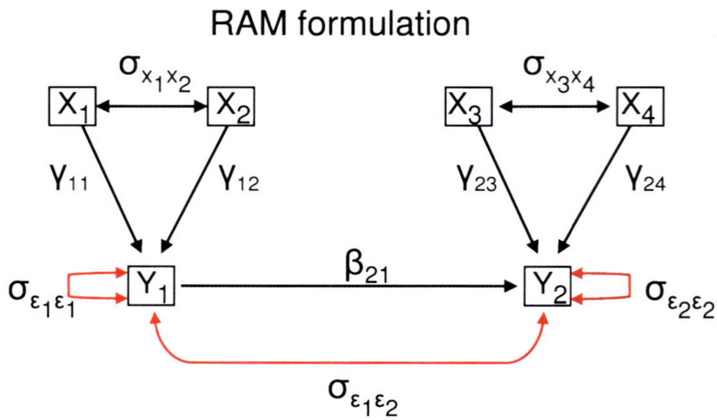

Fig. 11.2 RAM formulation of the path diagram reported in Fig. 11.1. Note that the (*red*) *double-headed arrows* encode variance and covariance parameters of the error terms

a convenient form for specifying and estimating SEMs. Each error variance and covariance is represented by a bidirectional arrow $\leftrightarrow$. Each structural coefficient is represented as a unidirectional arrow. For example, Fig. 11.2 represents the RAM formulation of the path diagram depicted in Fig. 11.2. Based on the RAM formulation, the causal model can be specified using the following R code which specifies each arrow, and assigns a name and possibly start value for the parameter. If `NA` is chosen as start value, the program automatically chooses the start value.

```
# load the sem library
library(sem)
CausalModelRAM=specify.model()
x1 -> y1, gamma11, NA
x2 -> y1, gamma12, NA
x3 -> y2, gamma23, NA
x4 -> y2, gamma24, NA
y1 -> y2, beta21, NA
x1 <-> x1, sigma.x1x1, NA
x2 <-> x2, sigma.x2x2, NA
x1 <-> x2, sigma.x1x2, NA
x3 <-> x3, sigma.x3x3, NA
x4 <-> x4, sigma.x4x4, NA
x3 <-> x4, sigma.x3x4, NA
y1 <-> y1, sigma.eps1eps1, NA
y2 <-> y2, sigma.eps2eps2, NA
y1 <-> y2, sigma.eps1eps2, NA

#Now we present R code for fitting an sem model
# and constructing a likelihood ratio test for evaluating its fit
semObject =sem(CausalModelRAM,S=S,N=m,maxiter=1000)
```

The following excerpt from the R output summarizes the sem model fit

```
> summary(semObject)

 Model Chisquare =  1.5385   Df =  7 Pr(>Chisq) = 0.98095
 Chisquare (null model) =  941.56   Df =  15
```

```
 Goodness-of-fit index =  0.9966
 Adjusted goodness-of-fit index =  0.98982
 RMSEA index =  0    90% CI: (NA, NA)
 Bentler-Bonnett NFI =  0.99837
 Tucker-Lewis NNFI =  1.0126
 Bentler CFI =  1
 SRMR =  0.022332
 BIC =  -33.536

 Normalized Residuals
     Min.    1st Qu.     Median       Mean    3rd Qu.       Max.
-9.15e-01 -1.88e-01 -6.96e-15 -1.12e-01  1.42e-02  1.85e-01

 Parameter Estimates
                 Estimate Std Error z value Pr(>|z|)
gamma11          1.05959  0.099051  10.6974 0.00000000 y1 <--- x1
gamma12          1.47564  0.065613  22.4903 0.00000000 y1 <--- x2
gamma23          2.47840  0.067153  36.9066 0.00000000 y2 <--- x3
gamma24          3.06989  0.055162  55.6519 0.00000000 y2 <--- x4
beta21           3.01463  0.048565  62.0742 0.00000000 y2 <--- y1
sigma.x1x1       0.81754  0.094741   8.6293 0.00000000 x1 <--> x1
sigma.x2x2       1.99479  0.231165   8.6293 0.00000000 x2 <--> x2
sigma.x1x2       0.36897  0.108898   3.3882 0.00070344 x2 <--> x1
sigma.x3x3       3.16646  0.366940   8.6294 0.00000000 x3 <--> x3
sigma.x4x4       4.70151  0.544828   8.6293 0.00000000 x4 <--> x4
sigma.x3x4       0.98262  0.326180   3.0125 0.00259091 x4 <--> x3
sigma.eps1eps1 1.22145  0.141564   8.6283 0.00000000 y1 <--> y1
sigma.eps2eps2 2.25345  0.266929   8.4421 0.00000000 y2 <--> y2
sigma.eps1eps2 0.56839  0.155560   3.6538 0.00025835 y2 <--> y1

 Iterations =  38
```

The `sem` summary output presents the maximum likelihood estimates (denoted by `estimate`) of the parameters. Note that in this case, the estimated parameters are close to the true values specified in the simulation model. The output also provides the standard errors, which are obtained from the inverse of the information matrix. As explained in many statistics text books, the information matrix is defined as the negative of the Hessian matrix of second-order partial derivatives of the log-likelihood evaluated at the parameter estimates. The ratio of each estimate to its standard error is known as the **Wald test statistic** (denoted `z value`) for testing the null hypothesis that the corresponding parameter is 0. An asymptotic p value can be calculated since for large m the Wald test statistic follows a standard normal distribution under the null hypothesis. In our case, all p values are zero, which indicates that each parameter is significantly different from 0. Note the highly significant error covariance corresponding to an error correlation of

$$\rho_{\varepsilon_1,\varepsilon_2} = \frac{0.56839}{\sqrt{1.22145 * 2.25345}} = 0.3425978.$$

At the top of the `sem` summary output, one can find several model fitting statistics: For example, the model fitting chi-square statistic (11.21) and the corresponding p value (denoted by `Pr(>Chisq)`). The `sem` function also outputs standard fitting indices that are implemented in the R package *sem* (Fox 2006) including the Root Mean Square Error of Approximation (RMSEA), Comparative Fit Index (CFI), Standardized Root Mean Square Residual (SRMSR), and BIC. Since a single fitting index reflects only a particular aspect of model fit, a favorable value of that index does not by itself demonstrate good model fit; it is important to assess the model fit based on multiple indices. We follow the following standard guidelines for interpreting these indices (Shipley 2000a; Loehlin 2004). Before calling an edge $A \rightarrow B$ causal, we recommend verifying that the corresponding causal model has a high model p value (say > 0.05), a low RMSEA score (say ≤ 0.05), a low SRMSR (say ≤ 0.10), a high CFI (say ≥ 0.90), and a significant Wald test p value (say $p \leq 0.05$). A more detailed explanation of these indices can be found in books on SEM models. Here we focus only on the model fitting chi-square statistic (11.21) and the corresponding model fitting p value, which can be obtained directly using the following R code:

```
ModelChisq=summary(semObject)$chisq
df= summary(semObject)$df
FittingtingPvalue=1-pchisq(ModelChisq,df)
ModelChisq # 1.538549
FittingtingPvalue #0.980945
```

For the remainder of this section, we will illustrate our matrix equations for calculating the model-based estimates of the variance–covariance matrix (11.9), (11.10), (11.12). This remainder is probably only interesting to readers who care about the calculations underlying the SEM model fit.

```
#maximum likelihood estimates of the model parameters
sigmaHat.x1x1=semObject$coeff[["sigma.x1x1"]]
sigmaHat.x2x2=semObject$coeff[["sigma.x2x2"]]
sigmaHat.x1x2=semObject$coeff[["sigma.x1x2"]]
sigmaHat.x3x3=semObject$coeff[["sigma.x3x3"]]
sigmaHat.x4x4=semObject$coeff[["sigma.x4x4"]]
sigmaHat.x3x4=semObject$coeff[["sigma.x3x4"]]
sigmaHat.eps1eps1=semObject$coeff[["sigma.eps1eps1"]]
sigmaHat.eps2eps2=semObject$coeff[["sigma.eps2eps2"]]
sigmaHat.eps1eps2=semObject$coeff[["sigma.eps1eps2"]]
betaHat21=semObject$coeff[["beta21"]]
gammaHat11=semObject$coeff[["gamma11"]]
gammaHat12=semObject$coeff[["gamma12"]]
gammaHat23=semObject$coeff[["gamma23"]]
gammaHat24=semObject$coeff[["gamma24"]]

# model based covariance matrix among exogenous variables
SigmaHat.xx=rbind(
c(sigmaHat.x1x1,sigmaHat.x1x2,0,0),
c(sigmaHat.x1x2,sigmaHat.x2x2,0,0),
c(0,0,sigmaHat.x3x3, sigmaHat.x3x4),
c(0,0,sigmaHat.x3x4,sigmaHat.x4x4))
```

```
# model based covariance among error variables
SigmaHat.eps=rbind(
c(sigmaHat.eps1eps1,sigmaHat.eps1eps2),
c(sigmaHat.eps1eps2,sigmaHat.eps2eps2))
# estimate of coefficient matrix B
BHat=rbind(
c(0,0),
c(betaHat21,0))

# estimate of coefficient matrix Gamma
GammaHat=rbind(
c(gammaHat11,gammaHat12,0,0),
c(0,0,gammaHat23,gammaHat24))
# to compute SigmaHat.yy, we define the following matrices

# identity matrix
I=diag(dim(datY)[[2]])
M1=solve(I-BHat)%*%GammaHat
M2=solve(I-BHat)
SigmaHat.yy=M1%*%SigmaHat.xx%*%t(M1)+M2%*%SigmaHat.eps%*%t(M2)
SigmaHat.yx=M1%*%SigmaHat.xx

# model-based covariance matrix for observed variables
SigmaHat=matrix(NA,nrow=q+p,ncol=q+p)
dimnames(SigmaHat)=dimnames(S)
SigmaHat[1:q,1:q]=SigmaHat.xx
SigmaHat[c((q+1):(q+p)),c((q+1):(q+p))]=SigmaHat.yy
SigmaHat[c((q+1):(q+p)),1:q]=SigmaHat.yx
SigmaHat[1:q,c((q+1):(q+p))]=t(SigmaHat.yx)
```

Let us now compare the model-based estimate of the variance–covariance matrix with the sample-based estimate.

```
> # model based estimate of the covariance matrix Sigma
> signif(SigmaHat,2)
     x1    x2    x3    x4   y1    y2
x1 0.82  0.37  0.00  0.00  1.4   4.3
x2 0.37  2.00  0.00  0.00  3.3  10.0
x3 0.00  0.00  3.20  0.98  0.0  11.0
x4 0.00  0.00  0.98  4.70  0.0  17.0
y1 1.40  3.30  0.00  0.00  7.6  24.0
y2 4.30 10.00 11.00 17.00 24.0 150.0

# sample covariance matrix S
> signif(S,2)
        x1     x2      x3     x4     y1    y2
x1  0.8200  0.370  0.0049 -0.150  1.400   3.8
x2  0.3700  2.000  0.0110  0.046  3.300  10.0
x3  0.0049  0.011  3.2000  0.980 -0.024  11.0
x4 -0.1500  0.046  0.9800  4.700 -0.180  16.0
y1  1.4000  3.300 -0.0240 -0.180  7.600  23.0
y2  3.8000 10.000 11.0000 16.000 23.000 150.0
```

Note that the model-based estimate $\Sigma(\hat{\theta})$ is very close to that of the sample covariance matrix S, which illustrates that the SEM provides an excellent fit to the data. In the following, we show how to calculate the model fitting criterion $FittingCriterion(\hat{\theta})$ (11.20), the fitting chi-square statistic (11.21), and the corresponding model fitting p value.

```
SigmaHatInverse=solve(SigmaHat)
no.rows.S=dim(S)[[1]] # same as q+p
FittingCriterion=
log(det(SigmaHat)/log(det(S)))+sum(S*t(SigmaHatInverse))-no.
   rows.S
# Fitting criterion can also be obtained from the sem object
FittingCriterion=semObject$criterion # 0.001101260
model.chisq=(m-1)*FittingCriterion
#number of unconstrained model parameters (count them)
t=14
# degrees of freedom for model fitting LRT test
df=(q+p)*(q+p+1)/2-t
# model fitting p-value
pvalue=1-pchisq(model.chisq,df) # 0.980945

#Let's calculate the model based likelihoods
loglike.overidentified=
-m/2*(log(det(2*pi*SigmaHat))+sum(S*t(SigmaHatInverse)))
loglike.justidentified=-m/2*(log(det(2*pi*S))+ no.rows.S)
LRT.statistic=2*(loglike.justidentified-loglike.overidentified)
LRT.statistic # 1.548874
model.chisq # 1.538549
#Note that model.chisq equals
LRT.statistic*(m-1)/m
```

11.3 Using Causal Anchors for Edge Orienting

As described above, SEMs can be used to evaluate how well the data fit a causal model. Here we are concerned how to infer (learn) a causal model from observational data. For theoretical and computational reasons, this turns out to be a very difficult task. To begin with, one can construct $4^{n(n-1)/2}$ different causal models between n random variables. Thus, over one million causal graphs can be constructed among $n = 5$ random variables. To complicate things, many of these causal graphs are "Markov" equivalent, which implies that they are indistinguishable with respect to observational data. Further, unobserved (latent) variables may complicate things further. The situation is already difficult for two random variables (denoted by A and B) which allow for 4 causal relationships. We denote the variables by A and B instead of X and Y since we are not sure which of them are endogenous or exogenous. The following four types of causal relationships can exist between A and B:

1. Variation in A directly causes variation in B (denoted by $A \rightarrow B$).
2. Variation in B directly causes variation in A (denoted by $B \rightarrow A$).
3. A and B directly cause each other (denoted by $A \leftrightarrow B$).
4. A and B have no direct causal relationship (denoted by $A \quad B$).

Fortunately, many applications involve learning the causal relationships between only two variables A and B. For example, in gene expression applications one may want to learn the causal relationship between the gene expression levels of a given gene (measured by variable A) and a clinical trait of interest (e.g., mouse body weight measured by variable B). Ideally, randomized trials or controlled experiments should be used to elucidate the relationships between the variables. But sometimes only observational data are available. Several algorithms for learning causal networks from observational data have been described in the literature (Cooper 1997), e.g., the IC/IC$*$ (Pearl 1988, 2000) and PC/PC$*$ (Spirtes et al. 2000) algorithms.

Sometimes causal inference is aided by external knowledge, which allows one to argue that some of the variables must be exogenous. We refer to such "sure" exogenous variables as **causal anchors** of a path diagram. For example, when studying causal relationships between genetic markers and gene expression data, genetic markers can often be interpreted as causal anchors (Schadt et al. 2005). We typically denote causal anchor variables by the letter "M". Assume that a causal anchor variable M is available which is associated with A or B or both variables. Since M is a causal anchor, a causal arrow may flow from M to A and/or from M to B. But there cannot be an arrow from A or from B to M since we are sure that M is an exogenous variable.

Recall that for each causal model one can calculate the model fitting p values (11.23). For example, $P(M \to A \to B)$ denotes the model fitting p value for the model in which causal anchor M causally affects trait A which in turn affects trait B. A small model p value (say $p < 0.05$) indicates that the causal model does not fit well. In contrast, a high p value indicates that the model fits the data well.

11.3.1 Single Anchor Local Edge Orienting Score

Here we will describe relative fitting indices (referred to as **local edge orienting scores**), which allow one to quantify how well the causal model $M \to A \to B$ fits relative to alternative causal models (Aten et al. 2008). Roughly speaking, the higher the value of the local edge orienting (LEO) score $LEO(A \to B)$, the stronger the evidence that the orientation $A \to B$ fits the data. We refer to models that involve two variables of interest (A and B) and additional causal anchors as **local structural equation models**. A local SEM that involves only a single causal anchor M is referred to as single anchor model. We only consider the five single anchor models depicted in Fig. 11.3a. Under the constraint that M is a causal anchor (graphically, arrows flow only from M and not into M), the five single anchor models exhaust all three possible variable models that both (1) explain significant correlations between $A - B$ and ($M - A$ or $M - B$) and (2) can be tested. The critical technical issue is that the chi-square statistics has nonzero degrees of freedom (d.f.) after estimating the model parameters. If the degrees of freedom are 0, the model p values cannot be calculated. Corresponding to the five single anchor models, one can

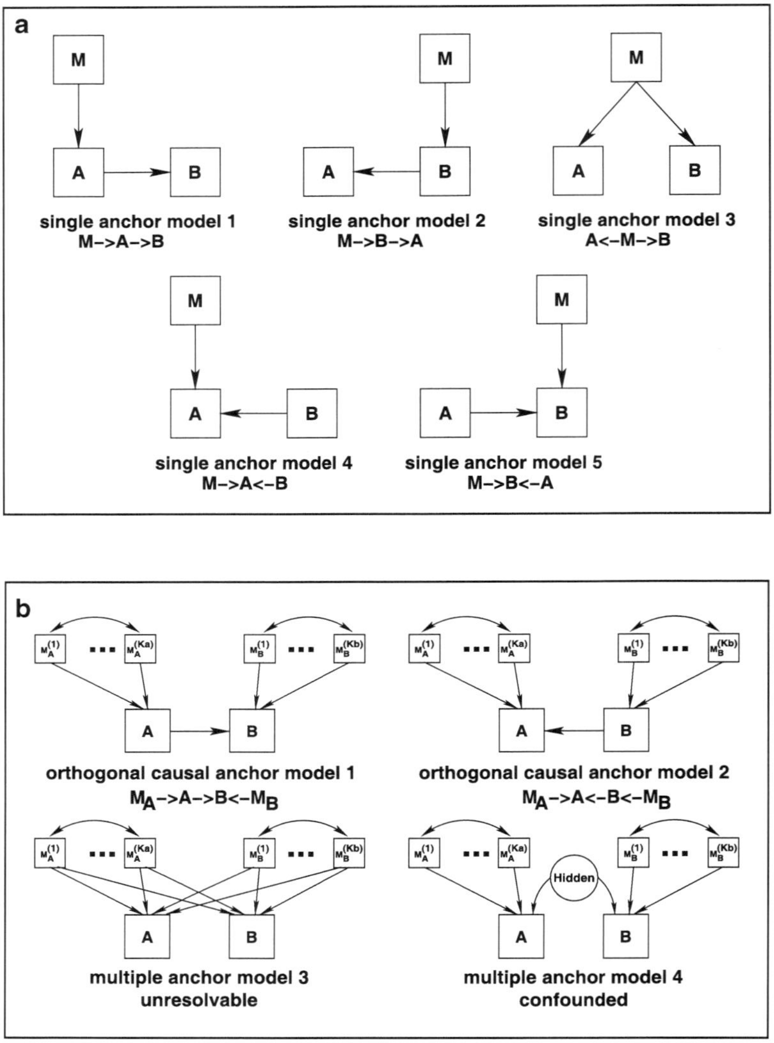

Fig. 11.3 Local SEMs involving causal anchors. (**a**) Single causal anchor models for calculating the LEO.SingleAnchor edge orienting score. The goal is to learn the causal model between variables A and B based on the causal anchor M. (**b**) Shows SEMs involving multiple causal anchors denoted by $M_A^{(i)}$ and $M_B^{(j)}$. By definition, $LEO(A \rightarrow B) = log_{10}(P(\text{model } 1))/(max_{i>1}(P(\text{model } i)))$, where the models in the definition are shown in (**a**) for single anchor LEO scores. (**b**) shows orthogonal causal anchor models used for the LEO.OCA score. The hidden confounder *Hidden* in model 4 is the causal parent of both A and B, i.e., $A \leftarrow H \rightarrow B$

calculate the following five model fitting p values: $P(M \rightarrow A \rightarrow B)$, $P(M \rightarrow B \rightarrow A)$, $P(A \leftarrow M \rightarrow B)$, $P(M \rightarrow A \leftarrow B)$, and $P(M \rightarrow B \leftarrow A)$. The model with the highest model fitting p value should be chosen as the inferred network since it is the best fitting model. We find it useful to report a relative fitting index for the orientation $A \rightarrow B$ (model 1) using a single number: the local edge orienting score. The LEO score is defined by dividing the model p value for $M \rightarrow A \rightarrow B$ by the p value of the best fitting alternative model, i.e., the best of models 2–5 in Fig. 11.3a. The model fitting p value of the best fitting alternative model is the maximum p value of the alternative causal models. Specifically, we define the single-anchor LEO score as follows:

$$
LEO.SingleAnchor(A \rightarrow B|M) = log_{10}\left(\frac{P(\text{model 1: } M \rightarrow A \rightarrow B)}{max\begin{pmatrix} P(\text{model 2: } M \rightarrow B \rightarrow A) \\ P(\text{model 3: } A \leftarrow M \rightarrow B) \\ P(\text{model 4: } M \rightarrow A \leftarrow B) \\ P(\text{model 5: } M \rightarrow B \leftarrow A) \end{pmatrix}}\right), \tag{11.24}
$$

where the logarithm has base 10. A positive $LEO.SingleAnchor(A \rightarrow B|M)$ score indicates that the p value in favor of model $M \rightarrow A \rightarrow B$ is higher than that of any of the competing models in Fig. 11.3a. For example, a LEO score of 1 indicates that $M \rightarrow A \rightarrow B$ fits the data ten times better than any competing model when the fit is measured using the p value. Based on simulations, we have found that 0.8 is often a reasonable threshold for $LEO.SingleAnchor(A \rightarrow B|M)$. A negative LEO score indicates that the $A \rightarrow B$ model is inferior to at least one alternative model. p-value based model fitting indices have several limitations, e.g., they depend on the sample size m (Kline 2005). Despite these limitations, we chose the model p value as the basis of the LEO scores (11.24) because it is the key ingredient of most, if not all, alternative fitting indices. Several alternative model fitting statistics are monotonic functions of the model fitting p value and are therefore threshold- or rank-equivalent to it. Alternative local edge orienting scores could be defined by replacing the model p value by another fitting index for which high values indicate good fit, see a related exercise.

11.3.2 Multi-Anchor LEO Score

Here we consider causal inference between A and B when multiple causal anchors are available. Let us first make the assumption that all causal anchors affect exactly one of the two variables of interest A and B. For example, we may assume that all causal anchors only affect A but not B. In this case, there is a set $M_{set} = \{M^{(1)}, M^{(2)}, \ldots\}$ of causal anchors. If the orientation $A \rightarrow B$ is correct, then each of the causal anchors has a pleiotropic effect by impacting first A and subsequently B, which is the reason why we refer to these anchors as **common pleiotropic anchors (CPA)** (Aten et al. 2008). In this case, the single anchor LEO score (11.24) can be generalized to the LEO.CPA score as follows:

$$LEO.CPA(A \to B|M_{set}) = log_{10}\left(\frac{P(\text{model 1: } M_{set} \to A \to B)}{max\begin{pmatrix} P(\text{model 2: } M_{set} \to B \to A), \\ P(\text{model 3: } A \leftarrow M_{set} \to B), \\ P(\text{model 4: } M_{set} \to A \leftarrow B), \\ P(\text{model 5: } M_{set} \to B \leftarrow A) \end{pmatrix}}\right). \quad (11.25)$$

Note that the multi-anchor models used in the definition of $LEO.CPA$ correspond to the single anchor models of Fig. 11.3a, with the single anchor M replaced by the set of anchors M_{set}.

Let us now assume that both A and B are affected by causal anchors. For clarity, we denote the sets of causal anchors affecting variables A and B by $M_A^{(1)}, M_A^{(2)}, \ldots, M_A^{(Ka)}$ and $M_B^{(1)}, M_B^{(2)}, \ldots, M_B^{(Kb)}$, respectively. We assume that M_A contains anchors that are more strongly correlated with A than with B. Similarly, M_B contains anchors more strongly correlated with B than with A. The set of anchors M_B are referred to as **orthogonal causal anchors (OCAs)** since the model $A \to B$ implies that these anchors impact B, but are independent of both A and M_A. The LEO.OCA score is a relative fitting index which contrasts the fit of the model $M_A \to A \to B \leftarrow M_B$ with that of the alternative models depicted in Fig. 11.3b. Specifically, we define

$$LEO.OCA(A \to B|M_A, M_B) = log_{10}\left(\frac{P(\text{model 1: } M_A \to A \to B \leftarrow M_B)}{max\begin{pmatrix} P(\text{model 2: } M_A \to A \leftarrow B \leftarrow M_B), \\ P(\text{model 3: } B \leftarrow M_A \to A; A \leftarrow M_B \to B), \\ P(\text{model 4: } M_A \to A \leftarrow H \to B \leftarrow M_B) \end{pmatrix}}\right). \quad (11.26)$$

Note that model 4 in the denominator involves a hidden confounder H. Model equivalence is also a key consideration in choosing which models to compare. Note that the multi-anchor models presented in Fig. 11.3b include a model with a hidden (latent) variable connecting A and B, and that no such model is included in the single anchor model comparisons. In the single anchor situation, such a latent variable model was found to be indistinguishable from single anchor models 4 and 5 since both of these models and the hidden variable models test for independence in the marginal relationship between the anchor and the more distal variable node.

In practice, one often observes strong dependence relationships between causal anchors. By allowing for correlations between anchors affecting one variable, LEO scoring allows multiple parents of a node to be correctly accounted for within each model. Simulations show that correlations between causal anchors can reduce the power of edge orienting scores. Removing anchors that are highly correlated with others may alleviate the loss of power. Therefore, the NEO R software automatically removes highly correlated causal anchors (Aten et al. 2008).

11.3.3 Thresholds for Local Edge Orienting Scores

For the single anchor score *LEO.SingleAnchor* and the *LEO.CPA* score, we often use a threshold of 0.8, which implies that the model p value of the causal model is $10^{0.8} = 6.31$-fold higher than that of the next best model. For the *LEO.OCA* score, we often use a lower threshold of 0.3 which implies that the causal model $A \rightarrow B$ fits the data $10^{0.3} = 2$ times better than the next best model. Using simulation, we found that these thresholds lead to false-positive rates that are often substantially below 0.05 (Aten et al. 2008), but we caution the reader that the LEO scores depend on the underlying sample size. For low values of m, lower thresholds may be more appropriate for generating hypotheses regarding causal relationships.

Permutation procedures and data-dependent schemes (e.g., based on the false discovery rate) may inform the user on how to pick a threshold for a particular application. Simulation studies can be used to determine the power and false-positive rates in different settings (sample size, causal signal, confounders, etc.).

11.4 Weighted Directed Networks Based on LEO Scores

A path diagram can be encoded in an unweighted, typically nonsymmetrical adjacency matrix where $A_{ij} = 1$ indicates a *direct* causal effect of variable i on variable j. It may be overly ambitious to infer direct causal relationships based on observational data. In practice, one is often interested in a less ambitious goal: learning whether any causal relationship between two variables exists irrespective of whether it is direct or indirect. Unweighted directed networks are often constructed by hard thresholding an association measure. As in the case of undirected networks, hard thresholding may lead to an information loss (preface). Therefore, it can be advantageous to consider weighted directed networks. We define a weighted directed network using an adjacency matrix, which is typically nonsymmetrical. Similar to the case of an undirected adjacency matrix (Sect. 11.4), we assume that each entry

A_{ij} is a number between 0 and 1 and diagonal elements equal 1. The adjacencies can be interpreted in many ways, e.g., high values of A_{ij} may indicate a strong (indirect or direct) causal effect of variable i on j or it may indicate strong evidence for such an effect. In general, we do not assume that $A_{ij} = 1$ indicates a *direct* causal effect.

Let us now outline some methods for constructing weighted or unweighted causal networks for a set of n numeric variables corresponding to the columns of a *datX*. Assume that we have a method for calculating a local edge orienting score $LEO_{ij} = LEO(x_i \rightarrow x_j)$. We do not assume any particular method for calculating LEO scores. The method may or may not make use of causal anchors. Our only assumption is that high values of $LEO_{i,j}$ indicate strong evidence that x_i causally affects x_j. $LEO_{i,j}$ can be interpreted as the i,jth element of an $n \times n$ dimensional matrix LEO, which is typically nonsymmetrical. Using a hard-threshold τ for dichotomizing the entries of LEO, one arrives at an unweighted directed network: $A_{ij} = 1$ if $LEO_{i,j} > \tau$ and 0 otherwise. To continuously transform LEO into a weighted directed adjacency matrix, one can make use of techniques and transformations described in Chap. 7. Several network concepts described in this book can easily be adapted to the case of a nonsymmetrical adjacency matrix, e.g., the incoming and outgoing connectivity are defined by

$$
\begin{aligned}
k_i^{incoming} &= \sum_{j \neq i} A_{ji} \\
k_i^{outgoing} &= \sum_{j \neq i} A_{ij}.
\end{aligned}
\tag{11.27}
$$

In an exercise, you are asked to adapt other types of network concepts to the case of a directed network.

11.5 Systems Genetic Applications

Causal inference approaches in systems genetics exploit genetic polymorphisms, e.g., quantitative trait loci (QTL), to infer causal relationships among phenotypes such as gene expression profiles or clinical traits. Several authors have argued that **genetic markers such as single nucleotide polymorphisms (SNPs) can serve as causal anchors for causal testing and causal inference** (Schadt et al. 2005; Smith 2006; Li et al. 2006; Kulp and Jagalur 2006; Chen et al. 2007b; Sieberts and Schadt 2007; Chen et al. 2007a; Emilsson et al. 2008).

Randomization of alleles during meiosis and the unidirectional influence of genetic polymorphisms on quantitative traits allow the inference of causal anchors for quantitative traits. For example, if a clinical trait or gene expression profile A is significantly associated with a genetic marker M, variation in M must be a cause of variation in A (denoted by $M \rightarrow A$) since the randomization of marker alleles during meiosis precedes their effect on trait A. Since the orientation of the edge between M

and A is unambiguous, M is a causal anchor of A. The causal relationships among quantitative traits can be inferred using the causal anchors, which allow one to distinguish between causal networks that would otherwise be equivalent with regard to their probability distribution. These causal anchors (genetic markers) provide significant statistical power and specificity for recovering directed edges (Zhu et al. 2007; Neto et al. 2008, 2010).

In Sect. 12.3, we provide a systems genetic analysis of the mouse expression data.

11.6 The Network Edge Orienting Method

The availability of hundreds of thousands of genetic markers poses new challenges: how to relate (anchor) gene expression profiles and clinical traits to multiple genetic markers, how to score the genetic evidence in favor of an edge orientation, and how to weigh the information from multiple markers. The Network Edge Orienting (NEO) method and software address the challenge of inferring unconfounded and directed gene networks from gene expression data using genetic marker data as causal anchors of local structural equation models. NEO can be used to address the following four **research goals**:

1. On the simplest level, NEO can be used to assign edge orienting scores to a single edge using manually chosen genetic markers.
2. When dealing with a single edge and multiple genetic markers, the NEO software can *automatically* select markers for edge orienting. Since the automatic marker selection entails certain parameter choices, we recommend carrying out a robustness analysis with respect to adding or removing genetic markers.
3. When dealing with a single trait A and manually selected genetic markers, the software can be used to screen for other traits that are causal or reactive to trait A.
4. When dealing with multiple edges, NEO can be used to arrive at a global directed network. This can be done as described in Sect. 11.4. If the resulting global network is acyclic (i.e., it does not contain loops), then d-separation (Pearl 1988) and standard SEM model fitting indices can be used to evaluate the fit of the global causal model to the data.

We now provide a detailed step-by-step description of a typical NEO analysis. An overview is also provided in Fig. 11.4.

11.6.1 Step 1: Combine Quantitative Traits and SNPs

NEO takes trait and genetic marker data as input. For simplicity, we assume that the genetic markers are single nucleotide polymorphisms (SNPs). For a given sample (e.g., a mouse), a biallelic SNP can take on one of three possible genotypes.

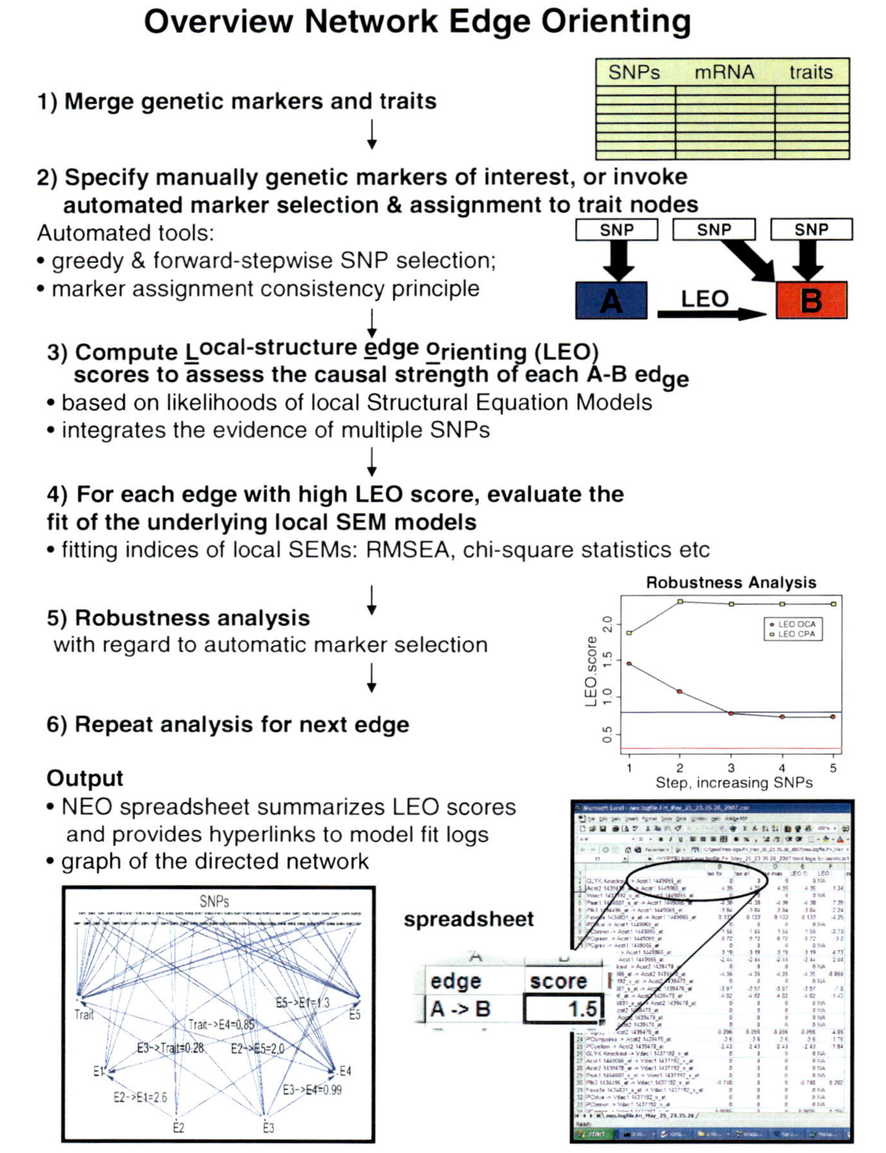

Fig. 11.4 Overview of the network edge orienting method. The steps of the network overview analysis are described in the text

By default, we assume an additive genetic effect and encode these genotypes as 0, 1, or 2, but alternative marker codings could also be considered. Ordinal variables are routinely used in path analysis and structural equation modeling

(Shipley 2000a; Bentler 2006). Variables A and B can include gene expression data, clinical phenotypes, or other quantitative variables (in mouse genetic applications, these variables as often referred to as traits). Each SNP or trait is a node in the network, and the NEO software evaluates and scores the edge between traits A and B if the absolute correlation $|cor(A,B)|$ lies above a user-specified threshold. For each edge $A - B$, NEO generates edge orienting scores for both possible orientations: $A \rightarrow B$ and $B \rightarrow A$. To allow the user to judge whether the existence of an edge is supported by the data, the NEO software outputs a Wald test statistic of the path coefficient, the corresponding p value, and the correlation between the two traits. If the Wald test p value is insignificant, orienting the edge may be meaningless.

11.6.2 Step 2: Genetic Marker Selection and Assignment to Traits

Edge orienting scores will only be generated for edges whose traits have been anchored to at least one genetic marker. Two basic approaches for anchoring traits to markers are implemented in the NEO software: a manual selection by the user or an automatic selection by the software itself.

11.6.2.1 Manual SNP Selection

NEO provides great flexibility to the user on how to anchor traits to markers. For example, the user can manually assign SNPs to the traits. This can also be carried out directly using the R code described in Sect. 11.8. This flexibility entails that the user carefully studies what constitutes a significant relationship between traits and markers and between the traits in the data set. The user may wish to anchor traits to SNPs that have been implicated by prior genetic analyses. For example, results from previous quantitative trait locus studies may implicate genetic markers associated with a trait. Multiple comparison issues are just starting to be addressed in the SEM literature (Cribbie 2000, 2007). Edge scores cannot be computed when an overly strict multiple testing control results in no causal anchors. On the other hand, an overly lax multiple testing control may result in spurious causal anchors which may lead to erroneous edge scores. We recommend that conservative measures of genome-wide QTL significance (Lander and Kruglyak 1995) and false discovery rate be applied when selecting the initial causal anchor(s). Once a causal anchor has been established as obtaining genome-wide significance, NEO can be used to evaluate the fit of different causal models.

11.6.2.2 Automatic SNP Selection

NEO can also be used to automatically relate (anchor) traits to SNPs. The automated SNP selection methods consider each trait A in isolation from the other traits when defining a *preliminary genetic marker set* (denoted by M'_A). Toward this end, the user can choose (1) a greedy approach based on univariate linear regression models, (2) a forward-stepwise approach based on multivariate linear regression models, or (3) both. The greedy SNP selection approach defines M'_A as the set of markers with the K highest absolute correlations with A. The greedy approach is equivalent to using univariate linear regression models to relate A to each marker separately and subsequently picking the K most significant markers.

For creating multivariate linear QTL models, NEO also implements forward-stepwise marker selection. The forward-stepwise marker selection method may avoid a pitfall that plagues the greedy SNP selection: if several genetic markers are located very close to each other (and are highly correlated), the greedy SNP selection may pick all of them before considering SNPs at other loci associated with the same trait. For this reason, we recommend combining greedy and forward-stepwise SNP selection methods.

Once the preliminary sets of markers M'_A and M'_B are obtained, NEO evaluates the consistency of each set. We use a *marker assignment consistency heuristic:* a genetic marker can only serve as causal anchor for one trait. To fulfill this heuristic, an SNP is moved from M'_A to M'_B if its correlation with B is stronger than that with trait A. We denote the resulting *consistent genetic marker sets* by M''_A and M''_B. The resulting consistent genetic marker sets may be comprised of dozens of SNPs. Therefore, it can be useful to further filter the SNPs according to their joint predictive power for the trait. Toward this end, we use the Akaike Information Criterion (AIC) in conjunction with multivariate regression models to select genetic markers from within the consistent genetic marker sets (Akaike 1973). Specifically, to define the *final genetic marker set* M_A for trait A, we use the AIC criterion to find a parsimonious multivariate regression model of A using predictors from within M''_A. The final sets of markers M_A and M_B are thus comprised of consistent genetic markers that according to the AIC criterion best predict their respective traits; we use these final sets as causal anchors in computing the edge orienting scores.

The forward-stepwise approach based on multivariate linear regression models is akin to a legal courtroom where two cases are built, weighed, and judged. Broadly, the strongest genetic support (multivariate eQTL models) for the genetic influence on A and B is built independently, using AIC-based halting criteria. After consistency checks, these multivariate eQTL models are weighed by embedding them in causal models (one principal causal model in favor of edge orientation $A \rightarrow B$ and alternative causal models), and models are then compared using SEM fitting indices. We consider automated SNP selection particularly useful when no prior evidence suggests causal anchors for the traits.

11.6.3 Step 3: Compute Local Edge Orienting Scores for Aggregating the Genetic Evidence in Favor of a Causal Orientation

Both LEO.CPA and LEO.OCA scores are computed for each edge orientation ($A \rightarrow B$ and $B \rightarrow A$). We recommend using the LEO.OCA score (11.26) as the primary edge orienting score if markers affect both A and B. However, if the results of the LEO.CPA score strongly disagree with those of the LEO.OCA score, the latter should not be trusted. As described in the next step, all fitting indices should be considered before calling an edge causal.

11.6.4 Step 4: For Each Edge, Evaluate the Fit of the Underlying Local SEM Models

Edges with high edge orienting scores may not necessarily correspond to causal relationships. Although edge orienting scores flag interesting edges, they are no substitute for carefully evaluating the fit of the underlying local SEMs. Since a LEO score is defined as a ratio of two model p values, it is advisable to check whether both p values are small, as this would indicate poor fit of either model. If the model p value of the confounded model $A \leftarrow C \rightarrow B$ is high, the correlation between A and B may be largely due to a hidden confounder C. NEO (using the underlying *sem* R package) also report a Wald test statistic for the path coefficient from $A \rightarrow B$. If the Wald test for an edge is significant, the data support its existence. The NEO software reports the standard SEM fitting indices (Shipley 2000a; Loehlin 2004) that are implemented in the R package *sem* (Fox 2006). Before calling an edge $A \rightarrow B$ causal, we recommend verifying that the corresponding causal model has a high model p value (say > 0.05), a low RMSEA score (say ≤ 0.05), a low SRMSR (say ≤ 0.10), a high CFI (say ≥ 0.90), and a significant Wald test p value (say $p \leq 0.05$).

11.6.5 Step 5: Robustness Analysis with Respect to SNP Selection Parameters

Since the edge orienting scores for an edge $A - B$ critically depend on the input genetic marker sets M_A and M_B, we also recommend carrying out a robustness analysis with respect to different marker sets. In particular, the automated SNP selection results should be carefully evaluated with regard to the threshold parameters that were used to define the marker sets. For example, when using a greedy SNP selection strategy, it is advisable to study how the LEO score is affected by altering the number of most highly correlated SNPs. For a given edge and a given edge orienting score (e.g., *LEO.OCA*), NEO implements a robustness analysis with respect

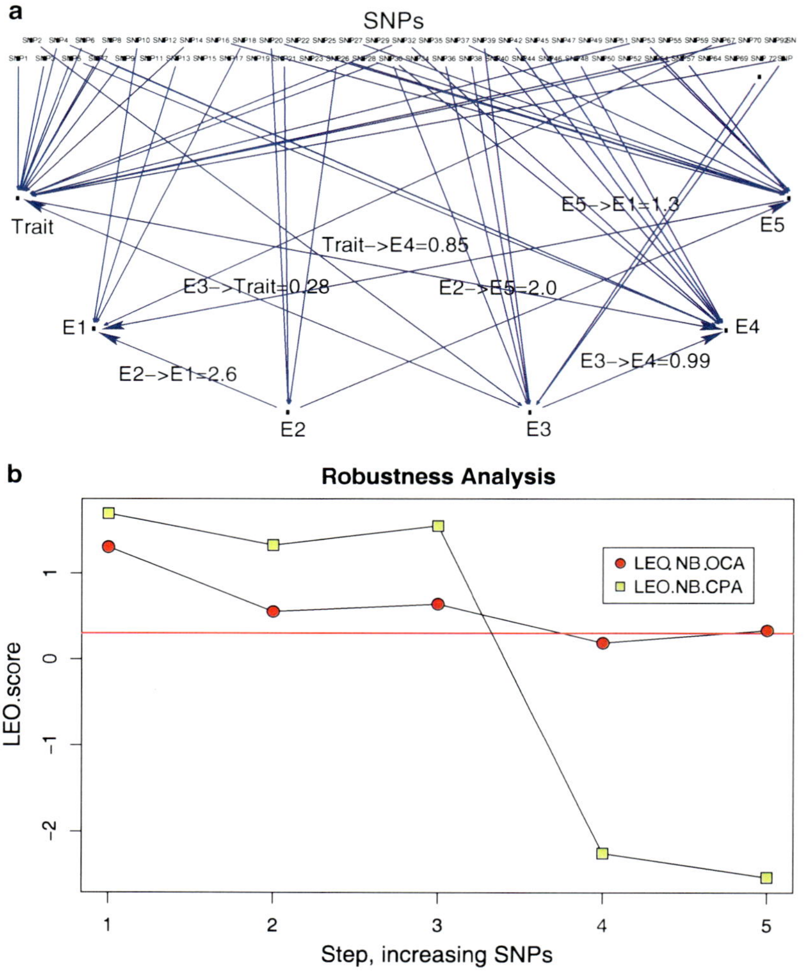

Fig. 11.5 Multi-edge simulation study involving five gene expression traits ($E1$-$E5$) and one clinical trait *Trait*. (**a**) shows an output graph of NEO: *blue edges* indicate significant correlations and a LEO.OCA score is added to each edges whose LEO.OCA score passes a user-supplied threshold. We find that all true causal edges are correctly retrieved at the recommended LEO.OCA threshold of 0.3. (**b**) shows the results of a robustness analysis for the LEO.OCA and LEO.CPA scores for the edge orientation $E4 \rightarrow Trait$. The LEO.OCA scores exceed the recommended threshold of 0.3 (*red horizontal line*), i.e., they retrieve the orientation correctly. Similarly, the LEO.CPA scores exceed the threshold of 0.8

to automatic marker selection (see Fig. 11.5b). A robustness plot shows how the LEO.OCA score (y-axis) depends on sets of automatically selected SNP markers (x-axis). When using the default SNP selection method (combined greedy and forward stepwise method), robustness step K corresponds to choosing the top K SNPs by greedy and forward selection for each trait. Since the greedy and forward SNP selection may select the same SNPs, step K typically involves fewer than $2K$ SNPs per trait. Ideally, the edge orienting results should be relatively robust with respect to different choices of K.

11.6.6 Step 6: Repeat the Analysis for the Next A–B Trait–Trait Edge and Apply Edge Score Thresholds to Orient the Network

NEO orients each edge separately in an undirected input trait network. The results are order-independent. For each edge, NEO repeats steps 1–3 until all edges have been assigned edge orienting scores. Once each edge has been scored, the user can generate a global, directed network by choosing an edge score (e.g., *LEO.OCA*) and a corresponding threshold (Fig. 11.5).

11.6.7 NEO Software and Output

The NEO software uses the `sem` R package by **John Fox** as computational engine and implements automatic methods (a) for finding causal anchors for quantitative traits, (b) for computing the local edge orienting (LEO) scores, e.g., the *LEO.SingleAnchor* (11.24), *LEO.CPA* (11.25), and *LEO.OCA* (11.26), and (c) for carrying out a statistical robustness analysis with regard to the causal anchor selection. The NEO software was originally implemented by **Jason E. Aten** but additional functions have been added over the years. Undirected correlation networks are oriented by considering each edge separately, thus reducing error propagation. To summarize the evidence in favor of a given edge orientation, LEO scores are considered. SEM fitting indices allow the user to assess local and overall model fit. R software tutorials, data, and supplementary material can be downloaded from: www.genetics.ucla.edu/labs/horvath/aten/NEO.

The primary output of NEO is an Excel spreadsheet which reports likelihood-based edge scores (LEO.CPA, LEO.OCA) and other edge scores that are described in the NEO manual. For each edge, the NEO spreadsheet also contains hyperlinks that allow the user to access the log file for each edge. The log file contains a host of information regarding computation of the edge orienting scores including SEM model p values, Wald test statistics for each path coefficient, and the SNP identifiers for the causal anchor sets M_A and M_B. Although the main output of NEO is scores for every edge orientation, one can construct a global directed network by thresholding an edge orienting score. The NEO software is documented in a series of separate tutorials that illustrate real data applications and simulation studies. These tutorials and the real data can be downloaded from our webpage.

11.6.8 Screening for Genes that Are Reactive to Insig1

We show that NEO is able to recover known causal relationships in the sterol homeostasis pathway using liver gene expression data from the mouse cross described in Sect. 5.5 and in Chap. 12. In this application, we illustrate that for a single trait (here *Insig1* gene expression) and manually selected genetic markers NEO can be used to screen for other traits (genes) that are reactive to the trait in question. We used genetic markers on chromosomes 8 and 16 as causal anchors for *Insig1*. For each gene expression trait B, we computed an LEO score for the edge $Insig1 \rightarrow B$. This allowed us to identify 23 genes that were downstream (reactive) to *Insig1* (Aten et al. 2008). Prior literature (Mounier and Posner 2006; Lusis 2006) suggests that 14 of the 23 genes are reactive to *Insig1* and are part of the well-studied sterol homeostasis pathway.

11.6.9 Discussion of NEO

Some R code is presented in Sect. 11.8. Additional R code for carrying out NEO analysis can be found at the following webpage: www.genetics.ucla.edu/labs/horvath/aten/NEO/.

The NEO R functions can use multiple causal anchors (e.g., genetic markers) to recover causal trait–trait relationships. NEO implements several edge orienting scores that measure the evidence in favor of a given edge orientation $A \rightarrow B$. Simulation studies show that orthogonal causal anchors lead to powerful edge scores that may outperform scores based only on candidate pleiotropic anchors (Aten et al. 2008). To afford flexibility to the user, the NEO software provides several options for anchoring the traits to causal anchors (manual versus automatic), computing local edge scores (LEO.CPA, LEO.OCA), and diagnosing poor model fit (RMSEA, CFI score, etc.). NEO provides multiple options for automatically anchoring a trait to causal anchors: greedy, forward, and combined (greedy and forward) anchor selection.

NEO's local, stepwise approach for orienting edges of a trait network allows one to orient networks involving hundreds or even thousands of traits. Since the calculation of edge orienting scores is based on local causal models, NEO is relatively robust with regard to mistaken orientation of some edges in the global network.

Although NEO performs well in simulation studies and the reported real data applications, we note that it has several limitations. The first limitation is that it requires the availability of causal anchors (genetic markers) that are significantly associated with at least one trait per edge. Spurious associations between the anchors and traits will result in meaningless edge orienting scores. Although the multi-anchor score (LEO.OCA) is quite robust to noise anchors in our simulations, false-positive input anchors will result in unreliable edge scores. The automatic anchor selection is particularly vulnerable to false positives, and its results should be carefully validated using biological experiments or causality analysis of independent data.

The second limitation is that the resulting global trait network may contain loops, i.e., it may be cyclic. In contrast, a directed acyclic graph (DAG) has no cycles. DAGs appear in models where it does not make sense for a trait to have a path to itself. While local DAGs are used for orienting individual edges, the reconstructed global trait network may no longer be acyclic. Acyclicity is theoretically desirable since it allows one to test causal predictions using Pearl's formalism of d-separation (Pearl 1988, 2000; Shipley 2000a). The constraint of acyclic graphs in many network learning algorithms is often more a mathematical convenience than reflective of biology; cycles may reflect feedback loops for maintaining homeostasis. When more and more edges are oriented, as in the IC/IC$*$ (Pearl 2000) and PC/PC$*$ (Spirtes et al. 2000) algorithms, an error in one part of the network can propagate and cause erroneous orientations in unrelated portions of the network. Most often these errors arise due to confusion between confounded and truly causal flows. To avoid being misled, NEO deliberately discards the evidence from correlated trait neighbors in the undirected graph during LEO scoring. By computing local edge orienting scores without regard to a global acyclicity constraint, the analysis is relatively robust to misoriented neighboring edges. NEO uses causal anchors for each edge separately and thus allows the genetic data to speak for themselves.

The LEO scores are local in that they orient one edge at a time without regard to the orientations of the other edges. The reconstruction of the global network should be taken with a grain of salt. While we report one simulation model where the global network was reconstructed correctly, future research should carefully evaluate the performance of the NEO approach for inferring global networks. A potential use of NEO is to use it for initializing an iterative edge orienting algorithms for large networks that maximizes a global SEM fitting index.

The third limitation is that the SEM-based edge orienting scores assume linear relationships between traits and causal anchors. This is mathematically convenient, but nonlinear effects are common and have been reported in the literature (Gjuvsland et al. 2007). The NEO approach works in the domain of linear graphical models since it is based on correlations and SEMs. Akin to the use of Pearson versus Spearman correlation, the software also offers the option of modeling monotone quantitative relationships in NEO by converting all data to ranks before further processing. NEO will not work for traits that satisfy non-monotone relationships.

Causal inference and structural equation modeling assume that relevant traits and causal anchors have been included in the causal model. Under-specified causal models, i.e., models that omit important variables, may mislead the user to detect spurious causal relationships.

Given all these potential limitations, it is reassuring that NEO performs well at retrieving known causal relationships in several real data applications (Aten et al. 2008; Farber et al. 2009; Presson et al. 2008; Plaisier et al. 2009; Inouye et al. 2010). Since NEO focuses on individual edges, we expect that NEO will be particularly useful for identifying traits that are causal for (or reactive to) a given trait. For example, we illustrate that NEO can be used to identify gene expression traits that are reactive to *Insig*1. The *sem* R package can be used to evaluate the global fit of an acyclic multi-trait network.

The NEO algorithm computes an edge score for each edge without regard to the information gained from neighboring edges. NEO aims to harness the power of the established upstream causal anchors (anchors) as fully as possible; thus, it is appropriate when genetic variations are a major source of the variation in the traits. To the extent the environment (e.g., diet) is also varied, the NEO approach may be less effective.

In our systems genetic applications (e.g., the mouse expression data in Sect. 12.3), we use SNP markers that capture only a limited amount of the sequence information of each individual. Since sequence information is likely to enhance the causal anchor assignment, sequence data may greatly improve the power of the NEO method. Apart from the common genetic variation that perturbs gene expression in mouse crosses, NEO can also be applied to orient edges on the basis of causal anchors that are not genetic markers.

We caution the reader that NEO gene screening and causal testing procedures based on observational data can only generate causal hypotheses. These should be validated using biological validation experiments. Biologic techniques for causal testing include transgenic modifications, viral-mediated overexpression, and chemical perturbation of genes. Edge orienting methods can also be based on various approaches that involve multiple perturbations, such as genetic and time series experiments (Geier et al. 2007).

11.7 Correlation Tests of Causal Models

SEMs are the method of choice for evaluating the fit of causal models. But sometimes one can define simple correlation tests for evaluating the fit of a causal model (Shipley 2000a). In the following, we provide a glimpse of how correlation tests can be used in case of very simple models. These techniques make use of path-analytic rules and a graph property called "d-separation" (introduced by **Judea Pearl**) (Pearl 1988, 2000). A detailed review of d-separation is beyond our scope but we provide the following terse description. Two variables A and B are d-separated in the graph by a set of variables S, then the two variables are independent given the variables in S. For example, $M \rightarrow A \rightarrow B$ implies that M and B are independent after conditioning on A. In particular, d-separation predicts the correlational consequences of conditioning in the causal graph. As usual, we quantify the linear relationship between variables M and A with a correlation coefficient $cor(M,A)$ (e.g., with the Pearson-, Spearman-, or biweight mid-correlation). To determine whether A mediates the effect of marker M on variable B ($M \rightarrow A \rightarrow B$), one can assess how conditioning on A affects the correlation between M and B. To quantify the linear relationship between M and B after conditioning on A, one can use the partial correlation coefficient (13.30):

$$cor(M,B|A) = \frac{cor(M,B) - cor(M,A)cor(B,A)}{\sqrt{(1-cor(M,A)^2)(1-cor(B,A)^2)}}.$$

According to path-analytic rules and additional assumptions, the causal model $M \rightarrow A \rightarrow B$ implies that the expected value of the partial correlation $cor(M,B|A)$ is zero. Fisher's Z transform or other transformations described in Sect. 10.3 can be used to compute p values for partial correlations. If the p value corresponding to $|cor(M,B|A)$ is high (nonsignificant), the data fit the assumed causal graph $M \rightarrow A \rightarrow B$. We also mention that path analytic rules applied to $M \rightarrow A \rightarrow B$ entail the following relationships among expected correlations

$$cor(M,B) = cor(M,A)cor(A,B). \tag{11.28}$$

The opposite (reactive) causal graph $M \rightarrow A \leftarrow B$ implies that the expected value of $cor(M,B)$ is zero since the causal paths collide at A. Conditioning on A activates this collider node, and the partial correlation $cor(M,B|A)$ may be nonzero. Similarly, one can show that the model $A \leftarrow M \rightarrow B$ implies that the expected value of $cor(A,B|M)$ is zero. These considerations illustrate that one can test the predicted correlational consequences of a causal model and thus evaluate its fit. By testing the correlational predictions, one can sometimes orient edges using observational data alone (Jordan 1998; Cooper 1997; Pearl 2000; Shipley 2000a,b; Korb and Nicholson 2004; Schaefer and Strimmer 2005; Opgen-Rhein and Strimmer 2007).

11.8 R Code for LEO Scores

11.8.1 R Code for the LEO.SingleAnchor Score

In the following, we will continue our analysis of the simulated model described in Sect. 11.2. Recall that we had simulated four exogenous variables $x1,\ldots x4$ and two endogenous variables $y1,y2$. In the following, assume that we want to infer the causal relationship between $A = y1$ and $B = y2$ based on the single causal anchor $M = x1$. Thus, we assume that the other exogenous variables have not been measured and cannot be used for edge orienting.

The following R code presents a function for calculating the *LEO.SingleAnchor* $(A \rightarrow B|M)$ (11.24) score.

```
library(sem)

# Now we specify the 5 single anchor models
# Single anchor model 1: M1->A->B
CausalModel1=specify.model()
A -> B, betaAtoB, NA
M1 -> A, gammaM1toA, NA
A <-> A, sigmaAA, NA
B <-> B, sigmaBB, NA
M1 <-> M1, sigmaM1M1,NA

# Single anchor model 2: M->B->A
CausalModel2=specify.model()
```

```
B -> A, betaBtoA, NA
M1 -> B, gammaM1toB, NA
A <-> A, sigmaAA, NA
B <-> B, sigmaBB, NA
M1 <-> M1, sigmaM1M1,NA

# Single anchor model 3: A<-M->B
CausalModel3=specify.model()
M1 -> A, gammaM1toA, NA
M1 -> B, gammaM1toB, NA
A <-> A, sigmaAA, NA
B <-> B, sigmaBB, NA
M1 <-> M1, sigmaM1M1,NA

# Single anchor model 4: M->A<-B
CausalModel4=specify.model()
M1 -> A, gammaM1toA, NA
B -> A, gammaBtoA, NA
A <-> A, sigmaAA, NA
B <-> B, sigmaBB, NA
M1 <-> M1, sigmaM1M1,NA

# Single anchor model 5: M->B<-A
CausalModel5=specify.model()
M1 -> B, gammaM1toB, NA
A -> B, gammaAtoB, NA
A <-> A, sigmaAA, NA
B <-> B, sigmaBB, NA
M1 <-> M1, sigmaM1M1,NA

# Function for obtaining the model fitting p-value
# from an sem object:
ModelFittingPvalue=function(semObject){
ModelChisq=summary(semObject)$chisq
df= summary(semObject)$df
ModelFittingPvalue=1-pchisq(ModelChisq,df)
ModelFittingPvalue}

# this function calculates the single anchor score
LEO.SingleAnchor=function(M1,A,B){
datObsVariables= data.frame(M1,A,B)
# this is the observed correlation matrix
S.SingleAnchor=cor(datObsVariables,use="p")
m=dim(datObsVariables)[[1]]

semModel1 =sem(CausalModel1,S=S.SingleAnchor,N=m)
semModel2 =sem(CausalModel2,S=S.SingleAnchor,N=m)
semModel3 =sem(CausalModel3,S=S.SingleAnchor,N=m)
semModel4 =sem(CausalModel4,S=S.SingleAnchor,N=m)
semModel5 =sem(CausalModel5,S=S.SingleAnchor,N=m)

# Model fitting p values for each model
p1=ModelFittingPvalue(semModel1)
p2=ModelFittingPvalue(semModel2)
```

```
p3=ModelFittingPvalue(semModel3)
p4=ModelFittingPvalue(semModel4)
p5=ModelFittingPvalue(semModel5)
LEO.SingleAnchor=log10(p1/max(p2,p3,p4,p5))

data.frame(LEO.SingleAnchor,p1,p2,p3,p4,p5)
} # end of function

# let's run the function
LEO.SingleAnchor(M1=x1,A=y1,B=y2)

# output
LEO.SingleAnchor p1 p2 p3 p4 p5
   4.306427 0.39657 4.550e-10 0 1.958e-05 5.9952e-14
```

Recall that the higher the model fitting p value, the better the model fits the data. Note that $LEO.SingleAnchor = 4.31$, which indicates that the (true) causal model $M \to A \to B$ fits $10^{4.31}$ times better than any competing model. Note that the value of the *LEO.SingleAnchor* score exceeds our recommended threshold of 0.8. Below, we show that considering additional causal anchors leads to more significant results.

In Sect. 12.3, we illustrate how to use the `LEO.SingleAnchor` function in a systems genetic analysis.

11.8.2 R Code for the LEO.CPA Score

Here we will continue our analysis of the simulated model described in Sect. 11.2. In the following, assume that we want to infer the causal relationship between $A = y1$ and $B = y2$ based on the two causal anchors $M1 = x1$ and $M2 = x2$. Now we present R code for calculating the $LEO.CPA(A \to B|M_A)$ (11.25) score.

```
#Specify the CPA models
# model 1
CausalModel1=specify.model()
A -> B, betaAtoB, NA
M1 -> A, gammaM1toA, NA
M2 -> A, gammaM2toA, NA
A <-> A, sigmaAA, NA
B <-> B, sigmaBB, NA
M1 <-> M1, sigmaM1M1,NA
M2 <-> M2, sigmaM2M2,NA
M1 <-> M2, sigmaM1M2,NA

#model 2
# A <- B
CausalModel2=specify.model()
B -> A, betaBtoA, NA
M1 -> B, gammaM1toB, NA
M2 -> B, gammaM2toB, NA
A <-> A, sigmaAA, NA
B <-> B, sigmaBB, NA
```

```
M1 <-> M1, sigmaM1M1,NA
M2 <-> M2, sigmaM2M2,NA
M1 <-> M2, sigmaM1M2,NA

#model 3
CausalModel3=specify.model()
M1 -> A, gammaM1toA, NA
M2 -> A, gammaM2toA, NA
M1 -> B, gammaM1toB, NA
M2 -> B, gammaM2toB, NA
A <-> A, sigmaAA, NA
B <-> B, sigmaBB, NA
M1 <-> M1, sigmaM1M1,NA
M2 <-> M2, sigmaM2M2,NA
M1 <-> M2, sigmaM1M2,NA

#model 4
CausalModel4=specify.model()
M1 -> A, gammaM1toA, NA
M2 -> A, gammaM2toA, NA
B -> A, gammaBtoA, NA
A <-> A, sigmaAA, NA
B <-> B, sigmaBB, NA
M1 <-> M1, sigmaM1M1,NA
M2 <-> M2, sigmaM2M2,NA
M1 <-> M2, sigmaM1M2,NA

#model 5
CausalModel5=specify.model()
M1 -> B, gammaM1toB, NA
M2 -> B, gammaM2toB, NA
A -> B, gammaAtoB, NA
A <-> A, sigmaAA, NA
B <-> B, sigmaBB, NA
M1 <-> M1, sigmaM1M1,NA
M2 <-> M2, sigmaM2M2,NA
M1 <-> M2, sigmaM1M2,NA

# function for calculating the LEO score
LEO.CPA=function(M1,M2,A,B){
datObsVariables= data.frame(M1,M2,A,B)
# this is the observed correlation matrix
S.CPA=cor(datObsVariables,use="p")
m=dim(datObsVariables)[[1]]

#fit the local SEM models
semModel1 =sem(CausalModel1,S=S.CPA,N=m)
semModel2 =sem(CausalModel2,S=S.CPA,N=m)
semModel3 =sem(CausalModel3,S=S.CPA,N=m)
semModel4 =sem(CausalModel4,S=S.CPA,N=m)
semModel5 =sem(CausalModel5,S=S.CPA,N=m)

p1=ModelFittingPvalue(semModel1)
p2=ModelFittingPvalue(semModel2)
```

```
p3=ModelFittingPvalue(semModel3)
p4=ModelFittingPvalue(semModel4)
p5=ModelFittingPvalue(semModel5)

# calculate the LEO.CPA score
LEO.CPA=log10(p1/max(p2,p3,p4,p5))

data.frame(LEO.CPA,p1,p2,p3,p4,p5)
} # end of function

# let's run the function
LEO.CPA(M1=x1,M2=x2,A=y1,B=y2)

#output
  LEO.CPA p1 p2 p3 p4 p5
1 5.192549 0.69898 0 4.4808e-06 2.220e-16 0
```

Clearly causal model 1 ($M_{set} \rightarrow A \rightarrow B$) fits best. The high value of the *LEO.CPA* score indicates that causal model 1 fits $10^{5.19} \approx 155,000$ times better than the next best model. Since the CPA score exceeds our threshold of 0.8, we conclude correctly that the edge orientation between A and B is given by $A \rightarrow B$.

11.8.3 R Code for the LEO.OCA Score

Here we will continue our analysis of the simulated model described in Sect. 11.2. In the following, assume that we want to infer the causal relationship between $A = y1$ and $B = y2$ based on the two causal anchors $MA1 = x1$ and $MA2 = x2$ associated with A and two causal anchors $MB1 = x3$ and $MB2 = x4$ associated with B. Now we present R code for calculating the $LEO.OCA(A \rightarrow B|M_A M_B)$ (11.26) score.

```
#Specify the causal models underlying the OCA score
CausalModel1=specify.model()
A -> B, betaAtoB, NA
MA1 -> A, gammaMA1toA, NA
MA2 -> A, gammaMA2toA, NA
MB1 -> B, gammaMB1toB, NA
MB2 -> B, gammaMB2toB, NA
A <-> A, sigmaAA, NA
B <-> B, sigmaBB, NA
MA1 <-> MA1, sigmaMA1MA1,NA
MA2 <-> MA2, sigmaMA2MA2,NA
MB1 <-> MB1, sigmaMB1MB1,NA
MB2 <-> MB2, sigmaMB2MB2,NA
MA1 <-> MA2, sigmaMA1MA2,NA
MB1 <-> MB2, sigmaMB1MB2,NA

CausalModel2=specify.model()
B -> A, betaBtoA, NA
MA1 -> A, gammaMA1toA, NA
MA2 -> A, gammaMA2toA, NA
```

```
MB1 -> B, gammaMB1toB, NA
MB2 -> B, gammaMB2toB, NA
A <-> A, sigmaAA, NA
B <-> B, sigmaBB, NA
MA1 <-> MA1, sigmaMA1MA1,NA
MA2 <-> MA2, sigmaMA2MA2,NA
MB1 <-> MB1, sigmaMB1MB1,NA
MB2 <-> MB2, sigmaMB2MB2,NA
MA1 <-> MA2, sigmaMA1MA2,NA
MB1 <-> MB2, sigmaMB1MB2,NA

CausalModel3=specify.model()
MA1 -> A, gammaMA1toA, NA
MA2 -> A, gammaMA2toA, NA
MA1 -> B, gammaMA1toB, NA
MA2 -> B, gammaMA2toB, NA
MB1 -> A, gammaMB1toA, NA
MB2 -> A, gammaMB2toA, NA
MB1 -> B, gammaMB1toB, NA
MB2 -> B, gammaMB2toB, NA
A <-> A, sigmaAA, NA
B <-> B, sigmaBB, NA
MA1 <-> MA1, sigmaMA1MA1,NA
MA2 <-> MA2, sigmaMA2MA2,NA
MB1 <-> MB1, sigmaMB1MB1,NA
MB2 <-> MB2, sigmaMB2MB2,NA
MA1 <-> MA2, sigmaMA1MA2,NA
MB1 <-> MB2, sigmaMB1MB2,NA

CausalModel4=specify.model()
MA1 -> A, gammaMA1toA, NA
MA2 -> A, gammaMA2toA, NA
MB1 -> B, gammaMB1toB, NA
MB2 -> B, gammaMB2toB, NA
A <-> B, sigmaHiddenConfounderAB, NA
A <-> A, sigmaAA, NA
B <-> B, sigmaBB, NA
MA1 <-> MA1, sigmaMA1MA1,1
MA2 <-> MA2, sigmaMA2MA2, 1
MB1 <-> MB1, sigmaMB1MB1,1
MB2 <-> MB2, sigmaMB2MB2,1
MA1 <-> MA2, sigmaMA1MA2,NA
MB1 <-> MB2, sigmaMB1MB2,NA

# define the function
LEO.OCA=function(MA1,MA2,MB1,MB2, A,B){
datObsVariables= data.frame(MA1,MA2,MB1,MB2,A,B)
# this is the observed correlation matrix
S.OCA=cor(datObsVariables,use="p")
m=dim(datObsVariables)[[1]]

#fit the causal models to the data
semModel1 =sem(CausalModel1,S=S.OCA,N=m)
semModel2 =sem(CausalModel2,S=S.OCA,N=m)
```

```
semModel3 =sem(CausalModel3,S=S.OCA,N=m)
semModel4 =sem(CausalModel4,S=S.OCA,N=m)

p1=ModelFittingPvalue(semModel1)
p2=ModelFittingPvalue(semModel2)
p3=ModelFittingPvalue(semModel3)
p4=ModelFittingPvalue(semModel4)
# calculate the OCA score
LEO.OCA=log10(p1/max(p2,p3,p4))

data.frame(LEO.OCA,p1,p2,p3,p4)
} # end of function

# let's run the function
LEO.OCA(MA1=x1,MA2=x2,MB1=x3,MB2=x4,A=y1,B=y2)

#output
LEO.OCA p1 p2 p3 p4
Inf 0.02727 0 0 0
```

Note that the *LEO.OCA* is infinite, i.e., causal model 1 ($M_A \rightarrow A \rightarrow B \leftarrow M_B$) fits infinitely better than the alternative models. Since our main interest is the causal orientation between A and B, the *LEO.OCA* score leads us to the correct conclusion regarding the edge orientation $A \rightarrow B$. Interestingly, the model p value for causal model 1 is significant ($p = 0.02727196$) which indicates poor fit of this model. This poor fit (significant p value) reflects that the causal model did not properly specify the effect of a hidden variable. Specifically, recall that in our original simulation model, we simulated two error variables errors ε_1 and ε_2 affecting $y1$ and $y2$, respectively. These error variables were simulated to have a nonvanishing covariance, i.e., $\sigma_{\varepsilon_1,\varepsilon_2} \neq 0$. The covariance (and correlation) between error variables can be interpreted as resulting from a hidden variable affecting both A and B. The presence of orthogonal causal anchors facilitates a powerful model fitting test that detects our failure to include hidden variable into the model specification. In contrast, an analysis that only uses CPAs fails to detect this model misspecification. Having said this, our main purpose is to orient the edge between A and B and the other variables only serve this purpose. By considering relative fit as opposed to absolute model fit, the *LEO.OCA* leads us to the correct conclusion regarding $A \rightarrow B$. This highlights the utility of using relative fitting indices.

11.9 Exercises

- Exercise regarding single anchor local SEM. Define a relative fitting index analogous to LEO.SingleAnchor that contrasts the model $A \rightarrow B \leftarrow M$ versus alternative models. Hint: Divide $P(A \rightarrow B \leftarrow M)$ by the maximum model fitting p value of the four remaining single anchor models. Log-transform the resulting ratio.

- Exercise regarding simulation studies. Change the coefficient values of the simulation model described in Sect. 11.2 as follows:

  ```
  beta21=2;gamma11=1;gamma12=1;gamma23=2;gamma24=1
  ```

  Run the R code for

  (i) Evaluating an SEM model (Sect. 11.2)
  (ii) Calculating the LEO.SingleAnchor score (Sect. 11.8.1)
  (iii) Calculating the LEO.CPA score (Sect. 11.8.2)
  (iv) Calculating the LEO.OCA score (Sect. 11.8.3)

- Exercise regarding the definition of alternative relative model fitting indices. Recall that the LEO.SingleAnchor score (11.24) is defined with respect to model fitting p values.

  (i) Develop analogous relative model fitting indices that contrast the fit of $M \rightarrow A \rightarrow B$ to models 2–5 in Fig. 11.3a. Hint: Recall that the summary command applied to an SEM object outputs several model fitting statistics, e.g., goodness-of-fit index, adjusted goodness-of-fit index, RMSEA index, Bentler–Bonnett NFI, Tucker–Lewis NNFI, Bentler CFI, SRMR, BIC. Read up on these indices. Choose one of them and determine whether high values indicate good or bad fit. If high values indicate bad fit, apply a monotonically decreasing function to it so that high values of the transformed index indicate good fit. Next replace the model p value in the LEO score by your chosen (and possibly transformed) fitting index.
  (ii) Simulate 100 causal models by varying the coefficient parameter values of the simulation model described in Sect. 11.2. Hint: For example, randomly sample the parameter values of $beta21, gamma11, gamma12, gamma23, gamma24$. For each simulated model, record the value of your relative fitting index and that of the LEO.SingleAnchor score.
  (iii) Create a scatter plot of your relative fitting index versus the LEO.SingleAnchor score across the different simulation models. What is the Spearman correlation between the fitting indices? Are the two fitting indices approximately threshold- or rank equivalent? Hint: Calculate the C-index.

References

Akaike H (1973) Information theory as the extension of the maximum likelihood principle. Akademiai Kiado, Budapest, Hungary, pp 267–281

Aten J, Fuller T, Lusis AJ, Horvath S (2008) Using genetic markers to orient the edges in quantitative trait networks: The NEO software. BMC Syst Biol 2(1):34

Bentler PM (2006) EQS 6 structural equations program manual. Multivariate Software, Inc, Encino, CA

Chen J, Xu H, Aronow BJ, Jegga AG (2007a) Improved human disease candidate gene prioritization using mouse phenotype. BMC Bioinform 8:392

Chen LS, EmmertStreib F, Storey JD (2007b) Harnessing naturally randomized transcription to infer regulatory relationships among genes. Genome Biol 8:R219

Cooper GF (1997) A simple constraint-based algorithm for efficiently mining observational databases for causal relationships. Data Min Knowl Discov 1:203–224

Cribbie RA (2000) Evaluating the importance of individual parameters in structural equation modeling: The need for type I error control. Pers Individ Dif 29:567–577

Cribbie RA (2007) Multiplicity control in structural equation modeling Struct Equ Model 14(1):98–112

Emilsson V, Thorleifsson G, Zhang B, Leonardson A, Zink F, Zhu J, Carlson S, Helgason A, Walters G, Gunnarsdottir S, Mouy M, Steinthorsdottir V, Eiriksdottir G, Bjornsdottir G, Reynisdottir I, Gudbjartsson D, Helgadottir A, Jonasdottir A, Jonasdottir A, Styrkarsdottir U, Gretarsdottir S, Magnusson K, Stefansson H, Fossdal R, Kristjansson K, Gislason H, Stefansson T, Leifsson B, Thorsteinsdottir U, Lamb J, Gulcher J, Reitman M, Kong A, Schadt E, Stefansson K (2008) Genetics of gene expression and its effect on disease. Nature 452(7186):423–428

Farber CR, vanNas A, Ghazalpour A, Aten JE, Doss S, Sos B, Schadt EE, IngramDrake L, Davis RC, Horvath S, Smith DJ, Drake TA, Lusis AJ (2009) An integrative genetics approach to identify candidate genes regulating bone density: Combining linkage, gene expression and association. J Bone Miner Res 1:105–16

Fox J (1984) Linear structural-equation models. In: Linear statistical models and related Methods, vol. 4. Wiley, New York

Fox J (2006) Structural equation modeling with the sem package in R. Struct Equ Model 13:465–486

Geier F, Timmer J, Fleck C (2007) Reconstructing gene-regulatory networks from time series, knock-out data, and prior knowledge. BMC Syst Biol 1:11

Gjuvsland A, Hayes B, Meuwissen T, Plahte E, Omholt S (2007) Nonlinear regulation enhances the phenotypic expression of trans-acting genetic polymorphisms. BMC Syst Biol 1(1):32

Inouye M, Silander K, Hamalainen E, Salomaa V, Harald K, Jousilahti P, Mannista S, Eriksson JG, Saarela J, Ripatti S, Perola M, van Ommen GJB, Taskinen MR, Palotie A, Dermitzakis ET, Peltonen L (2010) An immune response network associated with blood lipid levels. PLoS Genet 6(9):e1001113

Jordan MI (1998) Learning in graphical models. The MIT, Cabridge, MA

Kline RB (2005) Principles and practice of structural equation modeling. The Guilford, New York, NY

Korb KB, Nicholson AE (2004) Bayesian artifical intelligence. Chapman & Hall/CRC, Boca Raton, FL

Kulp DC, Jagalur M (2006) Causal inference of regulator-target pairs by gene mapping of expression phenotypes. BMC Genomics 7:125

Lander EJ, Kruglyak L (1995) Genetic dissection of complex traits: Guidelines for interpretation and reporting linkage results. Nat Genet 11:241–247

Li R, Tsaih SW, Shockley K, Stylianou IM, Wegedal J, Paigen B, Churchill GA (2006) Structural model analysis of multiple quantitative traits. PLos Genet 2(7):(e114) 1046–1057

Loehlin JC (2004) Latent variable models, 4th edn. Lawrence Erlbaum Associates, Mahwah, NJ

Lusis AJ (2006) A thematic review series: Systems biology approaches to metabolic and cardiovascular disorders. J Lipid Res 47(9):1887–1890

Mounier C, Posner BI (2006) Transcriptional regulation by insulin: From the receptor to the gene. Can J Physiol Pharmacol 84:713–724

Neto CE, Ferrara CT, Attie AD, Yandell BS (2008) Inferring causal phenotype networks from segregating populations. Genetics 179(2):1089–1100

Neto CE, Keller MP, Attie AD, Yandell BS (2010) Causal graphical models in systems genetics: A unified framework for joint inference of causal network and genetic architecture for correlated phenotypes. Ann Appl Stat 4(1):320–339

Opgen-Rhein R, Strimmer K (2007) From correlation to causation networks: A simple approximate learning algorithm and its application to high-dimensional plant gene expression data. BMC Syst Biol 1:37

Pearl J (1988) Probabilistic reasoning in intelligent systems: Networks of plausible inference. Morgan Kaufmann Publishers, Inc., San Francisco, CA

Pearl J (2000) Causality: Models, reasoning, and inference. Cambridge University Press, Cambridge, UK

Plaisier CL, Horvath S, Huertas-Vazquez A, Cruz-Bautista I, Herrera MF, Tusie-Luna T, Aguilar-Salinas C, Pajukanta P (2009) A systems genetics approach implicates USF1, FADS3, and other causal candidate genes for familial combined hyperlipidemia. PLoS Genet 5(9):e1000642

Presson AP, Sobel EM, Papp JC, Suarez CJ, Whistler T, Rajeevan MS, Vernon SD, Horvath S (2008) Integrated weighted gene co-expression network analysis with an application to chronic fatigue syndrome. BMC Syst Biol 2:95

Schadt EE, Lamb J, Yang X, Zhu J, Edwards J, GuhaThakurta D, Sieberts SK, Monks S, Reitman M, Zhang C, Lum PY, Leonardson A, Thieringer R, Metzger JM, Yang L, Castle J, Zhu H, Kash SF, Drake TA, Sachs A, Lusis AJ (2005) An integrative genomics approach to infer causal associations between gene expression and disease. Nat Genet 37(7):710–717

Schaefer J, Strimmer K (2005) An empirical Bayes approach to inferring large-scale gene association networks. Bioinformatics 21(6):754–764

Shipley B (2000a) Cause and correlation in biology, 2nd edn. Cambridge University Press, Cambridge, UK

Shipley B (2000b) A new inferential test for path models based on directed acyclic graphs. Struct Equ Model 7:206–218

Sieberts SS, Schadt EE (2007) Moving toward a system genetics view of disease. Mamm Genome 18(6):389–401

Smith GD (2006) Randomized by (your) god: Robust inference from an observational study design. J Epidemiol Community Health 60:382–388

Spirtes P, Glymour C, Scheines R (2000) Causation, prediction, and search, 2nd edn. The MIT, Cambridge, MA

Steiger JH, Fouladi RT (1997) What if there were no significance tests? Erlbaum, Mahwah, NJ

Zhu J, Wiener MC, Zhang C, Fridman A, Minch E, Lum PY, Sachs JR, Schadt EE (2007) Increasing the power to detect causal associations by combining genotypic and expression data in segregating populations. PLoS Comput Biol 3(4):0692–0703 (e69)

Chapter 12
Integrated Weighted Correlation Network Analysis of Mouse Liver Gene Expression Data

Abstract This chapter describes a case study and R code for carrying out an integrated weighted correlation network analysis of mouse gene expression, sample trait, and genetic marker data. It describes how to (a) use sample networks (signed correlation networks) for detecting outlying observations, (b) find co-expression modules and key genes related to mouse body weight and other physiologic traits in female mice, (c) study module preservation between female and male mice, (d) carry out a systems genetic analysis with the network edge orienting approach to find causal genes for body weight, and (e) define consensus modules between female and male mice. We also describe methods and software for visualizing networks and for carrying out gene ontology enrichment analysis.

The mouse cross is described in Sect. 5.5. The mouse gene expression data were generated by the labs of **Jake Lusis** and **Eric Schadt**. Here we used a mouse liver gene expression data set which contains 3,600 gene expression profiles. These were filtered from the original over 20,000 genes by keeping only the most variant and most connected probes. In addition to the expression data, several physiological quantitative traits were measured for the mice, e.g., body weight. The expression data are contained in the file "LiverFemale3600.csv" that can be found at the following webpages:
www.genetics.ucla.edu/labs/horvath/CoexpressionNetwork/Book
or
www.genetics.ucla.edu/labs/horvath/CoexpressionNetwork/Rpackages/WGCNA/.

Much of the following R code was created by **Peter Langfelder**.

12.1 Constructing a Sample Network for Outlier Detection

Here we use sample network methods for finding outlying microarray samples (see Sect. 7.7). Specifically, we define the **sample network** as canonical Euclidean distance based network $A_{uv} = 1 - ||S_u - S_v||^2/maxDiss$ (7.18). Next we use the standardized connectivity $Z.k_u = \frac{k_u - mean(k)}{\sqrt{var(k)}}$ (7.24) to identify array outliers. In the following, we present relevant R code.

S. Horvath, *Weighted Network Analysis: Applications in Genomics and Systems Biology*, DOI 10.1007/978-1-4419-8819-5_12, © Springer Science+Business Media, LLC 2011

```
# Change the path to the directory
# On Windows use a forward slash / instead of the usual \.
setwd("C:/Documents and Settings/SHorvath/My Documents/MouseData")
library(WGCNA)
options(stringsAsFactors = FALSE)

#Read in the female liver data set
femData = read.csv("LiverFemale3600.csv")
dim(femData)
[1] 3600 143
# note there are 3600 genes and 143 columns
# the column names are
names(femData)
# the output shows that the first 8 columns contain gene
      information
  [1] "substanceBXH" "gene_symbol" "LocusLinkID" "ProteomeID"
  [5] "cytogeneticLoc" "CHROMOSOME" "StartPosition" "EndPosition"
  [9] "F2_2" "F2_3" "F2_14" "F2_15"
 [13] "F2_19" "F2_20" "F2_23" "F2_24"
    ......................ETC (remainder of output omitted)

#Remove gene information and transpose the expression data
datExprFemale=as.data.frame(t(femData[, -c(1:8)]))
names(datExprFemale)=femData$substanceBXH
rownames(datExprFemale)=names(femData)[-c(1:8)]

# Now we read in the physiological trait data
traitData =read.csv("ClinicalTraits.csv")
dim(traitData)
names(traitData)
# use only a subset of the columns
allTraits=traitData[,c(2,11:15, 17:30, 32:38)]
names(allTraits)

# Order the rows of allTraits so that
# they match those of datExprFemale:
Mice=rownames(datExprFemale)
traitRows = match(Mice, allTraits$Mice)
datTraits = allTraits[traitRows, -1]
rownames(datTraits) = allTraits[traitRows, 1]
# show that row names agree
table(rownames(datTraits)==rownames(datExprFemale))

# sample network based on squared Euclidean distance
# note that we transpose the data
A=adjacency(t(datExprFemale),type="distance")
# this calculates the whole network connectivity
k=as.numeric(apply(A,2,sum))-1
# standardized connectivity
Z.k=scale(k)

# Designate samples as outlying
# if their Z.k value is below the threshold
thresholdZ.k=-5 # often -2.5
```

```
# the color vector indicates outlyingness (red)
outlierColor=ifelse(Z.k<thresholdZ.k,"red","black")

# calculate the cluster tree using flashClust or hclust
sampleTree = flashClust(as.dist(1-A), method = "average")
# Convert traits to a color representation:
# where red indicates high values
traitColors=data.frame(numbers2colors(datTraits,signed=FALSE))
dimnames(traitColors)[[2]]=paste(names(datTraits),"C",sep="")
datColors=data.frame(outlierC=outlierColor,traitColors)
# Plot the sample dendrogram and the colors underneath.
plotDendroAndColors(sampleTree,groupLabels=names(datColors),
colors=datColors,main="Sample dendrogram and trait heatmap")

# Remove outlying samples from expression and trait data
remove.samples= Z.k<thresholdZ.k | is.na(Z.k)
datExprFemale=datExprFemale[-remove.samples,]
datTraits=datTraits[-remove.samples,]
# Recompute the sample network among the remaining samples
A=adjacency(t(datExprFemale),type="distance")
# Let's recompute the Z.k values of outlyingness
k=as.numeric(apply(A,2,sum))-1
Z.k=scale(k)
```

Figure 12.1 shows the resulting cluster tree where each leaf corresponds to a mouse sample. The first color-band underneath the tree indicates which samples

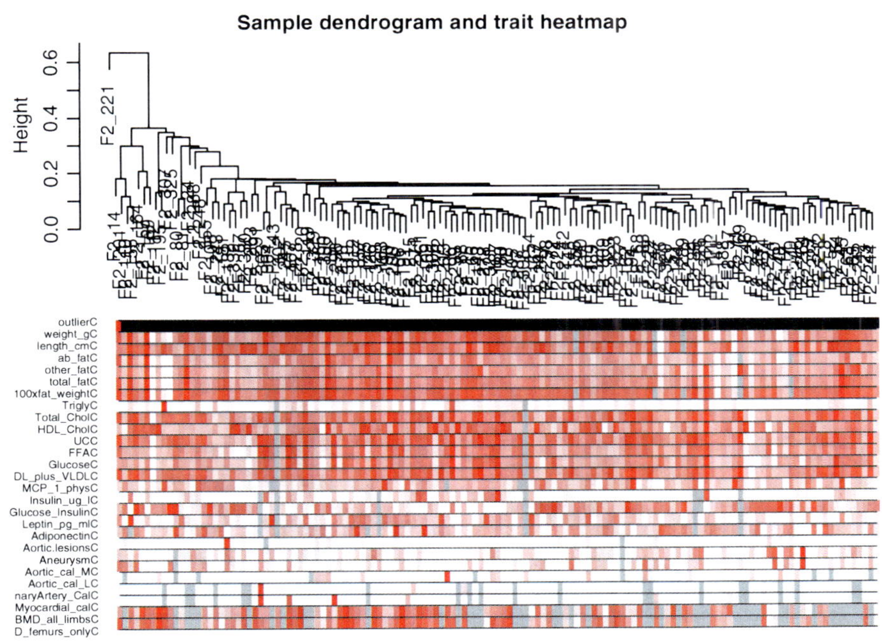

Fig. 12.1 Cluster tree of mouse liver samples. The leaves of the tree correspond to the mice. The first color band underneath the tree indicates which arrays appear to be outlying. The second color band represents body weight (*red* indicates high values). Similarly, the remaining color-bands color-code the numeric values of physiologic traits

appear outlying (colored in red) according to a low value. The mouse labeled F2_221 is highly outlying ($Z.k < -5$), which is why we remove it from the subsequent analysis. The other color bands color-code physiological traits. Note that the outlying samples do not appear to have systematically different physiologic trait values. Although the outlying samples are suspicious, we will keep them in the subsequent analysis.

12.2 Co-Expression Modules in Female Mouse Livers

12.2.1 Choosing the Soft Threshold β Via Scale-Free Topology

As detailed in Sect. 4.3, we propose to choose the soft threshold power β based on the scale-free topology criterion (Zhang and Horvath 2005), i.e., we choose the lowest β that results in approximate scale-free topology as measured by the scale-free topology fitting index (1.13)

$$R^2 = ScaleFreeFit = cor(log(p(dk)), log(BinNo))^2.$$

The following R code illustrates the use of the function `pickSoftThreshold` for calculating scale-free topology fitting indices R^2 corresponding to different soft thresholding powers β.

```
# Choose a set of soft thresholding powers
powers=c(c(1:10),seq(from=12,to=20,by=2))
# choose power based on SFT criterion
sft=pickSoftThreshold(datExprFemale,powerVector=powers)
# Plot the results:
par(mfrow=c(1,2))
# SFT index as a function of different powers
plot(sft$fitIndices[,1],
-sign(sft$fitIndices[,3])*sft$fitIndices[,2],
xlab="Soft Threshold (power)",ylab="Scale Free Topology
  Model Fit,
signed R^2",type="n",main=paste("Scale independence"))
text(sft$fitIndices[,1],
-sign(sft$fitIndices[,3])*sft$fitIndices[,2],
labels=powers,col="red")
# this line corresponds to using an R^2 cut-off of h
abline(h=0.90,col="red")
# Mean connectivity as a function of different powers
plot(sft$fitIndices[,1],sft$fitIndices[,5],type="n",
xlab="Soft Threshold (power)",ylab="Mean Connectivity",
main=paste("Mean connectivity"))
text(sft$fitIndices[,1],sft$fitIndices[,5],labels=powers,
  col="red")
```

The result is shown in Fig. 12.2. We choose the power $\beta = 6$ since this is where the curve reaches a saturation point. Also note that for this power $R^2 > 0.9$. Incidentally, $\beta = 6$ is the default choice for unsigned weighted networks. An advantage of weighted networks is that they are highly robust with regard to the power β, i.e., other choices would lead to very similar modules.

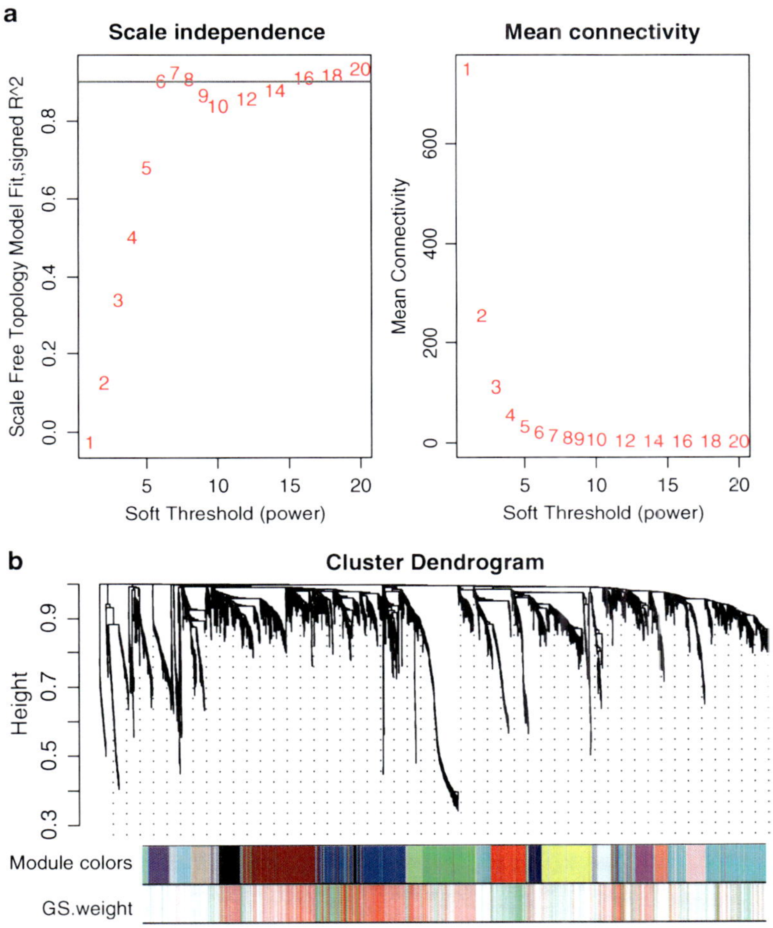

Fig. 12.2 Scale-free topology criterion and cluster tree of the female mouse liver co-expression network. (**a**) SFT plot for choosing the power β for the unsigned weighted correlation network. Left-hand side: the SFT index R^2 (y-axis) as a function of different powers β (x-axis). While R^2 tends to go up with higher powers, there is not a strictly monotonic relationship. Right-hand side: the mean connectivity (y-axis) is a strictly decreasing function of the power β (x-axis). (**b**) Hierarchical cluster tree (average linkage, dissTOM) of the 3,600 genes. The color bands provide a simple visual comparison of module assignments (branch cuttings) based on the dynamic hybrid branch cutting method. The first band shows the results from the automatic single block analysis and the second color band visualizes the gene significance measure: "*red*" indicates a high positive correlation with mouse body weight. Note that the *black*, *brown*, and *blue* modules contain many genes that have high positive correlations with body weight

12.2.2 Automatic Module Detection Via Dynamic Tree Cutting

The function `blockwiseModules` automatically implements all steps of module detection, i.e., it automatically constructs a correlation network, creates a cluster tree, defines modules as branches, and merges close modules. It outputs module colors and module eigengenes which can be used in subsequent analysis. Also one can visualize the results of the module detection.

The function `blockwiseModules` has many parameters, and in this example most of them are left at their default value. We have attempted to provide reasonable default values, but they may not be appropriate for the particular data set the reader wishes to analyze. We encourage the user to read the help file provided within the package in the R environment and experiment with changing the parameter values.

```
mergingThresh = 0.25
net = blockwiseModules(datExprFemale,corType="pearson",
maxBlockSize=5000,networkType="unsigned",power=6,minModuleSize=30,
mergeCutHeight=mergingThresh,numericLabels=TRUE,saveTOMs=TRUE,
pamRespectsDendro=FALSE,saveTOMFileBase="femaleMouseTOM")
moduleLabelsAutomatic=net$colors
# Convert labels to colors for plotting
moduleColorsAutomatic = labels2colors(moduleLabelsAutomatic)

# A data frame with module eigengenes can be obtained as follows
MEsAutomatic=net$MEs

#this is the body weight
weight = as.data.frame(datTraits$weight_g)
names(weight)="weight"
# Next use this trait to define a gene significance variable
GS.weight=as.numeric(cor(datExprFemale,weight,use="p"))
# This translates the numeric values into colors
GS.weightColor=numbers2colors(GS.weight,signed=T)
blocknumber=1
mColors=moduleColorsAutomatic[net$blockGenes[[blocknumber]]]
datColors=data.frame(mColors,GS.weightColor)
# Plot the dendrogram and the module colors underneath
plotDendroAndColors(net$dendrograms[[blocknumber]],
   colors=datColors,
groupLabels=c("Module colors","GS.weight"),dendroLabels=FALSE,
hang=0.03,addGuide=TRUE,guideHang=0.05)
```

The result can be found in Fig. 12.2b.

The parameter `maxBlockSize` tells the function how large the largest block can be. The default value is 5,000 which is appropriate for most modern desktops. Note that if this code were to be used to analyze a data set with more than 5,000 rows, the function `blockwiseModules` would split the data set into several blocks. We briefly mention the function `recutBlockwiseTrees` that can be applied to the cluster tree(s) resulting from `blockwiseModules`. This function can be used to change the branch cutting parameters without having to repeat the computationally intensive recalculation of the cluster tree(s).

12.2.3 Blockwise Module Detection for Large Networks

In this mouse co-expression network application, we work with a relatively small data set of 3,600 measured probes. However, modern microarrays measure up to 50,000 probe expression levels at once. Constructing and analyzing networks with such large numbers of nodes are computationally challenging even on a large server. We now illustrate a method, implemented in the WGCNA package, that allows the user to perform a network analysis with such a large number of genes. The basic idea is to use a two-level clustering. First, we use a fast, computationally inexpensive and relatively crude clustering method to pre-cluster genes into blocks of a manageable size. Second, perform a full network analysis in each block separately. Third, modules whose eigengenes are highly correlated are merged. The advantage of the block-wise approach is a much smaller requirement for memory (which is the main problem with large data sets on standard desktop computers), and a significant speed-up of the calculations. The tradeoff is that due to using a simpler clustering to obtain blocks, the blocks may not be optimal, causing some outlying genes to be assigned to a different module from that of a full network analysis. We will now pretend that even the relatively small number of genes, 3,600, that we have been using here is too large, and the computer we run the analysis on is not capable of handling more than 2,000 genes in one block. The automatic network construction and module detection function `blockwiseModules` can handle the splitting into blocks automatically; the user just needs to specify the largest number of genes that can form a block:

```
bwnet = blockwiseModules(datExprFemale, maxBlockSize = 2000,
power = 6, minModuleSize = 30,reassignThreshold = 0,
mergeCutHeight = 0.25, numericLabels = TRUE, saveTOMs = TRUE,
saveTOMFileBase = "femaleMouseTOM-blockwise",verbose = 3)
```

Below we will compare the results of this analysis to the results of Sect. 12.2.2 in which all genes were analyzed in a single block. To make the comparison easier, we relabel the block-wise module labels so that modules with a significant overlap with single-block modules have the same label:

```
# Relabel blockwise modules so that their labels
# match those from our previous analysis
moduleLabelsBlockwise=matchLabels(bwnet$colors,
     moduleLabelsAutomatic)
# Convert labels to colors for plotting
moduleColorsBlockwise=labels2colors(moduleLabelsBlockwise)

# measure agreement with single block automatic procedure
mean(moduleLabelsBlockwise==moduleLabelsAutomatic) #=0.8219444
```

The colorbands in Fig. 12.4 allow one to compare the results of the two-block analysis with those from the single block analysis. Clearly, there is excellent agreement between the two automatic module detection methods.

The hierarchical clustering dendrograms (trees) used for the module identification in the ith block are returned in `bwnet$dendrograms[[i]]`. For example, the cluster tree in the second block can be visualized with the following code:

```
blockNumber=2
# Plot the dendrogram for the chosen block
plotDendroAndColors(bwnet$dendrograms[[blockNumber]],
moduleColorsBlockwise[bwnet$blockGenes[[blockNumber]]],"Module
  colors",
main=paste("Dendrogram and module colors in block",blockNumber),
dendroLabels=FALSE,hang=0.03,addGuide=TRUE,guideHang=0.05)
```

The function `recutBlockwiseTrees` can be used to change parameters of the branch cutting procedure without having to recompute the cluster tree.

12.2.4 Manual, Stepwise Module Detection

Here we present R code that implements the following steps of module detection using the default WGCNA approach.

1. Define a weighted adjacency matrix A.
2. Define the topological overlap matrix (1.27) based dissimilarity measure $dissTOM_{ij} = 1 - TopOverlap_{ij}$.
3. Construct a hierarchical cluster tree (average linkage).
4. Define modules as branches of the tree.

```
# We now calculate the weighted adjacency matrix, using the
   power 6:
A = adjacency(datExprFemale, power = 6)
#define a dissimilarity based on the topological overlap
dissTOM =TOMdist(A)
#hierarchical clustering
geneTree = flashClust(as.dist(dissTOM),method="average")
```

Branches of the dendrogram group together densely interconnected, highly co-expressed genes. Modules are defined as branches of the cluster tree, i.e., module detection involves cutting the branches of the tree. There are several methods for branch cutting. As described in Sect. 8.6, the most flexible approach is the dynamic tree cutting method implemented in the `cutreeDynamic` R function (Langfelder et al. 2007). Although the default values of the branch cutting method work well, the user may explore choosing alternative parameter values. In the following R code, we choose a relatively large minimum module size of `minClusterSize=30`, and a medium sensitivity (`deepSplit=2`) to branch splitting.

```
# here we define the modules by cutting branches
moduleLabelsManual1=cutreeDynamic(dendro=geneTree,distM=dissTOM,
method="hybrid",deepSplit=2,pamRespectsDendro=F,
  minClusterSize=30)

# Relabel the manual modules so that their labels
# match those from our previous analysis
moduleLabelsManual2=
     matchLabels(moduleLabelsManual1,moduleLabelsAutomatic)
# Convert labels to colors for plotting
moduleColorsManual2=labels2colors(moduleLabelsManual2)
```

Clustering methods may identify modules whose expression profiles are very similar. More specifically, a module detection method may result in modules whose eigengenes are highly correlated. Since highly correlated modules are not distinct, it may be advisable to merge them. The following code shows how to create a cluster tree of module eigengenes and how to merge them (if their pairwise correlation is larger than 0.75).

```
# Calculate eigengenes
MEList=moduleEigengenes(datExprFemale,colors=moduleColorsManual2)
MEs = MEList$eigengenes
# Add the weight to existing module eigengenes
MET=orderMEs(cbind(MEs,weight))
# Plot the relationships among the eigengenes and the trait
plotEigengeneNetworks(MET,"",marDendro=c(0,4,1,2),
marHeatmap=c(3,4,1,2),cex.lab=0.8,xLabelsAngle=90)
```

Figure 12.3 shows the result of this analysis.

```
# automatically merge highly correlated modules
merge=mergeCloseModules(datExprFemale,moduleColorsManual2,
cutHeight=mergingThresh)
# resulting merged module colors
moduleColorsManual3 = merge$colors
# eigengenes of the newly merged modules:
MEsManual = merge$newMEs

# Show the effect of module merging by plotting the
# original and merged module colors below the tree
datColors=data.frame(moduleColorsManual3,moduleColorsAutomatic,
        moduleColorsAutomatic,GS.weightColor)
plotDendroAndColors(geneTree,colors=datColors,
groupLabels=c("manual hybrid","single block","2 block",
  "GS.weight"),
dendroLabels=FALSE,hang=0.03,addGuide=TRUE,guideHang=0.05)

# check the agreement between manual and automatic module labels
mean(moduleColorsManual3==moduleColorsAutomatic)
```

The result can be found in Fig. 12.4.

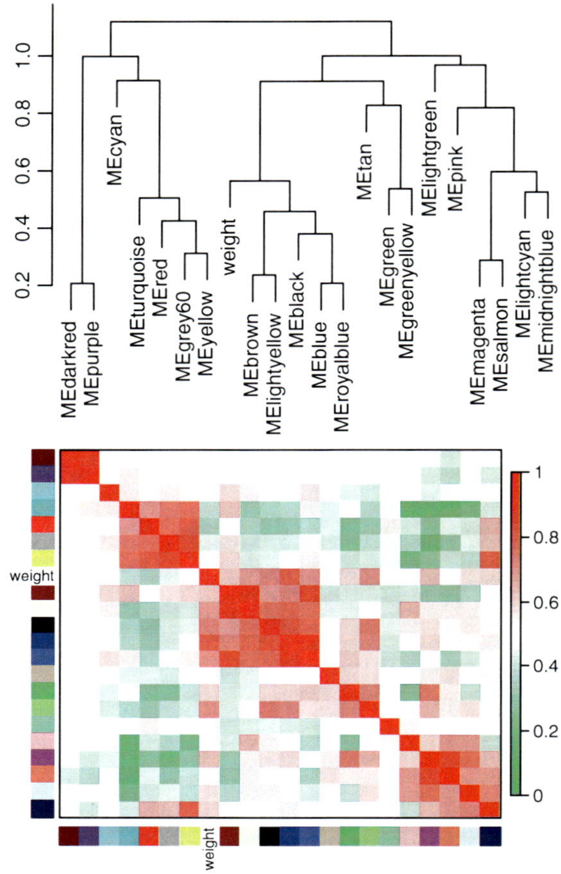

Fig. 12.3 Visualization of the eigengene network representing the relationships among the modules and the body weight. The *top panel* shows a hierarchical clustering dendrogram of the eigengenes based on the dissimilarity $diss(q_1, q_2) = 1 - cor(E^{(q_1)}, E^{(q_2)})$. The *bottom panel* shows the eigengene network, which is defined by the signed weighted correlation network $A_{q1,q2} = 0.5 + 0.5cor(E^{(q_1)}, E^{(q_2)})$

12.2.5 Relating Modules to Physiological Traits

Here we show how to identify modules that are significantly associated with the physiological traits. Since the module eigengene is an optimal summary of the gene

Fig. 12.4 (**a**) Cluster tree of 3,600 genes of the mouse liver co-expression network. The color bands provide a simple visual comparison of module assignments (branch cuttings) from three different branch cutting methods based on the dynamic hybrid branch cutting method. The first band shows the results from the manual (interactive) branch cutting approach, the second band shows the results of the automatic single block analysis, and the third band shows the results from the two block analysis. Note the high agreement among the methods. This illustrates that little is lost when using the blockwise approach. (**b**) Table of module-trait correlations and p values. Each cell reports the correlation (and p value) resulting from correlating module eigengenes (rows) to traits (columns). The table is color-coded by correlation according to the color legend

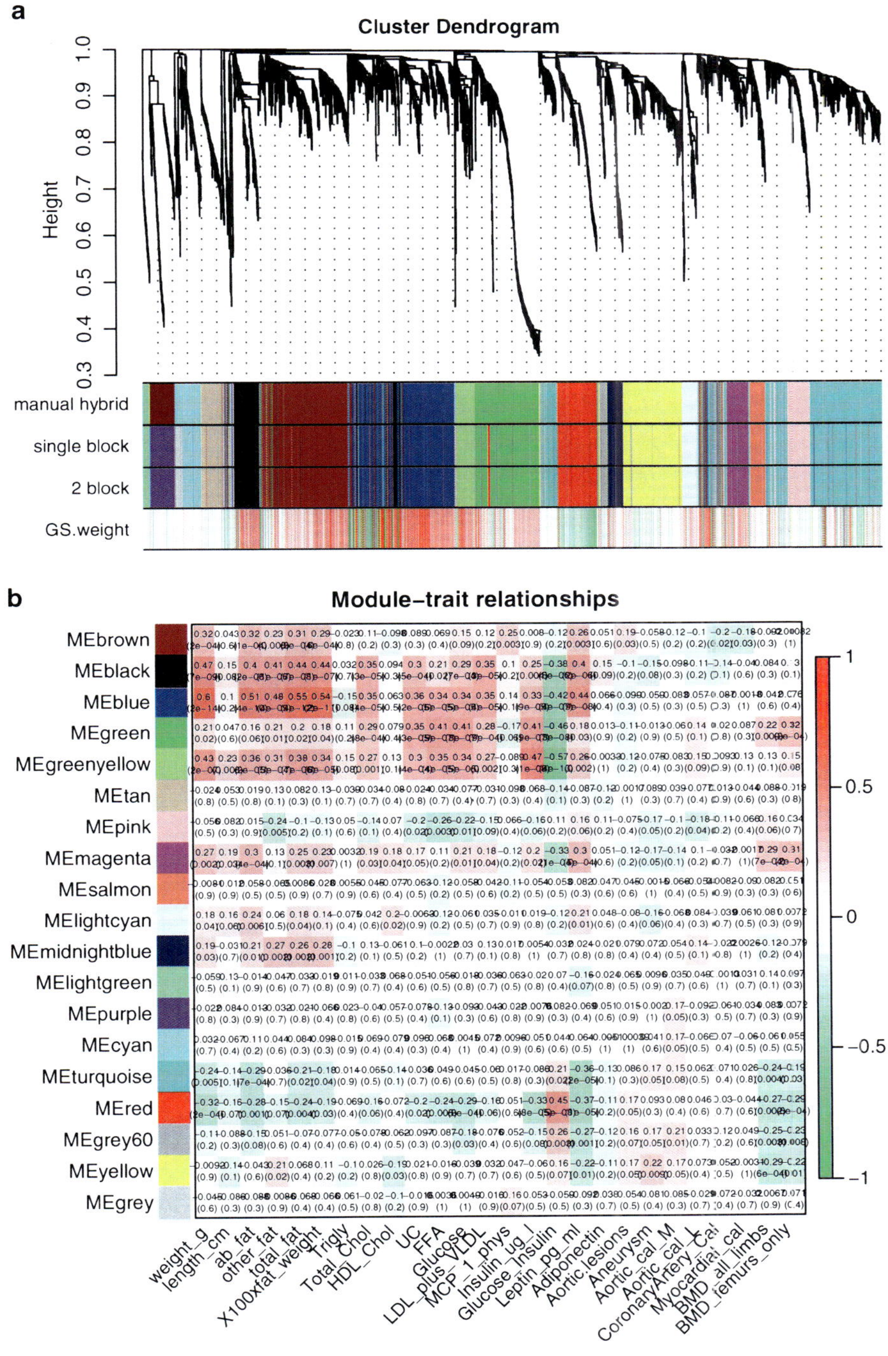
a
Cluster Dendrogram
Height
0.3
0.4
0.5
0.6
0.7
0.8
0.9
1.0
manual hybrid
single block
2 block
GS.weight
b
Module-trait relationships
MEbrown
MEblack
MEblue
MEgreen
MEgreenyellow
MEtan
MEpink
MEmagenta
MEsalmon
MElightcyan
MEmidnightblue
MElightgreen
MEpurple
MEcyan
MEturquoise
MEred
MEgrey60
MEyellow
MEgrey
1
0.5
0
−0.5
−1
weight_g
length_cm
ab_fat
other_fat
total_fat
X100xfat_weight
Trigly
Total_Chol
HDL_Chol
UC
FFA
Glucose
LDL_plus_VLDL
MCP_1_phys
Insulin_ug_l
Glucose_Insulin
Leptin_pg_ml
Adiponectin
Aortic.lesions
Aneurysm
Aortic_cal_M
Aortic_cal_L
CoronaryArtery_Cal
Myocardial_cal
BMD_all_limbs
BMD_femurs_only

expression profiles of a given module, it is natural to correlate eigengenes with these traits and to look for the most significant associations.

```
# Choose a module assignment
moduleColorsFemale=moduleColorsAutomatic
# Define numbers of genes and samples
nGenes = ncol(datExprFemale)
nSamples = nrow(datExprFemale)
# Recalculate MEs with color labels
MEs0 = moduleEigengenes(datExprFemale,moduleColorsFemale)
  $eigengenes
MEsFemale = orderMEs(MEs0)
modTraitCor = cor(MEsFemale, datTraits, use = "p")
modTraitP = corPvalueStudent(modTraitCor, nSamples)
#Since we have a moderately large number of modules and traits,
#a suitable graphical representation will help in reading
#the table. We color code each association by the correlation
  value:
# Will display correlations and their p-values
textMatrix = paste(signif(modTraitCor, 2), "\n(",
signif(modTraitP, 1), ")", sep = "")
dim(textMatrix) = dim(modTraitCor)
par(mar = c(6, 8.5, 3, 3))
# Display the correlation values within a heatmap plot
labeledHeatmap(Matrix = modTraitCor, xLabels = names(datTraits),
yLabels = names(MEsFemale), ySymbols = names(MEsFemale),
colorLabels=FALSE,colors=greenWhiteRed(50),textMatrix=textMatrix,
setStdMargins = FALSE, cex.text = 0.5, zlim = c(-1,1),
main = paste("Module-trait relationships"))
```

The resulting color-coded table is shown in Fig. 12.4b. The analysis identifies several significant module-trait associations. We will concentrate on weight (first column) as the trait of interest.

For each module, we also define a quantitative measure of module membership MM as the correlation of the module eigengene and the gene expression profile. This allows us to quantify the similarity of all genes on the array to every module.

```
# calculate the module membership values
# (aka. module eigengene based connectivity):
datKME=signedKME(datExprFemale, MEsFemale)
```

Intramodular analysis: identifying genes with high GS and MM Using the GS and MM measures, we can identify genes that have a high significance for weight as well as high module membership in interesting modules. As an example, we look at the blue module that the highest correlation with body weight. We plot a scatterplot of Gene Significance versus Module Membership in the blue module:

```
colorOfColumn=substring(names(datKME),4)
par(mfrow = c(2,2))
selectModules=c("blue","brown","black","grey")
par(mfrow=c(2,length(selectModules)/2))
for (module in selectModules) {
column = match(module,colorOfColumn)
```

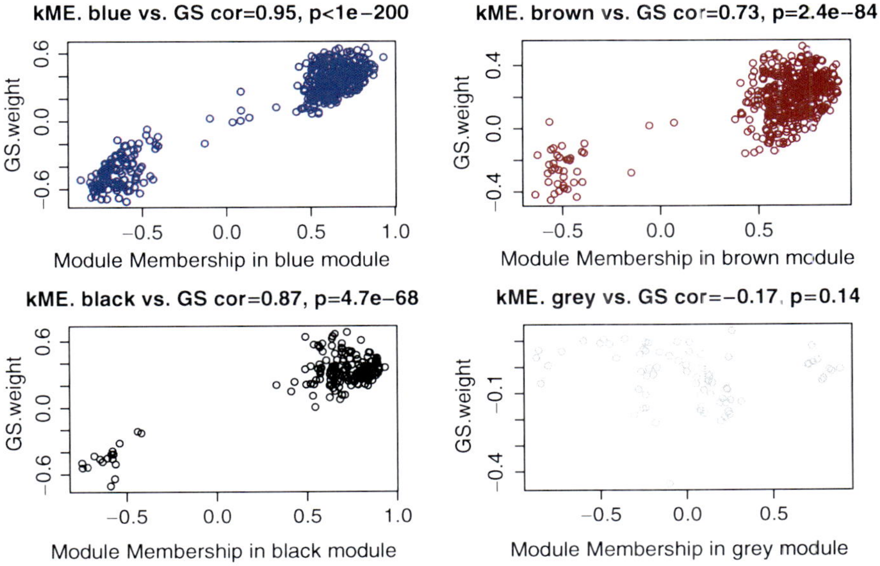

Fig. 12.5 Gene significance (GS.weight) versus module membership (kME) for the body weight related modules. GS.weight and MM.weight are highly correlated reflecting the high correlations between weight and the respective module eigengenes

```
restModule=moduleColorsFemale==module
verboseScatterplot(datKME[restModule,column],GS.weight
        [restModule],
xlab=paste("Module Membership ",module,"module"),
      ylab="GS.weight",
main=paste("kME.",module,"vs. GS"),col=module)
}
```

The result is presented in Fig. 12.5. We find that the blue, brown, and black modules contain genes that have high positive and high negative correlations with body weight. In contrast, the gray "background" genes show only weak correlations with weight.

12.2.6 Output File for Gene Ontology Analysis

Here we present code for merging the probe data with a gene annotation table (which contains locus link IDs and other information for the probes). Next, we show how to export the results. The list of locus link identifiers (corresponding to the genes inside a given module) can be used as input of gene ontology (GO) and functional enrichment analysis software tools such as David (EASE) (http://david.abcc.ncifcrf.gov/) or AmiGO, Webgestalt. The R software also contains relevant packages, e.g., GOSim (Frohlich et al. 2007). Most gene ontology-based functional enrichment analysis software programs simply take lists of gene identifiers as input. Ingenuity pathway analysis allows the user to input gene expression data or gene identifiers.

The following code shows how to add gene identifiers and how to output the network analysis results.

```
# Read in the probe annotation
GeneAnnotation=read.csv(file="GeneAnnotation.csv")
# Match probes in the data set to those of the annotation file
probes = names(datExprFemale)
probes2annot = match(probes,GeneAnnotation$substanceBXH)
# data frame with gene significances (cor with the traits)
datGS.Traits=data.frame(cor(datExprFemale,datTraits,use="p"))
names(datGS.Traits)=paste("cor",names(datGS.Traits),sep=".")
datOutput=data.frame(ProbeID=names(datExprFemale),
GeneAnnotation[probes2annot,],moduleColorsFemale,datKME,
datGS.Traits)
# save the results in a comma delimited file
write.table(datOutput,"FemaleMouseResults.csv",row.names=F,
  sep=",")
```

12.3 Systems Genetic Analysis with NEO

The identification of pathways and genes underlying complex traits using standard mapping techniques has been difficult due to genetic heterogeneity, epistatic interactions, and environmental factors. One promising approach to this problem involves the integration of genetics and gene expression.

Here, we describe a systems genetic approach for integrating gene co-expression network analysis with genetic data (Ghazalpour et al. 2006; Presson et al. 2008). The following steps summarize our overall approach:

1. Construct a gene co-expression network from gene expression data (see Sect. 12.2.2)
2. Study the functional enrichment (gene ontology, etc.) of network modules (see Sect. 12.2.6)
3. Relate modules to the clinical traits (see Sect. 12.2.5)
4. Identify chromosomal locations (QTLs) regulating modules and traits
5. Use QTLs as causal anchors in network edge orienting analysis (described in Chap. 11) to find causal drivers underlying the clinical traits

In step 4, we identify chromosomal locations (genetic markers) that relate to body weight and module eigengenes. The traditional quantitative trait locus (QTL) mapping approach relates clinical traits (e.g., body weight) to genetic markers directly. To relate quantitative traits to a single nucleotide polymorphism (SNP) marker, one can use a correlation test (Sect. 10.3) if the SNP is numerically coded. For example, an additive marker coding of an SNP results in trivariate ordinal variable, which takes on values $0, 1, 2$. Alternatively, one can carry out an LOD score analysis using the `qtl` R package (Broman et al. 2003). In practice, we find that results from a single locus LOD score analysis (with additive marker coding) are comparable to those from a correlation test analysis. Several authors have proposed a strategy that uses gene expression data to help map clinical traits (Jansen and Nap 2001, Schadt et al. 2003).

We used 1,065 single nucleotide polymorphism (SNP) markers that were evenly spaced across the mouse genome, to map gene expression values for all genes within each gene module (Ghazalpour et al. 2006). A **module quantitative trait locus** (mQTL) is a chromosomal location (e.g., an SNP) which correlates with many gene expression profiles of a given module. We found that the blue module, which was highly correlated with body weight, has nine mQTLs. These are located on chromosomes 2, 3, 5, 10 (middle), 10 (distal), 12 (proximal), 12 (middle), 14, and 19. In particular, the mQTL on chromosome 19 has a highly significant correlation with the blue module eigengene and mouse body weight. We use this mQTL (denoted $mQTL19.047$) as causal anchor to calculate Local Edge Orienting (LEO) scores for (a) determining whether the blue module eigengene has a causal effect on body weight, and (b) identifying genes inside the blue module that have a causal effect on body weight. For example, for the ith expression profile x_i in the blue module, we calculate the $LEO.SingleAnchor$ score (Sect. 11.8.1) for assessing the relative causal fit of

$$mQTL19.047 \rightarrow x_i \rightarrow weight.$$

The following assumes that the R code from Sect. 11.8.1 has been read into the R session.

```
# Get SNP markers which encode the mQTLs of the blue module
SNPdataFemale=read.csv("SNPandModuleQTLFemaleMice.csv")
# find matching row numbers to line up the SNP data
snpRows=match(dimnames(datExprFemale)[[1]],SNPdataFemale$Mice)
# define a data frame whose rows correspond to those of datExpr
datSNP = SNPdataFemale[snpRows,-1]
rownames(datSNP)=SNPdataFemale$Mice[snpRows]
# show that row names agree
table(rownames(datSNP)==rownames(datExprFemale))

SNP=as.numeric(datSNP$mQTL19.047)
weight=as.numeric(datTraits$weight_g)
MEblue=MEsFemale$MEblue

# evaluate the relative causal fit
# of SNP -> MEblue -> weight
LEO.SingleAnchor(M1=SNP,A=MEblue,B=weight)[[1]]
# output
 -2.823433
```

The negative LEO score indicates that there is no evidence for a causal relationship $MEblue \rightarrow weight$ if $mQTL19.047$ is used as causal anchor. However, we caution the reader that other causal anchors (e.g., SNPs in other chromosomal locations) may lead to a different conclusion. The critical step in a NEO analysis is to find valid causal anchors. While the module eigengene has no causal effect on weight, it is possible that individual genes inside the blue module may have a causal effect. The following code shows how to identify genes inside the blue module that have a causal effect on weight.

```
whichmodule="blue"
restGenes=moduleColorsFemale==whichmodule
datExprModule=datExprFemale[,restGenes]
attach(datExprModule)
LEO.SingleAnchorWeight=rep(NA, dim(datExprModule)[[2]] )
for (i in c(1:dim(datExprModule)[[2]]) ){
printFlush(i)
LEO.SingleAnchorWeight[i]=
LEO.SingleAnchor(M1=SNP,A=datExprModule[,i],B=weight)[[1]]}

# Find gens with a very significant LEO score
restCausal=LEO.SingleAnchorWeight>3 # could be lowered to 1
names(datExprModule)[restCausal]

# this provides us the corresponding gene symbols
data.frame(datOutput[restGenes,c(1,6)])[restCausal,]
#output
       ProbeID gene_symbol
7145 MMT00024300 Cyp2c40
8873 MMT00030448 4833409A17Rik
11707 MMT00041314
```

Note that the NEO analysis implicates three probes, one of which corresponds to the gene Cyp2c40. Using alternative mouse cross data, gene Cyp2c40 had already been found to be related to body weight in rodents (reviewed in (Ghazalpour et al. 2006)). Let us determine pairwise correlations

```
signif(cor(data.frame(SNP,MEblue,MMT00024300,weight),use="p"),2)
#output

              SNP    MEblue  MMT00024300  weight
SNP            1.00   0.17   -0.70         0.32
MEblue         0.17   1.00   -0.49         0.60
MMT00024300   -0.70  -0.49    1.00        -0.54
weight         0.32   0.60   -0.54         1.00
```

Note that the causal anchor (SNP) has a weak correlation ($r = 0.17$) with *MEblue*, a strong negative correlation ($r = -0.70$) with probe MMT00024300 of gene Cyp2c40, and a positive correlation with weight ($r = 0.32$). The differences in pairwise correlations explain why $MMT00024300$ is causal for weight while *MEblue* is not. Further, note that MMT00024300 has only a moderately high module membership value of $kMEblue = -0.49$, i.e., this one is *not* a hub gene in the blue module. This makes sense in light of our finding that the module eigengene (which can be interpreted as the ultimate hub gene) is not causal for body weight.

In Presson et al. (2008), we show how one can integrate module membership and causal testing analysis to develop systems genetic gene screening criteria.

12.4 Visualizing the Network

Module structure and network connections can be visualized in several different ways. While R offers a host of network visualization functions, there are also valuable external software packages.

12.4.1 Connectivity, TOM, and MDS Plots

A **connectivity plot** of a network is simply a heatmap representation of the connectivity patterns of the adjacency matrix or another measure of pairwise node interconnectedness. In an undirected network, the connectivity plot is symmetrical around the diagonal and the rows and columns depict network nodes (which are arranged in the same order). Often the nodes are sorted to highlight the module structure. For example, when the topological overlap matrix is used as measure of pairwise interconnectedness, the **topological overlap matrix (TOM) plot** (Ravasz et al. 2002) is a special case of a connectivity plot. The TOM plot provides a "reduced" view of the network that allows one to visualize and identify network modules. The TOM plot is a color-coded depiction of the values of the TOM measure whose rows and columns are sorted by the hierarchical clustering tree (which used the TOM-based dissimilarity as input).

As an example, consider Fig. 12.6a where red/yellow indicates high/low values of TOM_{ij}. Both rows and columns of TOM have been sorted using the hierarchical clustering tree. Since TOM is symmetric, the TOM plot is also symmetric around the diagonal. Since modules are sets of nodes with high topological overlap, modules correspond to red squares along the diagonal.

Each row and column of the heatmap corresponds to a single gene. The heatmap can depict topological overlap, with light colors denoting low adjacency (overlap) and darker colors higher adjacency (overlap). In addition, the cluster tree and module colors are plotted along the top and left side of the heatmap. Figure 12.6 was created using the following code.

```
# Set the diagonal of the TOM disscimilarity to NA
diag(dissTOM) = NA
# Transform dissTOM with a power to enhance visibility
TOMplot(dissTOM^7,geneTree,moduleColorsFemale,
main = "Network heatmap plot, all genes")
```

The TOM plot can be found in Figure 12.6.

Multidimensional scaling (MDS) is an approach for visualizing pairwise relationships specified by a dissimilarity matrix. Each row of the dissimilarity matrix is visualized by a point in a Euclidean space. The Euclidean distances between a pair of points is chosen to reflect the corresponding pairwise dissimilarity. An MDS algorithm starts with a dissimilarity matrix as input and assigns a location to each row (object, node) in d-dimensional space, where d is specified a priori. If $d = 2$, the resulting point locations can be displayed in the Euclidean plane. There are two major

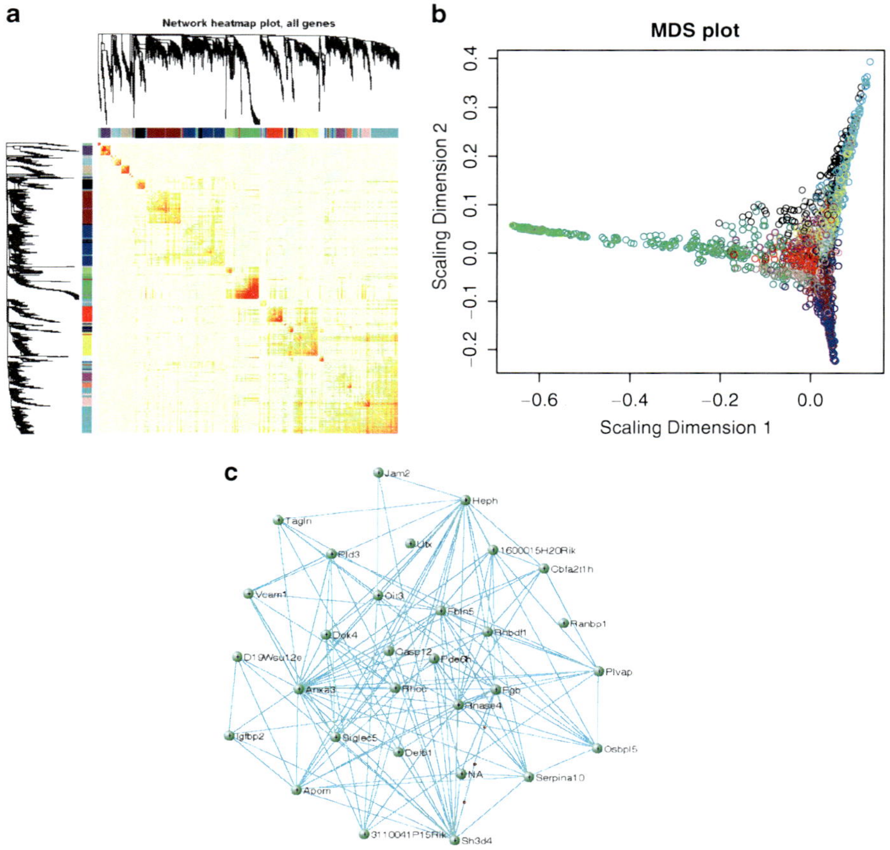

Fig. 12.6 Visualing the network connections. (**a**) Topological overlap matrix (TOM) plot (also known as connectivity plot) of the network connections. Genes in the rows and columns are sorted by the clustering tree. Light color represents low topological overlap and progressively darker red color represents higher overlap. Clusters correspond to squares along the diagonal. The cluster tree and module assignment are also shown along the left side and the top. (**b**) Classical multidimensional scaling plot whose input is the TOM dissimilarity. Each dot (gene) is colored by the module assignment. (**c**) VisANT plot of the connections among the top 30 most highly connected blue module genes of the female liver network. An edge threshold of 0.02 was chosen in the VisANT threshold

types of MDS: classical MDS (implemented in the R function `cmdscale`) and non-metric MDS (R functions `isoMDS` and `sammon`). The following R code shows how to create a classical MDS plot using $d = 2$ scaling dimensions.

```
cmd1=cmdscale(as.dist(dissTOM),2)
par(mfrow=c(1,1))
plot(cmd1,col=moduleColorsFemale,main="MDS plot",
xlab="Scaling Dimension 1",ylab="Scaling Dimension 2")
```

The resulting plot can be found in Fig. 12.6.

12.4.2 VisANT Plot and Software

VisANT (Hu et al. 2004) is a software framework for visualizing, mining, analyzing, and modeling multi-scale biological networks. The software can handle large-scale networks. It is also able to integrate interaction data, networks and pathways. The implementation of metagraphs enables VisANT to integrate multi-scale knowledge. The WGCNA R function `exportNetworkToVisANT` allows the user to export networks from R into a format suitable for ViSANT.

The following R code shows how to create a VisANT input file for visualizing the intramodular connections of the blue module.

```
# Recalculate topological overlap
TOM = TOMsimilarityFromExpr(datExprFemale, power = 6)
module = "blue"
# Select module probes
probes = names(datExprFemale)
inModule = (moduleColorsFemale==module)
modProbes = probes[inModule]
# Select the corresponding Topological Overlap
modTOM = TOM[inModule, inModule]
# Because the blue module is rather large,
# we focus on the 30 top intramodular hub genes
nTopHubs = 30
# intramodular connectivity
kIN = softConnectivity(datExprFemale[, modProbes])
selectHubs = (order(-kIN) <= nTopHubs)
vis = exportNetworkToVisANT(modTOM[selectHubs,selectHubs],
file=paste("VisANTInput-", module, "-top30.txt", sep=""),
weighted=TRUE,threshold = 0, probeToGene=
data.frame(GeneAnnotation$substanceBXH,GeneAnnotation$
      gene_symbol))
```

To provide an example of a VisANT visualization, we loaded the file produced by the above code in VisANT. For better readability, we varied the threshold for displaying a link between two nodes. The results are shown in Fig. 12.6c.

12.4.3 Cytoscape and Pajek Software

The WGCNA R function `exportNetworkToCytoscape` allows the user to export networks in a format suitable for Cytoscape (Shannon et al. 2003). Cytoscape is a widely used software platform for visualizing networks and biological pathways and integrating these networks with annotations, gene expression profiles, and other data. Although Cytoscape was originally designed for biological research, it is a general platform for complex network analysis and visualization. Cytoscape core distribution provides a basic set of features for data integration and visualization. Additional features are (freely) available as plugins. For example, plugins are available for network and molecular profiling analyses, scripting, and connection with databases.

Cytoscape allows the user to input an edge file and a node file, which allow one to specify connection strenghts (link weights) and node colors. The following R code shows how to create an input file for Cytoscape.

```
# select modules
modules = c("blue","brown")
# Select module probes
inModule=is.finite(match(moduleColorsFemale,modules))
modProbes=probes[inModule]
match1=match(modProbes,GeneAnnotation$substanceBXH)
modGenes=GeneAnnotation$gene_symbol[match1]
# Select the corresponding Topological Overlap
modTOM = TOM[inModule, inModule]
dimnames(modTOM) = list(modProbes, modProbes)
# Export the network into edge and node list files for Cytoscape
cyt = exportNetworkToCytoscape(modTOM,
edgeFile=paste("CytoEdge",paste(modules,collapse="-"),
     ".txt",sep=""),
nodeFile=paste("CytoNode",paste(modules,collapse="-"),
     ".txt",sep=""),
weighted = TRUE, threshold = 0.02,nodeNames=modProbes,
altNodeNames = modGenes, nodeAttr = moduleColorsFemale[inModule])
```

Inputting the network into Cytoscape is a bit involved since many options exist for interpreting edges and nodes. We refer the reader to Cytoscape documentation for the necessary details.

Another powerful network visualization software is called Pajek (de Nooy et al. 2005). This freely available software is widely used in many social network and biological applications.

12.5 Module Preservation Between Female and Male Mice

In the following, we present R code for studying the preservation of gene co-expression modules between female and male mouse liver networks. The mouse data are described in Sect. 5.5 and in Ghazalpour et al. (2006) and Fuller et al. (2007). In Sect. 12.2, we present R code for defining the modules as branches of a hierarchical clustering tree. This application demonstrates that module preservation statistics allow us to identify invalid, non-reproducible modules due to array outliers.

We use the `WGCNA` function `modulePreservation` to study module preservation of female modules in the male data.

```
maleData = read.csv("LiverMaleFromLiverFemale3600.csv")
names(maleData)
datExprMale=data.frame(t(maleData[,-c(1:8)]))
names(datExprMale)=maleData$substanceBXH

# We now set up the multi-set expression data
# and corresponding module colors:
setLabels = c("Female", "Male")
```

```
multiExpr=list(Female=list(data=datExprFemale),
        Male=list(data=datExprMale))
multiColor=list(Female=moduleColorsFemale)

# The number of permutations drives the computation time
# of the module preservation function.
# Set it to a low number (e.g., 3) if only the medianRank
  statistic
# and other observed statistics are needed.
# Permutations are only needed for calculating Zsummary
# and other permutation test statistics.
  nPermutations1=200
# set the random seed of the permutation test analysis
set.seed(1)
system.time({
mp = modulePreservation(multiExpr, multiColor,
referenceNetworks = 1, nPermutations = nPermutations1,
randomSeed = 1, quickCor = 0, verbose = 3)
  })
# Save the results of the module preservation analysis
save(mp, file = "modulePreservation.RData")
# If needed, reload the data:
load(file = "modulePreservation.RData")

# specify the reference and the test networks
ref=1; test = 2

Obs.PreservationStats= mp$preservation$observed[[ref]][[test]]
Z.PreservationStats=mp$preservation$Z[[ref]][[test]]
# Look at the observed preservation statistics
Obs.PreservationStats
# Z statistics from the permutation test analysis
Z.PreservationStats

# Let us now visualize the data.
modColors = rownames(Obs.PreservationStats)
moduleSize = Obs.PreservationStats$moduleSize
# we will omit the grey module (background genes)
# and the gold module (random sample of genes)
selectModules = !(modColors %in% c("grey", "gold"))
# Text labels for points
point.label = modColors[selectModules]

#Composite preservation statistics
medianRank=Obs.PreservationStats$medianRank.pres
Zsummary=Z.PreservationStats$Zsummary.pres

par(mfrow=c(2,2),mar = c(4.5,4.5,2.5,1))
# plot medianRank versus module size
plot(moduleSize[selectModules],medianRank[selectModules],col=1,
bg=modColors[selectModules],pch = 21,main="medianRank
  Preservation",
cex = 2, ylab ="medianRank",xlab="Module size", log="x")
```

```
labelPoints(moduleSize[selectModules],medianRank[selectModules],
         point.label,cex=1,offs=0.03)
# plot Zsummary versus module size
plot(moduleSize[selectModules],Zsummary[selectModules], col = 1,
bg=modColors[selectModules],pch = 21,
      main="Zsummary Preservation",
cex=2,ylab ="Zsummary", xlab = "Module size", log = "x")

labelPoints(moduleSize[selectModules],Zsummary[selectModules],
         point.label,cex=1,offs=0.03)
# Add threshold lines for Zsummary
abline(h=0); abline(h=2, col = "blue", lty = 2)
abline(h=10, col = "red", lty = 2)

whichmodule="lightgreen"
Eigengene=MEsFemale$MElightgreen
datExprModule=datExprFemale[,moduleColors==whichmodule]
# set the margins of the graphics window
# par(mar=c(0.3, 5.5, 3, 2))
# create a heatmap whose columns correspond to the arrays
# and whose rows correspond to genes
plot.mat(t(scale(datExprModule)),cex.axis=2,nrgcols=30,rlabels=F,
rcols=whichmodule,main=paste("heatmap",whichmodule,"module"))

#scatter plot between eigengene and sample network connectivity
verboseScatterplot(Eigengene,Z.k,
xlab=paste("ME",whichmodule,sep=""))
abline(h=-2,col="red",lwd=2)
```

The resulting plot is shown in Fig. 12.7. We find that most modules show strong evidence of preservation $Z_{summary} > 10$. The exception is the lightgreen module that has the second highest *medianRank* statistic implying that all but one of the other modules show stronger evidence of preservation. The lightgreen module also shows no evidence of preservation according to the $Z_{summary}$ statistic for which it obtains the lowest value of 2.00.

A heatmap of the lightgreen module expression values is shown in Fig. 12.7. Note that most genes are underexpressed in a single female mouse, which suggests that this module is due to an array outliers.

The lower right panel shows a scatter plot of MElightgreen versus the standardized sample network connectivity $Z.k$ (7.24), which is the measure of outlyingness described in Sect. 12.1. Note that the eigenvector takes on an extreme negative value for one female mouse sample (corresponding to the sample where most module genes are underexpressed). The outlying female mouse sample would have been removed if we had applied a stringent threshold of -2 to $Z.k$. This supports the golden rule of data preprocessing: when in doubt, throw it out.

In Sect. 8.7, we described several cluster quality statistics based on network concepts. In Table 8.1, we provide an overview of the cluster quality statistics (8.12). In the following, we present R code for evaluating the cluster quality of the modules inside the female mouse network.

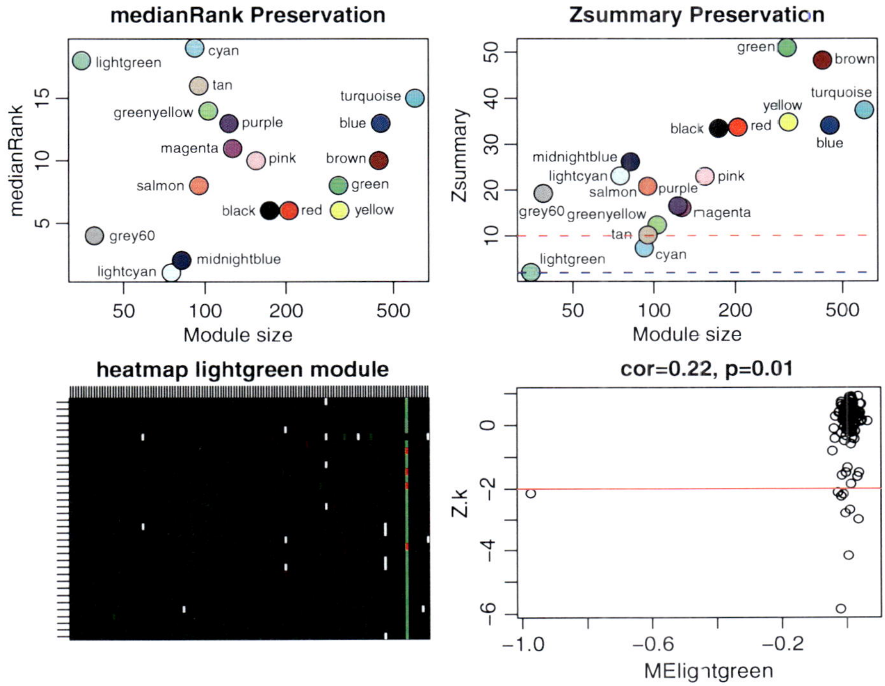

Fig. 12.7 Preservation of female mouse liver modules in male livers. The *upper left panel* shows the composite statistic *medianRank* (9.20) versus module size. The higher the *medianRank*, the less preserved is the module relative to other modules. Since *medianRank* is based on the observed preservation statistics (as opposed to Z statistics or p values), we find that it is much less dependent on module size. The *upper right panel* shows the composite statistic $Z_{summary}$ (9.1). If $Z_{summary} > 10$, there is strong evidence that the module is preserved (Langfelder et al. 2011). If $Z_{summary} < 2$, there is no evidence that the module preserved. Note that $Z_{summary}$ shows a strong dependence on module size. The lightgreen female module shows no evidence of preservation in the male liver network. The *lower left panel* shows a heatmap of the lightgreen module gene expressions (rows correspond to genes, columns correspond to female mouse tissue samples). The heatmap color-codes the scaled gene expression values: *red* corresponds to overexpression and green to underexpression. The barplots underneath the heatmap show the expression levels (y-axis) of the module eigengene across the mouse samples (x-axis). Note that the eigenvector takes on an extreme negative value (*low bar*) in one mouse sample where most module genes are underexpressed (*green*). This suggests that the lightgreen module is due to an array outliers. The *lower right panel* shows a scatter plot of the lightgreen module eigengene (x-axis) versus the standardized sample network connectivity $Z.k$ (7.24), which is the measure of outlyingness described in Sect. 12.1. Note that the module eigengene takes on an extreme negative value for one female mouse sample (corresponding to the sample where most module genes are underexpressed). The *red horizontal line* in the scatter plot corresponds to a $Z.k$ threshold of -2

```
# This calculates the quality of the modules in the female data
Obs.QualityStatsFemale= mp$quality$observed[[1]][[2]]
Z.QualityStatsFemale=mp$quality$Z[[1]][[2]]
QualityStats=data.frame(Obs.QualityStatsFemale[,1:2],
Zsummary.qual=Z.QualityStatsFemale$Zsummary.qual)
```

```
QualityStats["lightgreen",]
# Output
          moduleSize medianRank.qual Zsummary.qual
lightgreen 35 1 25.28073
```

According to the quality statistics in the female network, the lightgreen module looks fine (*medianRank.qual* = 1,*Zsummary.qual* = 25.28). This illustrates that no matter how good the quality statistics of a module in reference data set, there is no guarantee that it will be preserved in a test data set.

12.6 Consensus modules Between Female and Male Liver Tissues

In Sect. 12.5, we described how to assess module preservation between female and male mice. A related but distinct data analysis task is to construct modules that are present in several networks. In Sect. 7.11.1, we described how to construct "consensus" modules that are present in each of the underlying networks. For example, consensus modules in the female and male mouse liver networks are trivially preserved in each sex. In the following, we describe how to use the consensus dissimilarity $D^{Consensus,TOM}$ (7.40) for finding modules that are present in multiple correlation networks. Specifically, we find consensus modules that are present in both the female and the male mouse liver networks. We start out by defining the input of the automatic module detection function `blockwiseConsensusModules`.

```
# number of networks used in the consensus network analysis:
nSets=2
# Vector with descriptive names of the two sets.
setLabels=c("Female liver", "Male liver")
shortLabels=c("Female", "Male")
#Define a list whose components contain the data
multiExpr=vector(mode="list",length=nSets)
multiExpr[[1]] = list(data = datExprFemale)
names(multiExpr[[1]]$data) = names(datExprFemale)
rownames(multiExpr[[1]]$data) = dimnames(datExprFemale)[[1]]
multiExpr[[2]] = list(data = datExprMale)
names(multiExpr[[2]]$data) = names(datExprMale)
rownames(multiExpr[[2]]$data) = dimnames(datExprMale)[[1]]
# Check that the data has the correct format:
exprSize = checkSets(multiExpr)

# The variable exprSize contains useful information
# about the sizes of all of the data sets
# now we run automatic module detection procedure
netConsensus=blockwiseConsensusModules(multiExpr,
   maxBlockSize=5000,
power=6,minModuleSize=30,deepSplit=2,pamRespectsDendro=FALSE,
mergeCutHeight=0.25,numericLabels=TRUE,minKMEtoJoin=1,
minKMEtoStay=0,saveTOMs=TRUE)
```

```
# list of consensus module eigengenes
consMEs = netConsensus$multiMEs
# module labels
modLabCons0 = netConsensus$colors
# Relabel the consensus modules so that their labels match those
# from the automatic analysis in female mice only
modLabCons = matchLabels(modLabCons0,moduleLabelsAutomatic)
# check the agreement between consensus- and females-only modules
mean(modLabCons==moduleLabelsAutomatic) #=.655

# Convert the numeric labels to color labels
moduleColorsConsensus = labels2colors(modLabCons)
```

Similar to the function `blockwiseModules`, `blockwiseConsensusModules` has many parameters, and in this example most of them are left at their default value.

When dealing with many genes, it can be advantageous to carry out a blockwise consensus module detection analysis analogous to the case of a single network (see Sect. 12.2.3). The function blockwiseConsensusModules automatically partitions the genes into blocks of maximum size specified by the argument `maxBlockSize`. Since in our case the number of genes is smaller than `maxBlockSize`, the analysis amounts to a single block analysis. The consensus dendrogram (of the first block) is given by

```
blocknumber=1
consTree = netConsensus$dendrograms[[blocknumber]]
# plot the consensus dendrogram and module colors
datColors=data.frame(moduleColorsConsensus,moduleColorsFemale)
plotDendroAndColors(consTree,datColors,
c("Consensus module","female module"),dendroLabels=FALSE,
hang=0.03,addGuide=TRUE,guideHang=0.05,
main="Consensus gene dendrogram and module colors")
```

Figure 12.8 shows the consensus dendrogram, modules, and a comparison with the female modules.

12.6.1 Relating Consensus Modules to the Traits

In this section, we illustrate the use of module eigengenes to relate consensus modules to external microarray sample information such as clinical traits. We relate 26 physiological traits to consensus module eigengenes in each of the two sets. Each module is represented by two eigengenes: one in the female data set and another in the male data set. Similarly, we have the trait data separately for the female and for the male mice.

```
traitData = read.csv("ClinicalTraits.csv")
dim(traitData)
names(traitData)
```

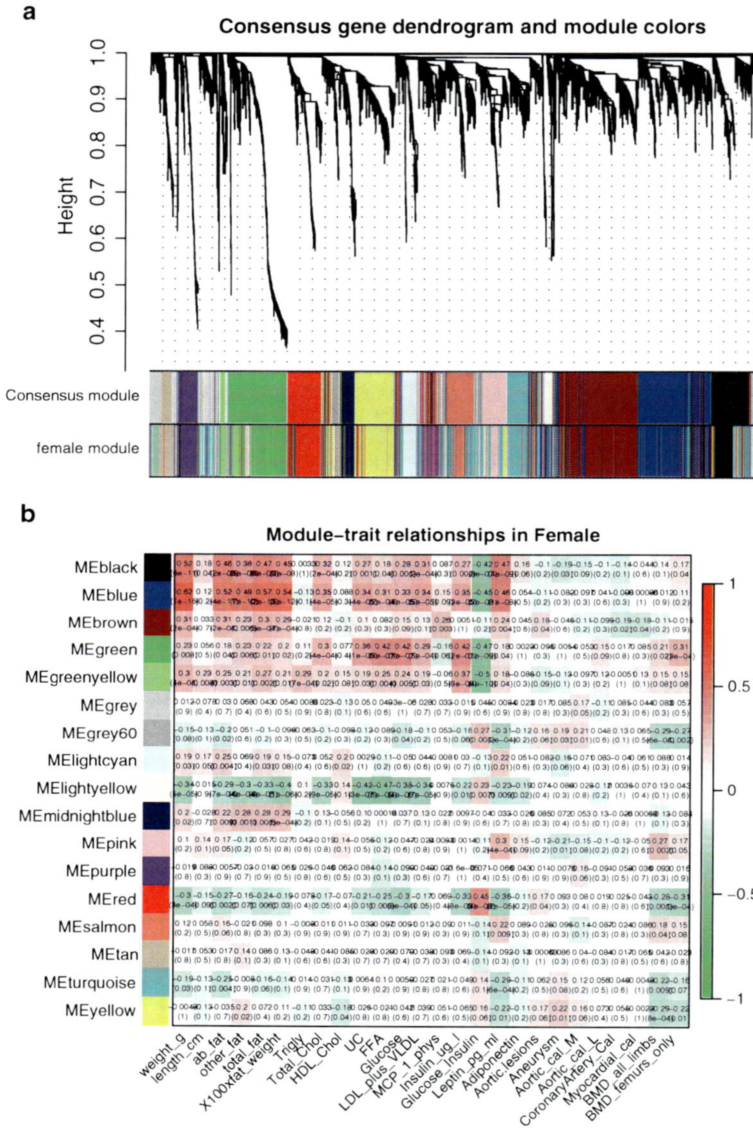

Fig. 12.8 (**a**) The consensus dendrogram for female and male mouse liver co-expression networks. The average linkage hierarchical clustering tree is based on the consensus dissimilarity $D^{Consensus,TOM}$ (7.40) of the female and male data. The first color-band shows the result of dynamic hybrid branch cutting of the consensus dendrogram. The second color-band shows the module colors based on the female mouse data alone. By visual inspection, it is clear that there is high agreement between the two module assignments, which reflects that many of the female mouse modules are also preserved in the male network. (**b**) Table of correlations and p values for studying the relationships between consensus module eigengenes (rows) and physiological traits (columns) in the female mice. An analogous table could also be created for the male mice

```
# remove columns that hold information we do not need.
allTraits=traitData[, -c(31, 16)]
allTraits=allTraits[, c(2, 11:36) ]
# Data descriptions
dim(allTraits)
names(allTraits)
allTraits$Mice


# Form a multi-set structure that will hold the clinical traits.
TraitsL = vector(mode="list", length = nSets)
for (set in 1:nSets)
{
setSamples = rownames(multiExpr[[set]]$data)
traitRows = match(setSamples, allTraits$Mice)
TraitsL[[set]] = list(data = allTraits[traitRows, -1])
rownames(TraitsL[[set]]$data) = allTraits[traitRows, 1]
}
collectGarbage()
# Define data set dimensions
nGenes = exprSize$nGenes
nSamples = exprSize$nSamples

# Set up variables to contain the module-trait correlations
modTraitCor = list()
modTraitP = list()
# Calculate the correlations
for (set in 1:nSets){
modTraitCor[[set]]=
   cor(consMEs[[set]]$data,TraitsL[[set]]$data,use="p")
modTraitP[[set]]=
   corPvalueFisher(modTraitCor[[set]],exprSize$nSamples[set])}
```

Now we will display the module–trait relationships using a color-coded table. The table reports the correlations and the corresponding p values. Entries are color-coded according to the p value.

```
MEColors = substring(names(consMEs[[1]]$data), 3)
MEColorNames = paste("ME", MEColors, sep="")
# Plot the module-trait relationship table for set number
set = 1 # 1 for females, 2 for males
textMatrix = paste(signif(modTraitCor[[set]], 2), "\n(",
signif(modTraitP[[set]], 1), ")", sep = "")
dim(textMatrix) = dim(modTraitCor[[set]])
par(mar = c(6, 8.8, 3, 2.2))
labeledHeatmap(Matrix = modTraitCor[[set]],
xLabels = names(TraitsL[[set]]$data),yLabels = MEColorNames,
ySymbols = MEColorNames,colorLabels = FALSE,
   colors=greenWhiteRed(50),
textMatrix=textMatrix,setStdMargins=FALSE,cex.text=0.5,
  zlim=c(-1,1),
main=paste("Module-trait relationships in", setLabels[set]))
```

Figure 12.8 presents the color-coded table. One can create an analogous figure based on the male data by choosing `set=2` in the above code.

12.6.2 Manual Consensus Module Analysis

Here we present a manual or interactive approach for finding consensus modules. Network construction starts by calculating the adjacencies in the individual sets

```
beta=6
# Initialize an appropriate array to hold the adjacencies
adjacencies=array(0,dim = c(nSets,nGenes,nGenes))
# Calculate adjacencies in each individual data set
for (set in 1:nSets)
{adjacencies[set,,]=abs(cor(multiExpr[[set]]$data,use="p"))^beta}

# Let's calculate corresponding Topological Overlap Matrices:
# Initialize an appropriate array to hold the TOMs
TOM = array(0, dim = c(nSets,nGenes,nGenes))
# Calculate TOMs in each individual data set
for (set in 1:nSets)
{TOM[set,,] = TOMsimilarity(adjacencies[set,,])}

#remove the adjacencies to free up space
rm(adjacencies)
```

Since the topological overlap matrices of the individual data sets may be differently calibrated (e.g., reflecting different sample sizes), it can be advantageous to calibrate them. For example, the TOM in the male data may be systematically lower than the TOM in female data. Since consensus is defined as the component-wise minimum of the two TOMs, poor calibration may lead to a bias. In Sect. 4.5, we describe how to use the power adjacency function AF^{power} (4.2) to calibrate weighted networks. We calibrate the male TOM such that its $prob = 0.95$ quantile equals that of the female TOM. The notation is explained in Sect. 4.5.

```
# Define the percentile (called prob) for defining the quantile
prob = 0.95
# Set random seed for reproducibility of sampling
set.seed(1)
#number of randomly samples entries of the TOM matrix
nSubset = 20000
# Choose the sampled TOM entries
subsetEntries = sample(nGenes*(nGenes-1)/2, size = nSubset)
TOMsubset = list()
# vector of quantiles of the individual TOM matrices
quantile.TOM = rep(1, nSets)
# Scaling powers to equalize reference TOM values
beta.prob = rep(1, nSets)
# Loop over sets
for (set in 1:nSets)
{# Select the sampled TOM entries
TOMsubset[[set]]=vectorizeMatrix(TOM[set,,])[subsetEntries]
# Calculate the quantile corresponding to prob
quantile.TOM[set] = quantile(TOMsubset[[set]],probs=prob,type = 8)
# calculate the power of the adjacency function
beta.prob[set] = log(quantile.TOM[1])/log(quantile.TOM[set])
# use the power adjacency function for calibrating the TOM
```

```
TOM[set, ,] = TOM[set, ,]^(beta.prob[set])
}
# let us look at the powers used for calibration
beta.prob
# output
[1] 1.000 0.982
# Pairwise relationship between uncalibrated matrices
par(mfrow=c(1,1))
verboseScatterplot(TOMsubset[[1]],TOMsubset[[2]],
  xlab="TOM[[1]]",ylab="TOM[[2]]"); abline(0,1)
```

By definition, the power β of the reference (female) network equals 1. Note that the power $\beta.95 = 0.98$ for the test (male) network is very close to 1, suggesting that even without calibration the two networks are very comparable.

Consensus module identification: We now calculate the consensus topological overlap by defining it using the parallel minimum of the TOMs in individual sets (7.36). We use the consensus TOM as input to hierarchical clustering, and identify modules in the resulting dendrogram using the Dynamic Tree Cut algorithm.

```
consensusTOM=pmin(TOM[1, , ],TOM[2, , ])
consTreeManual=flashClust(as.dist(1-consensusTOM),
  method="average")
# Module identification using dynamic tree cut:
unmergedLabels=cutreeDynamic(dendro=consTreeManual,
     distM=1-consensusTOM,
deepSplit=2,cutHeight=0.995,minClusterSize=30,
     pamRespectsDendro=FALSE)
unmergedColors=labels2colors(unmergedLabels)
```

The Dynamic Tree Cut may identify modules whose expression profiles are very similar. It may be prudent to merge such modules since their genes are highly co-expressed. To quantify co-expression similarity of entire modules, we calculate their eigengenes (MEs) and cluster them on their consensus correlation, that is the minimum correlation across the two sets:

```
# Calculate module eigengenes
unmergedMEs=multiSetMEs(multiExpr,colors=NULL,
     universalColors=unmergedColors)
# consensus dissimilarity of module eigengenes
consMEDiss=consensusMEDissimilarity(unmergedMEs)
# Cluster consensus modules
consMETree=flashClust(as.dist(consMEDiss), method = "average")
# The merging can be performed automatically:
merge = mergeCloseModules(multiExpr,unmergedLabels,cutHeight=0.25)
# resulting merged color
modLabConsManual0=merge$colors

# Relabel the manual consensus modules so that their labels match
    those from the automatic analysis
modLabConsManual=matchLabels(modLabConsManual0,modLabCons)
# Agreement between consensus modules and females-only modules
mean(modLabConsManual==modLabCons) #=0.9947222
```

```
# Convert the numeric labels to color labels
moduleColorsConsensusManual=labels2colors(modLabConsManual)

# Calculate consensus module eigengenes in each data set
consMEs=multiSetMEs(multiExpr,colors=NULL,
         universalColors=moduleColorsConsensusManual)

#Plot the gene dendrogram again:
plotDendroAndColors(consTreeManual,moduleColorsConsensusManual,
dendroLabels=FALSE,hang=0.03,addGuide=TRUE,guideHang=0.05)
```

The resulting plot is identical to that in Fig. 12.8, which was based on the automatic analysis implemented in the function `blockwiseConsensusModules`.

12.7 Exercises

1. Exercise. Reverse the role of the female and male mouse liver data sets and re-analyze the data. In particular, carry out the following tasks.
   (i) Construct a sample network based on the male mouse liver samples and identify potential outliers.
   (ii) Use different module detection procedures for finding co-expression modules in the male liver samples. Hint: dynamic hybrid method: single block, two block, and manual module analysis.
   (iii) Evaluate the agreement between the module detection methods.
   (iv) Relate the module eigengenes to physiological traits in male mice.
   (v) Evaluate module preservation of male modules in the female network.
   (vi) Discuss whether you have found something interesting. Is it publishable?
2. Exercise regarding visualizing a network. Use your favorite data set to define a weighted correlation network and modules (based on hierarchical clustering). Visualize the network and the modules using
   (i) A connectivity (TOM) plot
   (ii) A classical multidimensional scaling plot
   (iii) A ViSANT plot
   (iv) Which plots or software packages allow one to visualize a weighted network? Which tools work only for an unweighted network?
   (v) How can one visualize a weighted network with a tool that only visualizes unweighted networks? Answer: threshold the adjacency matrix
3. Exercise regarding systems genetic analysis with NEO. Recall that the analysis in Sect. 12.3 focused on the blue module. Here you are asked to analyze the brown module instead.
   (i) Show that the brown module correlates with body weight.
   (ii) Determine which SNP(s) correlate with the brown module eigengene. Hint: Run `cor(datSNP,MEbrown,use="p")`.

(iii) Use the most significant SNP to determine whether $SNP \rightarrow MEbrown \rightarrow weight$.
(iv) Which genes in the brown module show evidence of causally affecting body weight? Hint: use the SNP from (iii)
(v) How do the NEO results change if you use the second most significant SNP.
(vi) How do the results change if you use the LEO.CPA score (which uses the two most significant SNPs)? Hint: Use the R code from Sect. 11.8.2.

References

Broman KW, Wu H, Sen S, Churchill GA (2003) R/qtl: QTL mapping in experimental crosses. Bioinformatics 19(7):889–890

Frohlich H, Speer N, Poustka A, Beiszbarth T (2007) GOSim – an R-package for computation of information theoretic GO similarities between terms and gene products. BMC Bioinform 8(1):166

Fuller TF, Ghazalpour A, Aten JE, Drake T, Lusis AJ, Horvath S (2007) Weighted gene coexpression network analysis strategies applied to mouse weight. Mamm Genome 18(6–7):463–472

Ghazalpour A, Doss S, Zhang B, Plaisier C, Wang S, Schadt EE, Thomas A, Drake TA, Lusis AJ, Horvath S (2006) Integrating genetics and network analysis to characterize genes related to mouse weight. PloS Genet 2(2):8

Hu Z, Mellor J, Wu J, DeLisi C (2004) VisANT: An online visualization and analysis tool for biological interaction data. BMC Bioinform 5:17

Jansen RC, Nap JP (2001) Genetical genomics: The added value from segregation. Trends Genet 10(17):388+

Langfelder P, Zhang B, Horvath S (2007) Defining clusters from a hierarchical cluster tree: The Dynamic Tree Cut library for R. Bioinformatics 24(5):719–720

Langfelder P, Luo R, Oldham MC, Horvath S (2011) Is my network module preserved and reproducible? Plos Comput Biol 7(1):e1001057

de Nooy W, Mrvar A, Batagelj V (2005) Exploratory social network analysis with pajek (structural analysis in the social sciences). Cambridge University Press, Cambridge

Presson AP, Sobel EM, Papp JC, Suarez CJ, Whistler T, Rajeevan MS, Vernon SD, Horvath S (2008) Integrated weighted gene co-expression network analysis with an application to chronic fatigue syndrome. BMC Syst Biol 2:95

Ravasz E, Somera AL, Mongru DA, Oltvai ZN, Barabasi AL (2002) Hierarchical organization of modularity in metabolic networks. Science 297(5586):1551–1555

Schadt EE, Monks SA, Drake TA, Lusis AJ, Che N, Colinayo V, Ruff TG, Milligan SB, Lamb JR, Cavet G, Linsley PS, Mao M, Stoughton RB, Friend SH (2003) Genetics of gene expression surveyed in maize, mouse and man. Nature 422:297–302

Shannon P, Markiel A, Ozier O, Baliga NS, Wang JT, Ramage D, Amin N, Schwikowski B, Ideker T (2003) Cytoscape: A software environment for integrated models of biomolecular interaction networks. Genome Res 13(11):2498–2504

Zhang B, Horvath S (2005) General framework for weighted gene coexpression analysis. Stat Appl Genet Mol Biol 4:17

Chapter 13
Networks Based on Regression Models and Prediction Methods

Abstract Prediction methods (e.g., linear and nonlinear regression models) can be used to construct a network among a set of vectors $x_1, \ldots, x_n$. The construction follows three steps. First, the prediction method is used to predict $y = x_j$ on the basis of $x = x_i$. Second, a measure of predictive accuracy or statistical significance (e.g., a likelihood ratio test p value) is calculated. Third, the measure is transformed into an adjacency measure A_{ij} between the two vectors. Although general prediction and machine learning methods can be used, we focus on regression models for measuring nonlinear relationships (e.g., polynomial and spline regression models). Spline regression models are attractive since they provide a statistical framework for robustly capturing general nonlinear relationships. Generalized linear models allow one to model the relationship between binary variables, count data, and categorical variables. Multivariate regression models allow one to define a pairwise adjacency measure between variables, which conditions on additional covariates, which describe sample properties (e.g., batch effects). The partial correlation coefficient can be interpreted as a pairwise correlation measure which adjusts for other variables. Partial correlation networks can be used to define networks that encode direct relationships between variables.

13.1 Least Squares Regression and MLE

Assume we have two numeric vectors x and y representing samples (of size m) from corresponding random variables. Denote by $E(y)$ the expected value of y. We say that a linear relationship exists between $E(y)$ and x if numbers γ_0 and γ_1 can be found such that

$$E(y) = \gamma_0 1 + \gamma_1 x, \tag{13.1}$$

where $1 = (1, \ldots, 1)$. The numbers γ_0 and γ_1 are referred to as the **intercept** and **slope**, respectively. More generally, they are known as **regression coefficients**. In practice, the expected value $E(y)$ is unknown and the coefficient vector $\gamma = (\gamma_0, \gamma_1)$ must be estimated. In **least squares regression**, γ is estimated by minimizing the squared Euclidean distance (also known as error sums of squares)

S. Horvath, *Weighted Network Analysis: Applications in Genomics and Systems Biology*, DOI 10.1007/978-1-4419-8819-5_13,

$$Squared.Error(\gamma) = ||y - (\gamma_0 1 + \gamma_1 x)||^2 \quad (13.2)$$
$$= ||y - M\gamma||^2, \quad (13.3)$$

where $M = [1x]$ is the $m \times 2$ dimensional **model matrix**. One can prove that $\hat{\gamma}$ minimizes $Squared.Error(\gamma)$ if it satisfies the following **normal equation**:

$$M\hat{\gamma} = M(M^\tau M)^- M^\tau y$$
$$= ProjOp(M)y, \quad (13.4)$$

where

$$ProjOp(M) = M(M^\tau M)^- M^\tau \quad (13.5)$$

denotes the **projection operator**, $^-$ denotes the (pseudo) inverse, and $^\tau$ denotes the transpose of a matrix. Note that $ProjOp = ProjOp(M)$ is an $m \times m$ dimensional symmetric matrix. Recall that the **trace of the square** matrix denotes the sum of its diagonal elements, e.g.,

$$tr(ProjOp) = \sum_i ProjOp_{ii}.$$

In case of a projection operator, the trace $tr(ProjOp(M))$ equals the column-rank of M, which is defined as the number of independent columns of M. If the column-rank of M (or equivalently $tr(ProjOp(M))$) equals the number of columns of M (here 2), then the least squares estimates of the coefficients can be expressed as follows:

$$\hat{\gamma}_1 = cov(x,y)/var(x)$$
$$\hat{\gamma}_0 = mean(y) - \hat{\gamma}_1 * mean(x). \quad (13.6)$$

These results are derived in most introductory books on linear regression analysis. Both sides of the normal equation (13.4) define the least squares estimate of $E(y)$:

$$\hat{E}(y) = M\hat{\gamma} = M(M^\tau M)^- M^\tau y.$$

The linear model prediction $\hat{y}$ is defined as the least squares estimate of the expected value, i.e.,

$$\hat{y} = \hat{E}(y) = M\hat{\gamma} = M(M^\tau M)^- M^\tau y. \quad (13.7)$$

If the column-rank of M (or equivalently $tr(ProjOp(M))$) equals the number of columns of M then one can show that

$$\hat{y} = \hat{E}(y) = mean(y)1 + cor(x,y)\frac{\sqrt{var(y)}}{\sqrt{var(x)}}(x - mean(x)). \quad (13.8)$$

Based on (13.6), one can use a least squares regression model to find the optimal slope and intercept estimate for predicting $scale(y)$ on the basis of $scale(x)$. In this case, we find

$$\begin{aligned} \hat{\gamma}_1 &= cor(x,y) \\ \hat{\gamma}_0 &= 0, \end{aligned} \tag{13.9}$$

which implies that $\hat{scale}(y) = cor(x,y)scale(x)$.

13.2 R Commands for Simple Linear Regression

To illustrate the above results, we simulate two vectors x and y as follows:

```
x=rnorm(100,mean=2,sd=2)
y=x+rnorm(100)
```

To carry out least squares regression, one can use the following R command:

```
lm1=glm(y~x)
```

Since only one covariate x is used, the model is called a simple (univariate) linear regression model. To arrive at the coefficient values, one can use the following R command:

```
summary(lm1)$coefficients
# the intercept gamma_0 is given by
summary(lm1)$coefficients[1,1]
# the slope gamma_1 is given by
summary(lm1)$coefficients[2,1]
```

The least squares estimate $\hat{y}$ of the predicted value (and the expected value) is given by

```
yhat=as.numeric(predict(lm1))
```

The following code involves the model matrix $M = [1x]$ and the corresponding projection operator $ProjOp(M)$:

```
M=cbind(1,x)
ProjOp.M=M %*% solve(t(M)%*%M)%*% t(M)
#the trace of the projection operator is given by
sum(diag(ProjOp.M)) # equals 2
# estimate of the predicted value (and expected value)
yhat= ProjOp.M %*% y
```

Now we fit a linear model between the scaled variables $scale(y)$ and $scale(x)$ to verify (13.9):

```
lmScaled1=glm(scale(y)~ scale(x))
#compare the coefficient estimates with the correlation
summary(lmScaled1))
cor(x,y)
```

13.3 Likelihood Ratio Test for Linear Model Fit

Regression models allow one to test whether one model fits the data better than another model. To derive a likelihood-based test, assume that y represents a random sample from a normal distribution whose mean value is given by $\mu = \gamma_0 1 + \gamma_1 x$. Specifically, we assume that the likelihood of observing y (given x) is given by

$$likelihood(\gamma_0, \gamma_1, \sigma^2|x,y) = \frac{1}{(2\pi\sigma^2)^{m/2}} exp\left(-\frac{||y-(\gamma_0 1+\gamma_1 x)||^2}{2\sigma^2}\right). \quad (13.10)$$

The parameters of the normal distribution can be organized into a parameter vector $\theta = (\gamma_0, \gamma_1, \sigma^2)$. Then the logarithm of the likelihood function (13.10) is given by

$$log(likelihood(\theta|x,y)) = -\frac{m}{2} log\left(2\pi\sigma^2\right) - \frac{||y-M\gamma||^2}{2\sigma^2}, \quad (13.11)$$

where $M = [1x]$ is the model matrix and $\gamma = (\gamma_0, \gamma_1)$. **Maximum likelihood estimation (MLE)** is a widely used method for estimating parameter values when a likelihood function is available. For the fixed set of data x and y, maximum likelihood picks the values of the model parameters that make the data "more likely" than any other values of the parameters would make them. In our case, the maximum likelihood estimates $\hat{\gamma} = (\hat{\gamma}_0, \hat{\gamma}_1)$ and $\hat{\sigma}^2$ satisfy the following relationships:

$$\begin{aligned} M\hat{\gamma} &= ProjOp(M)y \\ \hat{\sigma}^2 &= \frac{||y-M\hat{\gamma}||^2}{m}. \end{aligned} \quad (13.12)$$

Note that the (13.12) satisfied by the MLE estimate $\hat{\gamma}$ equals that (13.4) satisfied by the least squares estimate. By plugging the MLEs into the log-likelihood, we arrive at the maximized log-likelihood of the linear regression model:

$$\begin{aligned} log(likelihood(\hat{\theta}|x,y)) &= -\frac{m}{2}(log(\hat{\sigma}^2) + log(2\pi) + 1) \\ &= -\frac{m}{2}(log\left(\frac{||y-M\hat{\gamma}||^2}{m}\right) + log(2\pi) + 1) \end{aligned} \quad (13.13)$$

Let us now derive the likelihood ratio test for comparing the simple linear regression model $E(y) = \gamma_0 1 + \gamma_1 x = M^{model1}\gamma$ versus the **intercept-only model** $E(y) = \gamma_0 1 = M^{model2}\gamma_0$, where

$$M^{model1} = [1x] \quad (13.14)$$

$$M^{model2} = [1]. \quad (13.15)$$

The maximized log-likelihoods (13.13) of the intercept only model is given by

$$log(likelihood(M^{model2})) = -\frac{m}{2}(log\left(\frac{||y - M^{model2}\hat{\gamma}_0||^2}{m}\right) + log(2\pi) + 1),$$

where the maximum likelihood estimate of the intercept-only model is given by $\hat{\gamma}_0 = mean(y)$. The likelihood ratio test statistic is twice the difference in these log-likelihoods:

$$\begin{aligned} LRT(model1, model2) &= -2(log(likelihood(M^{model1})) - log(likelihood(M^{model2}))) \\ &= -\frac{m}{2} log\left(\frac{||y - M^{model1}\hat{\gamma}||^2}{||y - M^{model2}\hat{\gamma}_0||^2}\right). \end{aligned} \quad (13.16)$$

If both models fit the data equally well, the LRT statistic follows approximately a chi-square distribution

$$LRT \sim \chi^2(df),$$

where *df* equals the difference in the number of parameters between model 1 and model 2 (here $df = 1$). For the case of two linear models, the degrees of freedom df can be found by forming the trace of the difference of the two projection operators:

$$df = tr(ProjOp(M^{model1}) - ProjOp(M^{model2})). \quad (13.17)$$

If the likelihood ratio statistic LRT takes on a large value (larger than a suitably chosen quantile of the chi-square distribution), then we reject the null hypothesis that model 2 fits as well as model 1.

To compute the LRT statistic and a corresponding chi-square test p value, one can use the following R code:

```
set.seed(1);m=100;x=rnorm(m);y=x+3*rnorm(m)
lmModel1=glm(y~x)
lmModel2=glm(y~1)
anova1=anova(lmModel2,lmModel1,test="Chisq")
pvalueLRT=anova1$"P(>|Chi|)"[[2]]
pvalueLRT # equals 0.0020396
# Alternatively, the following code can be used
lm1=glm(y~x)
anova1=anova(lm1,test="Chisq")
#Using terminology from a generalized linear model:
#LRT statistic=deviance/dispersion, i.e.,
LRT.statistic=
      anova1$Deviance[2]/summary(lm1)$dispersion
LRT.statistic #=9.5135
DegreesOfFreedom=anova1$Df[2] #=1
pvalueLRT=1-pchisq(LRT.statistic,DegreesOfFreedom)
```

13.4 Polynomial and Spline Regression Models

If x and y are drawn from statistically *independent distributions*, then their correlation is expected to be 0 but the converse is not always true. Only in the special case when x and y follow a joint normal distribution, does an expected correlation of zero imply independence of the corresponding random variables.

For example, $x = (-1, -0.9, -0.8, \ldots, 0.9, 1)$ and $y = x^2$ are highly dependent even though their Pearson correlation vanishes ($cor(x,y) = 0$). This quadratic relationship between x and y is a special case of a polynomial relationship.

We will present two major strategies for measuring possibly nonlinear relationships between x and y: the first is based on regression model methodology. Toward this end, we review polynomial regression and spline regression. The second approach is based on discretizing the quantitative variables and to measure the relationship between the corresponding categorical variable (see Chap. 14). In the following, we describe polynomial- and spline regression methods for estimating nonlinear relationships between x and y. These regression models allow one to model nonlinear dependencies between x and y as shown in Fig. 13.1. A **polynomial relationship of degree** d exists between $E(y)$ and x if a vector of numbers (parameters) $\gamma = (\gamma_0, \ldots, \gamma_d)$ can be found such that

$$\begin{aligned} E(y) &= \gamma_0 1 + \gamma_1 x + \gamma_2 x^2 \ldots + \gamma_d x^d \\ &= M\gamma, \end{aligned} \tag{13.18}$$

where

$$M = [1x \ldots x^d]. \tag{13.19}$$

Analogous to (13.12), one can show that the least squares estimate (and equivalently the maximum likelihood estimate) of the parameter vector $\hat{\gamma}$ satisfies

$$M\hat{\gamma} = ProjOp(M)y.$$

The fit of the quadratic model can be compared to that of an intercept-only model or to that of a simple linear regression model. Since our main interest lies in testing whether x has significant relationship to y, we describe how to test whether a quadratic model fits better than the intercept-only model. Toward this end, one can use the likelihood ratio test statistic LRT(model 1, model 2) (13.16) where $M^{model1} = [1xx^2]$ and $M^{model2} = [1]$. Under the null hypothesis of equally good fit, $LRT(model1, model2) \sim \chi^2(2)$. In contrast, the LRT statistic for comparing the quadratic fit versus a simple linear fit follows a chi-square distribution with 1 degree of freedom. If the LRT statistic takes on a large value, then the null hypothesis of equally good fit is rejected. Polynomial regression models have two potential shortcomings. First, polynomial regression models (with $d \geq 2$) can be inaccurate when it comes to predicting relationships beyond the observed range of x (extrapolation). Second, the model fit can be adversely affected by outlying observations. A single outlying observation (x_u, y_u) can "bend" the fitting curve into the wrong direction, i.e., adversely affect the estimate of γ.

To counter these disadvantages of polynomial regression, one can use **local polynomial regression models** (e.g., spline regression models) (Faraway 2002). Local refers to the fact that these models amount to fitting models on subintervals of the range of x. The boundaries of the subintervals are sometimes referred to as knots. The knots are typically prespecified, e.g., based on quantiles of x. To ensure that fit between y and x satisfies a continuous relationship, it is advantageous to implement local regression models by first transforming x using the **hockey stick function** $()_+$, which is defined as follows:

$$(s)_+ = \begin{cases} s & \text{if } s \geq 0 \\ 0 & \text{if } s < 0 \end{cases}. \qquad (13.20)$$

For example, $(-1)_+ = 0$. This function can also be applied to the components of a vector, e.g., $(x)_+$ denotes a vector whose negative components have been set to zero. Typically, the hockey stick function gets shifted by a number, which will be denoted by $knot1$, e.g., $(x - knot1)_+$ is a vector whose uth component equals $x[u] - knot1$ if $x[u] - knot1 \geq 0$ and 0 otherwise.

We are now ready to describe **linear spline model** (also known as **broken-stick regression**) for relating the expected value of y to a linear spline in x with two knots:

$$E(y) = \gamma_0 1 + \gamma_1 x + \gamma_2 (x - knot_1)_+ + \gamma_3 (x - knot_2)_+. \qquad (13.21)$$

The knot parameters (numbers) $knot_1, knot_2, \ldots$ are chosen before estimating the parameter values. To estimate the parameters γ, one can again use least squares regression. Figure 13.1 illustrates that the resulting fitting curve looks like a broken stick which is why *linear* spline regression is also known as broken stick regression (Faraway 2002). Linear spline regression localizes the influence of each data point to its particular subinterval (defined by the knots) which is good but it does not have the same smoothness afforded by polynomial regression. To combine the advantages of both methods (smoothness and local influence), one can fit higher order polynomials to the subintervals. For example, polynomial of degree 3 leads to a cubic spline regression model. The general form of a cubic spline with two knots is as follows:

$$E(y) = \gamma_0 1 + \gamma_1 x + \gamma_2 x^2 + \gamma_3 x^3 + \gamma_4 (x - knot_1)_+^3 + \gamma_5 (x - knot_2)_+^3. \qquad (13.22)$$

Obviously, the choice of the knots affects the model fit. It turns out that the values of the knots (i.e., their placement) is not as important as the number of knots. Let us review the following rule of thumb for the number of knots: if $m > 100$, use five knots, if $m < 30$, use three knots, otherwise use four knots.

Abstractly speaking, a spline regression model can be written as follows:

$$E(y) = \gamma_0 1 + \gamma_1 h_1(x) + \gamma_2 h_2(x) \ldots + \gamma_d h_d(x), \qquad (13.23)$$

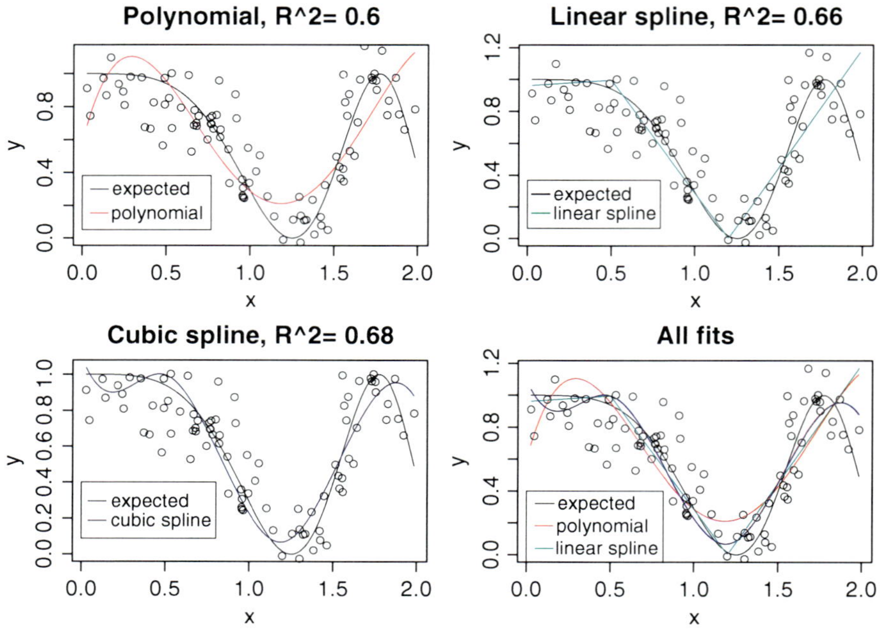

Fig. 13.1 Fitting polynomial and spline regression models to simulated data. The *black curve* shows the true expected value $E(y)$ given x. The red curve shows the fit of a polynomial regression model with degree $d = 4$. The *green curve* shows the fit of a linear spline regression model (13.21). The *blue curve* shows the fit of a cubic spline regression model (13.22), the dots show R code underlying this figure can be found in Sect. 13.5

where the exact form of the vectors $h_i(x)$ depends on the knots. The least squares estimate (and equivalently the maximum likelihood estimate) $\hat{\gamma}$ satisfies again $M\hat{\gamma} = ProjOp(M)y$ where $M = [1h_1(x)h_2(x)\dots h_d(x)]$. Also likelihood ratio tests can be used to determine whether the spline model fits better than an intercept-only model.

13.5 R Commands for Polynomial Regression and Spline Regression

The R functions adjacency.polyReg and adjacency.splineReg in the WGCNA R package allow one to construct adjacency matrices based on polynomial and spline regression models. Type help(adjacency.polyReg) for sample code and to learn more about this functions. The following R code shows how to fit an orthogonal polynomial regression model. The word "orthogonal" reflects that the corresponding columns of the model matrix are chosen to be orthogonal to each other, which has numerical advantages.

```
lmPolynomial=glm(y~poly(x,4))
yhatPolynomial=as.numeric(predict(lmPolynomial))
```

The following R code shows how to fit a natural cubic spline regression model. The word "natural" reflects that the splines are restricted to follow a linear model in the tails of the data. Instead of specifying knots, we specify the degrees of freedom which leads to an automatic choice of knots according to suitably defined quantiles of x.

```
library(splines)
lmSpline=glm(y~ns(x,df=4))
# this is the predicted value
yhatSpline=as.numeric(predict(lmSpline))
# model fitting index
Rsquared=cor(y,yhatSpline)^2
```

The following R code shows how to calculate a likelihood ratio test p value for measuring the association between y and a degree 4 polynomial in x:

```
lmPoly=glm(y~poly(x,degree=4))
lmIntercept=glm(y~1)
anova1=anova(lmIntercept,lmPoly,test="Chisq")
pvalueLRT=anova1$"P(>|Chi|)"[[2]]
# Alternatively, use the following code
anovaPoly=anova(lmPoly,test="Chisq")
LRT.statistic=
    anovaPoly$Deviance[2]/summary(lmPoly)$dispersion
DegreesOfFreedom=anovaPoly$Df[2]
pvalueLRT=1-pchisq(LRT.statistic,DegreesOfFreedom)
```

Here is the analogous code for testing the association between y and a spline model (with 5 degrees of freedom):

```
library(splines)
lmIntercept=glm(y~1)
lmSpline=glm(y~ns(x,df=5))
anova1=anova(lmIntercept,lmSpline,test="Chisq")
pvalueLRT=anova1$"P(>|Chi|)"[[2]]
#Alternatively, use the following code
anovaSp=anova(lmSpline,test="Chisq")
LRT.statistic=
      anovaSp$Deviance[2]/summary(lmSpline)$dispersion
DegreesOfFreedom=anovaSp$Df[2]
pvalueLRT=1-pchisq(LRT.statistic,DegreesOfFreedom)
```

The likelihood ratio statistic can be used as significance measure between x and y. The following R code shows how we created Fig. 13.1:

```
m=100; set.seed(1)
x=runif(m,min=0,max=2)
expected.y.fun=function(x) cos(x^2)^2
# expected outcome
expected.y=expected.y.fun(x)
# we define the observed outcome by random normal noise
y=expected.y+rnorm(m,sd=.2)
# now we generate another x vector whose values are sorted
x.sort=seq(from=min(x,na.rm=T),to=max(x,na.rm=T),length=200)
# here is the expected value of y corresponding to x.sort
```

```
expected.y.sort=expected.y.fun(x.sort)
#polynomial with 4 degrees
PolynomialModel=lm(y~x+I(x^2)+I(x^3)+I(x^4))
yhat.Polynomial=as.numeric(predict(PolynomialModel))
# model fitting index
Rsquared=signif(cor(y,yhat.Polynomial)^2,2)
# we will also predict the values for the following data
newdata1=data.frame(x=x.sort)
# this is the predicted value for x.sort
yhat.Polynomial.sort=
          as.numeric(predict(PolynomialModel,newdata=newdata1))
par(mfrow=c(2,2))
daty=data.frame(expected.y.sort,yhat.Polynomial.sort)
matplot(x.sort,daty,type="l",col=c(1,2),pch=21,lty=1,
main=paste("Polynomial, R^2=",Rsquared),xlab="x",ylab="y")
legend(0,.4,c("expected","polynomial"),col=c(1,2),lty=1)
points(x,y)
#Now we define hockey stick variables for the spline models
xknot1=ifelse(x-.5>=0,x-.5,0); xknot2=ifelse(x-1.2>=0,x-1.2,0)
# hockey stick variables corresponding x.sort
xknot1.sort=ifelse(x.sort-.5>=0,x.sort-.5,0)
xknot2.sort=ifelse(x.sort-1.2>=0,x.sort-1.2,0)
LinearSplineModel=lm(y~x+xknot1+xknot2)
# predicted value of the spline
yhat.LinearSpline=as.numeric(predict(LinearSplineModel))
CubicSplineModel=lm(y~x+I(x^2)+I(x^3)+I(xknot1^3)+I(xknot2^3))
yhat.CubicSpline=as.numeric(predict(CubicSplineModel))
newdata1=data.frame(x=x.sort,xknot1=xknot1.sort,
   xknot2=xknot2.sort)
yhat.LinearSpline.sort=
         as.numeric(predict(LinearSplineModel,newdata=newdata1))
yhat.CubicSpline.sort=
         as.numeric(predict(CubicSplineModel,newdata=newdata1))
Rsquared=signif(cor(y,yhat.LinearSpline)^2,2)
daty=data.frame(expected.y.sort,yhat.LinearSpline.sort)
matplot(x.sort,daty, type="l",pch=21,col=c(1,3),lty=1,
main=paste("Linear spline, R^2=",Rsquared),xlab="x",ylab="y")
legend(0,.4,c("expected","linear spline"),col=c(1,3),lty=1)
points(x,y)

Rsquared=signif(cor(y,yhat.CubicSpline)^2,2)
daty=data.frame(expected.y.sort,yhat.CubicSpline.sort)
matplot(x.sort,daty, type="l",pch=21,col=c(1,4),lty=1,
main=paste("Cubic spline, R^2=",Rsquared),xlab="x",ylab="y")
legend(0,.4,c("expected","cubic spline"),col=c(1,4),lty=1)
points(x,y)

daty=data.frame(expected.y.sort,yhat.Polynomial.sort,
yhat.LinearSpline.sort,yhat.CubicSpline.sort)
matplot(x.sort,daty, type="l",pch=21,col=c(1:4),lty=1,
main="All fits",xlab="x",ylab="y")
legend(0,.4,c("expect","polynomial","lin. spline","cubic spline")
col=c(1:4),lty=1)
points(x,y)
```

13.6 Conditioning on Additional Covariates

Thus far, we only considered two sample vectors x and y. Let us now assume that additional sample vectors $z1, z2, \ldots$ are available. For example, z_1 could represent a batch effect when dealing with gene expression data. For ease of notation, we assume two additional vectors that correspond to the columns of the data frame $datZ = (z1, z2)$. Further, assume that the expected value of y is given by

$$E(y) = \gamma_0 1 + \gamma_1 x + \delta_1 z1 + \delta_2 z2 = M^{model1}(\gamma, \delta)^{\tau}, \tag{13.24}$$

where the model matrix is given by $M^{model1} = [1, x, z1, z2]$. Under the null hypothesis that x does not improve the fit, we get that

$$E(y) = \delta_0 1 + \delta_1 z1 + \delta_2 z2 = M^{model2}(\delta), \tag{13.25}$$

where $M^{model2} = [1, z1, z2]$. Using the likelihood ratio statistic defined in (13.16), we are able to test whether x is significantly related to y after adjusting for the covariates $datZ$. Under the null hypothesis of equally good fit, the LRT statistic follows asymptotically a chi-square distribution with 1 degree of freedom.

Let us now present R code for fitting a linear model which adjusts the relationship between x and y:

```
lmAdjusted=glm(y~x+z1+z2)
```

The (training set) prediction $\hat{y}$ can be computed using the following R code:

```
yhat=as.numeric(predict(glm(y~.,data=datZ)))
```

A **likelihood ratio test statistic** and p value for testing the dependence of y on x given the other covariates can be computed as follows:

```
set.seed(1);m=100;x=rnorm(m);y=x+3*rnorm(m)
z1=rnorm(m);z2=x+5*rnorm(m);datZ=data.frame(z1,z2)
glmAdj=glm(y~.+x,data=datZ)
lmZonly=glm(y~.,data=datZ)
anova1=anova(lmZonly,glmAdj,test="Chisq")
pvalueLRT=anova1$"P(>|Chi|)"[[2]]
pvalueLRT # 0.001863
# Alternatively, use the following code
anovaZ=anova(glmAdj,test="Chisq")
no.Zvariables=dim(data.frame(datZ))[[2]]
LRT.statistic=
    anovaZ$Deviance[no.Zvariables+2]/summary(glmAdj)$dispersion
DegreesOfFreedom=anovaZ$Df[no.Zvariables+2]
pvalueLRT=1-pchisq(LRT.statistic,DegreesOfFreedom)
```

The following R code shows how to fit a multivariate spline regression model for regressing y on a natural spline in x (degree 5) that adjusts for additional variables which form the columns of a data frame (matrix) $datZ = (z1, z2)$:

```
glmAS=glm(y~.+ns(x,df=5),data=data.frame(datZ))
lmZonly=glm(y~.,data=data.frame(datZ))
anova1=anova(lmZonly,glmAS,test="Chisq")
pvalueLRT=anova1$"P(>|Chi|)"[[2]]
pvalueLRT #0.01620675
# Alternatively, use the following code
no.Zvariables=dim(data.frame(datZ))[[2]]
anovaAS=anova(glmAS,test="Chisq")
LRT.statistic=
    anovaAS$Deviance[no.Zvariables+2]/summary(glmAS)$dispersion
DegreesOfFreedom=anovaAS$Df[no.Zvariables+2]
pvalueLRT=1-pchisq(LRT.statistic,DegreesOfFreedom)
```

The LRT statistic (or the corresponding p value) can be used as a significance measure that adjusts for the conditioning variables.

13.7 Generalized Linear Models

To evaluate the fit of a regression model, one can use several **residual plots**. For example, by plotting the raw residual $residual = y - \hat{y}$ (or a scaled residual) versus y or $\hat{y}$, one can often detect violations of model assumptions. Any discernible pattern in a residual plot indicates poor fit (e.g., due to outliers, violation of the linearity assumption, serial correlation, unequal variances, etc.). Linear, polynomial, and spline regression models assume that the variance of the residuals does not depend on the predicted values. If this basic assumption is violated, a **generalized linear model** (GLM) may afford a better fit to the data. Generalized linear models unify various other statistical models, including linear regression, logistic regression, and Poisson regression under one framework. Describing generalized linear models is beyond our scope. But we briefly mention some R code that shows how to fit important generalized linear models. If y is a binary vector, one can fit the following **logistic regression model**:

```
glm1=glm(y~x,family=binomial(link="logit"))
```

If y represents count data (nonzero integers), then one can use the following R code for fitting a Poisson regression model and calculating a LRT-based p value:

```
set.seed(1);m=50;x=rnorm(m,sd=.5);y=rpois(m,lambda=exp(x))
glmPoisson=glm(y~x,family=poisson)
glmIntercept=glm(y~1,family=poisson)
anova1=anova(glmIntercept,glmPoisson,test="Chisq")
pvalueLRT=anova1$"P(>|Chi|)"[[2]]
pvalueLRT #equals 2.22e-08
# Alternatively, the following code can be used
anova1=anova(glmPoisson,test="Chisq")
LRT.statistic=anova1$Deviance[2]/summary(glmPoisson)$dispersion
DegreesOfFreedom=anova1$Df[2]
pvalueLRT=1-pchisq(LRT.statistic,DegreesOfFreedom)
```

13.8 Model Fitting Indices and Accuracy Measures

Let us now assume that we have two numeric vectors x and y and a prediction method *PredictionMethod* for predicting y on the basis of x. The prediction model could be a polynomial or spline regression model or it could a machine learning predictor (e.g., tree predictor, random forest predictors, neural network predictors, support vector machine). In general, the predictor will depend on parameters θ and/or additional covariates *datZ*. After fitting (a.k.a. "training") the prediction method to the data, one arrives at a prediction $\hat{y}$. To determine how well the predicted vector $\hat{y}$ approximates the "observed" vector y, one can use predictive accuracy measures. For example, to measure the extent of a linear relationship between $\hat{y}$ and y, one can use the fitting index

$$R^2 = cor(y, \hat{y})^2, \tag{13.26}$$

which is only useful if y is a numeric vector. If the prediction method is linear regression, then R^2 (13.26) is also known as **model fitting index** R^2. In the context of a regression model, R^2 is also known as **coefficient of determination** or as **proportion of variance explained** by the covariates. The latter name stems from the following relationship: $R^2 = \frac{var(\hat{y})}{var(y)}$. In an exercise, you are asked to prove that the fitting index of the simple linear (univariate) regression model $E(y) = \gamma_0 1 + \gamma_1 x$ is given by $R^2 = cor(y, \hat{y})^2 = cor(x, y)^2$. In this special case, the fitting index is symmetric in x and y. We use this relationship to define the **straight line fitting index** between x and y:

$$straightlineFitIndex(x, y) = R^2 = cor(y, \hat{y})^2 = cor(x, y)^2. \tag{13.27}$$

Similarly, one can use a polynomial regression model to define a polynomial fitting index between x and y $polyFitIndex(x, y) = cor(\hat{y}, y)^2$. In general, the polynomial fitting index is not symmetrical with regard to x and y.

Another measure of predictive accuracy (more precisely discriminatory power) is the concordance index $C.index(\hat{y}, y)$ (5.18), which estimates the "probability of concordance between predicted and observed responses". The C-index measures how well the model prediction $\hat{y}$ discriminates between different observed values y, i.e., it determines whether the predicted value is low for low observed values y and high for high observed values of y.

13.9 Networks Based on Predictors and Linear Models

Here we describe several strategies for using a prediction method (e.g., a regression model) to construct a network among a set of numeric vectors which correspond to the columns of $datX = [x_1 \ldots x_n]$.

The first strategy for constructing a network is based on p values. For each pair of variables $x = x_i$ to $y = x_j$ (and possibly conditioning variables *datZ*), a regression model allows one to define likelihood based p values (e.g., likelihood ratio test, Wald test, or score test p values). These p-values measure whether the variable $x = x_j$ is significantly related to the variable $y = x_i$. The resulting pairwise p values can be organized into an $n \times n$ dimensional matrix $M^{p.values} = (p.value_{ij})$. Note that this matrix will in general be nonsymmetric, and its entries take on large values if no relationship exists between i and j. Many approaches can be used to turn $M^{p.values}$ into an adjacency matrix (see Sect. 7.3). For example, the p values can be thresholded to arrive at a binary matrix. Next, this matrix can be symmetrized.

The second strategy for constructing a network is based on fitting indices. For each pair of variables $x = x_i$ and $y = x_j$, a prediction method allows us to calculate the fitting index $R^2 = cor(y, \hat{y})^2$ (13.26) or alternatively the concordance index $C.index = cor(y, \hat{y})^2$ (see Sect. 5.1.4). The resulting matrices can be turned into networks as described in Sect. 7.3. For example, a polynomial regression model can be used to construct a (nonsymmetric) polynomial fitting index between $x = x_i$ and $y = x_j$ as follows:

$$M_{ij}^{polynomial} = polyFitIndex(x, y, d) = cor(y, \hat{y})^2, \tag{13.28}$$

which depends on the degree $d \geq 1$. Since $M_{ij}^{polynomial}$ is defined as square of a correlation it takes on values between 0 and 1. In general, the matrix $M = (M_{ij})$ is not symmetrical and needs to be symmetrized to arrive at an adjacency matrix. The R functions adjaceny.polyReg and adjacency.splineReg in the WGCNA R package allow one to construct adjacency matrices based on polynomial and spline regression models. As another example, consider the $n \times n$ dimensional matrix $M^{C.index}$ of pairwise concordance indices (5.18):

$$M_{ij}^{C.index} = C.index(x_i, x_j). \tag{13.29}$$

Note that $M^{C.index}$ is a nonsymmetric matrix whose entries lie between 0 and 1. To define an adjacency matrix, one can symmetrize it (7.9).

13.10 Partial Correlations and Related Networks

We will now define the partial correlation between two vectors x and y after conditioning out the effect of additional variables that correspond to the columns of a data frame $datZ = (z1, z2)$. Based on the projection operator $ProjOp(datZ) = datZ(datZ^{\tau}datZ)^{-}datZ^{\tau}$, define the following residual vectors:

$$\begin{aligned} xResidualZ &= x - \hat{x} = (I - ProjOp(datZ))x \\ yResidualZ &= y - \hat{y} = (I - ProjOp(datZ))y, \end{aligned}$$

where I denotes the $m \times m$ dimensional identity matrix. Note that *xResidualZ* is a vector of residuals that results from the linear model that regresses x on the columns of *datZ*. The **partial correlation coefficient** $cor(x,y|z1,z2)$ is defined as the Pearson correlation between the residuals:

$$cor(x,y|datZ) = cor(xResidualZ, yResidualZ) \tag{13.30}$$

Thus, $cor(x,y|datZ)$ measures the extent of a linear relationship between vectors x and y, with the effect of the conditioning vectors removed.

Methods for calculating p values for correlations (described in Sect. 10.3) can be easily adapted to the case of the partial correlation coefficient. For example, a Student t-test statistic can be calculated by replacing $m-2$ in (10.4) with $m-2-N$, where N is the number of conditioning vectors in (13.30). If the partial correlation is statistically significant, the association between x and y is not due to the variation measured by the variables z_1, z_2.

In case of a single conditioning variable $z1$, the partial correlation coefficient can be expressed in terms of pairwise correlation coefficients:

$$cor(x,y|z1) = \frac{cor(x,y) - cor(x,z1)cor(y,z1)}{\sqrt{(1-cor(x,z1)^2)(1-cor(y,z1)^2)}} \tag{13.31}$$

In the following, assume we want to compute the pairwise partial correlations between $n \geq 3$ vectors by conditioning on the remaining vectors. Assume that the n vectors form the columns of a matrix (or data frame) $datX = (x_1 \ldots x_n)$. Denote by $datX_{-(i,j)}$ the matrix that results from *datX* after removing columns i and j. Thus, $datX_{-(i,j)}$ is an $m \times (n-2)$ dimensional matrix. Assume that the correlation matrix $cor(datX) = (cor(x_i, x_j))$ is invertible. The inverse of the correlation matrix is also known as **precision matrix**:

$$Precision = (cor(datX))^{-1} = [p_{ij}].$$

In this case, the following simple formula can be used to compute the partial correlations in terms of entries p_{ij} of the precision matrix (Whittaker 1990):

$$cor(x_i, x_j|datX_{-(i,j)}) = -\frac{p_{ij}}{\sqrt{p_{ii}p_{jj}}}, \tag{13.32}$$

where $datX_{-(i,j)}$ denotes the matrix that results after removing columns i and j from datX. Several authors have proposed to construct a correlation network on the basis of partial correlations (Magwene and Kim 2004; Schaefer and Strimmer 2005; Krainik et al. 2006; Opgen-Rhein and Strimmer 2007; Chen and Zheng 2009). Partial correlations $cor(x_i, x_j|datX_{-(i,j)})$ (see (13.32) and (13.30)) measure the similarity between pairs of node profiles x_i and x_j, given the set of remaining node profiles $datX_{-(i,j)}$. Unsigned and signed **partial correlation-based adjacency matrices** can be defined as follows:

$$\begin{aligned} A_{ij}^{partial} &= |cor(x_i, x_j|datX_{-(i,j)})| \\ signed.A_{ij}^{partial} &= \frac{1 + cor(x_i, x_j|datX_{-(i,j)})}{2}. \end{aligned} \tag{13.33}$$

While partial correlation networks are clearly very useful, we should point out that their construction is not robust with respect to noise. Relatively small changes in the conditioning variables can lead to very different networks. While standard correlation networks (based on the Pearson correlation or a robust correlation coefficient) can be less efficient than partial correlation networks, they tend to be much more robust with respect to noise.

13.11 R Code for Partial Correlations

Now we illustrate how to calculate a partial correlation coefficient:

```
xResidualZ=as.numeric(glm(x~.,data=datZ)$residuals)
yResidualZ=as.numeric(glm(y~.,data=datZ)$residuals)
# this is the partial correlation coefficient
cor(xResidualZ,yResidualZ)
```

where the dot after ~ indicates that all columns of *datZ* should be used as covariates.

The inverse of the correlation matrix (i.e., the precision matrix) can be computed with the following R command:

```
Precision=solve(cor(datX,use="p"))
```

The R function `cor2pcor` in the R library `corpcor` (Opgen-Rhein and Strimmer 2007) allows one to compute partial correlations from a correlation matrix. The R package `corpcor` also implements efficient estimation methods (and shrinkage estimators) of covariances and (partial) correlations.

13.12 Exercises

1. Exercise regarding the meaning of the model fitting index R^2. Prove that for a univariate regression model $E(y) = \gamma_0 + \gamma_1 x$) (involving a single covariate vector x) the following relationship holds $cor(y,\hat{y})^2 = cor(x,y)^2$ where $\hat{y}$ denotes the predicted value. Further, verify this statement numerically. Hint: Use (13.7). Use the following R commands:

   ```
   yhat=predict(glm(y~x));cor(yhat,y)^2;cor(y,x)^2
   ```

2. Exercise regarding polynomial and spline regression. Simulate two numeric vectors as follows:

   ```
   set.seed(1);x=rnorm(100);y=x^2+rnorm(100)
   ```

   (i) Use a polynomial regression model with degree 2 to regress y on x and calculate the corresponding model fitting index *R.squared.xy*. Hint. See Sect. 13.5.

(ii) Reverse the roles of x and y, i.e., use a polynomial regression model with degree 2 to regress x on y and calculate the corresponding fitting index *R.squared.yx*.

(iii) Do the two fitting indices have the same value? If not, suggest a way to combine the two fitting indices so that a symmetric measure results. Hint: e.g., $R.squared.symmetrized = (R.squared.xy + R.squared.yx)/2$.

3. Exercise regarding the concordance index. Simulate two numeric vectors as follows:

```
set.seed(1);x=rnorm(100);y=x^2+rnorm(100)
```

(i) Use a polynomial regression model with degree 2 to regress y on x and compute the corresponding predicted value $\hat{y}$. Next compute the concordance index between the predicted and the observed outcome. Hint: Sect. 5.1.4 and the following R code:

```
library(Hmisc)
C.index.xy=rcorr.cens(yhat,y,outx=TRUE)[[1]]
```

(ii) Which value of the C-index corresponds to no relationship? Hint: It is *not* 0.

(iii) Reverse the role of x and y (i.e., regress x on y and compute the resulting concordance index. Hint:

```
C.index.yx=rcorr.cens(xhat,x,outx=TRUE)[[1]]
```

(iv) Do the two C-indices have the same value? If not, suggest a way to combine the two C-indices so that the resulting measure is symmetrical. Hint:

```
C.index.mean=(C.index.xy+C.index.yx)/2
C.index.max=max(C.index.xy,C.index.yx)
```

4. Exercise networks based on measuring nonlinear relationships between variables. Consider the following data frame $datX = [x_1, \ldots, x_5]$.

```
library(splines)
library(Hmisc)
m=100;set.seed(1)
z1=rnorm(m,sd=.5)
# this is the true but unobserved vector
x1.true=rnorm(m)
# this is the measurement error associated with x1
epsilon1=rnorm(m,sd=.5)
# this is the observed vector x1 which is also affected by z1
x1=x1.true+z1+epsilon1
x2.true=(x1.true)^2+rnorm(m,sd=3); epsilon2=rnorm(m)
x2=x2.true+z1+epsilon2
x3.true=rnorm(m,mean=-1); epsilon3=rnorm(m,sd=.1);
x3=x3.true+epsilon3
x4.true=x3.true^2+rnorm(m,sd=.5); epsilon4=rnorm(m,sd=.1)
x4=x4.true+z1+epsilon4
x5.true=exp(-x3.true^2);epsilon5=rnorm(m,sd=.1);
x5=x5.true+z1+epsilon5
```

```
datX=data.frame(x1,x2,x3,x4,x5)
#pairwise scatterplots between observed vectors
pairs(datX)
#pairwise scatterplots between true underlying vectors
datX.true=data.frame(x1.true,x2.true,x3.true,x4.true,x5.true)
pairs(datX.true)
```

(i) For each combination of $x = x_i$ and $y = x_j$ fit a natural spline with 5 degrees of freedom and report the resulting model fitting index $R^2 = cor(\hat{y}, y)^2$, the C-index, the likelihood ratio test statistic for comparing the fit to the intercept-only model, and the corresponding p value. Hint:

```
no.variables=dim(datX)[[2]]
RsquaredMatrix=matrix(NA,nrow=no.variables,ncol=no.
  variables)
CindexMatrix=matrix(NA,nrow=no.variables,ncol=no.
  variables)
LRTMatrix=matrix(NA,nrow=no.variables,ncol=no.variables)
PvalueMatrix=matrix(NA,nrow=no.variables,ncol=no.
  variables)
for (i in 1:no.variables ){
for (j in 1:no.variables ){
x=datX[,i]
y=datX[,j]
lmSpline=glm(y~ns(x,df=5))
yhat=as.numeric(predict(lmSpline))
RsquaredMatrix[i,j]=cor(y,yhat)^2
CindexMatrix[i,j]=rcorr.cens(yhat,y,outx=T)[[1]]
anovaSp=anova(lmSpline,test="Chisq")
LRTMatrix[i,j]=anovaSp$Deviance[2]/
DegreesOfFreedom=anovaSp$Df[2]
PvalueMatrix[i,j]=1-pchisq(LRTMatrix[i,j],
  DegreesOfFreedom)summary(lmSpline)$dispersion
} # end of for loop over i
} # end of for loop over j
# Set the maximum value of LRT equal to 1000
LRTMatrix[LRTMatrix>1000]=1000
```

(ii) Visualize the relationships between elements of the matrices LRTMatrix, RsquaredMatrix, and CindexMatrix using pairwise scatterplots. Hint:

```
library(WGCNA)
vecRsquaredM=vectorizeMatrix(RsquaredMatrix)
vecLRTM=vectorizeMatrix(LRTMatrix)
vecPM=vectorizeMatrix(PvalueMatrix)
vecCM=vectorizeMatrix(CindexMatrix)
par(mfrow=c(1,2))
verboseScatterplot(vecLRTM,vecRsquaredM,
               xlab="LRT statistic",ylab="R^2")
verboseScatterplot(vecLRTM,vecCM,
               xlab="LRT statistic",ylab="C-index")
```

(iii) Dichotomize each of the matrices by choosing a suitable threshold value. Try to find thresholds so that the resulting binary matrices are similar.

```
tau.LRT=median(vecLRTM,na.rm=T)
tau.C=median(vecCM,na.rm=T)
tau.Rsquared=median(vecRsquaredM,na.rm=T)
LRTMatrixThresholded=ifelse(LRTMatrix>tau.LRT,1,0)
CindexMatrixThresholded=ifelse(CindexMatrix>tau.C,1,0)
RsquaredMatrixThresholded=ifelse(RsquaredMatrix>
  tau.Rsquared,1,0)
# Compare the off-diagonal elements of the difference
  between matrices
table(vectorizeMatrix(LRTMatrixThresholded-CindexMatrix
  Thresholded))
table(vectorizeMatrix(LRTMatrixThresholded-RsquaredMatrix
  Thresholded))
```

Message: If we define significant relationships between pairs of variables using the above thresholds, we arrive at the same results.

(iv) Use the thresholded matrices to define an unweighted adjacency matrix by symmetrizing the matrices with the maximum function. Hint:

```
ADJ.unweighted=
  pmax(RsquaredMatrixThresholded,
  t(RsquaredMatrixThresholded))
```

(v) Measure the extent to which the unthresholded matrix RsquaredMatrix threshold-implies the matrices LRTmatrix and CindexMatrix. Hint:

```
C.indexRsqtoLRT=rcorr.cens(vecRsquaredM,vecLRTM,outx=T)
  [[1]]
C.indexRsqtoC=rcorr.cens(vecRsquaredM,vecCM,outx=T)[[1]]
```

(vi) Show that minus the LRTMatrix threshold-implies the PvalueMatrix, i.e., for each threshold for the p values, one can find a corresponding threshold for the LRT matrix. Hint: Show that the C-index between the vectorized matrices equals 1.

```
rcorr.cens(-vecLRTM,vecPM,outx=T)[[1]]
```

(vii) Construct a weighted adjacency matrix on the basis of *RsquaredMatrix*.

```
ADJ.weighted=pmax(RsquaredMatrix,t(RsquaredMatrix))
```

(viii) Which adjacency matrix (weighted or unweighted) is a more accurate representation of the simulated true pairwise relationships? Hint: The answer depends on how we measure the simulated true relationships. I suggest to define the true network A^{true} using the following R code:

```
RsquaredMatrix.true=matrix(NA,nrow=no.variables,
  ncol=no.variables)
for (i in 1:no.variables ){
for (j in 1:no.variables ){
x=datX.true[,i];y=datX.true[,j]
lmSpline=glm(y~ns(x,df=5))
yhat=as.numeric(predict(lmSpline))
```

```
RsquaredMatrix.true[i,j]=cor(y,yhat)^2
} # end of for loop over i
} # end of for loop over j
ADJ.true=pmax(RsquaredMatrix.true,t(RsquaredMatrix.true))
```

Next correlate the vectorized version of this matrix with the vectorized versions of *ADJ.unweighted* and *ADJ.weighted*.

```
vA.true=vectorizeMatrix(ADJ.true)
vA.weighted=vectorizeMatrix(ADJ.weighted)
vA.unweighted=vectorizeMatrix(ADJ.unweighted)
cor(data.frame(vA.true,vA.weighted,vA.unweighted))
```

(ix) How do the results of the nonlinear weighted adjacency matrix compare with those from a standard correlation networks?

```
A.standard.weighted=abs(cor(datX))^6
vA.standard.weighted=vectorizeMatrix(A.standard.weighted)
datv=data.frame(vA.true,vA.weighted,vA.unweighted,
  vA.standard.weighted)
signif(cor(datv,use="p"),2)
```

(x) Recall that the simulation code included a variable $z1$ that affected the observed values of $x_1,\ldots,x_5$. Modify the above R code so that you arrive at association measures (R^2, C-index, LRT statistic, p value) that adjust for the covariates $z1$ and repeat tasks (i)–(viii) with the new adjusted measures. Hint: Define `lmSplineAdjusted=glm(y~z1+ns(x,df=5))`. Further, replace two lines of the R code by the following:

```
LRTMatrix[i,j]=
anovaSp$Deviance[3]/summary(lmSpline)$dispersion
DegreesOfFreedom=anovaSp$Df[3]
```

References

Chen L, Zheng S (2009) Studying alternative splicing regulatory networks through partial correlation analysis. Genome Biol 10(1):R3

Faraway JJ (2002) Practical regression and anova using R. R pdf file at http://cranr-projectorg/doc/contrib/Faraway-PRApdf

Krainik A, Duffau H, PelegriniIssac S, Lehericy J, Benali H (2006) Partial correlation for functional brain interactivity investigation in functional MRI. NeuroImage 32:228–237

Magwene P, Kim J (2004) Estimating genomic coexpression networks using first-order conditional independence. Genome Biol 5(12):R100

Opgen-Rhein R, Strimmer K (2007) From correlation to causation networks: A simple approximate learning algorithm and its application to high-dimensional plant gene expression data. BMC Syst Biol 1:37

Schaefer J, Strimmer K (2005) An empirical Bayes approach to inferring large-scale gene association networks. Bioinformatics 21(6):754–764

Whittaker J (1990) Graphical models in applied multivariate statistics. Wiley, Chichester

Chapter 14
Networks Between Categorical or Discretized Numeric Variables

Abstract Categorical variables take on non-numeric values, e.g., a discretized numeric variable can be interpreted as a categorical variable. Many association measures exist for measuring the statistical dependence between categorical variables (e.g., Pearson chi-square statistic, likelihood ratio test statistic, Fisher's exact test, mutual information). Association measures between two discretized numeric vectors have been used to measure nonlinear dependencies between them. We describe several approaches for defining weighted networks among categorical variables. In particular, mutual information networks are often used for constructing gene networks. The close relationship between mutual information and a likelihood ratio test statistic allows us to define a conditional measure of mutual information, which accounts for additional covariates. Estimating the mutual information between numeric variables is rather challenging and involves parameter choices. We argue that in many applications the mutual information measure can be approximated by a correlation-based measure. We review the ARACNE approach for constructing an unweighted mutual information network and generalize it to correlation networks and other association networks.

14.1 Categorical Variables and Statistical Independence

In R, categorical variables are also known as factor variables. The components of a categorical vector contain strings of characters. The possible values of a categorical variable are referred to as levels. For example, the levels of the variable *eyecolor* may be "brown", "black", and "blue". Assume that the categorical random variable DX takes on $no.lDX \geq 1$ different levels (character strings) $ldx_1, \ldots, ldx_{no.lDX}$ with probabilities $p_{DX}(ldx_1), \ldots, p_{DX}(ldx_{no.lDX})$, respectively. Similarly, assume that DY is another categorical random variable which takes on the values $ldy_1, \ldots, ldy_{no.lDY}$ with probabilities $p_{DY}(ldy_1), \ldots, p(ldy_{no.lDY})$, respectively. Denote the joint probability distribution between DX and DY by $p_{DX,DY}(ldx_r, ldy_c)$, i.e., $p_{DX,DY}(ldx_r, ldy_c)$ is the probability that DX takes on value ldx_r and DY takes on value ldy_c. Then the

S. Horvath, *Weighted Network Analysis: Applications in Genomics and Systems Biology*, DOI 10.1007/978-1-4419-8819-5_14,

random variables DX and DY are **independently** distributed if (and only if) for all values of r and c:

$$p_{DX,DY}(ldx_r, ldy_c) = p_{DX}(ldx_r)p_{DY}(ldy_c). \tag{14.1}$$

When (14.1) does not hold, i.e., when the two categorical random variables are dependent, we will informally say that they are associated with each other.

In practice, the frequency distribution of a categorical variable is unknown and must be estimated from a random sample. Assume that the vector $dx = (dx_1, \ldots, dx_m)$ represents a random sample (of size m) from DX. Let O_r denote the number of components of dx that equal ldx_r. The observed **relative frequency** of ldx_r is defined as

$$p(ldx_r) = \frac{O_r}{m}. \tag{14.2}$$

In other words, $p(ldx_r)$ is the proportion of components of dx with value ldx_r. The vector of **(observed relative) frequencies** is given by

$$p(dx) = (p(ldx_1), \ldots, p(ldx_{no.ldx})). \tag{14.3}$$

Assume now that $(dx_1, dy_1), \ldots, (dx_m, dy_m)$ represents a joint sample of the random variables DX and DY. Denote by O_{rc} the number of observations for which dx takes on the value ldx_r and dy takes on the value ldy_c. The **observed joint relative frequency** $p(ldx_r, ldy_c)$ is defined as proportion of observations for which dx equals ldx_r and dy equals ldy_c, i.e.,

$$p(ldx_r, ldy_c) = O_{rc}/m. \tag{14.4}$$

An alternative definition of the joint frequency is based on the **paste function**, which will be explained in the following. $paste(dx, dy)$ denotes a vector of length m whose uth component equals the concatenation of the strings specified by dx_u and dy_u. For example, if $dx_u =$ "$green$" and $dy_u =$ "tea", then $paste(dx, dy, sep = \text{""})_u =$ "$greentea$", $paste(dx, dy, sep = \text{","})_u =$ "$green, tea$". Let us assume that the default separator is ",". In general, $paste(dx, dy)$ is a categorical vector with $no.ldx * no.ldy$ different levels. The observed joint frequency $p(ldx_r, ldy_c)$ (14.4) is the proportion of components of $paste(dx, dy)$ which start out with the string specified by ldx_r and end with the string specified by ldy_c. We say that two categorical vectors dx and dy have (exactly) **independent frequency distributions** if their observed joint frequencies factor as follows:

$$p(ldx_r, ldy_c) = p(ldx_r)p(ldy_c). \tag{14.5}$$

Our definition of independence between observed frequency distributions (14.5) is analogous to the definition of statistical independence between random variables (14.1) (for which the theoretical frequency distributions are known).

The observed joint frequency $p(ldx_r, ldy_c, ldz_d)$ of three variables is the proportion of components of $paste(dx, dy, dz)$, which equal the combination of the three strings ldx_r, ldy_c, ldz_d. The three vectors have independent observed frequency distributions if their joint frequency factors as follows:

$$p(ldx_r, ldy_c, ldz_d) = p(ldx_r)p(ldy_c)p(ldz_d),$$

and if each pairwise joint distribution factors as well, e.g., $p(ldx_r, ldz_d) = p(ldx_r)p(ldz_d)$.

14.2 Entropy

Consider the categorical random variable DX for which the frequency of the rth level is given by $p(ldx_r)$. Then the entropy (Shannon 1948) of (the frequency distribution of) DX is defined as:

$$Entropy(DX) = -\sum_{r=1}^{no.lDX} p_{DX}(ldx_r)log(p_{DX}(ldx_r)), \tag{14.6}$$

where we set $0 * log(0) = 0$ when the probability equals 0. Consider the case when each level of DX has the same frequency (i.e., $p_{DX}(ldx_r) = 1/no.lDX$), then $Entropy(DX) = log(no.lDX)$. One can show that this is the maximum entropy across the set of categorical vectors with $no.lDX$ distinct values. We briefly mention Shannon's **defining properties of the entropy**. The entropy is the only real valued function that satisfies (Shannon 1948):

1. *Additivity*: $Entropy(DX, DY) = Entropy(DX) + Entropy(DY)$ if DX and DY are independent (14.1).
2. *Continuity*: $Entropy(DX)$ is a continuous function of the frequency distribution $p_{DX}(ldx_r)$.
3. *Maximality with respect to constant distributions*: it takes on the maximal value when $p_{DX}(ldx_r) = 1/no.lDX$ is constant.
4. *Order-independence*: $Entropy(DX)$ is invariant with respect to the order of the frequencies $p_{DX}(ldx_r)$.

The entropy is also an important concept in physics (thermodynamics) and chemistry where it is interpreted as the amount of information needed to exactly specify the state of a discrete system or the "potential for disorder" of the system. The entropy can be interpreted as a measure of uncertainty (Shannon 1948): the higher the entropy, the more uncertain one is about a randomly sampled value of DX. Assume again that the vector dx (of length m) represents a random sample of DX. Then $p_{DX}(ldx_r)$ can be estimated with the observed relative frequency $p(ldx_r)$ and the $Entropy(DX)$ can be estimated with

$$Entropy(dx) = -\sum_r p(ldx_r)log(p(ldx_r)). \tag{14.7}$$

Paraphrasing (Fraser and Swinney 1986; Steuer et al. 2002), $Entropy(DX)$ quantifies the amount of surprise one should feel upon finding out the value of a single component of dx. To illustrate this, consider the extreme case where DX can take on only a single value ldx_1. In this case, $p(ldx_1) = 1$ implies that $Entropy(DX) = 0$. Clearly, one would feel zero surprise upon learning that a randomly sampled value of DX equals ldx_1. As an opposite extreme case, assume that each component of dx takes on a distinct value, i.e., the number of levels $no.ldx = m$. In this case, $p(ldx_r) = 1/m$ and $Entropy(dx) = log(m)$. Since all components of dx take on distinct values, one would feel maximally surprised upon learning the value of a randomly sampled component of dx.

Assume now that a second categorical variable DY is available, which takes on the values $ldy_1, \ldots, ldy_{no.lDY}$ with probabilities $p_{DY}(ldy_1)$, ..., $p(ldy_{no.lDY})$, respectively. Denote the joint probability distribution between DX and DY by $p_{DX,DY}(ldx_r, ldy_c)$. The **joint entropy** of DX and DY is defined as:

$$Entropy(DX,DY) = -\sum_r \sum_c p_{DX,DY}(ldx_r, ldy_c) log(p_{DX,DY}(ldx_r, ldy_c)).$$

As described in Sect. 14.1, $paste(dx, dy)$ denotes a categorical vector (of length m) with $no.ldx * no.ldy$ different levels with relative frequencies $p(ldx_r, \ldots, ldx_c)$. The joint entropy $Entropy(dx, dy)$ is defined as the entropy of the vector $paste(dx, dy)$, i.e.,

$$\begin{aligned} Entropy(dx,dy) &= Entropy(paste(dx,dy)) \\ &= -\sum_{r=1}^{no.ldx} \sum_{c=1}^{no.ldy} p(ldx_r, ldy_c) log(p(ldx_r, ldy_c)) \end{aligned} \tag{14.8}$$

where $p(ldx_r, ldy_c)$ is the relative frequency of observations for which dx and dy take on the values ldx_r and ldy_c, respectively. Analogously, one can define the joint entropy between three categorical vectors dx, dy, and dz as follows:

$$Entropy(dx,dy,dz) = Entropy(paste(dx,dy,dz)).$$

In the following, we present R code for calculating the entropy based on a categorical vector dx: The entropy can be computed using the following R command

```
# vector of relative frequencies
p.ldx=table(dx)/length(dx)
Entropy.dx=-sum(p.ldx*log(p.ldx),na.rm=T)
```

14.2.1 Estimating the Density of a Random Variable

Usually the probability density distribution of a random variable X is unknown and must be estimated from a random sample. Assume that the components of the vector $x = (x_1, \ldots, x_m)$ contain a random sample of X.

Many methods have been developed for estimating the probability density. Here we review the **empirical estimation method** (also known as "naive", "plug-in", or "maximum likelihood" method) which is based on discretizing the range of x into subintervals. Assume that the components of x take on values between the lower bound $LowerB$ and upper bound $UpperB$. For example, we could define the lower and upper bound as the minimum and maximum value of x, respectively. The **equal-width discretization method** (also known as histogram method) partitions the interval $[LowerB, UpperB]$ into equal-width bins (subintervals). Specifically, the rth ($r = 1, 2, \ldots, no.bins$) bin is defined as follows:

$$bin_r = [LowerB + (r-1)width, LowerB + rwidth), \tag{14.9}$$

where the bin width is given by $width = \frac{UpperB - LowerB}{no.bins}$. The vector $discretize(x)$ has the same length as x but its lth component reports the bin number in which x_l falls:

$$discretize(x)_l = r \text{ if } x_l \in bin_r. \tag{14.10}$$

The number of bins $no.bins$ is the only parameter of the equal-width discretization method. Its default value equals the square root of the number of observations, i.e., $no.bins = as.integer(\sqrt{m})$. For example, if x has $m = 103$ components, then $discretize(x)$ is a categorical variable whose components take on one of the following values $1, 2, \ldots, 10$. The (observed) number of components of x that fall into the rth bin is denoted by O_r. The proportion p_r of components of x that fall into the rth bin is given by

$$p_r = O_r/m, \tag{14.11}$$

i.e., p_r is the **observed relative frequency** of the rth bin. A **histogram** (also known as **frequency plot**) can be used to visualize the estimated frequency distribution as a series barplots whose heights correspond to p_r. While it is straightforward to estimate the frequency distribution of a single variable, it is much more challenging to estimate the probability density of two or more variables.

We will now present relevant R code. Several approaches for discretizing a numeric vector are implemented by the R function `discretize` in the `infotheo` R package (see also the `minet` package) (Meyer et al. 2008). To discretize x into $no.bins = 10$, one can use the following R command:

```
dx=discretize(x,nbins=10,disc="equalwidth")
```

The vector of observed counts $O = (O_1, \ldots, O_{no.bins})$ can be calculated with the following R command

```
O=table(discretize(x,no.bins,disc="equalwidth"))
```

The vector of relative frequencies can be calculated with the following R command

```
p=table(discretize(x,no.bins,disc="equalwidth"))/length(x)
```

A histogram can be created with the following R command:

```
hist(x,breaks=no.bins)
```

14.2.2 Entropy of a Discretized Continuous Variable

The mutual information $\mathrm{MI}(dx,dy)$ between the discretized numeric variables $dx = discretize(x)$ and $dy = discretize(y)$ can be used to measure possibly nonlinear relationship between x and y (Cover and Thomas 1991). Note that $MI(discretize(x), discretize(y)) = 0$ implies that $dx = discretize(x)$ (14.10) and $dy = discretize(y)$ have independent frequency distributions (14.5). Unfortunately, the mutual information between discretized vectors is highly dependent on the discretization. For definiteness, we will use the equal-width discretization method (14.10) in the following. The resulting mutual information measure depends on the number of bins $no.bins$ used in the discretization step. As the default number of bins, we and others use the square root of the number of observations, i.e., $no.bins = \sqrt{m}$. Then the resulting mutual information measure depends on the vector length m. In the following, we will describe estimators for the entropy of a *discretized* continuous variable $DX = discretize(X)$. Many estimation procedures have been proposed in the literature (Beirlant et al. 1997; Darbellay and Vajda 1999; Paninski 2003; Daub et al. 2004; Paninski 2003; Nemenman 2004). Here we will briefly mention some of them since they will be relevant when discussing mutual information networks. Specifically, assume that the equal-width discretization method is used to split the range of X into $no.bins$. If the probability density p_X is known, then one can compute the true entropy $Entropy(DX)$. But in practice, p_X is unknown, and $Entropy(DX)$ must be estimated from a random sample, i.e., a discretized numeric vector $discretize(x)$. The equal-width discretization method (14.10) results in a vector of relative frequencies $p = (p_1, \ldots, p_{no.bins})$, where p_r denotes the frequency of the rth bin. Using these relative frequencies, the **empirical estimator** (also known as "naive", "plug-in" or "maximum likelihood" estimator) of $Entropy(DX, no.bins)$ is given by:

$$Entropy^{empirical}(DX) = -\sum_{r=1}^{no.bins} p_r log(p_r). \tag{14.12}$$

The empirical estimator of the entropy tends to be smaller than the true value $Entropy(DX)$. To counter this bias, several alternative methods have been developed. The **Miller–Madow estimator** is given by

$$Entropy^{MM}(DX) = Entropy^{empirical}(DX) + \frac{no.bins - 1}{2 * m}. \tag{14.13}$$

We briefly mention two other entropy estimators: the **shrinkage estimator** $Entropy^{shrinkage}$ uses shrinkage estimates of the relative frequencies and the **Schurmann–Grassberger** estimator $Entropy^{SG}$ uses the Dirichlet distribution. Most R functions for estimating the entropy (and the mutual information) allow the user to choose one of these methods.

As default method, we use the Miller–Madow estimation procedure which results in the Miller–Madow estimate of the entropy and the mutual information. It is also the default method in the `WGCNA` R function `mutualInfoAdjacency`. But an argument in the function allows one to choose alternative estimation methods.

14.3 Association Measures Between Categorical Vectors

If dx and dy represent random samples from corresponding random variables then the approximate relationship between the observed frequencies

$$p(ldx_r, ldy_c) \approx p(ldx_r)p(ldy_c) \tag{14.14}$$

supports the claim that the corresponding random variables are statistically independent. A better way of saying the same thing is to use the language of hypothesis testing: if the observed frequencies $p(ldx_r, ldy_c)$ are significantly different from $p(ldx_r)p(ldy_c)$ then we reject the null hypothesis of independence between the corresponding random variables DX and DY. In this case, we say that the two variables are associated with each other. To measure the difference between $p(ldx_r, ldy_c)$ and $p(ldx_r)p(ldy_c)$, one could define the sum of squared differences

$$SSq = \sum_{r=1}^{no.ldx} \sum_{c=1}^{no.ldy} (p(ldx_r, ldy_c) - p(ldx_r)p(ldy_c))^2$$

and use a permutation test (described in Sect. 10.2) to calculate a corresponding permutation p value. But statisticians prefer a relative difference measure known as **Pearson's chi-square statistic**:

$$Pearson.chisq(dx, dy) = m \sum_r \sum_c \frac{(p(ldx_r, ldy_c) - p(ldx_r)p(ldy_c))^2}{p(ldx_r)p(ldy_c)} \tag{14.15}$$

The higher the value of Pearson's chi-square statistic, the stronger the statistical evidence that the **two variables are associated with each other**. To compute a p value, one can make use of the fact that under the null hypothesis of independence between dx and dy, the chi-square statistic follows asymptotically (i.e., for large m) a chi-square distribution with $df = (no.ldx - 1) * (no.ldy - 1)$ degrees of freedom. The p value estimation is fairly accurate if the observed cell counts satisfy $O_{rc} \geq 5$. In case of low cell counts, it is advisable to use Fisher's exact test described below.

An alternative test statistic of independence is the **likelihood ratio test (LRT)** statistic

$$LRT.statistic(dx, dy) = 2m \sum_r \sum_c p(ldx_r, ldy_c) log \left(\frac{p(ldx_r, ldy_c)}{p(ldx_r)p(ldy_c)} \right), \tag{14.16}$$

where log denotes the natural logarithm (base e) and $0 * log(any.number) = 0$ for the terms of the sum. The larger the $LRT.statistic(dx, dy)$, the stronger the evidence that the corresponding random variables DX and DY are statistically dependent. Like the Pearson chi-square test statistic (14.15), the LRT is an association measure that follows a chi-square distribution with $(no.ldx - 1)(no.ldy - 1)$ degrees of freedom under the null hypothesis of independence. The name "likelihood ratio test" stems from the fact that it can be derived as a likelihood ratio test of a multinomial regression model, which is a generalized linear regression model.

The p values calculated using Pearson's chi-square statistic or the likelihood ratio test statistic are sometimes referred to as **asymptotic** p **values** since they are accurate in the limit of a large sample size m. For reasonably large m (say observed cell counts $O_{rc} \geq 5$), the LRT and the chi-square test will lead to the same conclusions. But the approximation to the theoretical chi-square distribution for the LRT test tends to be better. For small samples, **Fisher's exact test** is preferable to either Pearson's chi-square test or the likelihood ratio test. Fisher's exact test is called "exact" since the p value can be calculated exactly, rather than relying on an approximation that becomes exact in the limit as the sample size grows to infinity, as with asymptotic statistical tests. The p value from the test is computed as if the margins of the table are fixed which leads under a null hypothesis of independence to a hypergeometric distribution of observed cell counts O_{rc}. A drawback of Fisher's exact test is that it is computationally challenging if both $no.ldx$ and $no.ldy$ are larger than 2.

Alternatively, one can use the mutual information for contrasting the left-hand side of (14.14) with the right-hand side: the **mutual information** (MI) is defined as

$$MI(dx,dy) = \sum_{r=1}^{no.ldx} \sum_{c=1}^{no.ldy} p(ldx_r, ldy_c) log \left(\frac{p(ldx_r, ldy_c)}{p(ldx_r)p(ldy_c)} \right). \tag{14.17}$$

Here we will use a natural logarithm (base e) which implies that the MI is measured in units of "nats". It is straightforward to show that $MI(dx,dy) = 0$ if, and only if, the frequency distributions of the vectors are independent (14.5). Note that the following simple relationship exists between the mutual information (14.17) and the likelihood ratio test statistic (14.16):

$$MI(dx,dy) = \frac{LRT.statistic(dx,dy)}{2m}. \tag{14.18}$$

The relationship between mutual information and the LRT statistic has many uses including the following. First, it can be used to prove that the mutual information takes on nonnegative numbers. Second, it can be used to calculate an asymptotic p value for the mutual information. Toward this end, the following R code can be used:

```
p.valueMI=1-pchisq(MI.dxdy*2*m,df)
```

Equation (14.18) suggests a way for calculating an asymptotic p value which circumvents the computationally intensive calculation of a permutation test p value for MI (see Sect. 10.2). However, permutation test procedures are needed if dx and dy arise from discretizing numeric vectors.

The mutual information is well defined for categorical variables (14.17) but it is challenging to arrive at a useful definition in case of two continuous numeric variables X and Y. Most definitions and estimation approaches make use of the following relationship: $MI(X,Y) = Entropy(X) + Entropy(Y) - Entropy(X,Y)$ (14.23), which shows that $MI(X,Y)$ can be defined and estimated as long as the same holds for the (joint) entropy. While Shannon's entropy was originally defined for categorical variables, it is straightforward to define it for a continuous variable via integration $Entropy(X) = -\int p_X(x)log(p_X(x))dx$ (15.3). Although the entropy of a continuos numeric variable shares many of the properties of that of a discrete variable, it has two major shortcomings. First, it may be infinitely large or take on a negative value. Second, it is in general not invariant under monotonic transformations $x \longrightarrow increasingF(x)$. Because of these challenges, it is often preferable to define $MI(X,Y)$ in terms of $MI(discretize(X), discretize(Y))$ where $discretize(X)$, denotes a discretized version of the original continuous variable. Discretization turns numeric variables into categorical variables for which it is straightforward to calculate the (joint) entropies and hence the mutual information.

14.3.1 Association Measures Expressed in Terms of Counts

Let O_{rc} denote the observed number (count) of observations for which dx and dy equal ldx_r and ldy_c, respectively. The expected count $E_{rc} = E(O_{rc})$ under the null hypothesis of independence is given by

$$E_{rc} = \frac{O_{r\cdot}O_{\cdot c}}{O_{\cdot\cdot}},$$

where $O_{r\cdot} = \sum_c O_{rc}$ and $O_{\cdot\cdot} = m$. If the frequency distributions of dx is independent of that of dy (14.5), then $E_{rc} = O_{rc}$. Pearson's chi-square statistic (14.15), the likelihood ratio statistic (14.16), and the mutual information can be expressed in terms of observed and expected cell counts as follows:

$$\begin{aligned}
Pearson.chisq(dx,dy) &= \sum_{r=1}^{no.ldx}\sum_{c=1}^{no.ldy} \frac{(O_{rc}-E_{rc})^2}{E_{rc}} \\
LRT.statistic(dx,dy) &= 2\sum_r\sum_c O_{rc}log\left(\frac{O_{rc}}{E_{rc}}\right) \\
MI(dx,dy) &= \frac{1}{m}\sum_r\sum_c O_{rc}log\left(\frac{O_{rc}}{E_{rc}}\right).
\end{aligned} \tag{14.19}$$

14.3.2 R Code for Relating Categorical Variables

In the following, we present some R code. To turn any vector w into a categorical vector (whose components are strings), use `factor(w)`. The levels of a variable (vector) can be obtained with the following R command `levels(dx)`. The number of levels $no.ldx$ of dx can be computed using the following R command:

```
no.ldx=length(levels(dx))
```

The observed relative frequencies (14.3) can be calculated as follows:

```
p.dx=table(dx)/length(dx)
```

A table of counts O_{rc} between dx and dy can be computed the command `table(dx,dy)`. In particular, O_{rc} is given by `table(dx,dy)[r,c]`. The R function `paste` inputs two or more vectors of equal length and outputs a vector of combined character strings. The help file of the `paste` function describes several options for dealing with missing values. The default separator of the R function `paste` is the space " ". We find it useful to change the default separator to a comma, which can be accomplished using the following R code:

```
paste=function(x,...) base::paste(x,sep=",",...)
```

Then a vector of joint frequencies $p(ldx_r, ldy_c)$ can be calculated as follows:

```
p.ldx.ldy=table(paste(dx,dy))/m
```

Pearson's chi-square statistic (14.15) can be calculated as follows:

```
chisq.test(table(dx,dy),correct=F)$statistic
```

Pearson chi-square test p value is given by

```
chisq.test(table(dx,dy))$p.value
```

The multinomial likelihood ratio statistic (14.16) and a corresponding p value can be calculated as follows:

```
library(nnet)
glmMulti=multinom(factor(dy)~factor(dx))
glmMulti0=multinom(factor(dy)~1)
anova1=anova(glmMulti,glmMulti0)
LRT.statistic=anova1$"LR stat."[2]
DegreesOfFreedom=anova1[2,5]
pvalue.LRT=1-pchisq(LRT.statistic,DegreesOfFreedom)
```

Fisher's exact test p value can be calculated using the following R command:

```
fisher.test(table(dx,dy))$p.value
```

14.3.3 Chi-Square Statistic Versus Cor in Case of Binary Variables

In this section, we assume that the components of the categorical vectors $dx = (dx_1, \ldots, dx_m)$ and $dy = (dy_1, \ldots, dy_m)$ take on one of two possible values. A binary categorical variable can be easily recoded into a binary numeric variable, e.g., the variable gender can be coded to take on the value 1 for males and 0 for females. In case of two binary categorical variables, one can easily verify that Pearson's chi-square statistic $Pearson.chisq(dx, dy)$ (14.15) equals the square of the correlation $cor(x, y)^2$ times the number of observations:

$$Pearson.chisq(dx, dy) = m^* cor(x, y)^2. \tag{14.20}$$

This implies that $m^* cor(x,y)^2$ follows asymptotically a noncentral chi-square distribution with 1 degree of freedom under the null hypothesis of independence between *dx* and *dy* (which entails an expected correlation of 0 between *x* and *y*). Conversely, (14.20) suggests that a chi-square statistic can be defined for testing whether the correlation between two binary variables is zero.

The following R code turns a binary categorical variable *dx* into a binary numeric variable:

```
ldx1=levels(factor(dx))[[1]]
x=ifelse(dx==ldx1,1,0)
```

The following R command can be used for calculating a correlation test *p* value in case of two binary variables

```
pvalue.Corxy=1-pchisq(q=length(x)*cor(x,y)^2,df=1)
```

Note that this *p* value is identical to Pearson's chi-square test for testing the dependence of two categorical variables, i.e., it is equal to the *p* value calculated using the following R code:

```
pvalue.Chisq=chisq.test(table(x,y),correct=F)$p.value
```

We should point out that it is statistically advantageous to set the option `correct=T` in the previous R command. This results in a test statistic (and asymptotic *p* value) that uses a "continuity" correction (Agresti 2007), which amounts to subtracting 0.5 from all absolute differences $|O_{rc} - E_{rc}|$ in (14.19). Alternatively, one can use Fisher's exact test *p* value, which can be calculated as follows:

```
fisher.test(table(x,y))$p.value
```

14.3.4 Conditional Mutual Information

A major advantage of the likelihood ratio test formulation (14.18) of the mutual information is that it lends itself to generalizations. For example, we will describe how to define a **conditional mutual information measure** that adjusts for additional conditioning variables. Specifically, we describe how to define a (nonsymmetric) association measure between *dx* and *dy* that takes account of other variables corresponding to the columns of a matrix *datZ*. For example, if *dx* and *dy* represent discretized gene expression measurements, then *datZ* may contain a variable encoding a batch effect or other relevant information. Analogous to the likelihood ratio test statistic for testing the multivariate linear model described in Sect. 13.6, one can use a multivariate *multinomial regression model* for regressing *dy* on *dx* and the columns of *datZ*. The following R code describes how to calculate a corresponding LRT statistic and asymptotic *p* value:

```
library(nnet)
glmMulti=multinom(factor(dy)~.+factor(dx),data=data.frame(datZ))
glmMultiZ=multinom(factor(dy)~.,data=data.frame(datZ))
```

```
anova1=anova(glmMulti,glmMultiZ)
LRT.statistic=anova1$"LR stat."[2]
DegreesOfFreedom=anova1[2,5]
pvalue.LRT=1-pchisq(LRT.statistic,DegreesOfFreedom)
```

Up to a scaling factor of $2 * m$, the likelihood ratio test statistic can be interpreted as (nonsymmetric) measure of mutual information between dx and dy that adjusts for $datZ$. Strategies for symmetrizing a measure are described in (7.9).

14.4 Relationships Between Networks of Categorical Vectors

Assume now that we have a set of n categorical vectors $dx_1, \ldots, dx_n$ which form the columns of the matrix $datDX = [dx_1, dx_2, \ldots, dx_n]$. Any of the categorical significance measures in Sect. 14.3 can be used to relate $dx = dx_i$ to $dy = dx_j$. Thus, each of the categorical association measures gives rise to an $n \times n$ dimensional symmetric matrix S^{assoc} whose i, jth element measures the extent of an association between dx_i to dx_j. For example, Pearson's chi-square test statistic $Pearson.chisq(dx_i, dx_j)$ (14.15), the likelihood ratio test statistic $LRT.statistic(dx_i, dx_j)$ (14.16), and the mutual information $MI(dx_i, dx_j)$ (14.17) give rise to symmetric $n \times n$ dimensional matrices denoted by $S^{Pearson}$, S^{LRT}, and S^{MI}, respectively. The higher the value, the stronger the evidence for an association between the pairs of vectors. p values between pairs of variables also give rise to matrices, e.g., Pearson's chi-square test p value $S^{pvaluePearson} = (-log10(p_{ij}))$, the likelihood ratio test p value $S^{pvalueLRT}$), and Fisher's exact test p value $S^{pvalueFisher}$. Similarly, one could use the q-value (10.21) to define matrices. In Sect. 7.2, we describe several approaches for turning the resulting symmetric matrices into adjacency matrices. Let us now focus on the construction of an unweighted network, which can be accomplished using a (hard) threshold τ for any of the above-mentioned categorical association measure. Let us now investigate the relationships between the resulting unweighted networks using terminology from Sect. 10.11. Note that a matrix $S^{(1)}$ threshold-implies another matrix $S^{(2)}$ if, and only if, a nondecreasing function $nonDecreasingF$ can be defined such that

$$vectorizeMatrix(S^{(2)}) = nonDecreasingF(vectorizeMatrix(S^{(1)})).$$

Further, we say that $S^{(1)}$ is threshold-equivalent to $S^{(2)}$ if $nonDecreasingF$ is a strictly increasing function. Since m and n are fixed, it is straightforward to show (Exercise) that

- $S^{Pearson}$ threshold-implies $S^{pvaluePearson}$.
- S^{LRT} threshold-implies $S^{pvalueLRT}$.
- S^{LRT} and S^{MI} are threshold-equivalent.
- $S^{Pearson}$ is asymptotically threshold-equivalent to S^{LRT}, which means that they are approximately equivalent if m is large.

- $S^{pvaluePearson}$ is asymptotically (i.e., for large m) threshold-equivalent to $S^{pvalueLRT}$.
- $S^{pvaluePearson}$ is asymptotically threshold-equivalent to $S^{pvalueFisher}$.

Any matrix S^{pvalue} of p values threshold-implies a corresponding matrix $S^{qvalueLRT}$ of q-values since the q-values are a nondecreasing function (10.21) of the p values.

These results show that the mutual information measure is threshold-equivalent to other widely used statistical association measures. But since many bioinformaticians are trained in information theory, the mutual information measure is frequently used to define networks between categorical or discretized numerical variables (Chow and Liu 1968; Butte et al. 2000; Cheng et al. 2002; Steuer et al. 2002; Meyer et al. 2008) or for clustering data (Kraskov et al 2003).

14.5 Networks Based on Mutual Information

Let us now describe some strategies for defining a mutual information-based adjacency matrix between n categorical variables (or discretized numeric variables) $dx_1, dx_2, \ldots, dx_n$.

The definition of many (but not all) mutual information networks proceeds along the following two steps. First, one estimates the mutual information matrix (MI) whose i,jth element is given by

$$MI_{ij} = MI(dx_i, dx_j). \tag{14.21}$$

Second, the mutual information matrix is transformed into an adjacency matrix. This transformation is sometimes referred to as network inference or reconstruction algorithm. Unweighted mutual information networks can simply be defined by thresholding the mutual information. An incarnation of this approach in the context of gene expression data is known as relevance network (Butte and Kohane 2000; Butte et al. 2000). Let us now define approaches for defining weighted mutual information networks which are based on a matrix of upperbounds for MI, and then setting (7.4):

$$A_{ij} = \frac{MI_{ij}}{Upperbounds_{ij}}. \tag{14.22}$$

Upper bounds for the mutual information can be derived from the following relationship:

$$MI(dx, dy) = Entropy(dx) + Entropy(dy) - Entropy(dx, dy), \tag{14.23}$$

where $Entropy(dx, dy)$ denotes the joint entropy (14.8). In an exercise, you are asked to prove that the mutual information satisfies the inequalities

$$MI(dx,dy) \leq min(Entropy(dx), Entropy(dy)) \tag{14.24}$$
$$\leq \frac{Entropy(dx) + Entropy(dy)}{2}, \tag{14.25}$$
$$\leq max(Entropy(dx), Entropy(dy)). \tag{14.26}$$

Inequality (14.24) suggests $UpperBounds_{ij} = (Entropy(dx_i) + Entropy(dx_j))/2$. We define the **symmetric uncertainty based mutual information adjacency matrix** as follows:

$$A_{ij}^{MI,SymmetricUncertainty} = \frac{MI(dx_i, dx_j)}{(Entropy(dx_i) + Entropy(dx_j))/2}. \tag{14.27}$$

Note that the diagonal elements of $A^{MI,SymmetricUncertainty}$ equal 1 since $MI(dx_i, dx_i) = Entropy(dx_i)$. The Moebius adjacency transformation (4.3) of $A^{MI,SymmetricUncertainty}$ leads to another important adjacency matrix:

$$A_{ij}^{MI,UniversalVersion1} = AF^{Moebius}(A^{MI,SymmetricUncertainty}, a=1, b=0, c=-1, d=2)_{ij}$$
$$= \frac{A_{ij}^{MI,SymmetricUncertainty}}{-A_{ij}^{MI,SymmetricUncertainty} + 2}, \tag{14.28}$$

which we refer to as **universal mutual information-based adjacency matrix (version 1)**. The term "universal" reflects the fact that the adjacency-based dissimilarity $dissMI_{ij}^{UniveralVersion1} = 1 - A^{MI,UniversalVersion1}$ turns out to be a universal distance function . Roughly speaking, the universality of $dissMI_{ij}^{UniveralVersion1}$ implies that any other distance measure between dx_i and dx_j will be small if $dissMI_{ij}^{UniveralVersion1}$ is small (Kraskov et al 2003). The term "distance" reflects the fact that it satisfies the properties of a distance including the triangle inequality (7.11). Since the Moebius transformation is monotonically increasing, $A^{MI,SymmetricUncertainty}$ and $A^{MI,UniversalVersion1}$ are rank-equivalent (10.38) and hence threshold-equivalent. Roughly speaking, they lead to the same unweighted networks (if thresholds are suitably chosen).

Another adjacency matrix is based on the upper bound implied by inequality (14.26). We define **version 2 of a universal mutual information-based adjacency matrix** as follows:

$$A^{MI,UniversalVersion2} = \frac{MI(dx_i, dx_j)}{max(Entropy(dx_i), Entropy(dx_j))}. \tag{14.29}$$

The name reflects the fact that $dissMI^{UniveralVersion2} = 1 - A^{MI,UniversalVersion2}$ is also a universal distance measure. Strictly speaking, $A^{MI,UniversalVersion1}$ and $A^{MI,UniversalVersion2}$ are not rank-equivalent. But in many applications they are approximately rank-equivalent as one can verify by calculating C-indices and Spearman correlations between their vectorized versions.

We should point out that many alternative approaches exist for defining MI-based networks, e.g., the ARACNE approach is described in Sect. 14.7) (Margolin et al. 2006).

14.6 Relationship Between Mutual Information and Correlation

Assume that two numeric variables x and y which satisfy a linear relationship, e.g., x,y may represent random samples from a bivariate normal distribution (15.4). In this case, a correlation coefficient would be an appropriate association measure. Alternatively, the mutual information $MI(discretize(x), discretize(y))$ between the discretized vectors could be used as association measure. The two association measures are quite different for the following reasons. First, the estimated mutual information depends on parameter choices, e.g., the number of bins used in the equal-width discretization step. Second, the mutual information aims to measure general dependence-relationships, while the correlation only measures linear relationships. Third, the equations for the two measures are very different. Given these differences, it is surprising that a simple relationship can be derived between the two association measures under the following assumptions. First, x and y satisfy a linear relationship. Second, the equal-width discretization method with $no.bins = \sqrt{m}$ is used.

Under these assumptions, we argue below that $A^{MI,UniversalVersion2}$ (14.29) can be expressed as follows:

$$\begin{aligned} A^{MI,UniversalVersion2}(dx,dy) &= \frac{MI(dx,dy)}{max(Entropy(dx),Entropy(dy))}, \\ &\approx af^{cor-MI}(cor(x,y)) \end{aligned} \tag{14.30}$$

where the real value function

$$af^{cor-MI}(s) = \frac{log(1+\varepsilon-s^2)}{log(\varepsilon)}(1-\omega)+\omega \tag{14.31}$$

depends on the following two parameters

$$\begin{aligned} \omega &= 0.43m^{-0.30} \\ \varepsilon &= \omega^{2.2}. \end{aligned} \tag{14.32}$$

We call it the "cor-MI" function since it is useful for relating the correlation coefficient with the mutual information. In the following, we show that $af^{cor-MI}(s,\omega,\varepsilon) = \frac{log(1+\varepsilon-s^2)}{log(\varepsilon)}(1-\omega)+\omega$ is a monotonically increasing function that maps $[0,1]$ to $[0,1]$ as long as the two parameters ε and ω satisfy the following relationship:

$$0 < \varepsilon \leq \omega < 1. \tag{14.33}$$

Checking that the function is monotonically increasing is straightforward: (14.33) implies that the first derivative

$$\frac{d}{ds}af^{cor-MI}(s) = -\frac{2s(1-\omega)}{(1+\varepsilon-s^2)log(\varepsilon)}$$

is nonnegative on the unit interval. The minimum of $af^{cor-MI}(s)$ is

$$a_{min} = af^{cor-MI}(0) = \frac{log(1+\varepsilon)}{log(\varepsilon)}(1-\omega)+\omega$$

and the maximum is given by $a_{max} = af^{cor-MI}(1) = 1$. Condition 14.33 implies that $0 < a_{min} \leq 1$, i.e., $[a_{min}, 1]$ is a subinterval of $[0,1]$. The function $af^{cor-MI}(s)$ can be used to define the **correlation-mutual information (cor-MI) adjacency function** $AF^{cor-MI}(*, \varepsilon, \omega)$. Specifically, the i, jth element of $AF^{cor-MI}(A^{original}, \omega, \varepsilon)$ is defined as

$$AF^{cor-MI}(A^{original}, \omega, \varepsilon)_{ij} = \begin{cases} \frac{log(1+\varepsilon-(A_{ij}^{original})^2)}{log(\varepsilon)}(1-\omega)+\omega & \text{if } i \neq j \\ 1 & \text{if } i = j \end{cases} \quad (14.34)$$

Since condition (14.33) implies that the function is monotonically increasing, we find that AF^{cor-MI} is a rank-preserving adjacency function.

Let us now describe simulations that can be used to verify (14.30). Our simulations assume two normally distributed variables X and Y which satisfy a linear relationship specified by the correlation coefficient. The sample vectors x and y are used to estimate the correlation coefficient and the mutual information. The mutual information-based adjacencies are calculated in three steps. First, the two numeric vectors is discretized according to the equal-width discretization method with the default number of bins given by $no.bins = \sqrt{m}$, which results in $dx = discretize(x)$ and $dy = discretize(dy)$. Second, the mutual information $MI(dx, dy)$ is calculated between the discretized vectors using the Miller–Madow estimation method (but alternative approaches could also be used). Third, we transform the mutual information to the following three weighted MI-based adjacencies: $A^{MI,SymmetricUncertainty}(dx, dy)$ (14.27), $A^{MI,UniversalVersion1}(dx, dy)$ (14.28), $A^{MI,UniversalVersion2}(dx, dy)$ (14.29).

Figure 14.1 shows the relationships among the MI-based adjacency measures depend and the (observed) Pearson correlation. As can be seen from Fig. 14.1a,b, the cor-MI-based adjacency function with parameters specified in (14.32) provides a highly accurate prediction of $A^{MI,UniversalVersion2}$ (14.29) on the basis of $cor(x, y)$ and m. Figure 14.1d shows that $A^{MI,UniversalVersion2}$ is practically indistinguishable from $A^{MI,SymmetricUncertainty}$, which is why the cor-MI function can also be used to predict $A^{MI,SymmetricUncertainty}$ on the basis of the correlation measure. Figure 14.1c shows that $A^{MI,UniversalVersion1}$ and $A^{MI,UniversalVersion2}$ satisfy a monotonic relationship, which reflects the fact that these two measures are related by the Moebius adjacency function.

The following R code was used to produce Fig. 14.1.

```
library(minet)
library(entropy)
library(infotheo)
library(WGCNA)
# adjacency function cor-MI with default parameter values
```

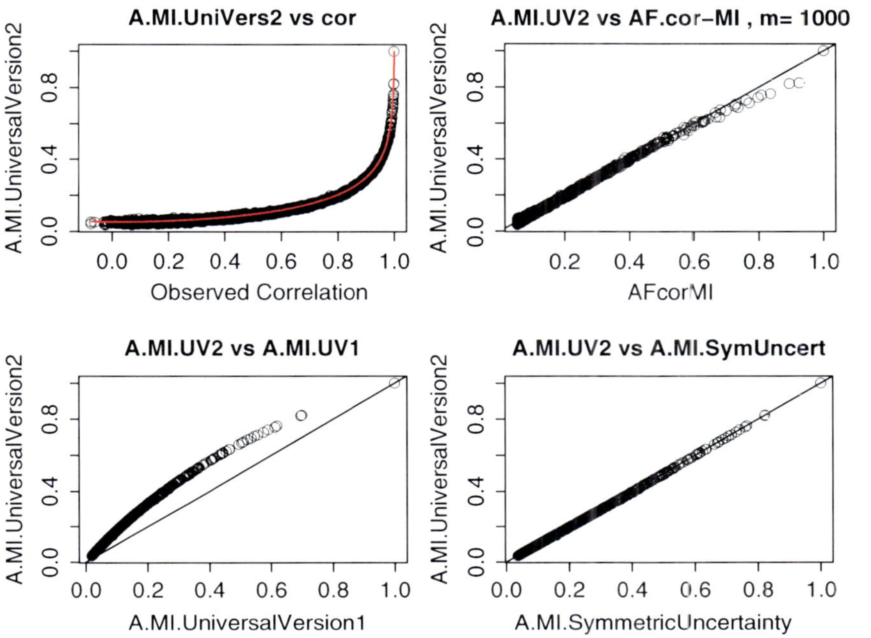

Fig. 14.1 Relating mutual information-based adjacencies to the Pearson correlation. Each point corresponds to a pair of numeric vectors x and y with length $m = 1,000$. The pair of vectors were simulated to exhibit different levels of correlations (see the R code in Sect. 14.6). The upper left panel shows the observed correlation (x-axis) and the corresponding MI-based adjacency matrix $A^{MI,UniversalVersion2}$ (14.29). The *red line* shows the predicted value according to the cor-MI adjacency function (14.31) with parameters described in (14.32). The upper right panel shows the observed $A^{MI,UniversalVersion2}$ (y-axis) versus its predicted value based on the cor-MI adjacency function. The *straight line* has slope 1 and intercept 0. The lower left panel shows the scatterplot between $A^{MI,UniversalVersion2}$ and $A^{MI,UniversalVersion1}$ (14.28). The lower right panel shows the scatterplot between $A^{MI,UniversalVersion2}$ and $A^{MI,SymmetricUncertainty}$ (14.27)

```
AFcorMI=function(cor,m,omega=0.43*m^(-0.30),epsilon=omega^2.2){
log(1+epsilon-cor^2)/log(epsilon)*(1-omega)+omega
} # end of function

#vector which contains true correlations
trueCor=seq(from=0,to=1,length=2000)
# vector with observed Pearson correlations
observedCor=rep(NA, length(trueCor))
A.MI.SymmetricUncertainty=rep(NA,length(trueCor))
A.MI.UniversalVersion1=rep(NA,length(trueCor))
A.MI.UniversalVersion2=rep(NA,length(trueCor))
for (i in 1:length(trueCor) ) {
set.seed(i)
# simulate two numeric vectors x and y that are correlated
m=1000; x=rnorm(m); y=trueCor[i]*x+sqrt(1-trueCor[i]^2)*rnorm(m)
observedCor[i]=cor(x,y)
```

```
# calculate mutual information adjacency matrices
MIinfo1=mutualInfoAdjacency(datE=data.frame(x,y),
discretizeColumns=TRUE,entropyEstimationMethod="MM",numberBins=
  NULL)
A.MI.SymmetricUncertainty[i]=MIinfo1$AdjacencySymmetric
Uncertainty[1,2]
  A.MI.UniversalVersion1[i]=MIinfo1$AdjacencyUniversal
  Version1[1,2]
A.MI.UniversalVersion2[i]=MIinfo1$AdjacencyUniversalVersion2[1,2]
} # end of for loop

par(mfrow=c(2,2))
plot(observedCor,A.MI.UniversalVersion2,xlab="Observed
  Correlation",
cex=1.5,cex.axis=1.5,cex.lab=1.5,cex.main=1.5,
main="A.MI.UniVers2 vs cor",)
lines(observedCor, AFcorMI(observedCor,m),col="red",lwd=2)
plot(AFcorMI(observedCor,m),A.MI.UniversalVersion2,cex=1.5,
cex.axis=1.5,cex.lab=1.5,cex.main=1.5,xlab="AFcorMI",
main=paste("A.MI.UV2 vs AF.cor-MI",", m=",m))
abline(0,1)
plot(A.MI.UniversalVersion1, A.MI.UniversalVersion2,
main=paste("A.MI.UV2 vs A.MI.UV1"),
cex=1.5,cex.axis=1.5,cex.lab=1.5,cex.main=1.5)
abline(0,1)
plot(A.MI.SymmetricUncertainty, A.MI.UniversalVersion2,
main=paste("A.MI.UV2 vs A.MI.SymUncert"),
cex=1.5,cex.axis=1.5,cex.lab=1.5,cex.main=1.5)
abline(0,1)
```

14.6.1 Applications for Relating MI with Cor

In Sect. 14.6 and Fig. 14.1, we used simulations to reveal a close relationship between mutual information-based adjacencies and the underlying correlation if the numeric vectors satisfy linear relationships. But even in real gene expression data sets where linear relationships may only be approximately satisfied, we and others have found close relationships between correlation and mutual information.

Lin Song compared the two measures on many gene expression data sets. Her studies confirm those of others who found that these two measures can be highly related when it comes to real gene expression data. For example, Fig. 14.2 shows the relationship between mutual information and correlation-based networks based on the female mouse liver example (described in Chap. 12). We find that for the vast majority of gene pairs, $A^{MI,UniversalVersion2}(dx_i,dx_j)$ can be approximated by $af^{cor-MI}(cor(x_i,x_j))$ if the parameters ω,ε satisfy (14.32). For the relatively few gene pairs where the two measures disagree (i.e., one measure detects a dependence while the other does not), scatterplots between x_i and x_j often show that the correlation-based measure is the more sensible choice since the mutual information appears to detect strange, nonbiological dependencies. Having said this, when m is large (say $m > 300$), then the mutual information can be valuable for detecting nonlinear dependencies.

We find that correlation networks based on the biweight midcorrelation (5.17) are particularly robust and often outperform mutual information-based networks. Figure 14.2 shows scatterplots involving gene pairs where the mutual information measure detects possibly nonbiological relationships but fails to detect linear relationships.

Our empirical studies and simulation studies lead us to the following.

Observation 14.1 (Relationship between MI- and correlation-based networks). *For many real data sets, $A^{MI,UniversalVersion1}$ (14.28) and $A^{MI,UniversalVersion2}$ (14.29) can be approximated by a monotonically increasing function (the cor-MI-based adjacency specified in (14.34) and (14.32) of the absolute pairwise correlations.*

Observation 14.1 entails that unweighted MI based network that result from thresholding a mutual information measure can be approximated by an unweighted correlation network that results from thresholding the Pearson correlation coefficient (or a corresponding p value). In the parlance of Sect. 10.15, the weighted correlation network $A = (|cor(x_i, x_j)|)$ approximately threshold-implies the MI-based networks (e.g., $A^{MI,UniversalVersion2}$) in many applications.

14.7 ARACNE Algorithm

The Algorithm for the Reconstruction of Accurate Cellular Networks (ARACNE) developed by **Andrea Califano**'s group is a mutual information-based algorithm for defining association networks between numeric variables (Margolin et al. 2006). ARACNE leads to an unweighted adjacency matrix whose edges represent "irreducible" statistical dependencies between numeric variables. In this sense, ARACNE pursues a similar goal as partial correlation network, conditioning approaches, and SEM models: an edge indicates a direct dependence (or association) which is not mediated by other variables.

In the first step, the ARACNE algorithm thresholds the mutual information $MI(x_i, x_j)$ between numeric variables using a hard threshold τ. The threshold could be chosen based on a significance level, which in turn can be estimated using permutation test p value (Sect. 10.2). Two variables x_i and x_j with no direct dependence relationship may have a high mutual information since they are connected via a third variable x_k. Thus in the second step, ARACNE aims to remove these indirect interactions using an information theoretic inequality, known as **data processing inequality** (DPI). The DPI applied to association networks states that if variables x_i and x_j interact only through a third variable x_k (i.e., if the interaction network is $x_i - \ldots - x_k - \ldots - x_j$ and no alternative path exists between x_i and x_j), then

$$MI(x_i, x_j) \leq min(MI(x_i, x_k), MI(x_k, x_j)). \quad (14.35)$$

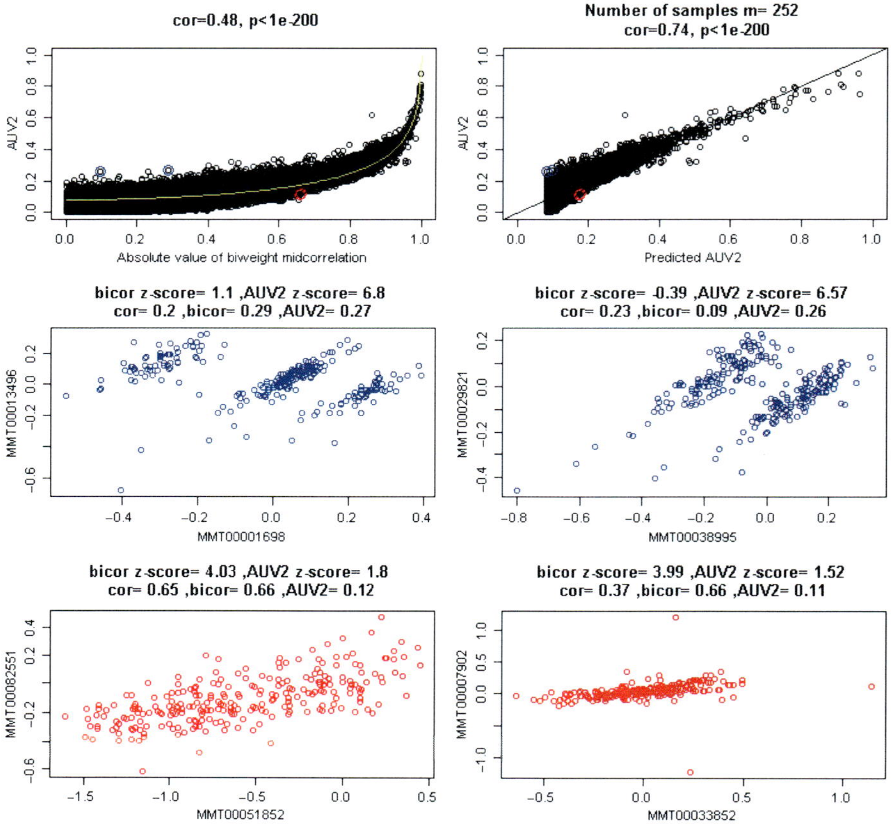

Fig. 14.2 Relationship between mutual information and correlation in female mouse liver tissues. The *upper left panel* shows the scatter plot of $AUV2 = A^{MI,UniversalVersion2}$ (14.29) versus the absolute value of the biweight midcorrelation. The *yellow line* visualizes the predicted values based on the function $af^{cor-MI}(s)$ (whose parameters ω, ε satisfy (14.32)). Each dot corresponds to a pair of gene expression profiles. The *upper right panel* shows the scatter plot of observed AUV2 values (y-axis) versus the predicted $AUV2$ values (x-axis). Dots colored in *blue* correspond to a pair of vectors with a relatively high $AUV2$ value but low biweight midcorrelation (5.17). By contrast, dots colored in red correspond to pairs of genes with high midcorrelation but low $AUV2$ value. The *middle panel* presents scatter plots for "*blue*" pairs of genes, i.e., two genes with high AUV2 but low correlation. Note that the relationships of a *blue* pair of genes appear highly unusual. AUV2 picks up a mathematically interesting association, but it is not clear whether it is biologically meaningful. The *lower panel* shows the scatter plot for "*red*" pairs of genes, i.e., two genes with high midcorrelation but relatively low $AUV2$ value. The *red* gene pair satisfies a linear relationship, which AUV2 measure barely detects

Although all three mutual information values $MI(x_i, x_j)$, $MI(x_i, x_k)$, and $MI(x_k, x_j)$ will likely be high (or significant), the DPI (14.35) suggest that no "direct" interaction exists between i and j, which suggests to remove the edge between x_i and x_j.

In summary, ARACNE starts with a network graph where each pair of nodes with $MI_{ij} > \tau$ is connected by an edge. The algorithm then examines each vector triplet for which all three MIs are greater than τ and removes the edge with the smallest value. Each triplet is analyzed irrespectively of whether its edges have been marked for removal by prior DPI applications to different triplets. Thus the network reconstructed by the algorithm is independent of the order in which the triplets are examined.

To minimize the impact of the variance of the MI estimator, a relaxed version of the DPI can be defined which depends on another threshold parameter called ε. In this case, the edge between i and j is only removed if the difference between $min(MI(x_i,x_k),MI(x_k,x_j))$ and $MI(x_i,x_j)$ lies above a threshold ε, i.e., the edge is removed if

$$MI(x_i,x_j) \leq min(MI(x_i,x_k),MI(x_k,x_j)) - \varepsilon. \tag{14.36}$$

The tolerance threshold ε could be chosen to reflect the variance of the MI estimator and should decrease with increasing sample size m. Using a nonzero tolerance $\varepsilon > 0$ can lead to the persistence of some 3-vector loops. A slightly modified version of the ARACNE algorithm is implemented as part of the `minet` R package (Meyer et al. 2008). Based on the ARACNE algorithm, we will define the ARACNE function which takes as input the mutual information matrix MI and outputs an $n \times n$ dimensional matrix whose i,jth element equals 0 if (14.36) is satisfied and it outputs $MI_{ij}/max(MI)$ otherwise. Specifically, we define

$$aracne(MI,\varepsilon)_{ij} = \begin{cases} \frac{MI_{ij}}{max(MI)} & \text{if } MI(x_i,x_j) > min(MI(x_i,x_k),MI(x_k,x_j)) - \varepsilon \\ 0 & \text{if } MI(x_i,x_j) \leq min(MI(x_i,x_k),MI(x_k,x_j)) - \varepsilon \end{cases} \tag{14.37}$$

Note that $aracne(MI,\varepsilon)$ defines a weighted adjacency matrix.

14.7.1 Generalizing the ARACNE Algorithm

The ARACNE algorithm can be generalized to general (non-mutual information based) networks in multiple ways. One approach of generalizing the ARACNE algorithm is based on expressing the ARACNE function (14.37) in terms of adjacency function. We have shown that the ARACNE algorithm can be expressed as follows:

$$aracne(MI,\varepsilon) = AF^{ultra}(\frac{MI}{max(MI)},\varepsilon), \tag{14.38}$$

where AF^{ultra} (7.33). Since $A = \frac{MI}{max(MI)}$ represents a weighted adjacency matrix, (14.38) suggests to generalize the ARACNE algorithm to the case of a general adjacency matrix A by simply setting $A.new = AF^{ultra}(A,\varepsilon)$.

An alternative approach for adapting the ARACNE algorithm to correlation networks is based on the insight that the data processing inequality (14.35) of the mutual information measure, which underlies the central step of the ARACNE algorithm, also holds for the following Association Measure:

$$A^{weighted.cor}(x_i,x_j) = |cor(x_i,x_j)|^{\beta},.$$

Based on $A^{weighted.cor}$, we propose the following **correlation-based DPI inequality**:

$$A^{weighted.cor}(x_i,x_j) \leq min(A^{weighted.cor}(x_i,x_k),A^{weighted.cor}(x_k,x_j)). \quad (14.39)$$

Like the MI-based DPI, the correlation-based DPI holds if variables x_i and x_j interact only through a third variable x_k (i.e., if $x_i - \ldots - x_k - \ldots - x_j$ is the only path connecting x_i with x_j). Under certain assumptions, one can prove that the least of the three adjacencies $A^{weighted.cor}(x_i,x_j), A^{weighted.cor}(x_i,x_k),\ and\ A^{weighted.cor}(x_k,x_j)$ can come from indirect interactions only. Using the notation from Sect. 15.2, checking against the correlation-based DPI may identify those variable pairs for which $\phi_{ij}(x_i,x_j) = 0$ even though $p_{X_i,X_j}(x_i,x_j) \neq p_{X_i}(x_i)p_{X_j}(x_j)$.

14.7.2 Discussion of Mutual Information Networks

Mutual information is an attractive choice for time series data since these data may exhibit nonlinear dependence relationships (reviewed in (Wiggins and Nemenman 2003)).

For categorical variables, the MI measure is (asymptotically) equivalent to other widely used statistical association measures (e.g., LRT statistic or the Pearson chi-square test, Sect. 14.4), and it is a reasonable choice. Having said this, even for categorical data it can be advantageous to use a likelihood ratio test statistic since it facilitates a straightforward approach for incorporating additional covariates into the analysis (see Sect. 14.3.4). In my humble opinion, the MI measure is often unattractive when it comes to relating continuous numeric variables. I realize that some readers will disagree with this opinion, which is why I will explain my reasoning in the following. On a conceptual level, it is difficult to define the mutual information between two continuous numeric variables. We described a widely used approach based on discretized variables, but alternative kernel-based approaches are often used. On a practical level, mutual information networks are often threshold-implied by correlation networks (Observation 14.1). For example, when it comes to gene expression data involving relatively few independent measurements (say $m < 100$), then we find that mutual information is often highly related to the absolute value of the correlation coefficient. Given this empirical finding, the sample correlation coefficient is an attractive alternative to the MI for the following reasons. First, the correlation can be accurately estimated with relatively few observations, and it does not require the estimation of the (joint) frequency distribution. Estimating the mutual information and the joint density typically requires larger sample sizes. Second, the correlation does not depend on hidden parameter choices. In contrast, most MI estimation approaches involve (hidden) parameter choices, e.g., the number of bins when a discretization step is being used. Third, the correlation allows one to quickly calculate p values and q-values since asymptotic tests are available (see Sect. 10.3). In contrast, it is computationally challenging to calculate a permutation test p value for the mutual information between two discretized vectors (Sect. 10.2). Fourth, the sign of the correlation allows one to distinguish positive from negative relationships. For example, signed correlation networks (see Sect. 5.3) have been found useful in biological applications (Mason et al. 2009). Fifth, modules comprised of highly correlated vectors can be effectively summarized with the module eigennode (the first principal component of scaled vectors). It is less clear, how to define a centroid for a MI-based module. Sixth, the correlation allows for a straightforward angular interpretation, which facilitates a geometric interpretation of network methods and concepts (see Chap. 6).

A widely noticed advantage of mutual information over correlation is its potential for detecting nonlinear relationships between numeric variables. In contrast, the Pearson correlation can only measure linear dependencies. Having said this, it is straightforward to generalize the correlation measure to model fitting indices, which can be used to measure nonlinear or monotonic relationships. Spline- or polynomial regression model-based approaches for measuring nonlinear relationships (Sect. 13.4) are statistically attractive alternatives to the MI since they allow one to calculate asymptotic p values, model fitting statistics, and to adjust the analysis for additional covariates. Further, the Spearman correlation coefficient measures any monotonic relationship, not just linear relationships.

14.7.3 R Packages for Computing Mutual Information

To construct mutual information networks, one can make use of several R packages, e.g., the R packages `infotheo` and `minet` provide a set of functions to infer mutual information networks (Meyer et al. 2008). Four different entropy estimators are made available in the package minet (empirical, Miller–Madow, Schurmann–Grassberger, and shrink). Also, the package integrates accuracy assessment tools, precision recall curves and ROC-curves to compare the inferred network with a reference one.

Another R package that allows one to estimate the entropy and mutual information is the `entropy` package (Hausser and Strimmer 2008). Based on functions implemented in the R packages `entropy` (Hausser and Strimmer 2008), `infotheo`, and `minet` (Meyer et al. 2008), we have implemented the function `mutualInfoAdjacency` in the `WGCNA` R package. The function `mutualInfoAdjacency` calculates $A^{MI,SymmetricUncertainty}$ (14.27), $A^{MI,UniversalVersion1}$ (14.28), and $A^{MI,UniversalVersion2}$ (14.29) based on an input data frame $datE$ whose rows correspond to samples (observations) and whose columns correspond to the vectors (for which the pairwise mutual info should be estimated). As a default, the `mutualInfoAdjacency` assumes that the input variables are numeric and need to be discretized. But if the input variables are already discrete random variables, the user can avoid discretizing by setting input option `discretizeColumns=FALSE`. The number of bins used in the discretization step can be specified using the input parameter `numberBins` (whose default value is $\sqrt{m}$). The function also computes the raw (unscaled) mutual information and entropy measures using natural logarithms (i.e., in units of nats).

The function `mutualInfoAdjacency` allows the user to choose the following estimation methods for mutual info (reviewed in Hausser and Strimmer (2008); Meyer et al. (2008)). The default input option `entropyEstimationMethod="MM"` leads to the Miller–Madow asymptotic bias corrected empirical estimator. Input option `entropyEstimationMethod="ML"` leads to the maximum likelihood estimator (also known as plug-in or empirical estimator). Input option `entropyEstimationMethod="shrink"` leads to the shrinkage estimator (Hausser and Strimmer 2008). Input option `entropyEstimationMethod="SG"` leads to the Schurmann–Grassberger estimator. For additional details, please consult the help file of the `entropy` or the `minet` package.

The function `mutualInfoAdjacency` outputs a list which includes the following components:

1. Output component `Entropy` is a vector whose components report entropy estimates of each column of *datE*. Using the notation from the Wikipedia entry (http://en.wikipedia.org/wiki/Mutual_information), this vector contains the values $Entropy(X)$ where X corresponds to a column in *datE*.
2. Output component `MutualInformation` is a symmetric matrix whose entries contain the pairwise mutual information measures $M(dx_i, dx_j)$ between the discretized columns of *datE*. In general, the entries of this matrix will be larger than 1, i.e., this similarity matrix is *not* an adjacency matrix. The diagonal of the mutual information matrix equals the entropy $Entropy(dx_i)$.
3. Output component `AdjacencySymmetricUncertainty` is the weighted adjacency matrix $A^{MI,SymmetricUncertainty}$ (14.27).
4. Output component `AdjacencyUniversalVersion1` is the adjacency matrix $A^{MI,UniversalVersion1}$ (14.28).
5. Output component `AdjacencyUniversalVersion2` is the adjacency $A^{MI,UniversalVersion2}$ (14.29).

14.8 Exercises

1. Exercise regarding Pearson's chi-square statistic. Use R code to verify numerically the equality of the count version (14.15) and the frequency version (14.15). Hint:

```
set.seed(1);m=30;dx=sample(c(1,2),m,replace=T)
dy=dx; dy[1:(m/2)]=sample(c(1,2),m/2,replace=T)
O=table(dx,dy)
RowSum=apply(O,1,sum)
ColSum=apply(O,2,sum)
```

```
E=outer(RowSum,ColSum,FUN="*")/m
chi.Counts=sum((O-E)^2/E,na.rm=T)
RowFreq=RowSum/sum(RowSum);ColFreq=ColSum/sum(ColSum)
ObsFreq=O/sum(O)
ExpFreq=outer(RowFreq,ColFreq,FUN="*")
chi.Freq=m*sum((ObsFreq-ExpFreq)^2/ExpFreq,na.rm=T)
chi.Counts; chi.Freq
chisq.test(table(dx,dy),correct=F)
```

2. Exercise regarding theoretical properties of the mutual information.
Consider two categorical vectors dx and dy. Denote by $p(ldx_r)$ the frequency of observations that take on the rth value ldx_r.

   (i) Prove that $MI(dx,dy) = 0$ if, and only if, the frequency distributions of the vectors are independent (14.5), i.e., if $p(ldx_r, ldy_c) = p(ldx_r) * p(ldy_c)$
   (ii) Prove that $MI(dx,dy) \geq 0$. Hint: The proof is trivial with the likelihood ratio test formulation of the MI (14.18). Since the MI is proportional to a likelihood ratio statistic, it must be nonnegative.

3. Exercise regarding permutation p values.

   (i) Simulate two numeric vectors x and y of length $m = 500$ with true underlying correlation of $\rho = 0.5$. Hint:

```
set.seed(1)
m=500;x=rnorm(m)
rho=0.5
y=rho*x+sqrt(1-rho^2)*rnorm(m)
```

   (ii) Define an R function called AssocMeasure that inputs a numeric matrix `datE` and outputs the pairwise mutual information between the discretized columns.

```
library(infotheo);library(minet)
AssocMeasure=function(datE){build.mim(datE,disc=
  "equalwidth")}
```

   (iii) For this association measure between x and y calculate permutation test p values using (10.1) and (10.2). Hint:

```
datX=data.frame(x,y)
M.observed=AssocMeasure(datX)[1,2]
no.perms.vec=c(10,50,100,200,500,1000)
p.permutation.vector=rep(NA,length(no.perms.vec))
index.no.perms=6
no.perms=no.perms.vec[index.no.perms]
M.perm=rep(NA,no.perms)
for (i in 1:no.perms ){
set.seed(i)
datX.perm=apply(datX,2,sample)
M.perm[i]=AssocMeasure(datX.perm)[1,2]
}
Z.perm=(M.observed-mean(M.perm,na.rm=T))/
sqrt(var(M.perm,na.rm=T))
```

```
p.Zbased.perm=pnorm(-Z.perm)
M.extended=c(M.observed,M.perm)
permutation.p.value=mean(M.extended>=M.observed,na.rm=T)
p.permutation.vector[index.no.perms]=permutation.p.value
```

(iv) Repeat the exercise with different numbers of permutations and save the resulting permutation p values in a vector. Create a plot that shows how the two permutation p values depend on the number of permutations *no.perms*. Hint: Change the values of index.no.perms in the above code.

```
plot(no.perms.vec,-log10(p.permutation.vector))
```

Message: The more the permutations are used, the more accurate is the estimate of the permutation test p value.

(v) Compare the permutation p values from ii) with those of the Student t-test. Hint: `p.Student=cor.test(x,y)$p.value`

(vi) Repeat (i) and (ii) with $m = 30$. Compare the resulting permutation p values with those of the Student t-test. Do you think that mutual information between discretized numeric vectors is a good association measure when dealing with small sample sizes m?

4. Exercise regarding mutual information and its relationship to Pearson's chi-square statistic.

(i) Use R code to verify numerically the equality between the count version (14.23) and the frequency version of the mutual information (14.23). Hint:

```
MI.Freq=sum(ObsFreq*log(ObsFreq/ExpFreq),na.rm=T)
MI.Count=1/m*sum(O*log(O/E),na.rm=T)
```

(ii) Use R code to verify that the mutual information can be expressed using the likelihood ratio test statistic of a multinomial regression model. Hint:

```
library(nnet)
glmMulti=multinom(factor(dy)~factor(dx))
glmMulti0=multinom(factor(dy)~1)
anova1=anova(glmMulti,glmMulti0)
LRT.statistic=anova1$"LR stat."[2]
```

(iii) Compute an asymptotic p value for the mutual information. Hint: Toward this end, you can use the following code:

```
no.ldx=length(levels(dx));no.ldy=length(levels(dy))
df1=(no.ldx-1)*(no.ldy-1)
p.valueMI=1-pchisq(MI.Freq*2*m,df=df1)
# or you can use the following R code:
DegreesOfFreedom=anova1[2,5]
pvalue.LRTVersion1=1-pchisq(LRT.statistic,DegreesOfFreedom)
pvalue.MI=1-pchisq(LRT.statistic,DegreesOfFreedom)
```

(iv) Write R code for calculating a permutation test p value for the mutual information. Hint: See the previous exercise.

(v) Compare the permutation test p value with that from Pearson's chi-square test $chisq.test(table(dx,dy),correct = T)$ and the LRT test.

(vi) Discuss the pros/cons of the permutation test procedure for calculating the p value (versus the asymptotic p values resulting from Pearson's chi-square test or the LRT test). Hints: Describe which tests make distributional assumptions. What happens if you have to study the pairwise relationships between tens of thousands of variables?

5. Exercise regarding the computational demands of the permutation test. Assume that you want to test whether pairwise relationships exist between $n = 10,000$ numeric vectors. Note that $no.tests = n*(n-1)/2$ pairs can be formed. Assume that you use the Bonferroni-corrected significance level $\alpha_{Bonferroni} = 0.05/no.tests$ for evaluating whether a pair of variables is significantly associated. How many permutations $no.perms$ will be needed if the pairwise permutation test p value is calculated according to (10.2). Hint: (10.3).
6. Exercise regarding the properties of the adjacency function AF^{cor-MI}.

   (i) Prove that it satisfies the properties of an adjacency function, i.e., that it maps an adjacency matrix to a new adjacency matrix.
   (ii) For which parameter values is the function rank-preserving (4.20)?

References

Agresti A (2007) An introduction to categorical data analysis (wiley series in probability and statistics), 2nd edn. Wiley, New York

Beirlant J, Dudewica EJ, Gyofi L, Meulen E (1997) Nonparametric entropy estimation: An overview. Int J Math Stat Sci 6(1):17–39

Butte AJ, Kohane IS (2000) Mutual information relevance networks: Functional genomic clustering using pairwise entropy measurments. Pac Symp Biocomput 5:418–429

Butte A, Tamayo P, Slonim D, Golub T, Kohane I (2000) Discovering functional relationships between RNA expression and chemotherapeutic susceptibility using relevance networks. Proc Natl Acad Sci USA 97:12182–12186

Cheng J, Greiner R, Kelly J, Bell D, Liu W (2002) Learning bayesian networks from data: An information-theory based approach. Artif Intell 137(1–2):43–90

Chow C, Liu C (1968) Approximating discrete probability distributions with dependence trees. IEEE Trans Inf Theory 14:462–467

Cover TM, Thomas JA (1991) Elements of information theory. Wiley, New York

Darbellay G, Vajda I (1999) Estimation of the information by an adaptive partitioning of the observation space. IEEE Trans Inf Theory 45:1315–1321

Daub CO, Steuer R, Selbig J, Kloska S (2004) Estimating mutual information using B-spline functions - an improved similarity measure for analysing gene expression data. BMC Bioinform 5(1):118

Fraser AM, Swinney HL (1986) Independent coordinates for strange attractors from mutual information. Phys Rev A 33(2):1134–1140

Hausser J, Strimmer K (2008) Entropy inference and the James-Stein estimator, with application to nonlinear gene association networks. J Mach Learn Res 10(July 2009):1469–1484

Kraskov A, Stogbauer H, andrzejak R, Grassberger P (2003) Hierarchical clustering based on mutual information CoRR q-bio.QM/0311037

Margolin AA, Nemenman I, Basso K, Wiggins C, Stolovitzky G, Favera RD, Califano A (2006) ARACNE: An algorithm for the reconstruction of gene regulatory networks in a mammalian cellular context. BMC Bioinform 7(Suppl. 1):S7

Mason M, Fan G, Plath K, Zhou Q, Horvath S (2009) Signed weighted gene co-expression network analysis of transcriptional regulation in murine embryonic stem cells. BMC Genomics 10(1):327

Meyer P, Lafitte F, Bontempi G (2008) minet: A R/Bioconductor package for inferring large transcriptional networks using mutual information. BMC Bioinform 9(1):461

Nemenman I (2004) Information theory, multivariate dependence, and genetic network inference. Technical Report. NSF-KITP-04-54, KITP, UCSB. arXiv: q-bio/0406015

Paninski L (2003) Estimation of entropy and mutual information. Neural Comput 15(6):1191–1253

Shannon CE (1948) A mathematical theory of communication. CSLI Publications, Stanford, CA

Steuer R, Kurths J, Daub CO, Weise J, Selbig J (2002) The mutual information: Detecting and evaluating dependencies between variables. Bioinformatics 18(Suppl 2):S231–S240

Wiggins C, Nemenman I (2003) Process pathway inference via time series analysis. Exp Mech 43(3):361–370

Chapter 15
Network Based on the Joint Probability Distribution of Random Variables

Abstract The chapter describes approaches for defining networks based on modeling the joint distribution between a set of random variables. Since it is notoriously difficult to estimate the joint probability function, simplifying assumptions are often made, e.g., multivariate normality. The joint probability distribution can be parameterized using structural equation models, Bayesian network models, or a partitioning function approach. The Kullback–Leibler divergence, which is closely related to the mutual information, can be used to measure the difference between an observed probability distribution and a model-based probability distribution. By minimizing the KL divergence, one can estimate parameter values. This chapter is rather theoretical and requires some background in calculus and probability theory.

15.1 Association Measures Based on Probability Densities

We start out reviewing some basic concepts from probability theory. Consider a continuous numeric variable X whose probability distribution (frequency distribution) $p_X(x)$ is known. The probability density $p_X(x)$ is a function that takes on non-negative values and its integral equals 1, i.e., $\int_{Domain} p_X(x)dx = 1$. Typically, the integration domain could be an interval, e.g., $Domain = [-\infty, +\infty]$. For notational convenience, we sometimes drop the reference to the integration domain. If the random variable X follows a **normal distribution** (also known as Gaussian distribution or bell curve) $N(\mu, \sigma^2)$, the probability distribution is given by

$$p_X(x) = \frac{1}{\sqrt{2\pi}\sigma} exp\left(-\frac{(x-\mu)^2}{2\sigma^2}\right). \quad (15.1)$$

For a continuous variable with distribution $p_X(x)$, the mean and variance are defined as integrals:

$$Mean(X) = \int_{Domain} x p_X(x)dx$$
$$Variance(X) = \int_{Domain} (x - Mean(X))^2 p_X(x)dx. \quad (15.2)$$

S. Horvath, *Weighted Network Analysis: Applications in Genomics and Systems Biology*,
DOI 10.1007/978-1-4419-8819-5_15,

If X follows a normal distribution (15.1), then $Mean(X) = \mu$ and $Variance(X) = \sigma^2$. For a continuous variable with distribution $p_X(x)$, the entropy is defined as follows:

$$Entropy(X) = -\int_{-\infty}^{+\infty} p_X(x) log(p_X(x)) dx. \tag{15.3}$$

For example, if X follows a normal distribution (15.1), then $Entropy(X) = 0.5 + log(\sqrt{2\pi}\sigma)$.

Assume that we have a second continuous random variable Y with probability distribution $p_Y(y)$. Further assume that X and Y have a joint probability distribution $p_{X,Y}(x,y)$. For example, if X and Y follow a **joint bivariate normal distribution**

$$p_{X,Y}(x,y) = \frac{1}{2\pi\sigma_X\sigma_Y\sqrt{1-\rho^2}} exp\left(-\frac{numerator(x,y)}{2(1-\rho^2)}\right), \tag{15.4}$$

where

$$numerator(x,y) = \frac{(x-\mu_X)^2}{\sigma_X^2} - \frac{2\rho(x-\mu_X)(y-\mu_Y)}{\sigma_X\sigma_Y} + \frac{(y-\mu_Y)^2}{\sigma_Y^2}.$$

Two continuous numeric random variables X and Y are **independent** if their joint probability distribution factors are as follows:

$$p_{X,Y}(x,y) = p_X(x) p_Y(y). \tag{15.5}$$

Informally, we say that two continuous numeric random variables are **associated** with each other, if they are *not* independent. The following statistics can be used to measure how two variables change together:

$$Covariance(X,Y) = \int_{-\infty}^{+\infty}\int_{-\infty}^{+\infty} (x - Mean(X))(y - Mean(Y)) p_{X,Y}(x,y) dxdy$$

$$Correlation(X,Y) = \frac{Covariance(X,Y)}{\sqrt{Variance(X)Variance(Y)}}.$$

If X and Y follow a joint bivariate normal distribution (15.4), then *Correlation* $(X,Y) = \rho$. If X and Y follow a bivariate normal distribution, then *Correlation* $(X,Y) = 0$ implies that the two variables are independent. But in general a vanishing correlation does not necessarily imply that the two variables are independent.

Information-theoretic quantities of dependence are given by the joint entropy:

$$Entropy(X,Y) = -\int_{-\infty}^{+\infty}\int_{-\infty}^{+\infty} p_{X,Y}(x,y) log(p_{X,Y}(x,y)) dxdy \tag{15.6}$$

and the mutual information:

$$MutualInfo(X,Y) = Entropy(X) + Entropy(Y) - Entropy(X,Y). \quad (15.7)$$

If X and Y follow a joint bivariate normal distribution (15.4), then $MutualInfo(X,Y) = -0.5log(1-r^2)$. The mutual information and joint entropy have several disadvantages when it comes to *continuous* numeric variables X and Y. For this reason, one typically calculates these measures for discretized numeric vectors as described in Chap. 14.

Let us now assume that X is a discrete numeric random variable which takes on values $x_1, x_2, \ldots$ with probabilities $p_X(x_1), p_X(x_2), \ldots$, respectively. For example, X could be a count variable. The above formulas for the mean, variance, and entropy for a continuous random variable can be easily adapted to the case of a discrete numeric random variable by replacing integrals by summations, e.g.,

$$Mean(X) = \sum_r x_r p_X(x_r)$$
$$Variance(X) = \sum_r (x_r - Mean(X))^2 p_X(x_r)$$
$$Entropy(X) = -\sum_r p_X(x_r) log(p_X(x_r)).$$

The entropy has the following intuitive quantitative meaning: it is the expected number of yes/no questions needed to guess the value of a randomly drawn observation when the optimal question-asking strategy is adopted (Cover and Thomas 1991; Latham and Roudi 2009). The optimal question-asking strategy is to divide the probability (determined by p_X) in half on each guess by asking binary questions such as "is it greater than τ?". Then the average number of yes/no questions it takes to guess the value correctly lies between $Entropy(X)$ and $Entropy(X)+1$.

If the probability density of a random variable is unknown, it is inconvenient to use the integration formulas described in (15.2) to estimate $mean(X)$ and $var(X)$ based on the observed relative frequency distribution. Instead, the random sample vector x is used directly for calculating the sample mean $mean(x)$ and the sample variance $var(x)$. Also it is inconvenient to use the integration formulas described in (15.6) to estimate $Correlation(X,Y)$ and $Covariance(X,Y)$ based on the observed relative frequency distribution. Instead, the random sample vectors x and y are used directly for calculating the sample correlation $cor(x,y)$ (5.10) and the sample covariance $cov(x,y)$.

15.1.1 Entropy(X) Versus Entropy(Discretize(X))

In Sect. 14.2.2, we described several estimators for the entropy of a discretized numeric variable $DX = discretize(X)$. In the following, assume that $Entropy(dx)$ denotes an estimator of $Entropy(DX)$ based on the discretized numeric vector

$dx = discretize(x, no.bins)$, e.g., the empirical estimator (14.13). Further, assume that the frequency distribution p_X of X is known so that $Entropy(X) = -\int p_X(x) log(p_X(x))dx$ (15.3) can be calculated. Then the following equation describes an approximate relationship between $Entropy(dx)$ and $Entropy(X)$:

$$Entropy(dx) \approx Entropy(X) - log\left(\frac{max(x) - min(x)}{no.bins}\right). \tag{15.8}$$

Note that $Entropy(dx)$ differs from $Entropy(X)$ by a term that depends on the number of bins used in the discretization step. We briefly outline a mathematical derivation of (15.8). Note that the bin width is given by $width = \frac{max(x)-min(x)}{no.bins}$. The relative frequency of the rth bin p_r can be approximated as follows:

$$p_r \approx p_X(midp_r)width, \tag{15.9}$$

where the rth bin midpoint is given by $midp_r = min(x) + (r - 0.5)width$. Using (15.9), we find that the empirical estimator of the entropy $Entropy(dx) = -\sum_r p_r log(p_r)$ can be approximated as follows:

$$\begin{aligned} Entropy(dx) &\approx -\sum_r p_X(midp_r)widthlog(p_X(midp_r)width) \\ &= -\sum_r p_X(midp_r)log(p_X(midp_r))width \\ &\quad -\sum_r p_X(midp_r)widthlog(width). \end{aligned}$$

Approximating the first term with

$$-\sum_r p_X(midp_r)log(p_X(midp_r))width \approx -\int p_X(x)log(p_X(x))dx = Entropy(X)$$

and the second term with

$$\sum_r p_X(midp_r)widthlog(width) \approx \int p_X(x)dxlog(width) = log(width)$$

results in (15.8). As numeric example, consider the case of a random variable X which follows a normal distribution $N(\mu, \sigma^2)$. The following R code verifies that the left-hand side $Entropy(X) = 0.5 + log(\sqrt{2\pi}\sigma)$ of (15.8) is approximately equal to the right-hand side when the empirical or the Miller–Madow estimator is used for $Entropy(dx)$.

```
library(infotheo)
sigma=3;Entropy.X=0.5+log(sqrt(2*pi)*sigma)
m=1000;x=rnorm(m,mean=5,sd=sigma)
no.bins=sqrt(m)
dx=discretize(x,no.bins,disc="equalwidth")[,1]
```

```
p.dx=table(dx)/length(dx)
EntropyEmp=-sum(p.dx*log(p.dx),na.rm=T)
EntropyMM=EntropyEmp+(no.bins-1)/(2*m)
RHS=Entropy.X-log((max(x)-min(x))/no.bins);
EntropyEmp; EntropyMM; RHS
```

15.1.2 Kullback–Leibler Divergence for Assessing Model Fit

The *Kullback-Leibler (KL) divergence* (Kullback and Leibler 1951; Kullback 1987) (aka. relative entropy information gain, information divergence) is a widely used tool in pattern recognition, e.g., it is often used in speech or image recognition. The KL divergence between two probability distributions $p_X()$ and $p_Y()$, which are defined on the same domain, is defined as the following integral:

$$KL(p_X, p_Y) = \int_{Domain} p_X(x) log\left(\frac{p_X(x)}{p_Y(x)}\right) dx. \tag{15.10}$$

The KL divergence is only defined if $\int_{Domain} p_X(x) = 1$ and $\int_{Domain} p_Y(y)dy = 1$ and if $p_Y(x) > 0$ for any x where $p_X(x) > 0$. The quantities $0log(0)$ and $0log\left(\frac{0}{0}\right)$ are interpreted as 0. For discrete numeric variables (or categorical variables), the integral in (15.10) is replaced by summation. For example, for two discrete numeric variables X and Y which take on the same values, the KL divergence is defined as

$$KL(p_X, p_Y) = \sum_i p_X(x_i) log\left(\frac{p_X(x_i)}{p_Y(x_i)}\right). \tag{15.11}$$

The KL divergence satisfies the following properties:

1. $KL(p_X, p_Y) \geq 0$ (positivity), which be proven using Gibbs' inequality.
2. $KL(p_X, p_X) = 0$ (self-similarity).
3. $KL(p_X, p_Y) = 0$ only if $p_X = p_Y$ (self-identification).
4. $KL(p_X, p_Y)$ is invariant under parameter transformations $x \to h(x)$.
5. In general, the KL divergence is nonsymmetrical, i.e., $KL(p_X, p_Y)$ is not necessarily the same as $KL(p_Y, p_X)$.

The KL divergence is often used in information-theoretic applications and is closely related to the mutual information: note that the mutual information $MI(X,Y)$ (15.7) equals the KL divergence between the joint probability distribution $p_{(X,Y)}$ and the product of the marginal distributions, i.e.,

$$MI(X,Y) = KL(p_{(X,Y)}, p_X p_Y).$$

An **information theoretic interpretation of the KL divergence** is that it is the information gain about the random variable X achieved if p_X can be used instead of p_Y. Kullback proposed to use the KL divergence for model updating and model inference. He proposed the **Principle of Minimum Discrimination Information (MDI)**: given new data, a new probability density p_X should be chosen which is as similar to the original distribution p_Y as possible, i.e., the new distribution p_X should be determined by minimizing $KL(p_X, p_Y)$. Incidentally, the principle of MDI can be interpreted as a generalization of Laplace's principle of insufficient reason, and the principle of maximum entropy of E.T. Jaynes.

The KL divergence has been used to measure the difference of a model (represented by p_Y) from real data (represented by p_X). Then $KL(p_X, p_Y)$ measures the difference between the observed real data (p_X) and the model-based estimates (p_Y).

When trying to fit parameterized models to data, there are various estimators which attempt to minimize Kullback-Leibler divergence, such as maximum likelihood estimators (Burnham and Anderson DR 2002) or maximum spacing estimators (Ranneby 1984).

In Sect. 11.1.6, we show that a maximum likelihood-based method for evaluating the fit of a structural equation model is equivalent to the KL divergence applied to multivariate normal distributions.

15.1.3 KL Divergence of Multivariate Normal Distributions

In general, the KL divergence (15.11) between two distributions is neither analytically tractable nor do efficient computational algorithms exist for computing it. Some approaches tackle this problem by replacing the KL divergence with other functions (e.g., a basis expansions) that can be computed efficiently (Hershey and Olsen 2007). Fortunately, in case of two multivariate normal distributions, one can derive closed form solutions as described in the following.

We start out reviewing the multivariate normal distribution of a **random vector** $\mathbf{X} = (X_1, \ldots, X_k)$. A random vector is a vector whose components are random variables. Note that we use boldface to distinguish the random **vector** $\mathbf{X}$ from the univariate random **variable** X_1.

The multivariate normal distribution (aka. multivariate Gaussian distribution) is a generalization of the one-dimensional (univariate) normal distribution to $k > 1$ dimensions. $\mathbf{X}$ is said to be multivariate normally distributed if there exists a k-vector μ and a symmetric $k \times k$ dimensional matrix Σ, such that the probability density function of $\mathbf{X}$ can be expressed as:

$$p_X(x) = \frac{1}{(2\pi)^{k/2} det(\Sigma)^{1/2}} exp\left(-\frac{1}{2}(\mathbf{x}-\mu)^{\tau}\Sigma^{-1}(\mathbf{x}-\mu)\right), \tag{15.12}$$

where $det(\Sigma)$ is the determinant of Σ. We assume that Σ is a positive-definite matrix (see Sect. 7.8), which implies that it is invertible.

The multivariate normal distribution of the k-dimensional random vector $\mathbf{X}$ is denoted by $N(\mu, \Sigma)$ with k-dimensional mean vector $\mu = (E(X_1), \ldots, E(X_k))$ and $k \times k$ dimensional variance covariance matrix Σ whose i, jth component is given by $cov(X_i, X_j)$.

Let us now study measures of differences between two multivariate normal probability density functions p_X and p_Y corresponding to two k-dimensional multivariate normal random vectors **X** and **Y**. Let us assume that **X** and **Y** follow multivariate distributions, i.e., $X \sim N(\mu_x, \Sigma_{xx})$ and $y \sim N(\mu_y, \Sigma_{yy})$. The Kullback–Leibler divergence $KL(p_X, p_Y)$ (15.10) between the distributions of $X \sim N(\mu_x, \Sigma_{xx})$ and $y \sim N(\mu_y, \Sigma_{yy})$ is given by

$$KL = \frac{1}{2}\left[log\left(\frac{det(\Sigma_{yy})}{det(\Sigma_{xx})}\right) + trace(\Sigma_{xx}\Sigma_{yy}^{-1}) - no.rows(\Sigma_{xx}) + (\mu_x - \mu_y)^{\tau}\Sigma_{yy}^{-1}(\mu_x - \mu_y)\right], \tag{15.13}$$

where $no.rows(\Sigma_{xx}) = no.rows(\Sigma_{yy}) = k$ is the number of rows of the variance covariance matrices. If $\mu_x = \mu_x = 0$, (15.13) can be rewritten as follows:

$$KL(N(0,\Sigma_{xx}), N(0,\Sigma_{yy})) = \frac{1}{2}\left[log\left(\frac{det(\Sigma_{yy})}{det(\Sigma_{xx})}\right) + trace(\Sigma_{xx}\Sigma_{yy}^{-1}) - no.rows(\Sigma_{xx})\right]. \tag{15.14}$$

Note that the right-hand side of (15.14) is the KL pre-dissimilarity (7.25) $PreDissim_{KL}(\Sigma_{xx}, \Sigma_{yy})$ between positive definite matrices Σ_{xx} and Σ_{yy}.

In practice, Σ_{xx} is unknown and must be estimated from random samples. We briefly provide some formulas to estimate it. Denote the uth sample of the random vector **X** by $\mathbf{x}_u$ (note the small caps). Note that $\mathbf{x}_u$ is a numeric vector with k components. Then the **sample mean** is a vector defined by $\hat{\mu}_x = \frac{1}{m}\sum_{u=1}^{m}\mathbf{x}_u$. The sample variance–covariance matrix S_{xx} is defined by

$$S_{xx} = \frac{1}{m-1}\sum_{u=1}^{m}(\mathbf{x}_u - \hat{\mu}_x)(\mathbf{x}_u - \hat{\mu}_x)^{\tau},$$

where $^{\tau}$ denotes the transpose. Note that S_{xx} is a $k \times k$ dimensional symmetric matrix.

15.1.4 KL Divergence for Estimating Network Parameters

Consider a random vector $\mathbf{X} = (X_1, \ldots, X_k)$ whose k components are random variables. Denote the corresponding joint probability density by $p_{\mathbf{X}}(x_1, \ldots, x_k)$. For example, a single gene expression measurement involving $k = 50,000$ probes can be interpreted as a random sample from a (joint) probability distribution involving k random variables X_i. It is notoriously difficult to estimate joint probability density for a large number of variables. With all due caution, let us make the unrealistic assumption that the joint density among the X_i can be estimated or is known. Further, assume that we have a network model which implies that the probability density function has a certain structure. For example, structural equation models (described in Chap. 11) lead to a model prediction (denoted $p_{\mathbf{Y}}$) of the joint density. Alternatively, Bayesian network models can be used to parameterize $p_{\mathbf{Y}}$. A Bayesian network is represented by a directed acyclic graph whose vertices (nodes) correspond to the random variables X_i and whose edges correspond to parent-child statistical dependencies among the variables (Pearl 1988; Heckerman 1996; Hartemink et al. 2001; Yu et al. 2002; Friedman 2004).

In general, the model prediction of the joint density involves parameters. The Kullback–Leibler divergence $KL(p_{\mathbf{X}}, p_{\mathbf{Y}})$ between the observed joint density $p_{\mathbf{X}}$ and the model prediction $p_{\mathbf{Y}}$ can be minimized (a) to estimate the model parameters, (b) to evaluate the fit of the network model, and (c) to choose among multiple network models.

15.2 Partitioning Function for the Joint Probability

The section describes an approach for defining a network between random variables based on (a) modeling the joint probability distribution with potentials and (b) using this model for defining a dependence concept. This approach follows methods of the Markov network literature (Pearl 1988) and in particular the ARACNE algorithm (described in Sect. 15.2) (Margolin et al. 2006).

The partitioning function approach models the joint probability $p_{\mathbf{X}}(x_1,\dots,x_n)$ as follows:

$$\begin{aligned} p_{\mathbf{X}}(x_1,\dots,x_n) &= \frac{1}{Z} exp\left(-\sum_{i=1}^{n}\phi_i(x_i) - \sum_{i=1}^{n}\sum_{j\neq i}^{n}\phi_{ij}(x_i,x_j) - \dots\right) \\ &= exp(-H(x_1,\dots,x_n)), \end{aligned} \tag{15.15}$$

where the normalization factor Z is called the **partition function**, the functions ϕ_i, ϕ_{ij} are referred to as **potentials**, and

$$H(x_1,\dots,x_n) = \sum_{i=1}^{n}\phi_i(x_i) + \sum_{i=1}^{n}\sum_{j\neq i}^{n}\phi_{ij}(x_i,x_j) + \dots$$

is sometimes referred to as **Hamiltonian** (energy). Within such a model, a pair of variables X_i and X_j can be defined to interact if and only if $\phi_{ij}(x_i,x_j) \neq 0$. Note that (15.15) does not define the potentials ϕ uniquely, and additional constraints are needed to avoid the ambiguity (Nemenman 2004; Margolin et al. 2006). Toward this end, one can invoke the maximum entropy approximation or the principle of minimum discrimination information (described in Sect. 15.1.2).

When the sample size m is small, inferring the exponential number of potential n-way interactions (potentials) of (15.15) is infeasible and a set of simplifying assumptions must be made about the dependency structure. Equation (15.15) provides a framework for introducing successive approximations. The simplest model is one where the variables are assumed to be independent, i.e.,

$$H_{independent}(x_1,\dots,x_n) = \sum_i \phi_i(x_i).$$

In this case, the first-order potentials $\phi_i(x_i)$ are determined by the marginal probabilities, $p_{X_i}(x_i)$, which in turn can be estimated from random samples (see Sect. 14.2.1). A sample size of $m > 100$ is sometimes sufficient to estimate pairwise potentials $\phi_{ij}(x_i,x_j)$.

For a sample size of $m < 1{,}000$, it is advisable to truncate the Hamiltonian (15.15) at the pairwise interaction level:

$$H_{pairwise}(x_1,\dots,x_n) = \sum_{i=1}^{n} \phi_i(x_i) + \sum_{i=1}^{n}\sum_{j\neq i}^{n} \phi_{ij}(x_i,x_j). \tag{15.16}$$

Based on $H_{pairwise}$, an unweighted network adjacency can be defined as follows:

$$A_{ij} = \begin{cases} 1 & \text{if } i=j \\ 1 & \text{if } i\neq j\ \phi_{ij}(x_i,x_j) > 0 \\ 0 & \text{if } i\neq j\ \phi_{ij}(x_i,x_j) = 0 \end{cases}. \tag{15.17}$$

Thus, an edge exists between variables X_i and X_j if the corresponding pairwise potential $\phi_{ij}(x_i,x_j)$ is nonvanishing. It is worth pointing out that statistical independence $p_{X_i,X_j}(x_i,x_j) = p_{X_i}(x_i)p_{X_j}(x_j)$ is not a sufficient condition for $\phi_{ij}(x_i,x_j) = 0$. Under certain assumptions about the network topology, one can define algorithms for reconstructing two-way interactions (see, e.g., the ARACNE algorithm).

15.3 Discussion

Constructing networks based on the joint probability density $p_{\mathbf{Y}}(y_1,\dots,y_n)$ between n random variables is very challenging for the following reasons. First, it is difficult to estimate a joint probability density without making some simplifying assumptions. For example, the parameters of a multivariate normal distribution $p_{\mathbf{Y}}$ can be easily estimated using maximum likelihood estimation.

Second, it is difficult to find a realistic parameterization of $p_{\mathbf{Y}}$. In Chap. 11, we review how to use structural equation models to parameterize $p_{\mathbf{Y}}$ based on the multivariate normal distribution. Structural equation models are also known as covariance regression models since they can be used to parameterize the variance–covariance matrix Σ_{yy}. Alternatively, Bayesian network models or the partitioning function-based approach (Sect. 15.2) can be used to parameterize $p_{\mathbf{Y}}$. Bayesian networks are implemented in several software packages, e.g., the LibB software (Friedman and Elidan 2011). However, Bayesian networks and SEMs have a hard time dealing with regulatory loops, and the resulting network construction methods can be highly sensitive to model assumptions and noise (Nemenman 2004).

Third, it is difficult to choose (and often to calculate) a suitable scoring metric for contrasting a model-based density $p_{\mathbf{Y}}$ with an observed density $p_{\mathbf{X}}$. Apart from the Kullback–Leibler divergence, several other "scoring metric" or model fitting functions have been proposed. For example, a Bayesian scoring metric penalizes complex graph models (Cooper and Herskovits 1992).

Fourth, it is difficult to estimate the parameters (and choosing a best fitting network) based on optimizing a scoring metric. For example, evaluating a scoring metric for each possible graph may be computationally infeasible, i.e., an NP-complete problem (Chickering 1996). In this case, heuristic optimization algorithms (e.g., greedy hill climbing or simulated annealing) can be used to search for locally optimal graph structures.

Fifth, it is difficult to define a network based on an estimated joint probability distribution $p_{\mathbf{Y}}$. A fundamental problem is that no general agreement exists on how to define a statistical dependency measure in the multivariate setting (Joe 1997; Nemenman 2004; Margolin et al. 2006). Instead of a universal definition of statistical dependence, standard statistical methods have produced a multitude of dependence concepts applicable in restricted contexts, such as normal noise, binary, bivariate, or numeric data (Joe 1997). For example, the notion of conditional (in)dependence in the form of **Bayesian networks** has been used in biological applications (Friedman 2004). Overall, it has been found that estimating the joint probability density to estimate pairwise dependencies is inferior to directly estimating pairwise dependencies (without first estimating the joint density) (Vapnik 1998; Nemenman 2004).

Because of these challenges, association networks based on pairwise association measures (e.g., the correlation coefficient or a p value) are widely used in biological applications.

References

Burnham K, anderson DR (2002) Model Selection and Multimodel Inference: A Practical Information-Theoretic Approach, Second Edition. Springer Science, NY

Chickering D (1996) Learning bayesian networks is np-complete. Learning from Data: Artificial Intelligence and Statistics pp 121–130

Cooper G, Herskovits E (1992) A bayesian method for the induction of probabilistic networks from data. Machine Learning 9:309–347

Cover T, Thomas J (1991) Elements of information theory. John Wiley Sons, New York

Friedman N (2004) Inferring cellular networks using probabilistic graphical models. Science 303(5659):799–805

Friedman N, Elidan G (2011) Libb 21 http://wwwcshujiacil/labs/compbio/LibB/

Hartemink A, Gifford D, Jaakkola T, Young R (2001) Using graphical models and genomic expression data to statistically validate models of genetic regulatory networks. Pac Symp Biocomput pp 422–433

Heckerman D (1996) A tutorial on learning with bayesian networks. Microsoft Research

Hershey J, Olsen P (2007) Approximating the kullback leibler divergence between gaussian mixture models

Joe H (1997) Multivariate models and dependence concepts

Kullback S (1987) Letter to the editor: The kullbackleibler distance. The American Statistician 41(4):340–341

Kullback S, Leibler R (1951) On information and sufficiency. Annals of Mathematical Statistics 22(1):79–86

Latham P, Roudi Y (2009) Mutual Information. Scholarpedia 4(1):1658

Margolin A, Nemenman I, Basso K, Wiggins C, Stolovitzky G, Favera R, Califano A (2006) Aracne: an algorithm for the reconstruction of gene regulatory networks in a mammalian cellular context. BMC Bioinformatics 7

Nemenman I (2004) Information theory, multivariate dependence, and genetic network inference. Tech Rep NSF-KITP-04-54, KITP, UCSB ArXiv: q-bio/0406015

Pearl J (1988) Probabilistic reasoning in intelligent systems: networks of plausible inference. San Francisco, CA: Morgan Kaufmann Publishers, Inc

Ranneby B (1984) The maximum spacing method an estimation method related to the maximum likelihood method. Scandinavian Journal of Statistics 11(2):93–112

Vapnik V (1998) Statistical Learning Theory. John Wiley Sons, New York

Yu J, Smith A, Wang P, Hartemink A, Jarvis E (2002) Using bayesian network inference algorithms to recover molecular genetic regulatory networks. 3rd International Conference on Systems Biology

Index

S. Horvath, *Weighted Network Analysis: Applications in Genomics and Systems Biology*, DOI 10.1007/978-1-4419-8819-5, © Springer Science+Business Media, LLC 2011

D

Printed by Publishers' Graphics LLC
MO20121214.19.21.8